What evidence is there for and against unified schemes for active galactic nuclei (AGN)? How do the AGN populations evolve over cosmological timescales? And what can the variability of their UV and X-ray emission tell us? These are just some of the exciting issues addressed in this volume of papers collected from the 33rd Herstmonceux conference in Cambridge.

AGN are among the most spectacular objects known to astronomy. Yet, despite years of intense and wide-ranging research, the debate continues – what is their fundamental source of power? Rapid progress has been made towards answering this question by a variety of large-scale, multi-wavelength monitoring campaigns and the latest generation of satellite-borne observations.

This volume provides a valuable overview and timely update of the exciting and rapidly developing field of AGN research – essential reading for graduate students and researchers.

The Nature of Compact Objects in Active Galactic Nuclei

Sponsored by:
> The Royal Greenwich Observatory
> The Institute of Astronomy
> The European Association for Research in Astronomy*
>
> *An association linking*
> *The University of Cambridge Institute of Astronomy*
> *The Institut d'Astrophysique de Paris*
> *Sterrewacht Leiden*

Organizing Committee:
> Brian Boyle
> Alice Duncan (secretary)
> Judith Perry
> Andrew Robinson
> Roberto Terlevich
> Simon White

The Nature of Compact Objects in Active Galactic Nuclei

Proceedings of the 33rd Herstmonceux conference,
held in Cambridge, July 6–22, 1992

Edited by
ANDREW ROBINSON
The Institute of Astronomy
University of Cambridge

and

ROBERTO TERLEVICH
The Royal Greenwich Observatory

Published by the Press Syndicate of the University of Cambridge
The Pitt Building, Trumpington Street, Cambridge CB2 1RP
40 West 20th Street, New York, NY 10011–4211, USA
10 Stamford Road, Oakleigh, Melbourne 3166, Australia

© Cambridge University Press 1994

First published 1994

Printed in Great Britain at the University Press, Cambridge

A catalogue record for this book is available from the British Library

ISBN 0 521 46480 3 hardback

Contents

I Evidence and Implications of Anisotropy in AGN

Evidence for Anisotropy and Unification 3
    R.A.E. Fosbury

Any Evidence against Unified Schemes? 13
    C.N. Tadhunter

Spectropolarimetry of Cygnus A 24
    N. Jackson & C.N. Tadhunter

Spectropolarimetery of the Ultraluminous Infrared Galaxy IRAS 110548–1131 . . . 28
    S. Young, J.H. Hough, J.A. Bailey, D.J. Axon & M.J. Ward

Are there Dusty Tori in Seyfert 2 Galaxies? 31
    T. Storchi-Bergmann, J.S. Mulchaey & A.S. Wilson

Imaging Spectrophotometry of Extended-Emission Seyfert Galaxies 36
    R.W. Pogge & M.M. DeRobertis

Spectroscopy of the Extended Emission Line Regions in NGC 4388 40
    I. Pérez-Fournon, B. Vila-Vilaró, J.A. Acosta-Pulido,
    J.I. González-Serrano, M. Balcells, A.S. Wilson & Z. Tsvetanov

Evidence and Implications of Anisotropy in Seyfert Galaxies 44
    J.A. Acosta-Pulido, B. Vila-Vilaró, I. Pérez-Fournon, A.S. Wilson &
    Z. Tsvetanov

Collimated Radiation in NGC 4151 46
    G. Kriss, I. Evans, H. Ford, Z. Tsvetanov, A. Davidsen & A. Kinney

A Dust Ring around the Nucleus of NGC 4151 48
    B. Vila-Vilaró, E. Pérez, A. Robinson, C. Tadhunter, I. Pérez-Fournon &
    R. González-Delgado

Evolution of Narrow Line Clouds 51
    P.A. Foulsham & D.J. Raine

Star Formation in NGC 5953 53
    R. González-Delgado & E. Pérez

Stellar Activity in the Seyfert Nucleus of NGC 1808 55
    D.A. Forbes

Direct Evidence for Anisotropy: Radio Maps and their Relation to Optical Morphology 58
    P. Alexander

The Radio-Optical Connection in AGN 66
    D.J. Axon, J.E. Dyson & A. Pedlar

Knots in Extragalactic Radio Jets 74
    E. Lüdke

Radio Emission and the Nature of Compact Objects in AGN 78
    L. Colina

The Radio Properties of Hidden Seyfert 1's: Implications for Unified Models 82
    E.C. Moran & J.P. Halpern

Anisotropic Optical Continuum Emission in Radio Quasars. 84
    J.C. Baker, R.W. Hunstead, V.K. Kapahi & C.R. Subrahmanya

The UV Component in Distant Radio Galaxies 88
    A. Cimatti, S. di Serego Alighieri & R.A.E. Fosbury

A Connection between BL Lacertæ Objects and Flat-Spectrum Radio Quasars? . . . 90
    P. Padovani

The Difference between BL Lacs and QSOs. 94
    P.A. Hughes, M.F. Aller, H.D. Aller & D.C. Gabuzda

The Evolutionary Unified Scheme and the θ–z Plane 96
    F. Vagnetti & R. Spera

II   Luminosity Functions and Continuum Energy Distributions

Radio Luminosity Functions of Active Galaxies 101
    J.A. Peacock

The Quasar Luminosity Function 110
    B.J. Boyle

UK ROSAT Deep & Extended Deep Surveys. 118
    L.R. Jones et al.

Luminosity Dependence of Optical Activity in Radio Galaxies 121
    J.S. Dunlop & J.A. Peacock

Modelling the Quasar Luminosity Function in Hierarchical Models for Structure Formation. 123
    M. Haehnelt

Active Galactic Nuclei in Clusters of Galaxies 125
    A.C. Edge

Clustering Properties of AGNs and their Contribution to the X-ray Background ... 129
   L. Toffolatti

Energy Distributions of AGN ... 131
   C. Impey

Absorption in the ROSAT X-ray Spectra of Quasars ... 139
   B.J. Wilkes, M. Elvis, F. Fiore & J. McDowell

Dust in AGNs ... 143
   A. Laor & B.T. Draine

First Simultaneous UBVRI Photopolarimetric Observations of a Sample of Normal Quasars ... 147
   L.O. Takalo, A. Sillanpää, M.R. Kidger & J.A. de Diego

Intermediate Resolution Spectropolarimetry of Three Quasars ... 150
   J.A. de Diego, E. Pérez, M.R. Kidger & L.O. Takalo

Active Galaxies which Emit Strongly at 25μm ... 154
   R.D. Wolstencroft, C.J. Lonsdale & Q.A. Parker

III The Broad Line Region: Variability and Structure

Emission Line and Continuum Variability in Active Galactic Nuclei ... 159
   B.M. Peterson

Results of the LAG Monitoring Campaign ... 167
   E. van Groningen & I. Wanders

A Relation Between the Profiles and Intensities of Broad Emission Lines ... 176
   G.M. Stirpe

Broad Line Profile Variability in NGC 4593 ... 180
   M. Dietrich & W. Kollatschny

Deconvolution of Variable Seyfert 1 Profiles ... 184
   D.M. Crenshaw

Ultra-violet Variability of AGN ... 188
   P.M. Rodríguez-Pascual

Broad-Line Variations in NGC 5548 ... 192
   W. Kollatschny & M. Dietrich

NGC 4593: A Low Luminosity Compact Seyfert 1 Nucleus ... 194
   M. Santos-Lleó, J. Clavel, P. Barr & I.S. Glass

UV Continuum Origin and BLR Structure in F-9 ... 197
   M.C. Recondo-González, W. Wamsteker, F. Cheng & J. Clavel

UV Emission Line Intensities and Variability: a Self Consistent Model for
Broad-Line Emitting Gas in NGC 3783 199
    A. Koratkar & G.M. MacAlpine

Non-Linear Anisotropic BLR Models 203
    P.T. O'Brien & M.R. Goad

Anisotropic Line Emission from Extended BLR's 207
    M.R. Goad & P.T. O'Brien

Active Galactic Nuclei and Nuclear Starbursts 209
    R.J. Terlevich, G. Tenorio-Tagle, J. Franco & B.J. Boyle

Rapidly Evolving Compact SNRs and the Nature of the Lag in AGNs 215
    G. Tenorio-Tagle, R.J. Terlevich, M. Rozyczka & J. Franco

Supernova Explosions in QSOs? - II 218
    I. Aretxaga, R. Cid Fernandes & R. Terlevich

High Metallicities in QSOs 220
    G.J. Ferland & F. Hamann

The Chemical Evolution of QSOs 227
    F. Hamann & G.J. Ferland

Non-Linearity of Ly α Response in Variable AGNs 231
    J.C. Shields & G.J. Ferland

Implications of Broad Line Profile Diversity among AGN 235
    A. Robinson

Emission Line Studies of AGN 244
    B.R. Espey

A Search for Velocity Shifts in QSO Broad Lines 248
    T.A. Small & W.L.W. Sargent

Broad Line Region Structure from Profile Shapes 251
    K.L. Thompson

IV   X-rays and Accretion Disks

X-ray Variability in AGN 257
    T.J. Turner

Thermal Reprocessing of X-rays in NGC 5548 265
    J. Clavel, K. Nandra, K. Pounds & W. Wamsteker

New Ginga Observation and Model of NGC 6814 Periodicity 270
    S. Tsuruta, K. Leighly & R. Sivron

Power Spectrum Fits to EXOSAT Long Looks 274
    I.E. Papadakis & A. Lawrence

Dramatic X-ray Spectral Variability of Mkn 841 278
    I.M. George, K. Nandra, A.C. Fabian, T.J. Turner, C. Done & C.S.R. Day

Thermal and Non-Thermal Emission from Accretion Disks. 280
    A.C. Fabian & R.R. Ross

Ultra-Soft X-ray Emission in AGN 291
    E.M. Puchnarewicz & K.O. Mason

Highly Ionized Gas in Seyfert Galaxies 295
    K. Nandra

EUV Observations of Seyfert 1 Galaxies and Quasars. 299
    P.M. Gondhalekar & B.J. Kellett

0.1–20 keV Spectra of 3C 273 and E1821+643 302
    R.D. Saxton, M.J.L. Turner & A. Lawson

Iron Lines from Ionized Discs 304
    G. Matt, A.C. Fabian & R.R. Ross

Reflection Effects in Realistic Discs 306
    C. Burigana

X-Ray Polarization Properties in the Two-Phase Model for AGN. 308
    G. Matt & F. Haardt

X-Ray Reprocessing and UV Continuum in NGC 4151 310
    G.C. Perola & L. Piro

Dense Clouds Near the Center of Active Galactic Nuclei 312
    R. Sivron & S. Tsuruta

Accretion Discs in AGN Context: Hints Toward Non-Standard Discs? 314
    S. Collin-Souffrin

Accretion Disk Instabilities. 323
    J.F. Hawley & S.A. Balbus

Compton-Heated Winds from Accretion Disks 332
    C.F. McKee, D.T. Woods, J.I. Castor, R.I. Klein & J.B. Bell

Determination of a Transonic Solution in a Stationary Accretion Disc 336
    J. Papaloizou & E. Szuszkiewicz

Black Holes and Accretion Disks 340
    F.H. Wallinder

Testing the "Disc X-ray Reprocessing" in UV-Optical Continuum and Line
Emission in NGC 5548.. 345
    E. Rokaki

Accretion Discs in Realistic Potentials................................. 348
    R.J.R. Williams & J.J. Perry

Test of the Accretion Disc Model and Orientation Indicator.............. 350
    M. Joly

Orientation Effects in QSO Spectra...................................... 354
    P.J. Francis

The Luminosity-Colour Distribution of Quasar Accretion Disks............ 356
    D.M. Caditz

V   Beams, Jets and Blazars

Magnetic Propulsion of Jets in AGN................................... 361
    M.C. Begelman

MHD Accretion-Ejection Model: X- and γ-rays and Formation of Relativistic
Pair Beams.. 368
    G. Pelletier, G. Henri & J. Roland

Relativistic Electron Beams in AGN: Construction of Transonic Solutions....... 372
    G. Henri & G. Pelletier

Properties of Relativistic Jets... 374
    S. Appl & M. Camenzind

A Massive Binary Black Hole in 1928+738?................................ 377
    N. Roos, J.S. Kaastra & C.A. Hummel

Gamma-Rays from Blazars: a Comparison of 3C 279, PKS 0537–441 and Mrk 421.
    L. Maraschi, G. Ghisellini & A. Boccasile   381

Microquasars in the Galactic Centre Region.............................. 385
    I.F. Mirabel

A Comparison of the Ultra-violet Continuum Variability Properties of Blazars
and Seyfert 1s.. 389
    R. Edelson, G. Pike, J. Saken, A. Kinney, M. Shull & J. Krolik

Simultaneous Optical and IR Monitoring of the Seyfert Nucleus NGC 7469...... 393
    D. Dultzin-Hacyan, A. Ruelas-Mayorga, R. Costero & M. Alvarez

Broad-Band Spectra and Polarization Properties of Variable Flat-Spectrum
Radio Sources... 397
    S.J. Wagner & A. Witzel

The Radio to Optical Variability of the BL Lac Object ON 231. 404
    E. Massaro, R. Nesci, G.C. Perola, D. Lorenzetti, L. Spinoglio, M. Felli & F. Palagi

January 1992 Microvariability Campaign of OJ 287. 406
    A. Sillanpää, L.O. Takalo, K. Nilsson, S. Kikuchi, Yu. S. Efimov, N.H. Shakhovskoy, D. Dultzin-Hacyan, R. Costero, E. Benitez, M.R. Kidger & J.A. de Diego

Blazar Microvariability: a Case Study of AO 0235+164. 408
    J.C. Noble & H.R. Miller

Timescales of the Optical Variability of the BL Lacertae Galaxy PKS 2201+044 . .
    J.W. Wilson, H.R. Miller, J.C. Noble & M.T. Carini

Dynamics of Quasar Variability 410
    S. Christiani & R. Vio

The Variability of a Large Sample of Quasars. 412
    I.M. Hook, R.G. McMahon, B.J. Boyle & M.J. Irwin

The Fate of Central Black Holes in Merging Galaxies. 420
    F. Governato, M. Colpi & L. Maraschi

Polarimetric Searching for Goldstone Bosons from AGNs 422
    Yu. N. Gnedin & S.V. Krasnikov

Concluding Talk

Unification of AGNs, and the Starburst Hypothesis 427
    A.V. Filippenko

Index of Participants

Acosta-Pulido, J.A.	Inst. de Astrofísica de Canarias, Spain	44
Albrecht, P.	Univ.-Sternwarte Göttingen, Germany	
Alexander, P.	MRAO, Univ. of Cambridge, UK	58
Appl, S.	Landessternwarte Heidelberg, Germany	374
Aretxaga, I.	Univ. Autónoma de Madrid, Spain	218
Axon, D.J.	Nuffield Radio Astronomy Lab., UK	28, 66
Begelman, M.C.	Univ. of Colorado, USA	361
Boisson, C.	Observatoire de Paris, France	
Boksenberg, A.	Royal Greenwich Observatory, UK	
Boyle, B.J.	IOA, Univ. of Cambridge, UK	110, 209, 416
Burigana, C.	Univ. di Padova, Italy	306
Caditz, D.M.	Montana State Univ., USA	356
Cavaliere, A.G.	Univ. di Roma, Italy	
Celotti, A.	IOA, Univ. of Cambridge, UK	
Cid Fernandes, R.	IOA, Univ. of Cambridge, UK	218
Clavel, J.	ESTEC, Holland	194, 197, 265
Colina, L.	Univ. Autónoma de Madrid, Spain	78
Collin-Souffrin, S.	Inst. d'Astrophysique de Paris, France	314
Colpi, M.	Univ. di Milano, Italy	420
Crenshaw, D.M.	Computer Sciences Corp., USA	184
Cristiani, S.	Univ. di Padova, Italy	412
Davidsen, A.F.	Johns Hopkins Univ., USA	46
de Diego, J.A.	Inst. de Astrofísica de Canarias, Spain	147, 150, 406
Dietrich, M.	Univ.-Sternwarte Göttingen, Germany	180, 192
Done, C.	Univ. of Leicester, UK	278
Dultzin-Hacyan, D.	UNAM, Mexico	393, 406
Dunlop, J.S.	Liverpool John Moores Univ., UK	121
Dyson, J.E.	Univ. of Manchester, UK	66
Edelson, R.	NASA/Goddard Space Flight Center, US	389
Edge, A.C.	IOA, Univ. of Cambridge, UK	125
Elvis, M.	Center for Astrophysics, USA	139
Espey, B.R.	Univ. of Pittsburgh, USA	244
Fabian, A.C.	IOA, Univ. of Cambridge, UK	278, 280, 304
Ferland, G.J.	Univ. of Kentucky, USA	220, 227, 231
Filippenko, A.V.	Univ. of California, Berkeley, USA	427
Forbes, D.A.	IOA, Univ. of Cambridge, UK	55
Fosbury, R.A.E.	ST-ECF, ESO, Germany	3, 88
Foulsham, P.A.	Univ. of Leicester, UK	51

xv

Francis, P.J.	Steward Observatory, USA	354
George, I.M.	NASA/Goddard Space Flight Center, USA	278
Gnedin, Yu.N.	Royal Greenwich Observatory, UK	422
Goad, M.R.	Univ. College London, UK	203, 207
Gondhalekar, P.M.	Rutherford Appleton Lab., UK	299
González-Delgado, R.	Inst. de Astrofísica de Canarias, Spain	48, 53
Haehnelt, M.	IOA, Univ. of Cambridge, UK	123
Hamann, F.	Ohio State Univ., USA	220, 227
Hawley, J.F.	Univ. of Virginia, USA	323
Henri, G.	Observatoire de Grenoble, France	368, 372
Hes, R.	Kapteyn Inst., Groningen, Holland	
Hook, I.	IOA, Univ. of Cambridge, UK	416
Horne, K.	Space Telescope Science Inst., USA	
Hughes, P.A.	Univ. of Michigan, USA	94
Hunstead, R.W.	Univ. of Sydney, Australia	84
Impey, C.D.	Steward Observatory, USA	131
Inglis, M.D.	Univ. of Hertfordshire, UK	
Jackson, N.	Sterrewacht Leiden, Holland	24
Jenkins, C.	Royal Greenwich Observatory, UK	
Joly, M.	Observatoire de Paris, France	350
Jones, L.R.	Univ. of Southampton, UK	118
Kaastra, J.S.	SRON, Leiden, Holland	377
Kellerman, K.I.	Nat. Radio Astronomical Obs., USA	
Kirk, J.	MPI für Kernphysik, Germany	
Kotilainen, J.	IOA, Univ. of Cambridge, UK	
Koratkar, A.	Space Telescope Science Inst., USA	199
Kriss, G.A.	Johns Hopkins Univ., USA	46
Lahav, O.	IOA, Univ. of Cambridge, UK	
Laor, A.	Inst. for Advanced Study, Princeton, USA	143
Lawrence, A.	QMWC London, UK	118, 274
Liu, R.	Mullard Space Science Lab., UK	
Lüdke, E.	Nuffield Radio Astronomy Lab., UK	74
Malkan, M.A.	Univ. of California, Los Angeles, USA	
Maraschi, L.	Univ. di Milano, Italy	371, 420
Masegosa, J.	Inst. de Astrofísica de Andalucía, Spain	
Massaro, E.	Univ. di Roma, Italy	404
Matt, G.	IOA, Univ. of Cambridge, UK	304, 308
McDowell, J.	Center for Astrophysics, USA	139
McKee, C.F.	Univ. of California, Berkeley, USA	332
Menon, T.K.	Univ. of British Columbia, Canada	
Miller, H.R.	Georgia State Univ., USA	408, 410
Mirabel, I.F.	Service d'Astrophysique, Saclay, France	385

INDEX OF PARTICIPANTS

Moran, E.C.	Columbia Univ., USA	82
Morganti, R.	Ist. di Radioastronomia, Bologna, Italy	
Nandra, K.	IOA, Univ. of Cambridge, UK	265, 278, 295
Nazarova, L.	Royal Greenwich Observatory, UK	
Ne'eman, Y.	Tel-Aviv University, Israel	
Norman, C.	Space Telescope Science Inst., USA	
O'Brien, P.T.	Univ. College London, UK	203, 207
Okoye, S.E.	IOA, Univ. of Cambridge, UK	
Padovani, P.	ESO, Germany	90
Papadakis, I.E.	QMWC London, UK	274
Peacock, J.A.	Royal Observatory Edinburgh, UK	101, 121
Pelletier, G.	Observatoire de Grenoble, France	368, 372
Pérez, E.	Inst. de Astrofísica de Canarias, Spain	48, 53, 150
Pérez-Fournon, I.	Inst. de Astrofísica de Canarias, Spain	40, 44
Perry, J.J.	IOA, Univ. of Cambridge, UK	348
Persaud, J.	IOA, Univ. of Cambridge, UK	
Peterson, B.M.	Ohio State Univ., USA	159
Petitjean, P.	IOA, Univ. of Cambridge, UK	
Phinney, E.S.	California Inst. of Technology, USA	
Piro, L.	Ist. Astrofisica Spaziale, Frascati, Italy	310
Pogge, R.W.	Ohio State Univ., USA	36
Puchnarewicz, E.M.	Mullard Space Science Lab., UK	291
Rauch, M.	IOA, Univ. of Cambridge, UK	
Recondo-González, M.	IUE Observatory, Spain	197
Rees, M.J.	IOA, Univ. of Cambridge, UK	
Robinson, A.	IOA, Univ. of Cambridge, UK	48, 235
Rodríguez-Pascual, P.	IUE Observatory, Spain	188
Rokaki, E.	Inst. d'Astrophysique de Paris, France	345
Roos, N.	Sterrewacht Leiden, Holland	377
Santos-Lleó, M.	Observatoire de Paris, France	194
Sanz, J.L.	IUE Observatory, Spain	
Saxton, R.	Univ. of Leicester, UK	302
Shields, J.	Ohio State Univ., USA	231
Sillanpää, A.	Univ. of Turku, Finland	147, 406
Small, T.A.	California Inst. of Technology, USA	248
Snijders, T.	Astronomisches Inst. Tübingen, Germany	
Solomon, P.	Stony Brook, New York, USA	
Stirpe, G.M.	Osserv. Astronomico di Bologna, Italy	176
Storchi-Bergmann, T.	Inst. de Fisica, Porto Alegre, Brazil	31
Szuszkiewicz, E.	QMWC London, UK	336
Tadhunter, C.N.	Univ. of Sheffield, UK	13, 24, 48
Takalo, L.O.	Univ. of Turku, Finland	147, 150, 406

Tenorio-Tagle, G.	Inst. de Astrofísica de Canarias, Spain	209, 215
Terlevich, R.J.	Royal Greenwich Observatory, UK	209, 215, 218
Thompson, K.L.	California Inst. of Technology, USA	251
Toffolatti, L.	Osserv. Astronomico di Padova, Italy	129
Turner, T.J.	NASA/Goddard Space Flight Center, USA	257, 278
Tsuruta, S.	Montana State Univ., USA	270, 312
Vagnetti, F.	Univ. di Roma, Italy	96
Van Groningen, E.	Uppsala Astronomical Obs., Sweden	167
Wagner, S.J.	Landessternwarte Heidelberg, Germany	397
Wallinder, F.H.	NORDITA, Denmark	340
Wanders, I.	Uppsala Astronomical Obs., Sweden	167
White, S.D.M.	IOA, Univ. of Cambridge, UK	
Wilkes, B.J.	Smithsonian Astrophysical Obs., USA	139
Williams, R.J.R.	IOA, Univ. of Cambridge, UK	348
Williger, G.	CTIO, Chile	
Wolstencroft, R.D.	Royal Observatory Edinburgh, UK	154
Yaqoob, T.	IOA, Univ. of Cambridge, UK	
Young, S.	Univ. of Hertfordshire, UK	28

Preface

This volume contains the proceedings of the 33rd Herstmonceux Conference, the latest in a venerable series initiated by the Royal Greenwich Observatory in its former home at Herstmonceux Castle. It is the second conference in the series to have been jointly organized by the RGO and the Institute of Astronomy at Cambridge. However, it also marks a beginning. Both the timing and the subject matter of the meeting in Cambridge were co-ordinated with a companion conference in Paris. Together, the two meetings marked the inauguration of the European Association for Research in Astronomy. This grouping links together the Institute of Astronomy, the Institut d'Astrophysique de Paris, and Leiden Observatory with the intention of encouraging scientific exchanges between the three laboratories and enhancing their collaborative research activities. The Paris conference, entitled *First Light in the Universe: Stars or QSO's*, took place during July 7–11, 1992 at the Institut d'Astrophysique and was concerned with the cosmological evolution of galaxies and quasars, with particular emphasis on the alternative rôles played by starbursts and active galactic nuclei. In Cambridge, our aim was to focus in detail on the sources which power active galactic nuclei themselves.

Active galactic nuclei (AGN) are undoubtedly the most spectacular objects known to astronomy yet the nature of the fundamental power source remains elusive, despite many years of intensive research. Indeed, the somewhat ambiguous conference title reflects the fact that the conventional black hole–accretion disk paradigm is now being strongly challenged by the starburst hypothesis. Nevertheless, recent years have seen rapid progress stimulated, in particular, by a variety of large-scale monitoring campaigns in several wavebands and by the latest generation of satellite-borne observatories, including IRAS, ROSAT, HST and most recently, GINGA. Many new results from these campaigns and instruments were presented during the conference. The scientific programme was organized around five general topics: anisotropic emission and unified schemes for AGN; AGN luminosity functions and their implications for the power source; the structure and dynamics of the broad emission line region; the variability and spectral energy distribution of the high-energy continuum and their implications for accretion disk models and lastly, theories and observed properties of jets in AGN. Each of these areas has recently seen important observational or theoretical advances which have fuelled the debate as to the nature of the "compact object".

On behalf of the Organizing Committee we would like to thank the staff and students of the Institute of Astronomy and the Royal Greenwich Observatory who contributed to the organization of the meeting. We are especially grateful to Alice Duncan, without whose efficiency in carrying out the many detailed tasks of organization and administration, the meeting could not have taken place. Particular thanks are also due to Dr. Michael Ingham and his staff, Sofia Bridgeman, Eileen Fenton, Margaret Harding and Judith Moss who coped admirably with an influx of over 100 visiting participants.

Finally, it is a pleasure to record a notable occasion during the course of the meeting; the unveiling, in the presence of its subject, of a statue of Professor Sir Fred Hoyle in the grounds of the Institute of Astronomy. Not only was Sir Fred a founder of the Institute and a pioneer in this field of astronomy (and in many others), but we can be sure that a good many of the participants were inspired to follow careers in astronomy by his popular books and articles.

A. Robinson
R.J. Terlevich
Cambridge, June 1993.

Evidence and Implications of Anisotropy in AGN

Evidence for Anisotropy and Unification

R.A.E. Fosbury[*]

Abstract

It is probable that the nuclei of all active galaxies radiate anisotropically either due to intrinsic beaming and/or due to extrinsic causes such as shadowing. Observations of scattered light and fluorescently excited extended line emission can be used to map the radiation pattern. The best established of the AGN unification schemes appears to be that for powerful radio sources. High redshift radio galaxies reveal their quasar/blazar nuclei clearly in their rest-frame ultraviolet radiation. Studies of local objects show in detail the physical process which are operating in these distant sources.

1. Introduction

The concept of radiation anisotropy has been remarkably successful in allowing the unification of various classes of active galaxy and quasar. Whilst the possibility of explaining the differing appearance of *all* the classes of active extragalactic sources simply in terms of orientation effects seems unrealistic now, it does appear that broad subclasses can be collapsed in this way. Thus, for example, the three subgroups of Seyfert 1s and 2s, Fanaroff/Riley class one (FR I) radio galaxies and BL Lac objects, and powerful radio galaxies and quasars, can be understood as entities having similar central engines and nuclear environments, but whose orientation with respect to the observer is the dominant variable determining the detailed appearance.

Fundamental differences in the nature of the central energy source may well exist and, if so, will certainly break the unification at some level. The impressive evidence for active nuclei fuelled by stellar rather than collapsed object processes presented by Terlevich and collaborators leaves us with little doubt that complete unification will never be achieved.

Direct observational evidence in support of radiation anisotropy has been accumulating steadily over the past couple of decades and we have now reached the point where the hypothesis of unification allows us to make strong predictions which are subject to observational test. Observational data can be used to map the nuclear radiation field in both frequency and angle and so provide some firm constraints on the nature of the central engine and its immediate environment. Since the processes which produce the anisotropy occur on spatial scales which are probably smaller than we can resolve, it is important to try to distinguish between *intrinsic* emission effects, e.g. relativistic beaming or emission from a non-spherical optically-thick surface, and *extrinsic* processes such as shadowing.

---

[*] Space Telescope – European Coordinating Facility, Karl-Schwarzschild-Str. 2, W–8046, Garching bei München, Germany. Affiliated to the Astrophysics Division, Space Science Department, European Space Agency.

While there is a great deal of evidence of obscuration and radiation anisotropy amongst the classical Seyfert galaxies—including the beautiful 'shadowing' ionization cones exemplified by NGC 5252 (Tadhunter & Tsvetanov 1989)—the evidence for intrinsic beaming is inconclusive except for a few well-studied cases. The attempts to unify the two classes by orientation shows some promise but it is unlikely to be as clean as in the case of the radio galaxies and radio quasars. For this reason, I shall concentrate on the radio sources in this talk.

2. High redshift radio galaxies (HZRG)

Ever since reasonably detailed optical observations of these faint objects became possible, HZRG have occupied a place of considerable interest in the study of galaxy formation and evolution. This is principally because it is so difficult to identify 'normal' galaxies at redshifts sufficiently high to be of historical interest. The fact that they play host to a nuclear engine which is capable of driving the powerful large scale radio structure was the cause of some discomfort to these historians. It was their hope, however, that the global colours would encode the message of the early stages of star formation rather than simply reflect the nuclear activity. This was justified by analogy with a few, rarely occurring, nearby powerful radio galaxies which are perturbed only slightly by emission lines and nuclear non-thermal continuum at optical wavelengths.

The discovery by McCarthy et al. (1987) and by Chambers, Miley & van Breugel (1987) that, above a redshift of 0.7 or so, the rest-frame ultraviolet images of the most powerful radio galaxies showed marked elongations which aligned with their radio axes, sent a shudder through the historical establishment. Are these high-z objects really different from their low-z counterparts, perhaps due to some evolutionary process in the galaxy or its environment, or is it simply that we view them at wavelengths different by a factor of about two? Whatever the cause of the elongation, this is clearly a crucial question to answer.

What do these elongated images have to do with unification? The answer demands a lot of intrinsically difficult observational work but I think we now have enough data to build a quite convincingly self-consistent picture. In sketching it out, I will try to point out the areas of particular difficulty and uncertainty and show that the picture, whilst very promising, is not yet very firmly established.

Let us suppose that the quasar–blazar–radio galaxy unified scheme described by Barthel (1989) is correct in its principal features. What does this imply for the appearance of the HZRG? In the scheme, the radio galaxies are quasars/blazars with radio axes making angles larger than about 40° to our line of sight such that we do not view their nuclei directly without obscuration. This explains why we can see the galaxy light directly without the disturbance of the glare from the quasar*. But what happens to the 'misdirected' quasar light? Is it really

---

* I shall assume that a quasar contains a beamed continuum source which, while it does not necessarily contain a large fraction of the AGN luminosity, would appear as an apparently bright blazar if directed towards the observer.

hidden?

It is well established that the ionizing radiation from the 'hidden' AGN can produce the huge extended nebulosities that are common around radio galaxies (e.g. Robinson *et al.* 1987) as the 'visible' AGN does in quasars. The photon deficit measured in the galaxies is due to the radiation anisotropy (Fosbury *et al.* 1992). But what happens to the visible and ultraviolet AGN continuum? Is it really hidden by obscuration and/or intrinsic anisotropy? If we want the clearest view of this continuum, it seems that the best place to look is in the ultraviolet where the galaxy starlight—unless the stellar population is very young of course—is faint. Simple calculations based on the observed Hubble diagram of, say, the 3C catalogue (Cimatti *et al.* in prep.) show that it is only necessary to scatter a few % of the quasar continuum to reproduce the aligned structures seen in the HZRG. How do the observations match up with this simple 'beaming/scattering' picture (Tadhunter *et al.* 1989) and can we find a suitable scattering medium to do the job? It is a powerful and easily refutable hypothesis since it makes clear predictions regarding the properties of the fluorescently excited line and scattered continuum radiation.

As far as the line radiation is concerned, there appears to be little controversy since the relative line intensities are consistent with hard-spectrum AGN-like photoionization and the nebulæ are sometimes seen beyond the radio lobes where any direct radio jet/ISM interaction could not be operating (McCarthy 1987). The strong prediction is that the scattered continuum should be linearly polarized with the electric vector—provided the source is unpolarized—always perpendicular to the photon path from the AGN to the scatterer. This is true whether the scatterers are electrons or dust grains although the scattered spectrum and the wavelength dependence of the polarization will be different in the two cases.

The current state of the polarization observations is encouraging but not entirely conclusive: see the recent papers by Tadhunter *et al.* 1992 and di Serego Alighieri, Cimatti & Fosbury 1992 for complete references). The weak points are to do with the spatial structure of the polarization. This is simply because the sources are very faint and imaging polarimetry is severely photon limited[‡]—as is the spatially resolved spectroscopy! In the very few cases where the polarization has been mapped (Scarrott, Rolph & Tadhunter 1990, di Serego Alighieri *et al.* 1992), the extranuclear polarization is indeed perpendicular and is as strong or stronger than the integrated polarization, but this really needs to be established for more objects before we feel comfortable. The other worry is to do with the secure identification of the polarization with a scattering rather than a non-thermal emission process. In general, the aligned extensions to these HZRG are not co-spatial with the radio emission but we have to be sure of this in the cases where we see extended polarization. Indeed, we can argue on the basis of the radio-optical spectral index that the optical extensions are not likely to be

---

[‡] The earliest imaging polarimetry of 3C368 by di Serego Alighieri, Fosbury, Quinn & Tadhunter (1989) was criticised by Meisenheimer & Hippelein (1992) on the grounds that they had underestimated the photon statistical errors by a factor > 2. After discussion, Meisenheimer & Hippelein now agree that their criticism was unfounded. On re-evaluation, however, di Serego Alighieri *et al.* have revised their error on the fractional polarization of 3C368 from 0.9% to 1.3%.

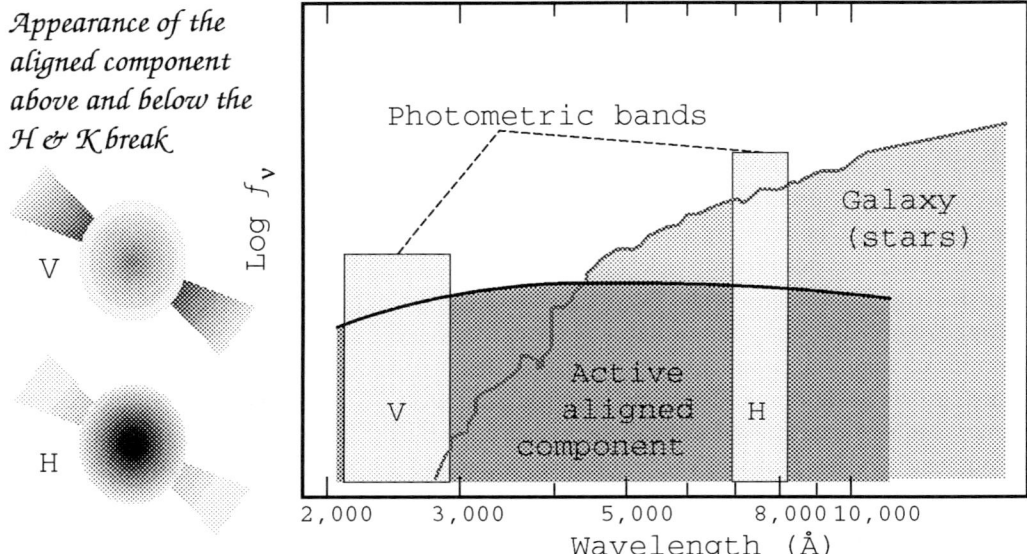

Figure 1. The spectral decomposition of a high redshift radio galaxy into a blue 'aligned' and a red 'symmetric' component. This cartoon is adapted from the data presented in Rigler et al. (1992). Given the very different energy distributions, imaging observations with filters below the rest-frame H & K break will be dominated by the aligned component.

the high frequency tail of the radio jets and hot-spots—but each individual case *has* to be examined.

I think that the strongest argument that we are on the right track comes from the analysis of the images by decomposition into two spectrally distinct components which has been performed most extensively by Rigler *et al.* (1992) using optical and infrared images of a sample of 3C sources. They showed that the aligned component is blue ($f_n \sim$ const) and contributes on average no more than about 10% to the more symmetric structures seen in the K-band ($\sim 1\mu m$ in the rest-frame). This result, by establishing the limit to the contribution of the 'active' aligned component to what is presumably a population of well-evolved stars, has removed a concern about the small scatter seen in the infrared Hubble diagram of powerful radio galaxies which has been pointed out by Lilly (1989).

If scattering is a significant contributor to this blue aligned component, the cartoon in figure 1 illustrates our expectation for the behaviour of the integrated fractional polarization as a function of the galaxy rest-frame wavelength of the filter passband used for the observation. Because of the very different energy distributions of the symmetric and the aligned components, their relative dominance of the integrated light will switch suddenly at a wavelength close to the H & K break near 4000Å. The observed polarization will 'switch on' at rest-frame wavelengths below the break. The plot in figure 2, adapted from the poster by Cimatti, di Serego Alighieri & Fosbury (this volume), shows just this behaviour for a collection of sources which have polarization measurements through an aperture which includes the whole galaxy rather than just the nuclear regions. Although we know that AGN-photexcited emission lines, which are presumably (but not demonstrably) unpolarized, contribute to the aligned component, this can be corrected for. In addition, a careful photometric subtraction of the (red, unpolarized) stellar component shows residual polarizations which are very high and leave little room for any other source of luminosity (Cimatti *et al.* in prep.).

This work shows that the beaming/scattering model is, so far, successful in explaining the photometric, polarimetric and morphological properties of the HZRG but that there are still several hurdles it needs to pass before we are confident. Perhaps the most pressing is the spectrum of the scattered light. There is no evidence yet for broad, quasar-like, emission lines from the aligned component. If this remains true with more sensitive, spatially-resolved spectroscopy then, in order to retain the basic idea, we will be forced to conclude either that the scattering is dominated by the intrinsically beamed blazar component which, on axis, swamps the BLR or that the scattering medium is composed of hot electrons which wash out the spectral features in velocity. This question demands the application of multi-band imaging polarimetry and sensitive spectroscopy on these very faint sources. An intriguing clue to the nature of the scattered light comes from recent work by Dunlop & Peacock (priv. comm.) which shows that the aligned component seen in the infrared, while weaker than at shorter wavelengths, is more *precisely* aligned with the radio axis. This suggests that the blazar beam may become increasingly dominant towards the red.

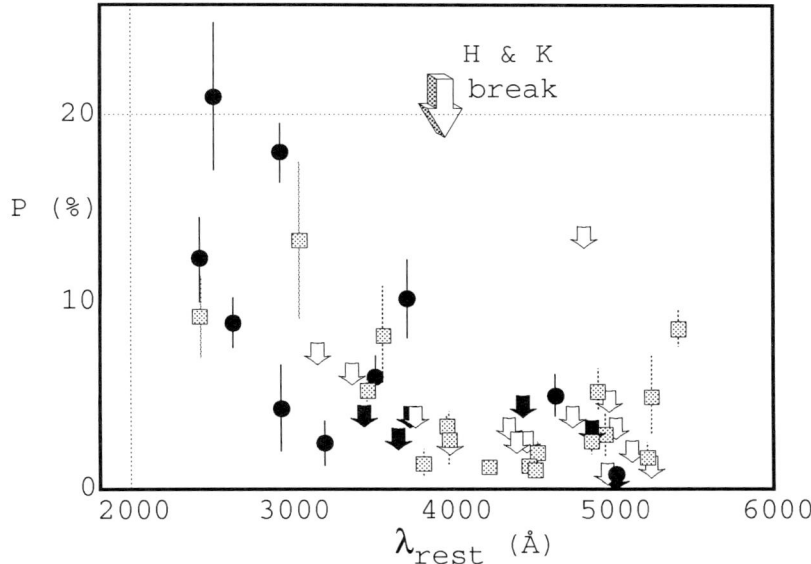

Figure 2 *A collection of polarization measurements of the spatially integrated light of radio galaxies collected by Cimatti, di Serego Alighieri & Fosbury (poster paper, this volume). They consist of measurements and upper limits by several observers. The fractional polarization is plotted against the wavelength of the filter passband transformed to the rest frame of the galaxy. This shows the polarization 'switching on' at wavelengths shortward of the H & K break in a manner which is readily understood if it is the blue, aligned component which is highly polarized.*

Figure 1 shows clearly why the appearance of an object is such a strong function of wavelength but we are still left with the question of whether the differences between the HZRG and local powerful radio galaxies is purely a 'K-correction' effect or whether there are real differences as a result of evolution of the galaxy or its environment. One obvious test is to use the HST to look at local powerful radio galaxies in the ultraviolet to see if they show any signs of the alignment effect. For technical reasons, this is not as easy as it sounds but I hope that it will be done soon. We can, however, use the local objects to map the anisotropy of the radiation field and investigate other details of the physical processes.

3. Nearby radio galaxies

I want to use a couple of local objects as examples of what we can do to investigate beaming anisotropy but also to point out a few dangers and uncertainties in the assumptions which are commonly made.

Perhaps the best evidence we have of extranuclear dust scattering from a beamed continuum source is the galaxy PKS 2152–69 (Tadhunter *et al.* 1987, 1988, di Serego Alighieri *et al.* 1988, Fosbury *et al.* 1990). Here we have a very highly excited cloud some 10kpc from the nucleus almost, but not quite, along the axis of the double radio source. What is remarkable is not so much the line emission—which is extreme with ions ranging up to Fe^{9+}—but the associated optical and ultraviolet continuum. This is very blue with a spectrum of $f_n \mu n^{+3}$ in the optical, turning down to become flat in the ultraviolet (di Serego Alighieri *et al.* 1988, Fosbury *et al.* 1990). The cloud as a whole is 10% linearly polarized in B with the electric vector perpendicular to the nuclear vector. The polarization becomes 12% if the emission line contamination is removed from the B-band. This is an assumption; we do not *know* that the emission lines are unpolarized but it seems clear that they are locally excited and not scattered. The question has been asked, however, by Meisenheimer & Hippelein (1992) following a detailed study of 3C368 in which they cannot find enough extended continuum to explain the polarization.

These properties are a classical signature of dust scattering and it is hard to think of an alternative mechanism which would produce such a blue, polarized source. It is argued that the source must be intrinsically beamed on two grounds. If the flux of radiation impinging on the cloud was emitted isotropically but absorbed in other directions, the infrared reradiation would be enormous and this is not seen by IRAS. Also, the galaxy is surrounded by emission line filaments which have a low state of ionization and are clearly not seeing the same nuclear luminosity as that seen by the cloud.

This object seems to be a good low-z model of the HZRG process, albeit on a small scale. An image in the UV would show the cloud continuum to be almost as bright as the host galaxy and the 'aligned continuum' would have approximately a flat spectrum (in f_n). It would be polarized and approximately aligned with the radio axis. At around 1μm—corresponding to K in the HZRG—the cloud is dominated by the galaxy starlight. The scattering in this case is well-modelled by dust (Fosbury *et al.* 1990) with relatively small grains producing a Rayleigh-like spectrum in the optical. Recent work by Binette & Magris (1992) has shown that an incident blazar spectrum can produce the line spectrum by photoionization and the continuum by dust scattering in a mutually consistent manner. They have also addressed the important question of the effect of dust on the Ly-a transfer in this type of nebulosity. Since we know that many of the HZRG are strong Ly-a emitters, it has been argued that this precludes the presence of significant quantities of dust in these objects since the resonantly scattered Ly-a photons would be too readily absorbed. It is important to realise, however, that the EELR geometry is quite different from internally ionized H II regions. In the radio galaxies, the extranuclear clouds are ionized by an *external* source and the Ly-a escape is strongly dependent on the viewing aspect. In their model of PKS 2152–69, Binette & Magris reproduce the observed Ly-a/Hb ratio in the presence of sufficient dust to scatter the continuum. This external illumination geometry also explains, perhaps, why we do not see much extinction—measured by the Balmer decrement—in EELR (Robinson *et al.* 1987). We have to look for dust in other ways such as the depletion of elements like calcium in nebulæ with an appropriate state of ionization (Ferland 1992).

The study of PKS 2152–69 explains my preference for dust as a scattering medium in the outer regions radio galaxies but I do not want to give the impression that Thomson scattering cannot play a rôle. In cluster cores with dense hot coronæ (Fabian 1989) and close to the nucleus, electrons can do the job. The case of NGC 1068 illustrates the complex mix of processes which can occur in the nuclear regions of a Seyfert (Miller, Goodrich & Mathews 1992).

Centaurus A is our closest radio galaxy although it is not intrinsically powerful on the scale of the high-z objects. According to our current understanding of the BL Lac parent population (Padovani & Urry 1990, 1991 Urry, Padovani & Stickel 1991), this—as a FR I source—should host this type of nucleus. The direct evidence is based partly on infrared and hard X-ray observations of the nucleus which can penetrate the obscuration of $A_v \sim 30$. Bailey *et al.* (1986) see a nucleus which is variable in flux but not in polarization and Fiegelson *et al.* (1981) and Morani *et al.* (1989) see variability by an order of magnitude in hard X-rays.

On the assumption that they are photoionized by the AGN, Morganti *et al.* (1991, 1992) have used the emission line filaments—which stretch out along the direction of the radio and X-ray jet to the northeast—as 'photon counters' to measure the nuclear photon luminosity in this direction. They deduce a value for the beamed nuclear luminosity which is close to that exhibited by BL Lac itself (it would be 4th magnitude in V if pointed towards us) and argue, on the basis of the far infrared luminosity and the direct X-ray and nearer infrared measurements of the nucleus, that the beaming must be intrinsic. While their assumption of AGN photoionization is based on the similarity of the filament spectra with other radio galaxy EELR when examined in line ratio diagnostic diagrams, there remains the alternative possibility that the gas is ionized locally by radiative shocks induced by the passage of the radio jet, This argument has been used by Sutherland, Bicknell & Dopita

(1992) who suggest that the existence of a chaotic velocity structure makes the presence of such shocks likely. The existence of two such different hypotheses to explain the same phenomenon illustrates one of the limitations of plasma diagnostics. It is very difficult to derive the detailed shape of the ionizing continuum from observations of a limited range of emission lines. As pointed out by Robinson *et al.* (1987), the diagnostic diagrams allow us to measure the *mean energy* of the ionizing photon spectrum but not really its *shape*. In this case the models do make different predictions for the ultraviolet lines but our most immediate hope may be to look at the locus traced by the observed lines from different filaments in the line ratio diagrams. The photon beam model predicts that the points scatter along the ionization parameter locus while I believe the shock model would, for a range of shock velocities, predict a spectrum shape locus which is approximately orthogonal. The data (Morganti *et al.* 1991) appear to favour the beam picture although this problem needs to be examined in more detail.

The Cen A story highlights a profound uncertainty in this type of analysis. This is the question of time variability and the existence of light echo effects (Ekers & Liang 1990). If we, as observers, sit within a BL Lac beam, we see a rapidly variable intensity and polarization. This does not, however, tell us much about the flux properties integrated over the beam profile, which are more relevant to the fluorescence and scattering observations. Also, we know little of the variability over periods of 10^{4-5} years or so which is the light travel time for the scales we are discussing. If there are dramatic variations in the AGN output on this timescale then the result will be strong side-to-side asymmetries in the appearance of the illuminated material. For example, if the inner Cen A filament—at a projected nuclear distance of a few kpc—is illuminated by a symmetric pulse from the nucleus and the jet is inclined at 30° to our line of sight, the counter jet would be seen much closer to the nucleus and possibly coincident with the peculiar 'knot-X' discussed by Phillips (1981). Is this why we don't see a counter-filament on a larger scale?

The beaming/scattering hypothesis implies that we observe a number of asymmetric properties. Attempts to find these will be promising lines of observational research. Dust scattering has strong forward/backward asymmetries which should be apparent for sources with axes inclined to the plane of the sky. This should show up—at least statistically—against the isotropic, AGN excited line emission.

4. Conclusions

However far the ideas of anisotropy and unification can be used to simplify our ideas about the structure of AGN it is clear that they have been most successful in stimulating a great deal of sharply directed work. Critics of unification argue that the differences in orientation do not explain everything. I agree: AGN have a complex structure. The recognition that they are not spherically symmetric was a clear signal that polarization observations of scattered light would be important. This has proved correct but these observations really do demand large telescopes and I don't think we will be able to go very much further with the high redshift objects before we gain access to them. The systematic errors in these measurements seem to be fairly well under control and the real limits are source photons and sky brightness.

Acknowledgments: I should like to thank the many collaborators who have worked with me on these topics over the years. In particular, I thank Andrea Cimatti and Sperello di Serego Alighieri for allowing me to discuss some of our polarization results here.

References

Bailey, J., Sparks, W.B., Hough, J.H. & Axon, D.J., 1986. Nature, 322, 150
Barthel, P., 1989. Astrophys. J., 336, 606
Binette, L. & Magris. G., 1992. In "The nearest active galaxies", Madrid, 11–14 May 1992. Ed. J. Beckman, in press
Chambers, K.C., Miley, G.K. & van Breugel, W., 1987. Nature, 329, 604
di Serego Alighieri, S., Binette, L., Courvoisier, T. J.-L., Fosbury, R.A.E. & Tadhunter, C.N., 1988. Nature, 334, 591
di Serego Alighieri, S., Cimatti, A. & Fosbury, R.A.E. 1992. Astrophys. J., in press
Ekers, R.D. & Liang, H., 1990. In: "Parsec scale radio jets", Ed. Zensus & Pearson, CUP, p333
Fabian, A.C., 1989. Mon. Not. R. astr. Soc., 238, 41p
Ferland. G., 1992. In "The nearest active galaxies", Madrid, 11–14 May 1992. Ed. J. Beckman, in press
Fiegelson, E.D., Schreier, E.J., Delavaille, J.P., Giacconi, R., Grindlay, J.E. & Lightman, A.P., 1981. Astrophys. J., 251, 31
Fosbury, R.A.E., di Serego Alighieri, S., Corvoisier, T. J.-L., Snijders, M.A.J., Walsh, J. & Wilson, W., 1990. In "Evolution in Astrophysics, IUE astronomy in the era of new space missions", Toulouse, France, 29 May–1 June 1990. ESA SP-310
Fosbury, R.A.E, Morganti, R., Robinson, R. & Tsvetanov, Z., 1992. In proc. 7th IAP astrophysics meeting: "Extragalactic radio sources - from beams to jets". Paris, 2-5 July 1991. Ed. Roland, Sol & Pelletier, CUP, p350
Lilly, S.J., 1989. Astrophys. J., 340, 77
McCarthy, P.J., 1987. PhD. Thesis, University of California
McCarthy, P.J., van Breugel, W., Spinrad, H. & Djorgovski, S., 1987. Astrophys. J., 321, L29
Meisenheimer, K. & Hippelein, H., 1992. Astron. Asrtrophys., in press
Miller, J.S., Goodrich, R.W. & Mathews, W.G., 1991. Astrophys. J., 378, 47
Morini, M., Anselmo, F. & Molenti, D., 1989. Astrophys. J., 347, 750
Morganti, R., Robinson, A., Fosbury, R.A.E., di Serego Alighieri, S., Tadhunter, C.N. & Malin, D.F., 1991. Mon. Not. R. Astr. Soc., 249, 91
Morganti, R., Fosbury, R.A.E., Hook, R.N., Robinson, A. & Tsvetanov, Z., 1992. Mon. Not. R. Astr. Soc., 256, 1p
Padovani, P. & Urry, C.M., 1990. Astrophys. J., 356, 75
Padovani, P. & Urry, C.M., 1991. Astrophys. J., 368, 373
Phillips, M.M., 1981. Mon. Not. R. Astr. Soc., 197, 659
Rigler, M.A., Lilly, S.J., Stockton, A., Hammer, F. & Le Fèvre, O., 1992. Astrophys. J., 385, 61
Robinson, A., Binette, L., Fosbury, R.A.E., & Tadhunter, C.N., 1987. Mon. Not. R. astr. Soc., 227, 97
Scarrott, S.M., Rolph, C.D. & Tadhunter, C.N., 1990. Mon. Not. R. astr. Soc., 243, 5p
Sutherland, R.S., Bicknell, G.V. & Dopita, M.A.D. 1992. In proc: "Astrophysical jets, a STScI symposium", May 12–14, 1992. in press
Tadhunter, C.N.& Tsvetanov, Z., 1989. Nature, 341, 422
Tadhunter, C.N., Fosbury, R.A.E., Binette, L.A., Danziger, I.J. & Robinson, A. 1987. Nature, 325, 504
Tadhunter, C.N., Fosbury, R.A.E., di Serego Alighieri, S., Bland, J., Danziger, I.J., Goss, W.M.,

McAdam, B. & Snijders, M.A.J., 1988. Mon. Not. R. astr. Soc., 235, 403
Tadhunter, C.N., Fosbury, R.A.E. & di Serego Alighieri S., 1988. In: "BL Lac objects – 10 years after: Proceedings of the Como conference 1988". Eds: Maraschi, Maccacaro & Ulrich., Springer-Verlag, p79
Tadhunter. C.N., Scarrott, S.M., Draper, P. & Rolph, C. 1992. Mon. Not. R. astr. Soc., 256, 63p
Urry, C.M., Padovani, P. & Stickel, M., 1991, Astrophys. J., 382, 501

Any Evidence against Unified Schemes?

Clive N. Tadhunter [*]

Abstract

Various observational tests of unified schemes for radio sources are reviewed critically. It is shown that, although some of the results might raise doubts, there exists no definitive evidence against these schemes.

1. Introduction

There is now considerable evidence to support the idea that the radiation from active galactic nuclei (AGN) is emitted anisotropically. For radio-loud objects much of the evidence has been gathered at radio wavelengths, and includes the measurements of superluminal motions in radio cores (e.g. Zensus 1988), the statistics of core and lobe dominated radio sources (e.g. Orr & Browne 1982), and the polarization asymmetry in the extended radio structures (e.g. Garrington et al. 1988, Laing 1988). For the radio-quiet objects the beautiful emission line images of radiation cones (e.g. Pogge 1989, Tadhunter & Tsvetanov 1989), and the detection of scattered broad lines within the extended regions (e.g. Miller et al. 1991), are also very strong evidence in favour of anisotropy.

These observations are comforting for the proponents of unified schemes who use anisotropy to explain the relationships between certain classes of AGN in terms of orientation effects. However, not all of the evidence is so positive, and in this paper I review some recent results which are less supportive of unified schemes.

2. Types of unified schemes

There is not just one unified scheme based on anisotropy/orientation effects, but several, and it is important to be clear about what we are discussing. Fig. 1 summarises the major types. Most are based on anisotropies produced by relativistic beaming and/or by extinction in a large-scale dusty disc or torus.

I will concentrate on unified schemes for radio-loud active galaxies and, in particular, on the proposal that radio galaxies and radio-loud quasars are the same thing viewed from different directions (e.g. Peacock 1987, Scheuer 1987, Barthel 1989). However, many of the comments that I will make also apply to the other unified schemes.

---

[*]Department of Physics, University of Sheffield, Sheffield, S3 7RH

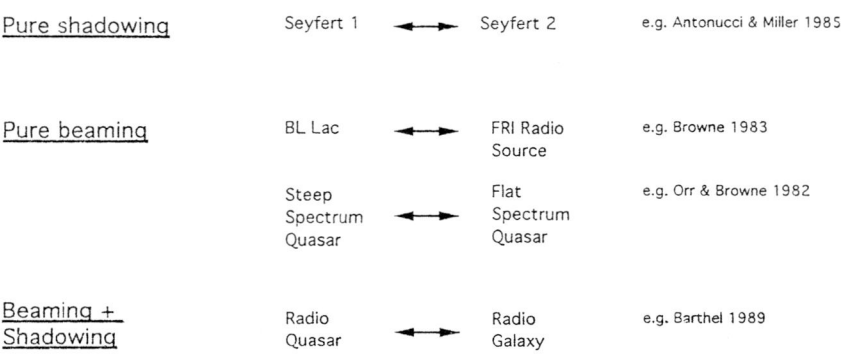

Figure 1. The main types of unified scheme based on anisotropy/orientation.

The radio galaxy/radio-loud quasar (hereafter RGQ) unification is perhaps the most ambitious yet proposed. It requires elements of both relativistic beaming and disc obscuration: the relativistic beaming to explain the relationship between the flat spectrum compact and steep spectrum extended radio quasars/galaxies, and the disc obscuration to "hide" the optical quasar broad lines and blue bump continuum in the radio galaxies. The general idea is that when the line of sight is within an angle θ_1 to the radio axis, the beamed component dominates and the object appears as a flat-spectrum compact quasar with a highly polarized superluminal core; at intermediate angles to the radio axis ($\theta_1 < \theta < \theta_2$) the object appears as a steep spectrum radio quasar; and at large angles to the radio axis ($\theta_2 < \theta < 90$) the optical quasar nucleus is completely obscured by the dusty disk/torus and the object appears as a narrow line radio galaxy.

It is important to emphasise that RGQ is a "two cone" unified scheme with separate cone opening half-angles θ_1 and θ_2 defined, respectively, by the relativistic beaming Lorentz factor and by the details of the disc obscuration. Measurements of the relative extended radio diameters, of the relative numbers of the different types, and of superluminal motions in the cores of the quasars support values of the beaming cone opening half-angle in the range $5 < \theta_1 < 15$, and values of the disc obscuration cone opening half-angle in the range $30 < \theta_2 < 45$ (Orr & Browne 1984, Browne 1983, Barthel 1989, Padovani & Urry 1992).

3. Tests of the unified schemes

When we test the unified schemes it is important to ensure that we are comparing like with like. For the radio galaxies this is usually done by selecting objects using their extended radio properties under the assumptions that the extended emission is emitted isotropically and is also a good indicator of the overall level of activity in the AGN[1]. The 3CR catalogue is useful for this purpose, since it is selected at low frequencies, and most 3CR objects are dominated by extended steep-spectrum radio emission.

A full list of tests that have been made of the radio-loud unified schemes is presented in Figure 2. It would be impossible to do full justice to all of these tests in a short review, so I concentrate below on a limited selection of tests made in the EUV–optical–infrared wavelength range, although I note that interesting results are also emerging at radio wavelengths (see Alexander 1992, this volume).

3.1 Relative numbers

The first questions we might ask about the RGQ unification are: do *all* radio galaxies have hidden quasars in their cores, and can we apply the same model parameters (e.g. same θ_1, θ_2) to radio galaxies across the range of radio power?

Partial answers to these questions have been provided by Lawrence (1991), who has considered how the relative numbers of broad line (BL) and narrow-line (NL) objects in the 3CR catalogue change as a function of radio power. If all powerful radio galaxies have luminous BL nuclei, and the unification model parameters are the same at all radio powers, then the relative frequency of BL and NL objects should not change as a function of power. *In fact, as Lawrence has shown, no broad line nuclei have been detected in the lowest power objects, whereas at the high power end $\sim 50\%$ or more of the 3CR objects are quasars.*

There are observational selection effects which may influence this result. In particular, if the quasar luminosity is a strong function of extended radio power, then the quasar broad lines and continuum might be very weak in the low power objects, and consequently difficult to detect against the diluting narrow lines and stellar continuum. Balancing this to some extent is the fact that the broad lines are easiest to detect at $H\alpha$, but $H\alpha$ falls outside the observable range for many of the high power/high redshift objects (Lawrence 1991). Personally, I believe that the former effect will dominate, and that we may be missing several weak BL nuclei at the low power end. A concerted effort is clearly required to search for weak broad lines in the lower power radio sources.

If we ignore the observational selection effects then there are two possible explanations for the change in the BL/NL frequency with power:

---

[1] N.B. Neither of these assumptions has ever been proved beyond doubt.

Extended Optical Emission	Alignments	Baum & Heckman 1989; McCarthy et al. 1987
	Ionizing Fluxes	Morganti et al. 1991
	Scattered Fluxes	Tadhunter 1990
Orientation Invariants	Far-IR	Heckman et al. 1992
	[OIII]	Jackson & Browne 1990; Lawrence 1991
	Ext. Radio	Antonucci & Ulvestadt 1985
Relative frequency of different types	NL/BL	Lawrence 1991
	SSQ/FSCD/Gal.	Orr & Browne 1982 Padovani & Urry 1992
	BL Lac/FRI	Browne 1983
Quasar nuclear properties	BL width vs. R	Wills & Browne 1986
	BL EW vs. R	Jackson et al. 1989
	Cont. vs. R	Browne & Murphy 1987
Extended radio	Projected sizes	Barthel 1989
	Depolarization Asymmetry	Garrington et al. 1988 Laing 1988
	Spectral-index Asymmetry	Liu & Pooley 1991
	Jet/Counterjet Ratio	Davis et al. 1991
Environments	Host galaxies	Hutchings 1987
	Clusters	Prestage & Peacock 1988 Hill & Lilly 1991
Scattered light in NL objects	Broad lines	Antonucci 1984
	Continuum Shape	Jackson & Tadhunter 1992

Figure 2. A list of observational tests of unified schemes for radio-loud objects.

1. The properties of the obscuration change with radio power e.g. the quasar nuclei are more strongly obscured at the low power end (see the discussion in Lawrence 1991).

2. Not all of the radio galaxies have powerful quasar nuclei all of the time, and the quasar nuclei occur much less frequently at the low radio powers. For example, the nuclei may be more highly variable at the low power end and spend more time in an "off" state. In at least one relatively low power radio source, the broad line activity has been found to switch from an "off" to an "on" state on a timescale of a few years, after apparently spending several years in the "off" state (Ulrich 1981).

In either case the results imply that the RGQ scheme in its simplest form cannot be applied in the same way to radio galaxies of all powers. This is not surprising in view of the fact that the properties of the extended radio sources change markedly from low radio powers — where the radio morphologies are predominantly of the FRI class — to high radio powers — where the sources are generally classified as FRII type[2]. There is now much evidence to support the idea that the FRI radio galaxies are the parent population of BL Lac objects rather than quasars (see Browne 1989 and references therein).

3.2 Orientation invariants — I. The narrow line luminosities.

Another way to approach the problem is to assume that the narrow emission lines are emitted isotropically by an extended region outside the central obscuring disk/torus, and then to compare the emission line luminosities of radio galaxies and quasars of similar extended radio power.

If the unified schemes are correct, and the assumption of isotropic NLR radiation holds, then the ranges of luminosities covered by the two groups should be the same, but *Baum & Heckman (1989), Jackson & Browne (1990) and Lawrence (1991) have shown that quasars have roughly a factor of 4 — 10 greater NL luminosities than powerful radio galaxies of similar extended radio power.*

Although this could mean that the quasars are intrinsically different from the radio galaxies, it is more plausible that the assumption of isotropic NL emission is wrong, and that the NLR is obscured to some extent in the radio galaxies. We can envisage two extreme modes of obscuration:

1. **Weak obscuration**: the entire NLR undergoes weak obscuration by a veil of obscuring material along the line of sight. It should be possible to test this

---

[2]However, Lawrence (1991) claims that the increase in relative BL frequency with radio power is present if the FRII sources are considered separately, although it is not clear whether the effect is as significant as in the combined sample.

idea by examining the narrow line ratios: the galaxies should show signs of stronger reddening (e.g. in the Balmer decrements) than the quasars. Jackson & Browne (1990) have already pointed out that NL of many radio galaxies are strongly reddened, but little is known about the NL reddening in the quasars.

2. **Strong obscuration**: a large fraction of the NLR is completely hidden by the central obscuring disk/torus in the radio galaxies. This idea might be tested by comparing the NL profiles and physical conditions of quasars and radio galaxies: if the NL are emitted closer to the AGN in the quasars, this should be reflected in more extreme emission line kinematics (e.g. broader narrow lines) and physical conditions (e.g. higher temperatures).

Of course, the reality may prove to be somewhere between these two. The degree of obscuration will depend on the scale of the main NLR region, and in this context it is notable that, while most radio galaxies have extended emission line regions, the bulk of the NL emission is concentrated in an unresolved region centred on the nucleus (Baum et al. 1988). Furthermore, direct evidence for a compact NLR is provided by the observations of strong NL variability in the BLRG 3C390.3 (Clavel & Wamstekar 1987; Perez et al. 1992), which suggest a NLR scale of 10 light years or less. Such a compact NLR would be easy to "hide" by a disc with the same scale height as that observed in Centaurus A, for example, and if most radio-loud objects have similarly compact NLR, then the strong obscuration picture is likely to be closest to the truth.

3.3 Orientation invariants — II. Far-IR luminosities

The far-IR continuum luminosities can be used in a similar way to the NL luminosities: that is, comparing the far-IR luminosities of radio galaxies and quasars of similar extended radio power, under the assumption that the far-IR radiation is emitted isotropically and represents re-processed AGN light — in this case optical/UV/X-ray continuum which has been absorbed by dust in the obscuring disk, then thermally re-radiated in the far-IR.

Although few radio galaxies and steep spectrum quasars have been detected directly by IRAS, Heckman et al. 1992 have investigated the relative far-IR properties by combining IRAS scans for large samples of 3CR radio galaxies and quasars. For this work, a redshift limit of $z > 0.3$ was chosen to ensure that the radio galaxies and quasars have similar distributions of extended radio power and redshift.

The comparison between the median far-IR fluxes is most reasonable, since these are least affected by the presence of a few extreme data points in the samples. Heckman et al. find that the *3CR quasars have median far-IR fluxes that are a factor $\sim 2\pm-0.7$ larger than the median far-IR fluxes of radio galaxies in the matched $z > 0.3$ samples.* The error bars are large, but the fact that the same result has been obtained in three IRAS bands suggests that it is significant. The difference is even larger if the mean

IRAS fluxes are considered — quasars have ~5× the mean far-IR fluxes of radio galaxies — but the mean fluxes are highly susceptible to a few extreme objects in the quasar sample.

At first sight these results appear to provide strong evidence against the unified schemes, but again we must question the assumptions underlying the test.

First, could a large fraction of the far-IR light in the 3CR quasars be non-thermal, blazar-like continuum rather than thermal re-processed radiation? This point is controversial. Sanders et al. 1989 have shown that the spectral energy distributions (SEDs) of some radio-loud PG quasars have radio to far-infrared slopes that are steeper than a "mean" blazar continuum. The same authors point out that the radio-loud quasars show the same "universal minimum" in their SEDs at $\sim 1\mu m$ as radio-quiet quasars.

A close examination of the SEDs in Impey & Neugebauer (1989) shows, however, that there exists a considerable variety in the SEDs of blazars, and many have similar SEDs to the steep spectrum 3CR quasars. Indeed, one of the few 3CR quasars in the sample of Heckman et al. 1992 which has been detected by IRAS — 3CR351 — is included in the Impey & Neugebauer (1989) paper as a blazar.

Thus it is plausible that a large fraction of the far-IR emission in some steep-spectrum quasars is non-thermal blazar-type emission. Presumably the non-thermal radiation would be beamed, and would act to boost the far-IR fluxes of the quasars relative to the radio galaxies. At optical wavelengths the highly polarized blazar continuum would be strongly diluted by the thermal "blue-bump" continuum component.

There is also the question of the isotropy of the far-IR emission, even if it does prove to be thermal dust emission: are the columns in the obscuring discs enough to produce a significant optical depth in the far-IR? Assuming gas/dust mix typical of our galaxy, a neutral hydrogen column depth of $N(HI) \sim 5 \times 10^{23}\ cm^{-2}$ is required to produce a $50\mu m$ dust optical depth of $\tau_{50\mu m} \sim 1$ (e.g. Heckman et al. 1992). Estimating the column depths in NLRG is fraught with difficulties, because the "real" nucleus is rarely observed at wavelengths shorter than the mm-radio range. However, Djorgovski et al. have detected the IR nucleus of the extremely powerful low-z radio galaxy Cygnus A, and have used the comparison between infrared and radio observations to deduce a visual nuclear extinction of $A_v \sim 30 - 90$ magnitudes. For typical Galactic ISM conditions this translates into a hydrogen column density of $N(HI) \sim 0.7 - 1.7 \times 10^{23}\ cm^{-2}$, which compares well with the estimate $N(HI) \sim 0.5 - 1.8 \times 10^{23}\ cm^{-2}$ obtained from an analysis of the X-ray spectrum by Arnaud et al. 1987. Note that these may actually *underestimate* the true column in Cygnus A, since, on the one hand, the IR measurements may be contaminated by hot dust emission from the circumnuclear regions, and on the other, the X-ray measurement may be contaminated by scattered hard X-ray radiation. It is also worth pointing out that direct measurements of the hard X-ray emission in Seyfert 2 galaxies give columns in the range N(HI)$\sim$ 0.5 – 7 $\times 10^{23}\,cm^{-2}$ (Mulchaey et al. 1992). Even larger columns (N(HI)$\sim 10^{25}\,cm^{-2}$) are suspected for the Seyfert 2 galaxies which have not been detected at hard X-ray wavelengths. If such large columns are also common in the radio galaxies, then the resultant anisotropies in the far-IR radiation fields would be sufficient to explain all of the effect noted by Heckman et al. 1992.

3.4 Extended emission line distributions

Extended emission line regions have provided some of the best evidence to support the idea of anisotropy (see Fosbury 1992 this volume, and references therein), but how do the extended emission line distributions match up with the detailed predictions of the unified schemes?

Assuming that the same radiation cone opening angles apply at all wavelengths, then, given an isotropic distribution of ISM, the distribution of extended emission line gas in the radio galaxies should reflect the angular dependence of the anisotropic ionizing radiation field of the hidden AGN.

Unfortunately, the ISM distributions in radio galaxies are both sparse and inhomogeneous, with large scale blobs, arcs and filaments rather than a regular, smooth screen of material (e.g. Baum et al. 1988). This perhaps explains why no clearly defined cone-like structures have been observed in the extended emission line regions of radio galaxies, while several have been detected in the Seyfert galaxies, which have less sparse ISM distributions.

It is reasonable nonetheless to use the statistics of the alignments between the radio axis and the extended emission line regions in an attempt to map the radiation fields. The most extensive work of this type has been done by Baum & Heckman (1989) for low redshift $z < 0.2$ radio galaxies, and by McCarthy et al. (1987) for higher redshift ($z > 0.5$) objects. Both studies show a tendency for the radio axis to line up with the extended emission lines (i.e. zero position angle difference) and, if we examine the distributions of the radio-emission line position angle differences, then for the low-z radio galaxies the width of the distribution is $\Delta\theta \sim 45°$, while for the high-z radio galaxies the width is $\Delta\theta \sim 20°$. Thus, the alignments appear to present strong evidence in favour of anisotropy, although, especially in the low-z case, the radiation cones must be broad (opening half-angles $\sim 45°$) — suggesting that the emission line distributions map the broader cone caused by the disk obscuration.

An interesting feature of these results is that, interpreted purely in terms of the angular dependence of the radiation field, they appear to provide evidence for a narrowing of the radiation cones as the redshift/radio power increases. This is in apparent contradiction with the finding that the relative number of broad line objects increases with redshift, which suggests the opposite result (Lawrence 1991, see section 3.1).

But could factors other than the radiation field influence the emission line alignments?

One potentially important mechanism is interaction between the ambient ISM and the radio-emitting plasma. There is direct observational evidence in some cases that the jets may excite the emission line regions (e.g. van Breugel et al. 1985), but perhaps more seriously, it has been proposed that the radio jets can cause density enhancements in the warm gas (e.g. de Young 1989, Rees 1989). The high velocity gas components observed in many powerful radio galaxies provide strong evidence for the disturbing influence of the radio plasma on the ISM (e.g. Tadhunter 1991). Thus the emission line/radio plasma alignments may have more to do with anisotropies in the gas distributions, induced by jet/ISM interactions, than with anisotropies in the radiation fields.

The overall message is that extreme caution is required when interpreting the radio/optical alignments: the ISM in radio galaxies may not just be sparse and inhomogeneous, it may be sparse, inhomogeneous *and* anisotropic.

3.5 Scattered broad lines

Many of what I have called "tests" of the unified schemes are little more than consistency arguments: determining whether a certain property in a NL object is consistent with a certain property in a BL object; whether there is evidence for anisotropy of the right sort for the unified schemes; whether the numbers of objects in the different classes are consistent with the model parameters.

However, what we really want to determine is whether or not all NL objects contain luminous BL nuclei. None of the tests that I have considered so far are capable of doing this. The only hope is to search for broad lines in the scattered light from the extended regions surrounding the nuclei — a technique which has been applied successfully to a number of Seyfert 2 galaxies (Antonucci & Miller 1985, Miller & Goodrich 1990).

The problem with this approach for the radio galaxies is that few of the optically bright, nearby radio galaxies show evidence for scattered light in the form of large linear polarization. This dearth of highly polarized objects is probably due, at least in part, to the sparsity of the scattering ISM in the radio galaxies. There is also the problem that, at low redshifts, the scattered light at optical wavelengths is strongly diluted by the continuum light from the old stellar populations in the host galaxies. It is only in a few extreme cases that the scattered light dominates.

The few existing spectropolarimetric studies of powerful radio galaxies have provided mixed results for the unified schemes. In the case of the highly polarized object 3C234, Antonucci (1984) has found strong evidence for a scattered broad Hα line, while in Cygnus A — a somewhat less highly polarized object — sensitive spectropolarimetry observations have failed to reveal any evidence for a scattered broad line (Jackson & Tadhunter 1992, these proceedings). The case of Cygnus A illustrates many of the pitfalls of this type of work, with a strong dilution of the scattered light by the old stellar continuum, and the possibility of a contribution to the polarization by mechanisms other than scattering.

Despite the difficulties I believe that the spectropolarimetry observations have much potential. There is evidence that the proportion of highly polarized 3C radio galaxies increases with redshift (Tadhunter et al. 1992), and it is becoming feasible to search for scattered broad lines in the fainter, higher redshift radio galaxies with the new generation of sensitive spectropolarimeters. If scattered broad lines are detected in a significant fraction of the highly polarized radio galaxies, then this will provide strong evidence for the RGQ scheme. On the other hand, the failure to detect the scattered broad lines is not definitive evidence against the unified schemes: it could just mean that the scattering is by hot electrons which doppler broaden the broad lines and reduce their chances of detection.

4. Conclusions

By their very nature unified schemes are slippery. We have seen that there exists no conclusive evidence against the radio galaxy/quasar unification, and there seems to be no single test capable of providing a definite yes/no answer (the search for scattered broad lines comes closest).

On the positive side, the observational tests are leading to a refinement in our knowledge of the central regions of active galaxies, and particularly in our knowledge of the geometry of the emitting regions. This refinement process is likely to continue as more tests are made in the future.

Acknowledgements. I thank Neal Jackson and Bob Fosbury for several stimulating discussions.

References

Antonucci, R.R.J., 1984. *Astrophys.J.*, **278**, 499.
Antonucci, R.R.J. & Miller, J.S., 1985. *Astrophys.J.*, **297**, 621.
Antonucci, R.R.J., & Ulvestadt, J.S., 1985. *Astrophys.J.*, **294**, 158.
Arnaud, K., Johnstone, R. Fabian, A., Crawford, C., Nulsen, P., Shafter, R. & Mushotzky, R., 1987. *Mon.Not.R.astr.Soc.*, **227**, 241.
Barthel, P.D., 1989. *Astrophys.J.*, **336**, 606.
Baum, S.A., Heckman, T., Bridle, A., van Breugel, W. & Miley, G., 1988. *Astrophys.J. Suppl. Ser.*, **68**, 643.
Baum, S.A. & Heckman, T., 1989. *Astrophys.J.*, **336**, 702.
Browne, I.W.A., 1983. *Mon.Not.R.astr.Soc.*, **213**, 23p.
Browne, I.W.A., 1989. In: *Proceedings of the Como Conference on BL Lac Objects*, L. Maraschi et al. (eds), Springer-Verlag, p401.
Browne, I.W.A. & Murphy, D.W., 1987. *Mon.Not.R.astr.Soc.*, **226**, 601.
Clavel, J. & Wamstekar, W., 1987. *Astrophys.J.*, 320, L9.

Davis, R.J., Unwin, S.C. & Muxlow, T.W.B., 1991. *Nature*, **354**, 374.
de Young, D.S., 1989. *Astrophys.J.Lett.*, **342**, L59.
Djorgovski, S., Weir, N., Matthews, K. & Graham, J.R., 1991. *Astrophys.J.*, **372**, L67.
Garrington, S.T., Leahy, J.P., Conway, R.G. & Laing, R.A., 1988. *Nature*, **331**, 147.
Miller, J.S. & Goodrich, R., 1990. *Astrophys.J.*, **355**, 456.
Heckman, T.M., Chambers, K.C. & Postman, M., 1992. *Astrophys.J.*, **391**, 39.
Hill, G.J. & Lilly, S.J., 1991. *Astrophys.J.*, **367**, 1.
Hutchings, J.B., 1987. *Astrophys.J.*, **320**, 122.
Impey, C.D. & Neugebauer, G., 1988. *Astron.J.*, **95**, 307.
Jackson, N., Browne, I.W.A., Murphy, D.W. & Saikia, D.J., 1989. *Nature*, **338**, 485.
Jackson, N. & Browne, I.W.A., 1990. *Nature*, **343**, 43.
Laing, R.A., 1988. *Nature*, **331**, 149.
Lawrence, A., 1991. *Mon.Not.R.astr.Soc.*, **252**, 586.
Lawrence, A. & Elvis, M., 1982. *Astrophys.J.*, **256**, 410.
Liu, R. & Pooley, G., 1991. *Mon.Not.R.astr.Soc.*, **249**, 343.
McCarthy, P., van Breugel, W., Spinrad, H. & Djorgovski, S., 1987. *Astrophys.J.*, **321**, L29.
Morganti, R., Robinson, A., Fosbury, R.A.E., di Serego Alighieri, S., Tadhunter, C.N. & Malin, D.F., 1991. *Mon.Not.R.astr.Soc.*, **249**, 91.
Miller, J.S., Goodrich, R. & Mathews, W., 1991. *Astrophys.J.*, **378**, 47.
Mulchaey, J.S., Mushotzky, R.F. & Weaver, K.A., 1992. *Astrophys.J.*, **390**, L69.
Orr, M.J.L. & Browne, I.W.A., 1982. *Mon.Not.R.astr.Soc.*, **200**, 1067.
Padovani, P. & Urry, C.M., 1992. *Astrophys.J.*, **387**, 449.
Peacock, J.A., 1987. In: *Astrophysical Jets and their Engines*, Kundt, W. (ed.), p185, D. Reidel, Dordrecht, Holland.
Perez et al., 1992. *Mon.Not.R.astr.Soc.*, submitted.
Pogge, R., 1989. *Astrophys.J.*, **345**, 730.
Prestage, R.M. & Peacock, J.A., 1988. *Mon.Not.R.astr.Soc.*, **230**, 131.
Rees, M.J., 1989. *Mon.Not.R.astr.Soc.*, **239**, 1p.
Sanders, D.B., Phinney, E.S., Neugebauer, G., Soifer, B.T. & Matthews, K., 1989. *Astrophys.J.*, **347**, 29.
Scheuer, P.A.G., 1987. In: *Superluminal Radio Sources*, Zensus, J.A. & Pearson, T.J. (eds), p104, Cambridge University Press.
Tadhunter, C.N., 1990. In: *New Windows on the Universe, Proceedings of the XIth ERAM Meeting, Vol.2*, Sanchez & Vazquez (eds), Cambridge University Press, p175.
Tadhunter, C.N., 1991. *Mon.Not.R.astr.Soc.*, **251**, 46p.
Tadhunter, C.N. & Tsvetanov, Z., 1989. *Nature*, **341**, 422.
Tadhunter, C.N., Scarrott, S.M., Draper, P. & Rolph, C., 1992. *Mon. Not.R. astr.Soc.*, **256**, 53p.
Ulrich, M-H., 1981. *Astron.Astrophys.*, **103**, L1.
van Breugel, W., Miley, G., Heckman, T., Butcher, H. & Bridle, A., 1985. *Astrophys.J.*, **290**, 496.
Wills, B.J. & Browne, I.W.A., 1986. *Astrophys.J.*, **302**, 56.
Zensus, J.A., 1988. In: *Proceedings of the Como Conference on BL Lac Objects*, L. Maraschi et al. (eds), Springer-Verlag, p3.

Spectropolarimetry of Cygnus A

N. Jackson [*]   C. N. Tadhunter [†]

Abstract

We fail to detect a polarized broad line in the 2% polarized narrow line radio galaxy Cygnus A. The possible explanations are that a diluting dichroic component is present (in which case more sensitive observations should show the polarized broad line), that a hidden broad line nucleus is being scattered by electrons which smear out the broad line, or that Cygnus A really is a narrow line radio galaxy and contains no "hidden quasar".

1. Introduction

There is now considerable evidence that some apparently narrow emission line objects contain hidden broad line regions. The first discovery in this field was the observation (Antonucci and Miller 1985) that the Seyfert 2 (narrow line) galaxy NGC 1068 had a broad line region in polarized light, which they interpreted as a scattered image of a broad line nucleus whose direct line of sight was blocked by an obscuring torus. Further such discoveries in Seyfert galaxies (Miller and Goodrich 1990), in one broad line radio galaxy (3C234, Antonucci 1984) and in one low–power narrow line radio galaxy (Inglis et al 1992) lend support to the idea that at least some narrow line objects would appear as broad line objects if seen from a suitable angle.

Most of the objects so far shown to contain polarized broad lines have been low–luminosity objects. Proposals that narrow–line and broad line high–luminosity, radio loud objects can be unified by a scenario such as that of Antonucci and Miller have been made over the last few years (Scheuer 1987, Peacock 1987, Barthel 1989). The obvious experiment to do in order to provide a powerful plausibility argument is to do spectropolarimetry of narrow line radio galaxies in order to try to find a hidden broad line – quasar – nucleus.

We chose the radio galaxy Cygnus A for this purpose. It is nearby, relatively bright and is known to contain continuum polarization (Tadhunter et al. 1990). Moreover, it contains a bright infra–red nucleus which has been proposed as a dust–enshrouded hidden quasar (Djorgovski et al. 1991, Ward et al 1991). We have been conducting spectropolarimetry of this object with the intention of finding – or at least setting limits on – any polarized broad line and also to investigate the nature of the polarized

---

[*]Sterrewacht Leiden, Postbus 9513, 2300RA Leiden, Netherlands
[†]Department of Physics, University of Sheffield, Sheffield S3 7RH, England

continuum. Results on 40 minutes' data were presented by us earlier (Jackson and Tadhunter 1992). Here we present the preliminary results from a much larger quantity of data (Jackson and Tadhunter, in preparation).

2. Observations and results

The observations were taken on the William Herschel Telescope on La Palma using the ISIS spectrograph and spectropolarimetry apparatus (for details see Tinbergen and Rutten 1992) at a resolution of 9Å in order to cover the range from Hβ to Hα. The total integration time was 12 hours over three nights (1991 July 8,9,10).

Figure 1 presents the spectrum in Stokes Q and U, and figure 2 presents the polarization percentage spectrum. Several things are clear from these figures. First, there is no immediately obvious broad line in the Stokes spectra. Second, there is a substantial dilution at the position of the narrow lines in the blue, but probably less in the red. Third, there is a decrease in the polarization level from blue to red.

3. Discussion and conclusions

Suppose Cygnus A contained a quasar nucleus, and suppose further that the continuum polarization observed by Tadhunter et al. (1990) was due to this nucleus being scattered by dust particles. We would then expect to see a broad line in the polarized spectrum of similar width and equivalent width to those of ordinary quasars. This we do not see: with the signal–to–noise in our polarized spectrum we would be able to detect a typical Hα broad line of equivalent width 400Å and width 80Å.

One alternative is that at least some of the continuum polarization comes from dichroic absorption in the host galaxy of Cygnus A. (It is probably not from our own galaxy: a nearby galaxy which also fell into our slit has a much lower polarization in a different direction from that of Cygnus A). It is particularly plausible that the red region of the spectrum – where the polarization falls to $\simeq 1\%$ – is mostly polarized in front of the narrow line region, as here the narrow lines do not cause much dilution of the overall polarization level. In this case we would expect observations a factor of 3 more sensitive to detect the polarized broad line (although such observations are a job for an 8m telescope).

The second alternative is that hot electrons are scattering the hidden quasar nucleus and smearing out the broad line (Fabian 1989). This is energetically quite plausible even though the scattering cross section is very small (the mass of gas required in the scattering region is $\leq 10^8 M_\odot$). It requires a reasonably large stellar contribution in order to lower the high degrees of polarization expected from electron scattering to the observed 2%. The major constraint on electron scattering models is that the electron scattering function does not depend strongly on wavelength, and so the polarized spectrum would be expected to resemble that of a quasar. Depending on

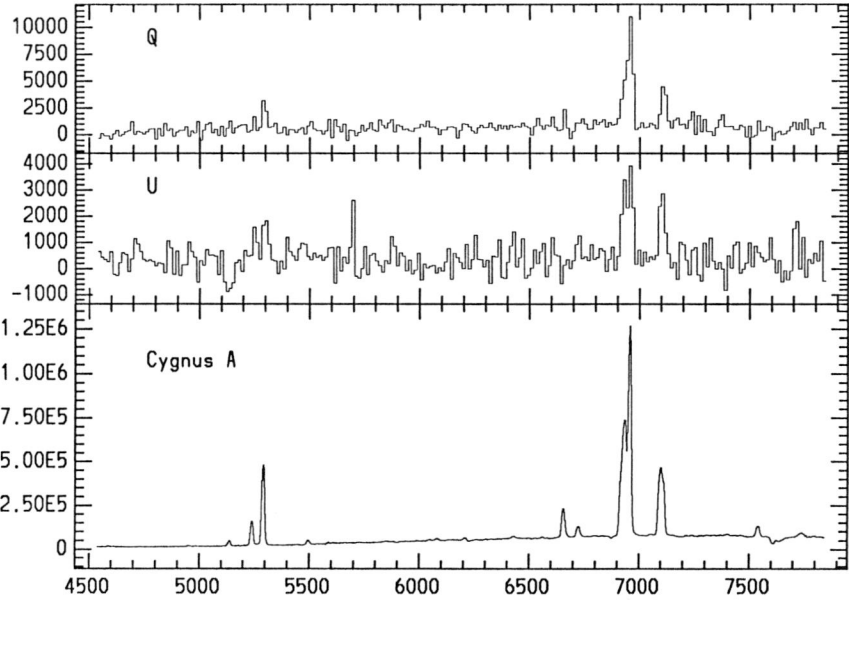
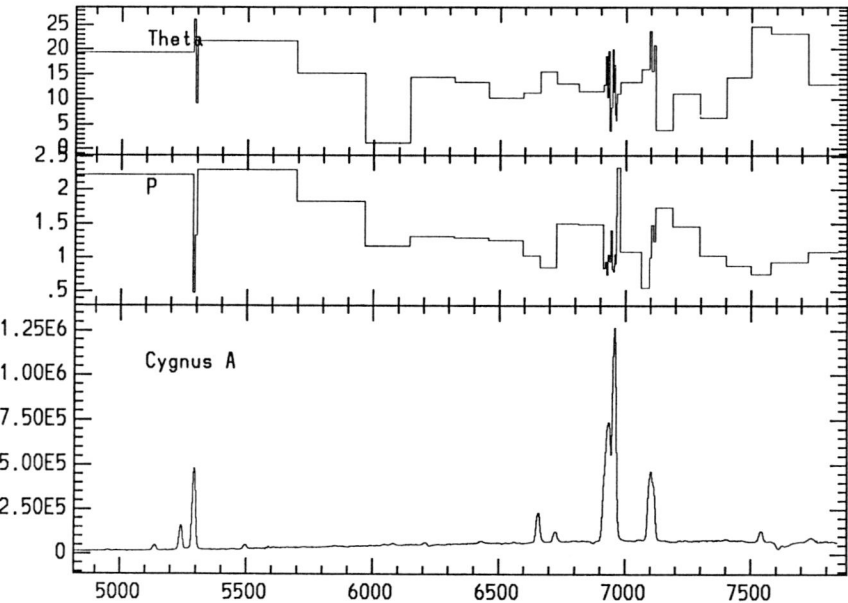

Figures 1 (top) and 2. Polarization spectra of Cygnus A (see text).

the exact reddening correction applied, this may just be possible, but our best current estimate is that the polarized spectrum is rather bluer than that of a typical quasar. If there really is substantial contamination by dichroic polarization, whose wavelength dependence is a Serkowski law (Serkowski, Mathewson and Ford 1975) the observed polarized spectrum will then become much bluer than that of a quasar and favour dust rather than electron scattering.

The third possibility is simply that, although Cygnus A contains a hidden nucleus (Djorgovski et al.) this nucleus is a radio galaxy nucleus and not a quasar. We note in this regard that no high power narrow line radio galaxy has yet been shown to contain polarized broad line emission indicative of a hidden broad line object. A search is under way to find such an object.

Acknowledgements. The William Herschel Telescope is operated on the island of La Palma by the Royal Greenwich Observatory at the Spanish Observatorio del Roque de los Muchachos of the Instituto de Astrofísica de Canarias. NJ thanks the Dutch SRON and ASTRON for financial support.

References

Antonucci, R. R. J., 1984, ApJ 278, 499
Antonucci, R. R. J. and Miller, J. S., 1985, ApJ 297, 621
Barthel, P. D., 1989, ApJ 336, 606
Djorgovski, S., Weir, N., Matthews, K., and Graham, J. R., 1991, ApJ 372, 67
Fabian, A. C., 1989, MNRAS 238, 41P
Inglis, M. et al., 1992, Mon. Not. R. astr. Soc.: submitted.
Jackson, N., and Tadhunter, C. N., 1992, in "Physics of Active Galactic Nuclei", proc. Heidelberg conference 1990, ed. S. Wagner, Springer–Verlag.
Miller, J. S. and Goodrich, R. W., 1990, ApJ 355, 456
Peacock, J. A., 1987, in W. Kundt (ed.), Astrophysical Jets and their Engines, p. 185, D. Reidel, Dordrecht, Holland
Scheuer, P. A. G., 1987, in J. A. Zensus and T. J. Pearson (eds.), Superluminal Radio Sources, p. 104, Cambridge University Press
Serkowski, K., Mathewson, D. S., and Ford, V. L., 1975, ApJ 196, 261
Tadhunter, C. N., Scarrott, S. M., and Rolph, C. D., 1990, MNRAS 246, 163
Tinbergen, J., and Rutten, R., 1992, Measuring polarization with ISIS on the WHT, RGO, Cambridge
Ward, M. J., Blanco, P. R., Wilson, A. S., and Nishida, M., 1991, ApJ 382, 115

Spectropolarimetry of the Ultraluminous Infrared Galaxy IRAS 110548-1131

S. Young * J.H. Hough* J.A. Bailey † D.J.Axon ‡
M.J.Ward §

Abstract

From spectropolarimetric observations of the galaxy, IRAS 110548-1131 we report strong, broad Hα emission (FWHM 7600 $km\ s^{-1}$) in the polarized flux spectrum. This suggests that IRAS 110548-1131 has an obscured broad line region, whose radiation is scattered into our line-of-sight by scatterers outside the obscured region.

1. Introduction

Since the discovery by Antonucci & Miller (1985) and its confirmation by Bailey et al. (1988) that the type II Seyfert galaxy NGC 1068 has broad hydrogen Balmer lines in it's polarized flux spectrum, and therefore has a Seyfert I type nucleus, it has been possible to construct a physical model in which the two types of Seyfert are in fact the same. Whether an object is seen as a type I or II depends upon the orientation of the galaxy and obscuration of the broad line region, probably in the form of a dusty torus, although the universality of such a model is an open question. Radiation from the BLR can escape along the axis of the torus and then be scattered into the line-of-sight by electrons and/or dust. The scattered component of these lines should then be observable in polarized flux.

In recent years there has been considerable interest in those galaxies identified from the IRAS survey which are very luminous in the far infrared region. Some evidence that highly luminous IRAS galaxies contain obscured QSO nuclei comes from the work of Hough et al. (1991), who from a broad band optical and IR polarization study of IRAS 23060+0505 proposed that the polarization is mainly due to the scattering of radiation from a QSO nucleus.

For our spectropolarimetry programme a selection of warm IRAS galaxies was chosen from the list due to deGrijp, Miley & Lub (1987). These are active galaxies based

---

*Division of Physical Sciences, University of Hertfordshire, Hatfield, Herts AL10 9AB.
†Anglo-Australian Observatory, PO Box 296, Epping, NSW 2121, Australia.
‡University of Manchester, Nuffield Radio Laboratories, Jodrell Bank, Macclesfield, Cheshire, SK12 9DL.
§University of Oxford, Nuclear and Astrophysics Laboratory, Keble Road, Oxford OX1 3RH.

on their colours at 25, 60 and 100 μm. Although the work is not yet complete we report the results on one of our objects IRAS 110548-1131, because of its extreme characteristics.

2. Observations

Observations of IRAS 110548-1131 were carried out at the Anglo-Australian telescope on 1992 March 7, 8 & 9 using the Hatfield waveplate modulator in conjunction with the RGO spectrograph and the 25cm camera with the 600R grating. A two aperture dekker, with each aperture 1.3 by 2.7 arcseconds in size with a projected separation of 23 arcseconds, was used to provide object and sky measurements. The object was switched between the two apertures every 1000 seconds, with a total observation time of 12,000 s centred on Hα and 8,000 s in the [O III], Hβ region.

The instrumental polarization was measured to be less than 0.1%; for flux calibration, the Oke standard L745-46 was used. The data were reduced using the STARLINK TSP package.

3. Comments

IRAS 110548-1131 has been classified as a Seyfert II galaxy with a Hα emission line of FWHM 260 $km\ s^{-1}$. Previous broad band polarimetry by Brindle et al. (1990) gave an average polarization of about 3% at a position angle of around 120 degrees between the B and H bands, in a 6" aperture. The object has extended [O III] emission with a bipolar structure along position angles 16 and 213 degrees (van Heerde 1988).

Our spectropolarimetry (Fig. 1), shows extremely broad Hα emission in the polarized flux. Fitting a gaussian to the broad line profile gives a FWHM of 7600 $km\ s^{-1}$ and FWZI of 16800 $km\ s^{-1}$, far in excess of the 260 kms^{-1} FWHM for the narrow Hα line. The measured flux of the broad line is 9 $10^{-15} ergs\ cm^{-2}\ s^{-1}$, and a luminosity 5.5 $10^{40}\ ergs\ s^{-1}$ ($z = 0.055$, $H_0 = 75\ km\ s^{-1}\ Mpc^{-1}$). If the scattering is optically thin then following Miller & Goodrich (1990), we calculate the total broad Hα luminosity as 6.9 $10^{43}\ ergs\ s^{-1}$, which also makes it one of the brighter Seyfert I galaxies; the scattered broad lines are also wider than for most Seyfert Is. The peak of the broad line is red-shifted by (900$\pm$400) $km\ s^{-1}$ with respect to the narrow Hα emission. This shift could be produced by radial outflow of the scatterers.

The average polarization for 110648-1131 is measured to be 3.23$\pm$0.09% at 118.6$\pm$0.8 degrees in very good agreement with the results of Brindle et al. (1990). This position angle is perpendicular to the axis of the extended narrow line emission, in accord with the polarization being due to scattering in this region.

Acknowledgements. SY is in receipt of a SERC Ph.D. studentship.

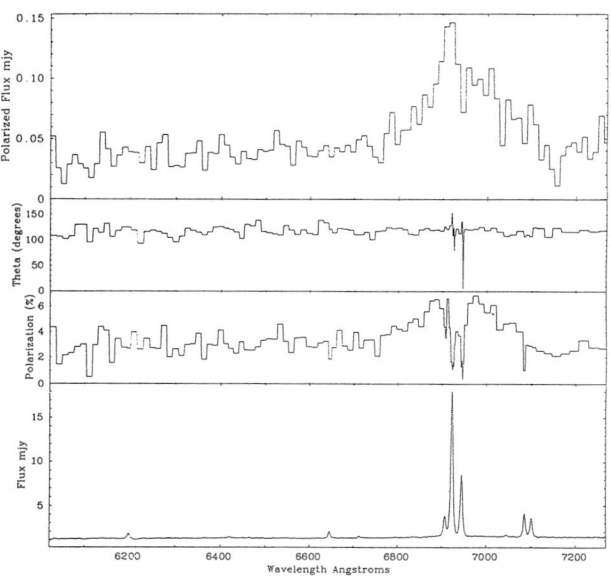

Figure 1. Spectropolarimetric data for IRAS 110548-1131.

References

Antonucci R.R.J., and Miller J.S., 1985, *ApJ*, **297**, 621

Bailey J.A., Axon D.J., Hough J.H., Ward M.J., McLean I. and Heathcote S.R., 1988, *MNRAS*, **234**, 899

Brindle C., Hough J.H., Axon D.J., Ward M.J., Sparks W.B., McLean I.S., 1990, *MNRAS*, **244**, 577

deGrijp M.H.K., Miley G.K., Lub J., 1987, *A&AS* **70**, 95

van Heerde G.M., 1988, *A&A*, **203**, 255

Hough J.H., Brindle C., Wills B.J., Wills D., and Bailey J., 1991, *ApJ*, **372**, 478

Miller J.S., Goodrich R.W., 1990, *ApJ*, **355**, 456

Are there Dusty Tori in Seyfert 2 Galaxies?

Thaisa Storchi-Bergmann *
John S. Mulchaey and Andrew S. Wilson [†]

Abstract

Recent work has revealed that a number of Seyfert 2 galaxies exhibit conically-shaped regions of gas apparently illuminated by a collimated, nuclear ionizing source. In this work, we test one model for this collimation, namely that the cones result from shadowing of a compact nuclear continuum source by a thick, dusty torus. From the emission-line ratios measured for gas within the cones, we have calculated the number of ionizing photons emitted by the compact nucleus. Then, on the assumption that the nuclear source radiates isotropically, we have found the power incident on the torus, which is expected to be reradiated in the infrared. Given the uncertainties in the calculation, and the fact that the torus may be somewhat anisotropic in the infrared, we find the observed IRAS luminosities are consistent with the torus model in 9 objects with sufficient data to perform the calculation.

1. Introduction

There is now considerable evidence that the ionizing photons escape from the nuclear source anisotropically in many Seyfert 2 galaxies (see Wilson 1992 for a recent review). For some objects, this anisotropy becomes evident in the elongated morphologies detected in images taken through narrow-band filters centred in high excitation lines (e.g. [OIII]λ5007) or through "ionization maps" (Pogge 1988a,b) – ratios between the continuum subtracted images in [OIII]λ5007 and in Hα+ [NII]$\lambda\lambda$6548,6583.

In this work, we combine new with existing data for a sample of nine Seyfert 2 galaxies with known "ionization cones" in order to test whether the collimation is the result of shadowing of radiation from a small, isotropic, nuclear source by a thick dusty torus, as suggested in the "Unified Scheme" for Seyfert galaxies (Antonucci & Miller 1985; Krolik & Begelman 1986). From long-slit spectroscopic data, the emission-line properties of the gas in the cones are obtained and used to calculate the luminosity of the ionizing source. Narrow-band images are used to obtain the opening angles of the cones which together with the luminosity of the source are used to predict

---
*Instituto de Fisica, UFRGS, Campus do Vale, 91500 Porto Alegre, RS, BRASIL, and Astronomy Department, University of Maryland, College Park, MD 20742

[†]Astronomy Department, University of Maryland,College Park, MD 20742, and Space Telescope Science Institute, 3700 San Martin Drive, Baltimore, MD 21218

2. The Sample

Our sample comprises galaxies with Seyfert 2 nuclei and ionization cones, for which long-slit spectra with good spatial resolution are available. The galaxies are listed in the first column of Table 1. Basic data, as well as references for the detection of the conical morphology and long-slit spectra used in the calculations of the predicted torus luminosity can be found in table 1 of Storchi-Bergmann, Mulchaey and Wilson (1992). Although some of the galaxies have star forming regions near the nucleus, we have selected only the spectra avoiding these regions.

3. The Calculations

Our goal is to predict the luminosity of the torus assuming the ionizing source is an isotropic radiator, the ionization cones represent the free solid angle, and the torus absorbs all photons in the wavelength range 100Å–1μm. The torus is expected to reradiate this energy in the infrared (Krolik & Lepp 1989). We first calculate the number of ionizing photons emitted from the central source using the emission line ratios in the cone to get the ionization parameter and gas density at a measured distance from the nucleus. Then, by assuming a spectral shape $L_\nu = A\nu^{-\alpha}$, the total source luminosity in the 100Å–1μm band can be calculated, as well as the fraction incident on the torus, using the narrow-band images to get the covering factor of the ionization cone. The slope α of the power-law continuum was obtained from Kinney et al. (1991), for the galaxies in their sample and assumed to be 1.5 for the other ones. The luminosity incident on the torus (hereafter called L_P) is assumed to be reradiated isotropically and can then be compared with the infrared luminosity determined from the IRAS fluxes.

Before comparing L_P with the infrared luminosity, we have to estimate the contribution of the galaxy to the far-infrared flux, because of the large IRAS beam. This was done by comparing ground-based 12μm observations from Roche et al. (1991; beam diameter of $\sim$ 5''), for five of our sample galaxies with the corresponding IRAS observations and assuming that the ground based observations are isolating the nuclear emission. For the galaxies without sufficient ground based data, we adopt a correction factor of 0.6, the median flux ratio between the ground based and IRAS 12μm flux that Roche et al. (1991) find for Seyfert 2's and LINERS. The observed IR luminosity (hereafter called L_{IR}) was corrected by this factor. Details of all the above calculations are described in Storchi-Bergmann, Mulchaey & Wilson (1992).

If a dusty torus is indeed present, the observed IR luminosity L_{IR} should be about equal to the calculated luminosity L_P. If L_P is significantly larger than L_{IR}, then the simple model is inconsistent with the data. In this case, the ionizing source could be intrinsically anisotropic (e.g. Miller, Goodrich and Mathews 1991; Madau 1988; Acosta-Pulido et al. 1990) or the torus itself could be optically thick even at mid-infrared wavelengths (Pier & Krolik 1992). We have adopted $H_0 = 75\,km\,s^{-1}Mpc^{-1}$ in calculating the luminosities; the astronomically important ratio L_P/L_{IR} is, of course, independent of distance.

Table 1. Measured and calculated properties

GALAXY	r(pc)	−Log(U)	n(cm^{-3})	α	θ	L_P	L_{IR}
NGC1068	4144	3.0	150	1.6	40	5.2×10^{11}	9.7×10^{10}
		2.9-3.1	50-250			$(0.1 - 1.1) \times 10^{12}$	
NGC1365	370	3.1	150	1.5	90	2.1×10^{9}	9.8×10^{9}
		2.9-3.4	50-250			$(0.4 - 5.6) \times 10^{9}$	
NGC2110	508	2.7	130	1.5	30	1.1×10^{10}	8.4×10^{9}
		2.5-2.9	50-200			$(0.3 - 2.8) \times 10^{10}$	
NGC3281	770	2.3	350	1.5	70	1.6×10^{11}	2.0×10^{10}
		2.2-2.4	300-400			$(1.1 - 2.2) \times 10^{11}$	
NGC4388	487	2.6	250	1.3	92(2)	1.6×10^{10}	4.6×10^{9}
		2.3-2.9	125-400		50(3)	$(0.4 - 5.2) \times 10^{10}$	
NGC5506	398	2.8	400	1.6	90	1.5×10^{10}	9.5×10^{9}
		2.7-2.9	200-600			$(0.6 - 2.8) \times 10^{10}$	
NGC5728	879	2.9	250	1.5	50	3.9×10^{10}	1.4×10^{10}
		2.8-3.0	100-400			$(1.2 - 7.8) \times 10^{10}$	
MKN78(1)	5061	2.5	≤100	2.0	50	$\leq 3.2 \times 10^{12}$	5.7×10^{10}
		2.5-3.0					
MKN573	934	2.9	800	0.9	80	5.9×10^{10}	1.6×10^{10}
		2.7-3.2	600-1000			$(0.2 - 1.2) \times 10^{11}$	

[1]In the low density limit of the [SII]$\lambda\lambda 6716,6731$ density diagnostic.
[2]SW cone (Pogge 1989)
[3]NE cone (Pogge 1989)

4. Results

Table 1 shows, for each sample galaxy, the relevant quantities and the results of the calculations: distance r from the nucleus at which the line ratios were measured (column 2); the negative of the logarithm of the ionization parameter $-log(U)$ obtained from the line ratios (column 3); gas density n at the distance r (column 4); slope α of the power-law continuum (column 5); observed opening angle of the ionization cone (column 6); the predicted torus luminosity L_P (column 7) and the observed infrared luminosity L_{IR} (Mulchaey et al. 1992) corrected for galactic contributions (column 8).

We also show in table 1, some error estimates for the parameters obtained from the emission-line ratios. For $-log(U)$, the second line for each entry shows the range of values that it can assume considering both the [OIII]λ5007/Hβ and [NII]λ6583/Hα ratios. For n, we also show the possible range of gas densities considering the errors affecting the [SII]λ6716/λ6731 ratio. Finally, we show the resultant maximum range for L_P due to these variations.

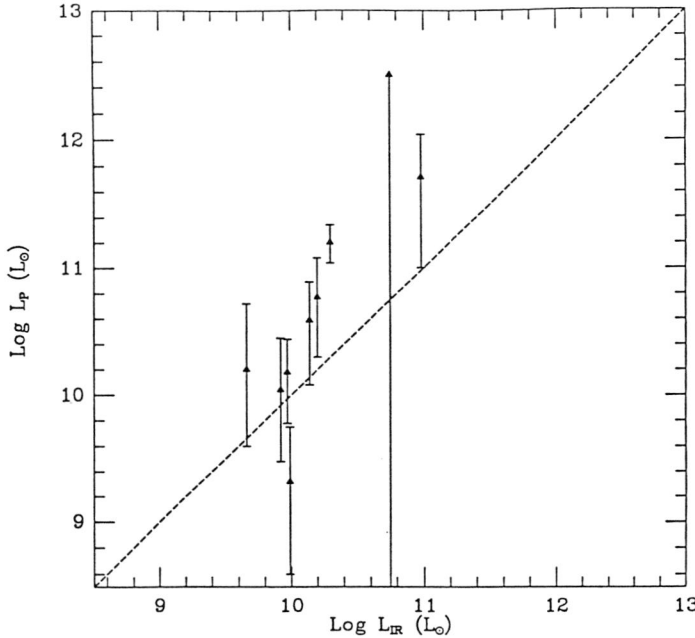

Figure 1. A logarithmic plot of L_P versus L_{IR}.

5. Discussion

A comparison of L_P and L_{IR} from Table 1 shows that for six of the sample galaxies the agreement between these two quantities is very good. A plot of L_P versus L_{IR} (Fig. 1) is largely consistent with a linear relation between these two quantities. For most galaxies, the nominal value L_P is larger than L_{IR}, but when the errors are taken into account, a significant discrepancy exists for only a few. Such an excess of L_P over L_{IR} may be consistent with the torus model if the torus emission is significantly anisotropic in the mid-infrared. Pier and Krolik (1992) have modelled the reradiation from dusty tori and indeed find that in edge-on systems the observed infrared flux may be significantly less than the true torus flux. The amount of anisotropy depends on the torus' optical thickness and shape, but a factor of a few should be expected in all cases. Given this, the considerable uncertainty in the predicted torus luminosity and estimating the nuclear component to the IRAS fluxes, values of L_P that are a few times larger than L_{IR}, as observed in our sample, must be considered consistent with the torus model.

The fact that all the selected galaxies present ionization cones indicates that they are oriented in such a way that the torus must be seen nearly edge on. We conclude, therefore, that the data are generally consistent with collimation and reradiation by a dusty torus.

References

Acosta-Pulido, J. A., Pérez-Fournon, I., Calvani, M. & Wilson, A. S., 1990. *ApJ*, **365**, 119.
Antonucci, R. R. J., & Miller, J. S., 1985. *ApJ*, **297**, 621.
Kinney, A. L., Antonucci, R. R. J., Ward, M. J. Wilson, A. S. & Whittle, M., 1991. *ApJ*, **377**, 100.
Krolik, J. H. & Begelman, M. C., 1986. *ApJ*, **308**, L55.
Krolik, J. H. & Lepp, S., 1989. *ApJ*, **347**, 179.
Madau, P., 1988. *ApJ*, **327**, 116.
Miller, J. S, Goodrich, R. W., & Mathews, W. G., 1991. *ApJ*, **378**, 47.
Mulchaey, et al., 1992. in preparation.
Pier, E. and Krolik, J. H., 1992, *ApJ*, in presss.
Pogge, R. W., 1988a. *ApJ*, **328**, 519.
Pogge, R. W., 1988b. *ApJ*, **332**, 702.
Roche, P. F., Aitken, D. K., Smith, C. H. & Ward, M. J., 1991. *MNRAS*, **240**, 838.
Storchi-Bergmann, T., Mulchaey, J. S. & Wilson, A. S., 1992. *ApJ*, **395**, L73.
Wilson, A. S., 1992, in *Physics of Active Galactic Nuclei*, eds. S. J. Wagner and W. J. Duschl, Springer-Verlag, in press.

Imaging Spectrophotometry of Extended-Emission Seyfert Galaxies

Richard W. Pogge [*‡] *Michael M. DeRobertis* [†‡]

Abstract

We present highlights of a multi-band optical imaging study of the ENLRs in seven Seyfert galaxies. Flux-calibrated narrowband emission-line and continuum images obtained in subarcsecond seeing conditions with the 3.6m CFHT reveal a remarkable wealth of emission-line and optical continuum structures. We show an example from our data of an apparent mechanical interaction between radio plasma ejecta and the ENLR, and describe an extended near-UV continuum component, which appears to be scattered nuclear continuum light, that is seen in all of the five Seyfert 2s, but is conspicuously absent in the two extended-emission Seyfert 1s.

1. Introduction

Many Seyfert nuclei are surrounded by kpc-scale regions of highly ionized gas known as the "extended narrow-line region" (ENLR). The energetics, ionization state, and morphology of the ENLR have been one of our most powerful probes of the interaction between a Seyfert nucleus and its host galaxy environment. Traditionally, the ENLR in Seyferts has been mapped using CCD images taken through narrow interference filters isolating the bright emission lines of Hα+[N II] and [O III]λ5007. These lines bracket a wide range of ionization potential, and so provide a simple, qualitative diagnostic of the ionization state of the gas. Two-band imaging surveys have had a fair amount of success so far at revealing extended emission in Seyferts and other AGN [1,2]. Often, however, one wants more quantitative information such as a particular line ratio which is diagnostic of a nebular parameter. Such data have been obtained by following the imaging work with conventional long-slit spectroscopy [1,3].

To bridge the gap between the semi-quantitative two-band technique and conventional spectroscopy, we have extended our narrowband filter imaging to include key diagnostic emission lines. In addition to Hα+[N II], we take images in bands isolating the [O I]λ6300, [O II]λ3727, and [O III] emission-lines which span a wide range

---

[*]Department of Astronomy, Ohio State University, Columbus, OH 43210, USA.
[†]Department of Physics and Astronomy, York University, North York, Ontario M3J 1P3, Canada
[‡]Visiting Astronomers, Canada-France-Hawaii Telescope, operated by the National Research Council of Canada, the Centre National de la Recherche Scientifique de France, and the University of Hawaii.

of ionization potential for a single ion, and the [S II]$\lambda\lambda$6716+31 line blend which is sensitive to density enhancements and ionization edges. With careful flux calibration, these data allow us to map the 2-D ionization structure of the ENLR, and to explore the radio and optical gas interaction with respect to line morphology and line ratios. Emission-line imaging techniques require the use of images taken in adjacent line-free continuum bands to remove the underlying stellar component of the host galaxy. In our study, we used three 150–200Å wide continuum bands in the near-UV (3600Å), Green (5200Å), and Red (6100Å) regions. In addition to providing high-quality host galaxy subtraction templates, these bands also span a wide enough wavelength range to allow us to address host galaxy continuum properties. In particular, we can search for changes in the stellar content of the galaxies, and the presence (or absence) of circumnuclear star formation regions as revealed by continuum colours.

Thus far, we have obtained emission-line images of seven Seyfert galaxies using the 3.6-m CFHT with median seeing of 0.5″ during photometric conditions. The complete data set is too large to treat within the scope of this contribution, so we present two interesting results. The first illustrates our ability to compare optical and radio data at the same spatial scale to address the interaction between the radio plasma ejecta and the ENLR with respect to both gas morphology and maps of the line ratios in the region. The second introduces maps of the near-UV/Red continuum colour. In our five Seyfert 2s, these maps have revealed a diffuse, extended near-UV excess component that is aligned with the extended radio/line-emission axis, but which is conspicuously absent in the two extended-emission Seyfert 1 galaxies we observed.

2. Radio & Optical Interaction in Mrk 573

Figure 1a shows the 6cm radio flux contours of the Seyfert 2 galaxy Mrk 573 [4] superimposed on our [O III]λ5007 emission-line image. Both images have a spatial resolution of 0.6″, corresponding to a spatial scale of $\sim 145 h_{100}^{-1}$ pc. We see the well-known alignment between the axis of the extended emission and the radio ejection axis, but it is clear that there is considerable optical emission substructure present. While aligned in the plane of the sky, there is no direct knot-for-knot correlation between the radio plasma and the line-emitting gas, except at the location of the unresolved active nucleus. Plotting along the radio axis, the peaks in the [O III] emission lie at larger radii than the peaks in the radio plasma (Fig 1b). The [O I]/[O III] line ratio cut (Fig. 1c) peaks "downstream" of the radio ejecta. This ratio is sensitive to the ionization parameter, and shows that the ionization parameter drops in the downstream peaks, possibly indicating enhanced density. The appearance of the emission-line gas as a set of two nested crescents of emission, the line ratio distribution, and the relative location of the radio plasma knots relative to this gas are all very suggestive that what we are seeing is a bow shock as the radio plasma ploughs into the circumnuclear ISM.

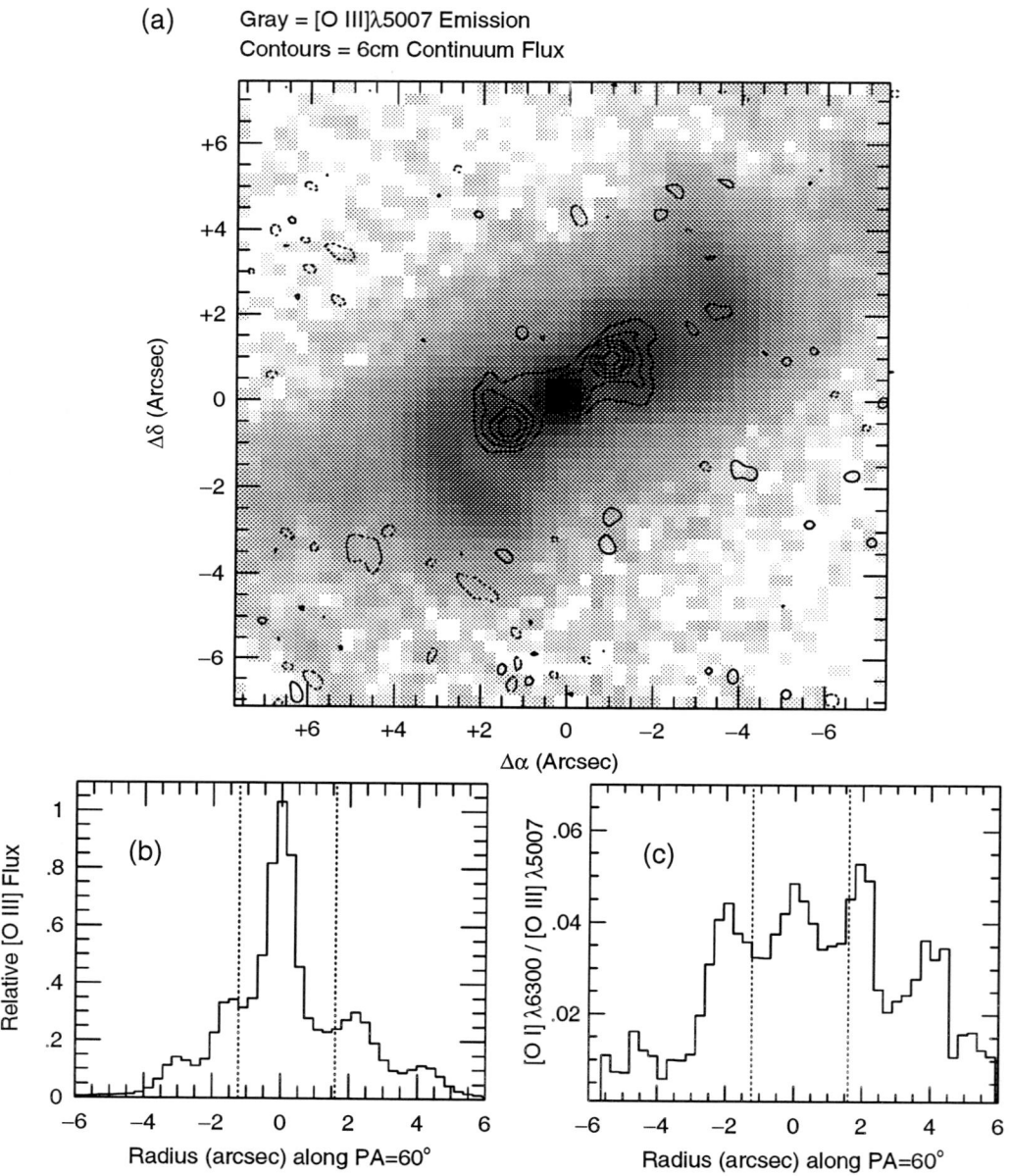

Figure 1. (a) Radio and [O III] Maps of Mrk 573, (b) Cuts along the radio axis in [O III] emission, and (c) [O I]/[O III] line ratio. The peaks of the off-nuclear radio knots are indicated by the dashed lines.

3. Extended Near-UV Continuum Emission

Using our near-UV and Red continuum images, we formed $I(UV)/I(Red)$ colour maps. These maps should emphasize blue OB associations (if present) and the blue semi-stellar active nucleus. Sensitivity to extinction between 3600Å and 6100Å should also emphasize regions of dust extinction. The colour maps met these expectations by detecting OB associations and dust lanes previously known to exist in the galaxies, but they also revealed unexpected circumnuclear regions with a UV colour excess in all five of the Seyfert 2s (Mrk 3 & 573, NGC 1068, 2110, & 3081). The results for three Seyfert 2s with the most complete supplementary data are discussed elsewhere [5]. In summary, the extended near-UV excess take two forms: discrete knots spatially coincident with the high-surface brightness ionized gas regions, and a diffuse component spatially coextensive with the diffuse [O III] emission. The discrete UV knots are all "fuzzy" in the same sense as the associated emission-line regions, and do not show the same sharp core structure, or the same colours, as the OB associations. The diffuse UV continuum emission in three cases [5] shares the striking conical (NGC 1068) or bi-conical (Mrk 3 & Mrk 573) morphology of the [O III] "ionization cones" in these galaxies. The other two do not show a conical morphology. The correlation of the extended UV-excess in NGC 1068 with the extended optical continuum polarization component [6], its non-correlation with the radio continuum structure, and its generally diffuse character suggests to us that we are seeing scattered nuclear continuum light. This is perhaps the same scattered light that in typical 4" diameter spectropolarimeter apertures has the polarized flux spectrum of a "hidden" Seyfert 1 nucleus. Both NGC 1068 and Mrk 3 are classic polarized broad-line Seyfert 2s [7,8].

Curiously, in the two Seyfert 1 galaxies (NGC 3516 & 7469), both noted for their extended emission regions, we do not see an extended UV-excess component. Rather, we see only a point-like blue nucleus (to our seeing limit) and essentially normal galaxy colours outside this, except in NGC 7469 where we see the OB associations in the well-known circumnuclear starburst ring.

References

[1] Pogge, R.W. 1989, ApJ, 345, 730
[2] Haniff, C.A., Ward, M.J., & Wilson, A.S. 1988, ApJ, 334, 584
[3] Storchi-Bergmann, T., Wilson, A.S., & Baldwin, J.A. 1992, ApJ, in press
[4] Ulvestad, J.S. & Wilson, A.S. 1984, ApJ, 278, 544
[5] Pogge, R.W. & DeRobertis, M.M. 1993, ApJ, 404, in press
[6] Miller, J.S., Goodrich, R.W., & Mathews, W.G. 1991, ApJ, 378, 47
[7] Antonucci, R.R.J. & Miller, J.S. 1985, ApJ, 297, 621
[8] Miller, J.S., & Goodrich, R.W. 1990, ApJ, 335, 456

Spectroscopy of the Extended Emission Line Regions in NGC 4388

Ismael Pérez-Fournon [*] *Baltasar Vila-Vilaró* [*]
José A. Acosta-Pulido [*] *J. Ignacio González-Serrano* [†]
Marc Balcells [‡] *Andrew S. Wilson* [§] *Zlatan Tsvetanov* [¶]

Abstract

We present and discuss the results of long-slit spectroscopic observations of the extended emission line regions (EELR) in NGC4388, a Seyfert 2 galaxy in which extended, off-nuclear broad Hα emission (FWZI $\approx$ 4000 km s^{-1}) has been reported (Shields & Filippenko 1988). These features have been interpreted as scattered radiation from a Seyfert 1 nucleus that is obscured along our line of sight. Our spectroscopic observations cover a large fraction of the inner part of the EELR, including some of the positions where the presence of broad lines has been claimed. Broad wings in the Hα + [NII] $\lambda\lambda$6548, 6583 complex are also present in our data but they can be explained by the superposition of several narrow components. However, we cannot, at present, exclude the possibility that an intrinsically broad component to Hα exists at some locations. The implications of our results for unified models of Seyfert galaxies are briefly discussed.

1. Introduction

NGC4388 is a high-inclination Seyfert 2 galaxy with bright EELR (Colina et al. 1987, Pogge 1988, Corbin et al. 1988). Radio observations reveal a double-peaked source close to the apparent optical nucleus and more extended emission aligned roughly perpendicular to the galaxy disk (Stone et al. 1988, Hummel & Saikia 1991). The axis of the cone-like, high-excitation gas is almost perpendicular to the galaxy disk and close to that of the extended radio emission (Pogge 1988, Corbin et al. 1988), suggesting that the EELR are photoionized by nuclear radiation which escapes preferentially along and around the radio axis. A broad component to the extranuclear Hα emission has been reported by Shields and Filippenko (1988) at some positions

---

[*] Instituto de Astrofísica de Canarias, 38200, La Laguna, Tenerife, Spain
[†] Departamento de Física Teórica, Universidad de Cantabria, Santander, Spain
[‡] Observatorio del Roque de los Muchachos, La Palma, Tenerife, Spain
[§] Space Telescope Science Institute, 3700 San Martin Drive, Baltimore, MD 21218, U.S.A.
[¶] Department of Physics and Astronomy, Johns Hopkins University, Baltimore, MD 21218, U.S.A.

in the EELR on the basis of total intensity spectroscopic observations. They suggest scattering of the emission from an obscured Broad Line Region (BLR) as the origin for the off-nuclear broad lines; a similar model has been proposed for the broad lines detected in the nuclei of Seyfert 2 galaxies by spectropolarimetry (Antonucci & Miller 1985). In the case of NGC4388, however, the scattering particles would be dust and the scattering regions located outside the nucleus. Thus, a detailed study of these regions is particularly interesting to test unified models of AGNs. In this paper we present the results of spectroscopic observations of the EELR of NGC4388, concentrating on the nature of the extranuclear broad emission lines.

2. Observations and Data Reduction

We have observed the EELR of NGC4388 using the Cassegrain Twin Spectrograph at the 3.5m telescope of the German-Spanish Astronomical Center, Calar Alto Observatory, and the ISIS Spectrograph at the 4.2m William Herschel Telescope (WHT) of the Observatorio del Roque de los Muchachos, La Palma[1]. Intermediate dispersion long-slit spectra were obtained at several positions in the EELR with the slit axis located approximately perpendicular to the radio axis. The seeing during the observations was about 1 arcsec FWHM. The observations were reduced using standard techniques with the FIGARO and IRAF reduction packages. Due to the good quality of the data, line profiles could be analyzed at the full spatial resolution.

3. Results

Broad wings to the Hα + [NII] $\lambda\lambda$6548, 6583 complex are present in our data in several locations in the EELR, confirming the discovery by Shields and Filippenko (1988). If these wings are interpreted as broad Hα emission, then the FWHM is of the order of 2000 km s^{-1}, in agreement with Shields and Filippenko. However, in the positions where these broad wings are present we detect several distinct kinematical components in narrow emission lines, such as [OI] λ6300 and [SII] $\lambda\lambda$6716, 6731. Unlike the Hα + [NII] $\lambda\lambda$6548, 6583 complex, the wings on such forbidden lines cannot originate in the BLR. In particular, at the slit position shown in Figure 1 the [SII] doublet can be well fitted with four Gaussian components for each of the two lines at our resolution of about 4 Å (Figure 2a). Applying precisely the same kinematical model (4 Gaussian components for each line), a good fit can be obtained to the Hα + [NII] $\lambda\lambda$6548, 6583 complex (Figure 2b). This implies that no intrinsically broad Hα emission is required to explain the line profiles observed at this location.

---

[1]The William Herschel Telescope is operated on the island of La Palma by the Royal Greenwich Observatory in the Spanish Observatorio del Roque de los Muchachos of the Instituto de Astrofísica de Canarias

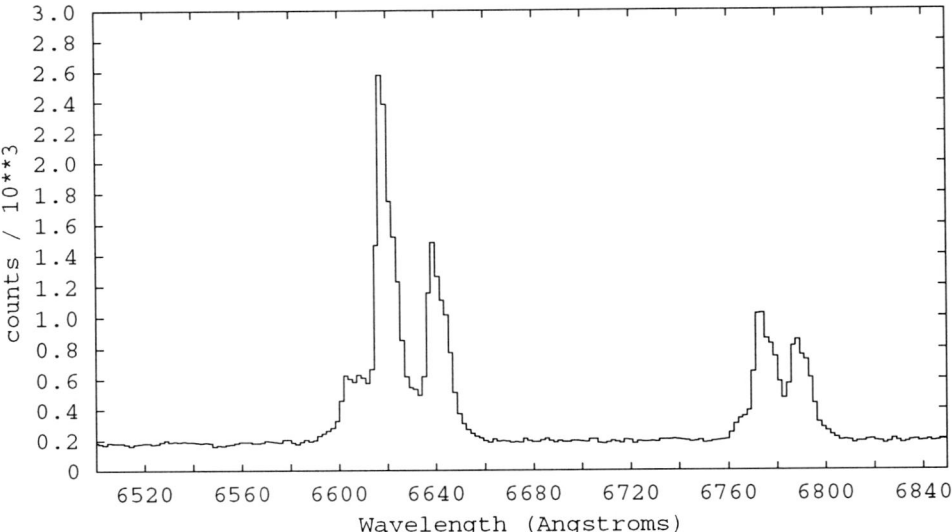

Figure 1. Red spectrum obtained with the 3.5m telescope at Calar Alto: average over 2.5 arcsec² centred approximately 4 arcsec south and 2 arcsec east of the apparent optical nucleus of NGC4388.

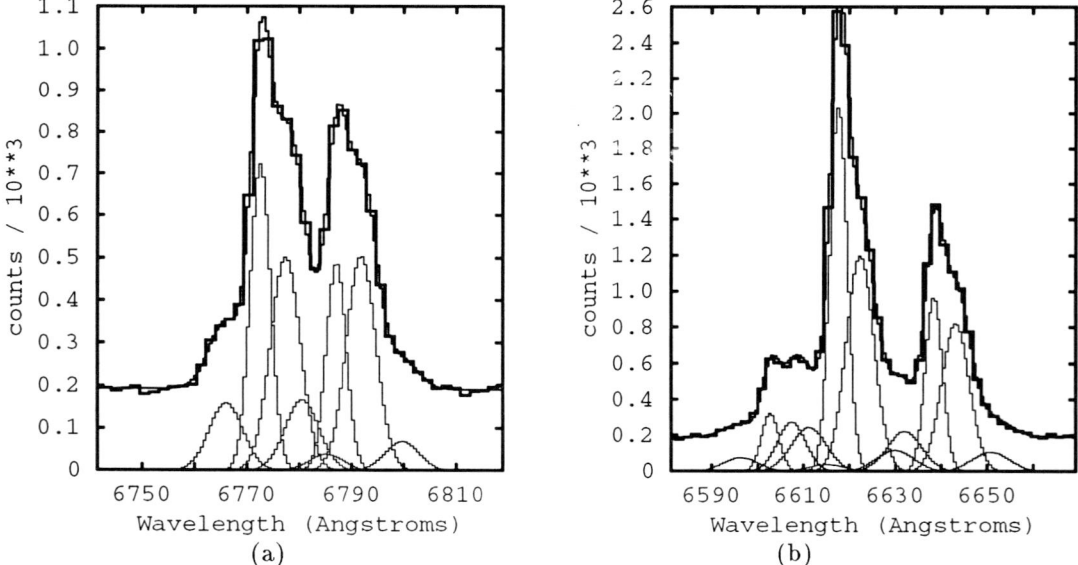

Figure 2. a) The [SII] $\lambda\lambda6716, 6731$ doublet for the same region as in Figure 1 showing the fitted Gaussian components and their superposition. b) Same as (a) for the Hα + [NII] $\lambda\lambda6548, 6583$ complex.

4. Discussion

We have shown that the apparent presence of broad wings in the Hα + [NII] $\lambda\lambda$6548, 6583 blend at some locations outside the nucleus of NGC4388 can be naturally explained by the superposition of several kinematical components which are also present in forbidden lines, such as [OI] λ6300 and [SII] $\lambda\lambda$6716, 6731. However, since our long-slit spectra do not cover all the regions where broad Hα emission is found by Shields and Filippenko (1988), we cannot rule out its presence at some locations.

Although we find no evidence for extranuclear scattering of broad lines emitted in a hidden BLR, we note that the line profiles at some locations in the EELR resemble those at the nucleus. Both the extended and nuclear emission lines exhibit a strong red-wing asymmetry (Colina et al. 1987, Ayani & Iye 1989, Veilleux 1991). Therefore, we suggest that some of the kinematical components of the emission lines could represent scattered radiation from the circumnuclear NLR, which is partially hidden from our direct view by large amounts of dust. This conclusion is consistent with the finding that the radio and IR nuclei of this galaxy are displaced about 2 arcsec north of the apparent optical nucleus.

References

Antonucci, R.R.J., Miller, J.S., *1985, Ap. J. 297, 621.*
Ayani, K., Iye, M., *1989, A. J. 97, 686.*
Colina, L., Fricke, K.J., Kollatschny, W., Perryman, M.A.C., *1987, Astr. Ap. 186, 39.*
Corbin, M.R., Baldwin, J.A., Wilson, A.S., *1988, Ap. J. 334, 584.*
Hummel, E., Saikia, D.J., *1991, Astr. Ap., 249, 43.*
Pogge, R.W., *1988, Ap. J. 332, 702.*
Shields, J.C., Filippenko, A.V., *1988, Ap. J. (Letters) 332, L55.*
Stone, Jr., J.L, Wilson, A.S., Ward, M.J., *1988, Ap. J. 330, 105.*
Veilleux, S., *1991, Ap. J. Suppl. 75, 357.*

Evidence and Implications of Anisotropy in Seyfert Galaxies

José A. Acosta–Pulido * *B. Vila–Vilaró* * *I. Pérez–Fournon* *
A. S. Wilson † *Z. Tsvetanov* ‡

Abstract

We present long–slit spectroscopy across the ENLR in the Seyfert 2 galaxy NGC 5252. Possible implications for anisotropic nuclear models are discussed.

1. Introduction

The most direct evidence for anisotropy of the optical/UV continuum in Seyfert galaxies comes from the observed high ionization cones (Tadhunter & Tsvetanov 1989). Several plausible models have been proposed to explain the origin of this anisotropy: **a** - A large obscuring torus at intermediate distances between the BLR and the NLR. **b** - A geometrically thick accretion disk. **c** - A blazar beam emitted by a relativistic jet.

Acosta–Pulido (1992) has predicted the angular variation of the ENLR spectrum for the ionizing continuum of each of the above models. In order to discriminate between these different models for the ionization cone in NGC 5252 (Tadhunter & Tsvetanov 1989) we have commenced a program of mapping out emission lines ratios within the cone using long–slit spectroscopy.

2. Observations and Results

We made long–slit observations of the ENLR of the Seyfert 2 galaxy NGC 5252. The slit was oriented perpendicular to the cone axis (PA=165°) at a distance of 15" SE from the nucleus. We used the ISIS Spectrograph at the Cass. focus of the 4.2m William Herschel telescope in La Palma[1]. The data were reduced using the IRAF software.

---

*Instituto de Astrofísica de Canarias, 38200 La Laguna, Spain.
†Space Telescope Institute, Baltimore, USA.
‡Johns Hopkins University, Baltimore, USA
[1]The Isaac Newton Group of telescopes are operated on the Island of la Palma by the Royal Greenwich Observatory in the Spanish Observatorio del Roque de los Muchachos of the Instituto de Astrofísica de Canarias.

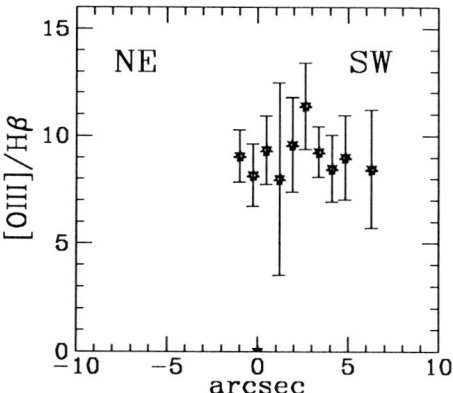 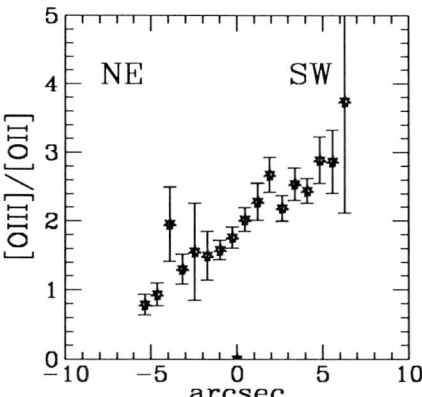

The extension of the line emission does not extend across whole opening angle of the ionization cone (*i.e.* 75°) as deduced from the outer shells, probably indicating a non-uniform distribution of the emitting gas. Furthermore, the line flux (*e.g.* [OIII] $\lambda5007$ and [OII] $\lambda3727$) is asymmetrically distributed with respect to the cone axis, showing a slow flux decrease towards the major axis of the host S0 galaxy (*i.e.* SW of the cone axis). On the contrary, towards the NE, there is a steep fall followed by a plateau of emission. The variation along the slit of the line ratios [OIII] $\lambda5007/\mathrm{H}\beta$ and [OIII] $\lambda5007/$[OII] $\lambda3727$ is presented in the figure. These show a very interesting behaviour: while [OIII] $\lambda5007/\mathrm{H}\beta$ is essentially constant, [OIII] $\lambda5007/$[OII] $\lambda3727$ increases towards the major axis of the galaxy. There are several possible explanations for this gradient: **a** - The reddening increases towards the major axis of the galaxy; however there is no observational support for this conjecture from either the continuum images (Tadhunter & Tsvetanov 1989) nor from our spectroscopy. **b** - Optically thin clouds more concentrated towards the SW. **c** - The ionization parameter increases by a factor ~ 3 towards the SW, but the constancy of [OIII] $\lambda5007/\mathrm{H}\beta$ poses a problem.

In conclusion, although line ratios can be used to map out the angular dependence of ionizing radiation escaping from the nucleus, the interpretation can be complicated by non–uniform gas and dust distributions.

References

Acosta–Pulido J.A., *Proceedings of the conference The Nearby Active Galaxies, Madrid 1992 (in press)*.
Tadhunter C. and Tsvetanov Z., *Nature, Vol. 341 (1989), p. 422*.

Collimated Radiation in NGC 4151

G. Kriss [*]   I. Evans [†]   H. Ford [*†]   Z. Tsvetanov [*]
A. Davidsen [*]   A. Kinney [†]

Abstract

We present a [O III] $\lambda 5007$ image of the nuclear region of the Seyfert 1 galaxy NGC 4151 obtained with the Planetary Camera on the Hubble Space Telescope. The [O III] image shows a striking biconical structure centred on the bright, unresolved central source. Simple geometric arguments place our line of sight *outside* the cone of ionizing radiation. Since we have a nearly unobstructed view of the UV continuum and broad-line region, an optically thick molecular torus cannot be the source of the collimation. Lower column density material visible in UV spectra is largely transparent at UV and optical wavelengths, but opaque beyond the Lyman limit. It can collimate the ionizing radiation field without obscuring our view of the central engine.

The simplest unified models of Seyfert 1 and Seyfert 2 galaxies invoke a dense, optically thick, molecular torus surrounding the central engine and broad-line region which serves as both a shield which prevents us from viewing the centres of Seyfert 2's and as the source of the collimation for the ionizing radiation. The difference between Seyfert 1's and Seyfert 2's is then due entirely to the opening angle of the torus and its orientation relative to our line of sight. The observations presented here and by Evans et al. [1] present a direct challenge to this simple view.

Figure 1 shows the raw image obtained with the Planetary Camera on the Hubble Space Telescope through the F502N filter in a 1608 s exposure. A striking biconical structure is revealed along PA = $60° \pm 10°$ with a projected opening angle of $75° \pm 10°$. The "string" of highly ionized emission-line knots visible in ground-based images as the ENLR of NGC 4151 [4] extends to 30" from the nucleus and lies wholly within the biconical structure defined by our HST images. The string lies in the plane of the galaxy, which is nearly face on, at an angle of 98° to our line of sight [4]. The geometry which places our line of sight closest to the axis of the ionization cone puts the string along the furthest edge of the cone from our line of sight. Allowing a projected opening angle of 75° for the cone and solving for the projection effects gives an angle of 65° between our line of sight and the cone axis. Hence, as shown in

---

[*]Department of Physics and Astronomy, Johns Hopkins University, Baltimore, MD 21218.
[†]Space Telescope Science Institute, 3700 San Martin Drive, Baltimore, MD 21218.

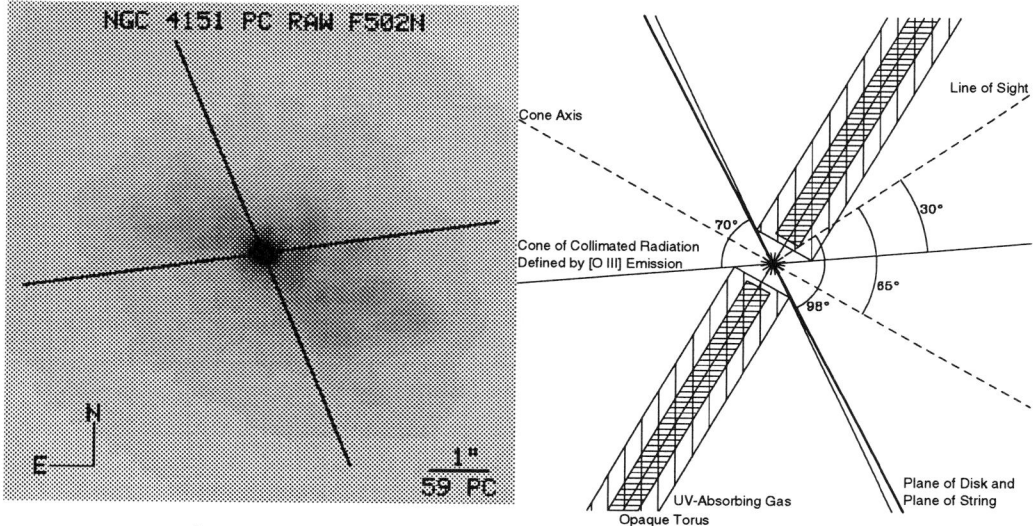

Figure 1. The left panel shows the raw [O III] image obtained with the Planetary Camera. The solid lines outline the ionization cones defined by the [O III] emission. The schematic in the right panel illustrates our conception of the collimation process. The opaque torus cannot block the line of sight to the observer, but lower column density gas surrounding the torus can effectively collimate the ionizing radiation without obstructing our view of the continuum and broad line regions at near UV and visible wavelengths.

Figure 1, we are looking at an angle of 30° from the edge of the cone.

The clear biconical structure of the narrow-line region of NGC 4151 implies that our line of sight lies *in the shadow* of the collimating material, and that we must look through it to see the broad-line and continuum regions of NGC 4151. (A line of sight interior to the cone would project emission from the narrow-line region onto the sky at all position angles surrounding the nucleus.) Yet the optical and UV continua of NGC 4151 show low extinction, and the broad lines are easily visible. Thus the collimation cannot be provided by an optically thick molecular torus. The lower column density and UV-absorbing gas observed in NGC 4151 with HUT [2] and IUE [3] may lie just outside the torus and play an important role in collimating the ionizing radiation field in NGC 4151. This geometry then makes the special absorption properties of NGC 4151 a rare occurrence since the observer's line of sight must lie between the cones defined by the opaque torus and the lower column density gas surrounding it.

References

[1] Evans, I. N., et al. 1993, in preparation
[2] Kriss, G. A., et al. 1992, ApJ, 392, 485
[3] Penston, M. V., et al. 1981, MNRAS, 196, 857
[4] Perez-Fournon, I., & Wilson, A. S. 1990, ApJ, 356, 456

A Dust Ring around the Nucleus of NGC4151*

Baltasar Vila-Vilaró [†]   *Enrique Pérez* [†]   *Andy Robinson* [‡]
Clive Tadhunter [§]   *Ismael Pérez-Fournon* [†]
Rosa González-Delgado [†]

1. Introduction

Several authors have proposed that one of the mechanisms for fuelling active galactic nuclei (AGN) may involve the presence of a stellar bar in the galaxy (Shlosman *et al.*, 1989 and references therein). The bar potential produces an overall instability that drives the ambient gas of the disk inwards, forming a gaseous disk sorrounding the active nucleus at a radius of several hundred parsecs. Under certain circumstances the gas can dissipate its angular momentum and fall into the centre. Numerical simulations show that the movement of the gas towards the nucleus would be preferentially along the stellar bar in two arms (Athanassoula 1992). The shocked material in the flow would be delineated by dust lanes that, in regions close to the nucleus might be seen as a ring-like structure or "inner spiral arms". We present here evidence of such a structure surrounding the nucleus in the Seyfert 1 galaxy NGC4151.

2. Results

Broad-band images of the nucleus of NGC4151 were obtained in the U,B,V,I and Z filters with the 1m JKT telescope in La Palma. The images were calibrated in magnitudes and then subtracted in pairs to produce a series of colour maps. The B and V images are clearly contaminated by extended emission lines. However, a line-free colour map was obtained by combining a continuum image obtained by Pérez *et al.* (1989) in a filter centred at 5700Å, with the I-band image. This reveals a red excess surrounding the nucleus and most of the extended narrow line region in a patchy elliptical ring (see Figure) with a major axis $\sim 6\,kpc$ ($H_0 = 50\,{\rm km\,s^{-1}\,kpc^{-1}}$). Assuming that the red excess is due to extinction by dust with similar mean properties to that in our Galaxy, we calculated A_V at several positions around the "ring". After allowing for the colours of the bulge population, the estimated visual extinction ranges from 0.1–0.35 magnitudes. Differential magnitudes measured at the same locations

---

*This work is based on data obtained by the **LAG** collaboration. LAG is a consortium of mainly European astronomers which was established to study active galaxies using International Time at the Canary Islands' observatories operated under the auspices of the Comité Científico Internacional.

[†] Instituto de Astrofísica de Canarias, 38200, La Laguna, Tenerife, SPAIN
[‡] Institute of Astronomy, Madingley Road, Cambridge CB3 0HA, UK.
[§] Department of Physics, University of Sheffield, Sheffield S3 7RH, UK.

Top. Line-free colour map showing red excess (lighter tones) enclosing the nucleus of NGC4151. A contour map of the extended [OIII]λ5007 emission (Pérez *et al.*) is overplotted and an ellipse of major axis $\approx 1'$ is also shown for comparison.

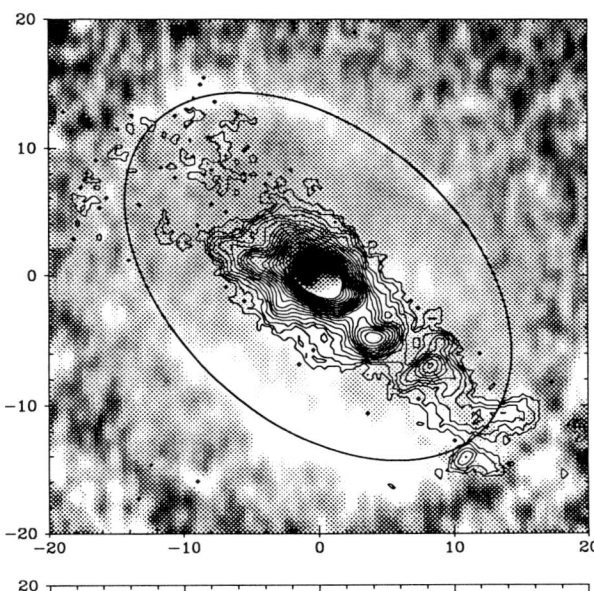

Bottom. I-band polarization map from Draper *et al.*.

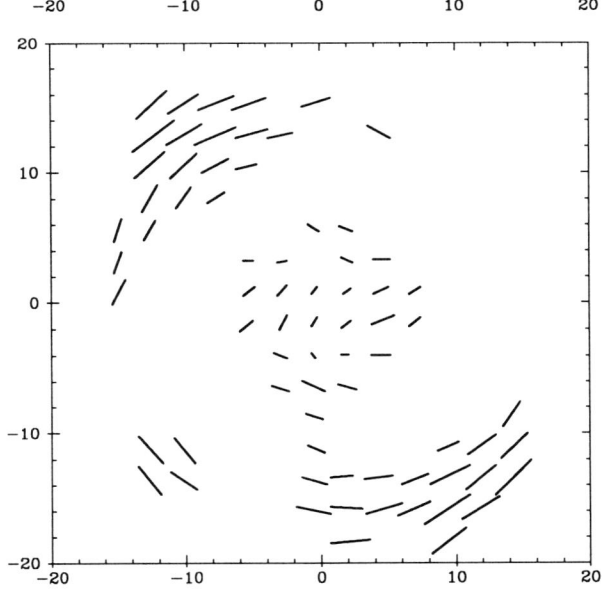

in the remaining colour maps (e.g., B–V; V–I etc.) are consistent with reddening corresponding to this range in A_V. Thus, we conclude that the elliptical structure seen in the colour maps is the result of dust extinction. Further evidence that this is indeed the case comes from recent broad-band polarimetry by Draper *et al.* (1992) whose I-band map shows polarization around the nucleus which is roughly co-spatial with the structure in our colour maps (see Figure) This feature is interpreted by the authors as dichroic absorption of the underlying stellar population by dust.

If the dust lies in the plane of the disk, its morphology is strikingly similar to the predictions of Athanassoula's (1992) simulations. On larger scales, these simulations also predict that gas in the galactic bar flows inwards towards the "inner spiral arms" in two symmetric curved lanes. Similar features have been found in NGC4151 in recent HI 21 cm line observations (Pedlar *et al.*, 1992). An alternative scenario is that the observed ellipse is actually a circular dust ring inclined with respect to the line of sight but roughly perpendicular to the stellar bar. This kind of structure has been detected in CO in some galaxies (Dereveux *et al.*, 1992, Ishizuki *et al.*, 1990) and indeed, from the ratio of major to minor axes we would deduce an inclination of $60°$ to the plane of the sky, which is compatible with the ring being perpendicular to the axis of the stellar bar.

Acknowledgements This work was partly supported by the *Acciones Integradas* programme of the British Council and the Spanish Ministerio de Educación y Ciencia.

References

Draper, P.W. *et al.* 1992, *MNRAS*, **257**, 310.
Dereveux, N. A., Kenney, J. D. P., and Young, J. S. 1992, *A. J.*, **103**, 784.
Athanassoula, E. 1992, *MNRAS.*, **259**, 345.
Pedlar, A., Howley, P., Axon, D. J., and Unger, S. W. 1992, **259**, 369.
Pérez, E., González-Delgado, R. M., Tadhunter, C. N., and Tsvetanov, Z. 1989, *MNRAS*, **241**, 31p.
Shlosman, I. *et al.* 1989, *Nature*, **338**, 4.
Ishizuki, S. *et al.* 1990, *Nature*, **334**, 224.

Evolution of Narrow Line Clouds

P. A. Foulsham[*]   D. J. Raine[*]

Cloud velocities in the NLR appear to be related mainly to the host galaxy (Whittle, 1992). This is compatible with evidence that the clouds are predominantly *infalling* (De Robertis & Shaw, 1990), with only a minor component of the line emission arising from jet induced emission. Either these infalling clouds contribute significantly to the BLR, connecting this to the NLR smoothly through an intermediate zone, or they are destroyed. We expect that the cloud geometry and dynamics will be reflected in the line emission ratios and profiles. We look here at the emission from clouds that are destroyed in an outflowing supersonic wind from the central nucleus which we assume to be in pressure balance with the ambient ISM (Smith, 1984; Mobasher & Raine, 1987)

We model the hydrodynamic evolution of two clouds each having an initial density of 10^4 cm^{-3} and a temperature of 10^4 K with a free-fall velocity of 2×10^7 cm s^{-1}. The smaller cloud has a mass of 6.6×10^{-4} M$_\odot$ ($r_c = 25 \times 10^{15}$ cm) and the larger one a mass of 6.6×10^2 M$_\odot$ ($r_c = 25 \times 10^{17}$ cm). The clouds fall under gravity into a supersonic wind of Mach number 1.5 with a density that increases as r^{-2}. The initial distance from the continuum source is 10^{21} cm and its luminosity is $\sim 10^{44}$ erg s^{-1}.

For the hydrodynamic simulation we employ a code utilising a first order Godunov scheme (Godunov, 1959) developed by R. Hillier of Imperial College London and adapted for astrophysical use at Leicester University by M. Dubal and P. Foulsham. The code solves the equations of non-self-gravitating non-adiabatic hydrodynamics in a cylindrical coordinate system. The heating-cooling function used is that suggested by Mathews & Doane (1990).

To calculate the emission from the evolved states we use the photoionisation code CLOUDY (Ferland & Truran, 1981) in a series of 1-D runs. The density and temperature at each position is set by the output of the hydrodynamics.

Table 1. shows the line ratios we obtain from the evolved clouds at ~ 1 sound crossing time compared with those from a slab and sphere. The line ratios from the clouds, except for the low ionisation sulphur lines, are in reasonable agreement with observation. Looking at the [OIII] $\lambda5007$ emission in figure 1. we see that it is predominantly coming from the face of the cloud and will, therefore, be almost independent of the cloud size. However, the [SII] $\lambda6718$ emission is coming from the whole cloud and will be dependent on cloud size. Thus, choosing a cloud of intermediate column density will allow us to fit the sulphur lines to observation also.

[*]Dept. of Physics and Astronomy, University of Leicester, Leicester LE1 7RH, UK

	Slab $n_H = 10^4 cm^{-3}$ $n_{col} = 10^{22} cm^{-2}$	Sphere $n_H = 10^4 cm^{-3}$ $R = 25 \times 10^{17} cm$	Small Cloud	Large Cloud	Observed
[OII]λ3727	0.32	7.0	0.14	0.15	0.30±0.15
[NeIII]λ3869	0.12	0.21	0.081	0.093	0.13±0.04
[OIII]λ4363	0.0062	0.0068	0.0055	0.0052	0.019±0.010
[Hβ]λ4861	0.10	0.14	0.085	0.085	0.093±0.025
[NII]λ6583	0.27	53	0.12	0.34	0.27±0.09
[SII]λ6718	0.036	16	0.020	0.39	0.14±0.05
[SII]λ6733	0.062	33	0.031	0.36	0.14±0.20

Table 1. Line ratios with respect to [OIII]λ5007.

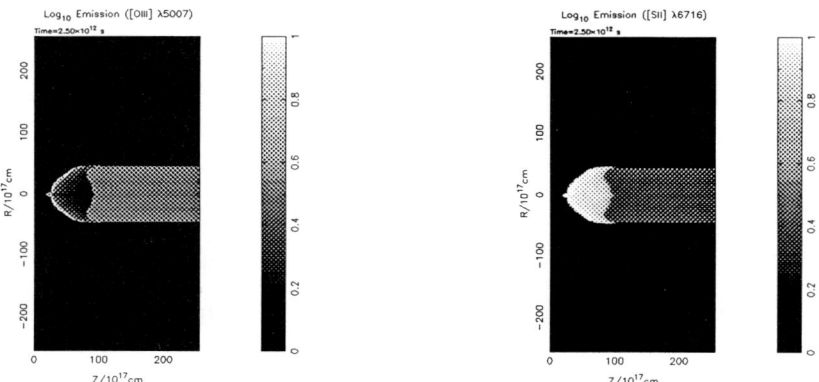

Figure 1. Cloud emission of [OIII]λ5007 and [SII]λ6718.

References

De Robertis, M. M. & Shaw, R. A., 1990. *Astrophys. J.*, **348**, 421.
Ferland, G. J. & Truran, J. W., 1981. *Astrophys. J.*, **244**, 1022.
Godunov, S. K., 1959. *Mat. Sb.*, **47**, 357.
Mathews, W. G. & Doane, J. S., 1990. *Astrophys. J.*, **352**, 423.
Mobasher, B. & Raine, D. J., 1987. *Mon. Not R. astr. Soc.*, **228**, 159.
Smith, M. D., 1984. *Mon. Not. R. astr. Soc.*, **209**, 913.
Whittle, M., 1992. *Astrophys. J.*, **387**, 121.

Star Formation in NGC 5953

R.M. González-Delgado and E. Pérez *

Abstract

We present results from an analysis of Hα, [OIII] images and long slit spectroscopy of the Seyfert 2 galaxy NGC5953. These show that the nucleus is extended in the northeast direction and surrounded by a vigorous burst of recent star formation.

1. Introduction

Some interacting Seyfert galaxies show strong circumnuclear emission associated with recent star formation, and one good example of these objects is NGC 5953. This Sa galaxy has a Seyfert 2 nucleus and is in interaction with the late type Scd galaxy NGC 5954 which has a LINER type nucleus, shows distorted spiral arms and is located 44" to the northeast of NGC 5953.

The radio continuum map at 1.4 GHz shows a diffuse structure over the main body of the galaxy and an enhanced extended component NE of the nucleus. There is a local maximum at p.a. 90°, 5" from the nucleus (Jenkins 1984); this component has been identified with a supergiant HII region (Rafanelli et al., 1990)

We present narrow band images in Hα, [OIII]λ5007 and the nearby continuum, that were obtained with the 4.2m William Herschel Telescope, using the Taurus II box in imaging mode with the f/4 camera. The spatial resolution was 0.27"/pixel. In addition we obtained long-slit spectroscopy with the IDS spectrograph at the 2.5m Isaac Newton Telescope. We used an EEV CCD detector with a spatial resolution of 0.6"/pixel, covering the spectral regions 3300–5200 Å and 5000–6900 Å. A slit of width 1.5" was oriented in the position angle 44°.

2. Results

The continuum image shows a central part separated in two sources, the nucleus and a field star 3" to the west. The Hα and [OIII] maps show an extended nucleus at p.a. 30° to the northeast. The nucleus is surrounded by a vigorous burst of recent star formation. The total luminosity in the Hα light is log L(Hα)=41.41 erg s^{-1}, and the nucleus represents 30% of the total emission. Eleven HII regions have been

---

*Instituto de Astrofísica de Canarias, 38200 La Laguna, Tenerife, Spain

identified, all of which are supergiant with luminosities between 40.7 and 39.81, and sizes between 200 and 450 pc. The star formation rate and the SFR/area in NGC 5953 is very high for its early morphological type, Sa. The cumulative luminosity function is well represented by a power law with a slope of -2.1 ± 0.1, which makes it shallower than the mean LF for the Sa morphological type (Kennicut et al., 1989).

The 2D spectra show four different knots: the nucleus, two HII regions to the southwest (A and B), at 7 and 3 arcsec, and an extended emitting region (C) to the northeast of the nucleus, which extends up to a distance of 10 arcsec. The [OIII] emission (as we can see in the [OIII] map) is more extended and brighter in this direction than in the southwest direction, and more so than Hα. Using the diagnostic diagrams of Veilleux and Osterbrock (1987), the A and B components are characterised by line ratios typical of low excitation HII regions. However, component C falls at the border line between HII regions and AGN. The spatial extent of the line ratio [OIII]/[OII] is larger towards the northeast, confirming the suggestion from the [OIII] map that this may be an HII region contaminated by extra ionization from the nucleus.

The rotation curves from Hα and [OIII]λ5007 towards the SW look very similar; however to the NE, where the higher excitation is seen in the spectra and the images, the [OIII] velocity is $\sim$ 40 km/s larger than that of Hα. This adds further evidence for an interaction between the radio plasma and the NLR, as has been found in other objects.

These regions are in the low density limit (100–300 cm^{-3}) and have electron temperatures between 5500 and 6600 K. The chemical abundances indicate an overabundance of O and N, and a deficiency of S; however in the C component the O abundance is lower than in the other regions; it is possible that some fresh material is falling onto this part of the disc from the companion galaxy NGC5954.

References

Jenkins, C.R., 1984. *Astrophys. J.*, **277**, 501.
Kennicut, Edgar, K., Hodge, P.W, 1989. *Astrophys. J.*, **337**, 761.
Rafanelli, P., Osterbrock, D.E., Pogge, R.W., 1990. *Astrophys. J.*, **99**, 53.
Veilleux, S., Osterbrock, D.E., 1987. *Astrophys. J. Suppl. Ser.*, **63**, 295.

Stellar Activity in the Seyfert Nucleus of NGC 1808

Duncan A. Forbes [*]

Abstract

Although claimed to possess a Seyfert nucleus, NGC 1808 reveals radio properties, line widths and ratios that are consistent with a few, albeit powerful, SNRs. There is as yet no compelling evidence for Seyfert activity, and this galaxy should be reclassified as a starburst galaxy (it shows many similarities to NGC 253 and M82).

1. Introduction

NGC 1808 is a 'hotspot' or Sersic–Pastoriza galaxy at a distance of 16.4 Mpc ($1'' =$ 80 pc). Although this galaxy shows optical hotspots, these hotspots have largely disappeared at near–infrared wavelengths. It is a highly inclined and dusty Sbc spiral, with evidence for a burst of star formation in the circumnuclear region about 5×10^7 yrs ago. Two pieces of evidence suggested that NGC 1808 habours a Seyfert nucleus:
- High resolution spectra revealed broad lines (Veron–Cetty & Veron 1985).
- The 6cm radio luminosity of the compact nucleus is 500 times that of the most luminous Galactic supernova remnant (SNR), suggesting a non–stellar nucleus (Saikia *et al.* 1990).

2. Line Ratios

We have obtained a red spectrum of NGC 1808 (see Forbes, Boisson & Ward 1992 for details) containing the [SIII] lines at 9069 and 9532Å. Fig. 1 shows a diagnostic diagram based on the sulphur line ratios. We also show the mixing curve between the location of HII regions and Galactic SNRs. This line ratio diagram (and those for [OI]6300Å, [NII]6583Å, [SII]6717+6731Å) are consistent with a high abundance HII region and a $\sim 20\%$ contribution from SNRs, i.e. no Seyfert nucleus is required to explain the line ratios. The ionization mechanism would be from hot OB stars combined with shocks associated with SNRs. The broad lines seen by Veron–Cetty & Veron (1985) could therefore be due to an expanding SNR shell. A similar conclusion

---

[*] Institute of Astronomy, Madingley Road, Cambridge CB3 0HA, England and Lick Observatory, University of California, Santa Cruz, CA 95064, USA.

was reached by Filippenko (1989), based on observations of SN 1987F located in an HII region, namely star formation could account for the optical spectra of some low–luminosity AGN.

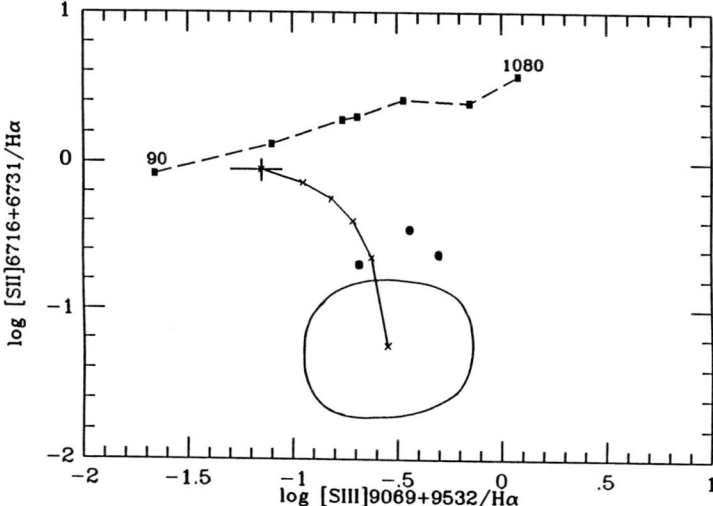

Figure 1. [SIII]9069+9532/Hα vs. [SII]6717+6731/Hα diagnostic diagram. The long dashed line represents the shock models with the extreme shock velocities indicated. The enclosed region represents observed line ratios for HII regions. The solid line represents a mixing curve from this region to the location of SNRs. The crosses indicate increments of 20% in the line ratio. Also shown (filled circles) are the line ratios for the nucleus and two locations south of the nucleus.

3. Radio Maps

High resolution radio maps have revealed a population of steep spectrum compact sources in the nuclear region, i.e. SNRs (Saikia et al. 1990). Saikia et al. measure a radio luminosity some 500 times Cas A. This is too powerful for a single SNR and so a Seyfert activity was claimed. However this radio emission could still be due to a small number of powerful SNRs. One example is an off–nuclear source in NGC 1808 itself. In the 20cm map of Saikia et al. this source is some 100 times more luminous than Cas A. Another example is 41.9+58 in M82, and 120 times more luminous than Cas A. Both of these sources lie some distance from the nucleus and cannot be AGN. Most recently, support for the SNR origin came from 2cm observations by Axon et al. (1992) in which the nucleus is resolved ($\sim 0.3''$). This effectively rules out a black hole and accretion disk as the source of the radio emission. We conclude stellar activity and gaseous processes in the nucleus of NGC 1808 can explain its 'Seyfert–like' properties, namely optical line ratios and radio emission. A similar situation may be present for other low–luminosity 'Seyfert' galaxies (see Terlevich et al. 1992).

References

Axon, D., *et al.* 1992. *Hot Gas in the Halo*, R.A.L. conference, in press.
Forbes, D., Boisson, C., & Ward, M., 1992. *M.N.R.A.S.,* in press.
Filippenko, A., 1989. *A. J.,* **97,** 726.
Saikia, D., *et al.* 1990. *M.N.R.A.S.,* **245,** 397.
Terlevich, R., *et al.* 1992. *M.N.R.A.S.,* **255,** 713.
Veron–Cetty, M., & Veron, P., 1985. *A. & A.,* **145,** 425.

Direct Evidence for Anisotropy: Radio Maps and their Relation to Optical Morphology

P. Alexander *

Abstract

Double radio sources provide perhaps the most graphic evidence for anisotropy in AGN, and observations of their large-scale structure can potentially give us a direct means of measuring the emission axis of the AGN itself. Attempts to unify radio sources using orientation and doppler boosting arguments alone have been relatively successful, however recent discoveries of correlated asymmetries on the kpc- to Mpc-scale in the radio, optical and IR may pose serious problems for such models.

1. Introduction

At first sight a paper about the large-scale (kpc – Mpc) structure in radio galaxies appears to have little direct relevance to the nature of the compact object in AGN. If we are considering evidence for anisotropy however then radio sources provide the most graphic evidence for such emission from the AGN – for example 4C74.26 (Fig. 1) shows all the principal features: an essentially linear structure with a one-sided jet linking the core with the southern lobe, the alignment of core and hotspots and a one-sided VLBI jet aligned with the large-scale jet. Potentially, radio sources could provide us with one of the best ways of determining the orientation of the emission axis of the AGN, however the recent discovery of large-scale correlated asymmetries, may not only make this a difficult task, but may also provide strong evidence *against* unified models of AGN phenomena in terms of orientation dependencies.

So as to limit my discussion I will consider only high-power objects with P(178 MHz) $\geq 10^{25.5}$ W Hz sr^{-1}. Unified models for such objects, taken together as a single class, seem to work well linking narrow line radio galaxies (NLRG), broad line radio galaxies (BLRG) steep-spectrum radio quasars (SSRQ) and flat-spectrum radio quasars (FSRQ) in a sequence with decreasing angle of the radio axis to the line-of-sight (θ). This unification, which relies upon bulk relativistic motions in the jets, explains the occurrence of one-sided jets and luminous radio cores in radio quasars as a consequence of doppler enhancement. Unification schemes of this sort for the radio population therefore provide direct support for models of AGN which associate

---

*Mullard Radio Astronomy Observatory, Cavendish Laboratory, Madingley Road, Cambridge, England, CB3 0HE

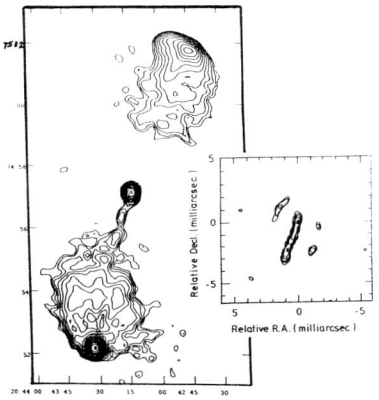

Figure 1. 4C74.26 at L-band, from Riley & Warner (1990), the inset shows the VLBI image (Pearson et al. 1992) of the core of 4C74.26 – note the scale of the VLBI image and the alignment of the VLBI jet with the one-sided kpc-scale jet.

quasars with those AGN aligned near to the line-of-sight. The discoveries which may challenge this picture are: (1) that the radio-axis tends to be aligned with the optical, IR and Extended-Narrow-Line regions (ENLR); and (2) the correlated asymmetries in the large-scale radio structures and ENLR.

2. Arguments in favour of orientation dependencies

If jets are relativistic then we *expect* effects such as superluminal motion, doppler boosting and hence we would expect to see one-sided jets and a strong orientation dependence. Can we therefore estimate the jet speed from radio maps? We can relate the jet-speed to the advance speed of the hotspot by assuming ram-pressure confinement of the hotspot (see Fig. 2) giving $v_j \approx v_h (1 + \rho_0/\rho_j)^{1/2}$; assuming a filling factor of unity for the lobe material and all acceleration to occur in the hotspots, we can relate the pressures and densities via adiabatic expansion, $(\rho_l/\rho_h) \approx (p_h/p_l)^{-1/\Gamma}$ where $\Gamma = 4/3$ is the adiabatic index for a relativistic plasma. The jet terminates in the hotspot via a strong shock, implying $\rho_h = (\Gamma + 1/\Gamma - 1)\rho_j$. Putting these together and conserving material flow along the jet and in the lobe we get $\frac{v_j}{v_h} \approx \left[\left(\frac{r_l}{r_j}\right) - 1\right]\left(\frac{\Gamma+1}{\Gamma-1}\right)\left(\frac{p_l}{p_h}\right)^{1/\Gamma}$. Observations give $r_l/r_j \geq 10$, $p_l/p_h \approx 0.1$ from minimum energy arguments and hence $v_j/v_h \geq 30$; since $0.05c \leq v_h \leq 0.2c$ from spectral-aging measurements (Alexander & Leahy 1987) we estimate a highly relativistic jet speed (and hence we should really re-do the above calculation). Observationally, the strongest evidence for bulk relativistic motion in jets comes from MERLIN/VLBI observations of superluminal motion and observations on all scales of the ubiquitous one-sided jets in radio quasars. For example, the observations of 3C273 by Davis, Unwin & Muxlow (1991) reveal not only an apparent expansion speed of between 2.2 c and 4.8 c, but also a limit on the jet to counter-jet ratio of 5300:1, these data are (just) consistent with $\beta \geq 0.95$. Bulk relativistic motions are also invoked to explain the high brightness-temperatures of the cores (e.g. Quirrenbach et al. 1989) and the lack of self-compton X-rays (Marsher 1986).

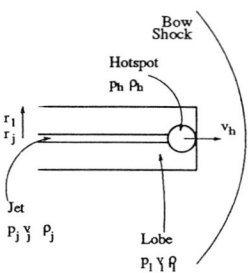

Figure 2. A simple model of a radio source; the subscript h refers to the hotspot, j to the jet and l to the lobe.

low $\theta \longrightarrow$ high θ	References	Comments
Radio loud QSO $\longrightarrow$ Radio quiet QSO	Scheuer & Readhead (1979)	Problem: lack of low-frequency extended structure in radio-quiet QSOs
FSRQ $\longrightarrow$ SSRQ	Orr & Browne (1982)	Problem: what about the large θ objects and the Broad-line objects?
FSRQ $\longrightarrow$ SSRQ $\longrightarrow$ BLRG $\longrightarrow$ NLRG	Barthel (1989) Antonucci (1989) Scheuer (1987) Peacock (1987)	Supported by: ability to fit all the population, linear-size distributions, source counts and luminosity functions

Table 1. Summary of unification schemes

With the assumption of bulk relativistic motion in the jets a number of unification schemes have been proposed, see Table 1 and the review by Antonucci (1993). Of these schemes the third, which unifies the whole radio-source population in a single model, is perhaps the most appealing; it is supported by the work of Padovani & Urry (1992) who have shown that the luminosity function of the supposedly low θ objects may be derived from the high θ population (the NLRGs) assuming a simple source model including a doppler-boosted jet. Further support comes from the work of Rawlings & Saunders (1991) who show that the narrow-line luminosity is strongly correlated with radio jet power for all sources – this is consistent with the unified model provided the narrow line luminosity is orientation independent.

3. The alignment effect

Despite the apparent success of unified models there is an increasing body of data which (at the very least) requires careful explanation and indeed may challenge all unification schemes. The first of these is the alignment effect. Since 1987 many groups have reported the alignment of radio and optical axes. For example, Chambers, Miley & van Breugel (1987) and Mc Carthy et al. (1987) find alignments in approximately 80% of their objects that are better than 20 degrees between the radio axis and the major axis of the optical continuum (BVR) and extended emission-line gas ([OII] λ3727).

Line ratios for eleven southern radio galaxies studied by Robinson et al. (1987) indicate photoionization of the emission-line gas, probably by the AGN. The origin of the continuum is more problematic. For the best-studied case, 3C368, (Djorgovski et al. 1987) there is alignment in all wave bands and the radio and optical emission are nearly co-spatial; the optical emission is polarized (Scarrott, Rolph & Tadhunter 1990) and the alignment is seen to be better at shorter wavelengths – this is consistent with the general trend observed by Rigler et al. (1992). Eales & Rawlings (1990) examined in detail the alignment at K-band (2.2 μm) in 3C356 and dismissed Rayleigh scattering, inverse compton and electron scattering as possible causes for the alignment, preferring triggered star formation as the most likely mechanism. The polarization of the extended emission in objects such as 3C368, however, argues strongly in favour of electron scattering. Recently, Eales (1992), in an attempt to reconcile these problems, has suggested that the alignment effect is caused by a strong selection effect – he argues that if a radio source propagates into a denser medium it will, for the same mechanical energy in the jet, have a significantly larger radio luminosity; in a flux limited sample therefore we will tend to select those objects which follow the mass distribution and hence we will see an alignment effect since the optical/IR light also traces the mass. Although the alignment effect in itself does not cast serious doubt on unification models it does provide evidence that the interaction of radio sources with their environment is of considerable importance.

4. Large-scale asymmetries

The first large-scale asymmetry to be observed was the jet-sidedness–depolarization asymmetry. The observed asymmetry is in the sense that in those sources with one-sided jets, the side/lobe of the radio source into which the jet propagates is less depolarized (Laing 1988; Garrington et al. 1988; Garrington et al. 1991; Garrington & Conway 1991). This result can be explained in terms of a doppler boosting model if we assume that the *one-sided* jet results from relativistic enhancement and hence the lobe into which the jet propagates is nearer to us. If the depolarization occurs in a magneto-active medium surrounding the radio source (e.g. a standard hot cluster with a cluster-scale magnetic field) then the distant lobe (non-jet side) will have a longer line-of-sight through this medium than the nearer-, jet-lobe, and hence the former will be more depolarized.

In addition to the jet-sidedness–depolarization asymmetry one also finds that in those objects which have asymmetrical structures the lobe nearest the nucleus is more depolarized shows patchy depolarization and is often associated with, and in some cases co-spatial with, extended emission-line gas (Pedelty et al. 1989; Mc Carthy et al. 1992; Liu & Pooley 1991b; Fig. 3). Mc Carthy et al. (1991) also show that the asymmetry in the emission-line gas distribution is strongest at higher redshifts,

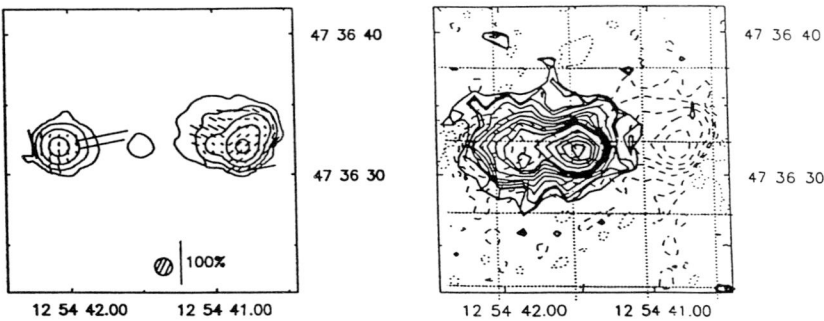

Figure 3. The polarization–emission-line-gas asymmetry in 3C280 from Liu & Pooley (1991b).

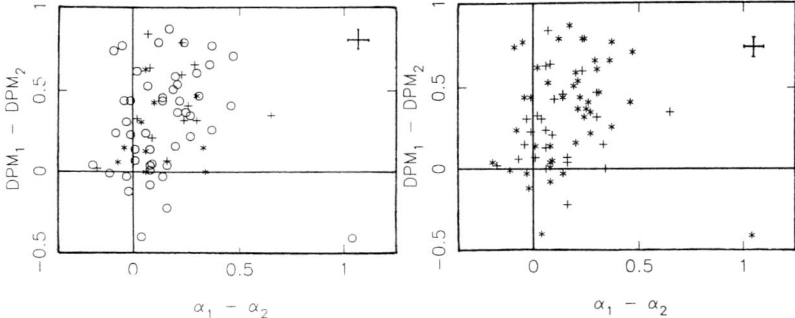

Figure 4. **a.** The depolarization–spectral-index asymmetry for the samples of Liu & Pooley (1991a) '+', Garrington et al. (1991) 'o' and Pedelty et al. (1989) '∗'; **b.** The depolarization–spectral-index asymmetry for FR II radio galaxies '+' and quasars '∗'.

and that the radio-structure is most asymmetric when the radio-source size and the size of the extended emission-line gas region are comparable. Taken together, these results again suggest that the dominant factor is the interaction of the radio source with its environment.

The third asymmetry, and the one which provides the most stringent constraints on models and unification schemes, is the spectral-index–depolarization asymmetry reported by Liu & Pooley (1991a) and mentioned in Garrington & Conway (1991). This asymmetry is in the sense that the lobe which is most depolarized also has a steeper spectrum. Such an asymmetry is most easily explained in terms of an intrinsic asymmetry perhaps in terms of the kinetic power of the two jets, the strength of the magnetic field in the lobes, or an asymmetric environment. In Fig. 4a I have constructed a plot showing the spectral-index–depolarization asymmetry using data from Liu & Pooley (1991a), Garrington et al (1991) and Pedelty et al. (1989) – the correlation holds strongly for all of these samples and therefore also implies a

	Near & NLR	Near & NOT NLR	Near & NLR?
Jet	1	0	17
NOT Jet	6	0	26
Jet?	13	3	–

Figure 5. The correlation between the narrow-emission-line gas regions (NLR), the structure asymmetry (Near?) and the jet-sidedness (Jet?) for all the available data taken from the references used to construct Fig. 4.

spectral-index–jet-sidedness correlation. The few objects not following the correlation do not show any significant differences from the remainder in terms of galaxy type (see Fig. 4) luminosity or linear size. How are these three asymmetries related to the distribution of emission-line gas? Unfortunately the overlap is rather poor between those galaxies which have emission-line images and those with good multi-frequency radio observations. The current situation can be summarized as follows: (i) the spectral-index–depolarization, depolarization–jet-sidedness correlations are very strong and essentially always hold and hence imply a strong spectral-index–jet-sidedness correlation; (ii) in those objects with an extended emission-line region there is a strong depolarization–structural–emission-line correlated asymmetry which is most apparent at high redshift; (iii) there are no strong correlations between jet-sidedness, structural asymmetries and emission-line gas distribution.

5. Explaining the asymmetries?

Can we explain all the observed asymmetry correlations in the framework of a unification scheme with orientation dependencies and relativistic bulk motions? As discussed above the jet-sidedness–depolarization asymmetry can be explained in this way if the depolarizing medium is a magnetized cluster gas. What of the other asymmetries? From light-travel-time arguments one would expect that the jet-side of the source should be longer since it is nearer to us and therefore seen at a later epoch – this does not work (see previous section on the lack of any jet-sidedness–structural correlation). The correlation with emission-line gas may be explained if the emission-lines come from the surface of optically thick clouds which are photoionized from the nucleus – the line emission on the far (non-jet) side would therefore be greater since we would observe the photoionized faces of the clouds directly leading to the observed correlation between depolarization and emission-line gas. The depolarization–spectral-index correlation poses a more severe problem. One possibility is that for those sources close to the line-of-sight the distant lobe is seen at an earlier epoch and if we postulate that the magnetic field decreases with source age then we would expect to see a depolarization–spectral-index correlation since the electron loss rate in the distant lobe will be greater. Such an explanation does not work for the whole sample however as we would predict a strong dependence on projected linear size and an almost complete absence of the effect in sources seen at more than a few degrees to the line-of-sight.

Of the other possible models I think we can dismiss the possibility of intrinsically one-sided jets in *all* objects since they do not provide a natural explanation for the depolarization–spectral-index correlation; a more serious objection still is the lack of radio-luminous quasar-like objects without jets which would be those members of the population with their jets directed away from us.

The final possibility which is strongly suggested by the *data* is the existence of an asymmetric environment – I believe we can explain all of the observed correlations as well as the alignment effect with such a model. Simple arguments can be used to show that we expect smaller lobes and hence higher energy densities on the side of the source with a greater external density. As a consequence of the higher density we will have more line-emission and greater depolarization; furthermore if the magnetic-field is in equipartition with the relativistic particles we expect a higher magnetic field strength on the denser side and hence a faster loss rate for the electron population leading to a steeper spectrum. The one remaining correlation, that of the jet-sidedness, is explained by some recent work by Fraix-Burnett (1992) who argues that the denser environment will lead to a thermalization of the jet kinetic power and hence we will only see the jet on the lower-density side where boundary-layer effects are dominant leading to efficient particle acceleration.

6. Conclusions

Although unified models are very attractive they are being challenged by discoveries of correlated asymmetries in the large-scale structure of radio galaxies and quasars. There are two main effects which I believe must play a role: (A) Asymmetric environments which can explain all the observed correlations and the alignment effect; (B) Doppler-boosting/orientation models which *may be able* to explain the observed correlations at least for objects close to the line-of-sight. At present my best guess is that (A) dominates, but that (B) becomes important for quasars. This may then salvage some of the ideas of unified models, but in a somewhat watered down form. Perhaps most importantly the possibility of very strong selection effects should not be ruled out. To finish on a positive note it is welcoming to think that the above models are testable and we may have the answers in the not too distant future.

References

Alexander P., Leahy J.P., 1987, MNRAS, 225, 1

Antonucci R., 1989, in 14th Texas Symposium of Relativistic Astrophysics, E. Fenyves ed., NY: NY Academy Science Press
Antonucci R., 1993, ARAA, 31
Barthel P.D., 1989, ApJ, 336, 606
Chambers K.C., Miley G.K., van Breugel W., 1987, Nature, 329, 604
Davis R.J., Unwin S.C., Muxlow T.W.B., 1991, Nature, 354, 374
Djorgovski S., Spinrad H., Pedelty J., Rudnick L., Stockton A., 1987, AJ, 93, 1307
Eales S.A., Rawlings, S., 1990, MNRAS, 241, 1P
Eales S.A., 1992, ApJ, in press
Fraix-Burnet D., 1992, A&A, in press
Garrington S.T., Leahy J.P., Conway R.G., Laing R.A., 1988, Nature, 331, 147
Garrington S.T., Conway R.G., 1991, MNRAS, 250, 198
Garrington S.T., Conway R.G., Leahy J.P., 1991, MNRAS, 250, 171
Laing R.A., 1988, Nature, 331, 149
Liu R., Pooley G.G., 1991a, MNRAS 249, 343
Liu R., Pooley G.G., 1991b, MNRAS 253, 669
Marsher A.P., 1986, in Continuum Emission in AGN, ed. M.L. Sitko, Tucson, NOAO, p. 143.
Mc Carthy P.J., van Breugel W., Spinrad H., 1987, ApJ, 321, L29
Mc Carthy P.J., van Breugel W., Kapahi V.K., 1991, ApJ, 371, 478
Orr M.J.L., Browne I.W.A., MNRAS, 1982, 200, 1067
Padovani P., Urry C.M., 1992, ApJ
Peacock J.A., 1987, in Astrophysical Jets and Their Engines, W. Kundt ed., Dordrecht, Reidel, p. 185
Pearson T.J., Blundell K.M., Riley J.M., Warner P.J., 1992, MNRAS in press
Pedelty J.A., Rudnick L., Mc Carthy, P.J., Spinrad H., 1989, AJ, 97, 647
Quirrenbach A., Witzel A., Krichbaum T., Hummel C.A., Alberdi A., Schalinski C., 1989, Nature, 337, 442
Rawlings S., Saunders R.D.E., 1991, Nature, 349, 138
Rigler M.A., Lilly S.J., Stockton A., Hammer, F., Le Fèvre O., 1992, ApJ. 385, 61
Riley J.M., Warner P.J., MNRAS, 246, 1P
Robinson A., Binette, L., Fosbury R.A.E., Tadhunter C.N., 1987, MNRAS, 227, 97
Scarrott S.M., Rolph C.D., Tadhunter C.N., 1990, MNRAS 243, 5P
Scheuer P.A.G., Readhead A.C.S., 1979, Nature, 303, 26
Scheuer P.A.G., 1987, in Superluminal radios sources, J.A. Zensus & T.J Pearson eds, CUP, p. 104

The Radio-Optical Connection in AGN

David J. Axon *   *J. E. Dyson* [†]   *Alan Pedlar* [‡]

Abstract

We review the evidence that ejected radio material plays a fundamental role in the formation and kinematics of the Narrow Line region and the extended emission line regions associated with radio jets in radio galaxies and QSO's. In Seyfert galaxies, the key observation is the existence of high-velocity (several hundred $km\,s^{-1}$ from the systemic velocity of the galaxy) emission line components which are found systematically closer to the nucleus that the radio emission peaks. We describe how this result can be explained with a high speed bowshock model. In radio galaxies, the strong shock created by the jet results in a surrounding hot cocoon of gas expanding away from the jet axis. These expanding cocoons are visible in the form of double velocity structure in high resolution optical spectra and have now been detected in 3C120, 3C 171, 3C405 and 3C265. The velocity separation between the components can be as high as several thousand $km\,s^{-1}$ We briefly discuss how these cocoons can be used to verify the relativistic beaming hypothesis in systems with strong one-sided jets.

1. Introduction

Extended emission line regions (EELR) closely aligned with the radio structure have been found in Seyferts [20] and radio galaxies at both low [3] and high redshift [10]. The physical conditions and kinematics of these EELR provide a probe of both the radiation field of the AGN and the role played by ejected material in exciting the emission [13]. Wilson [25] has recently reviewed at length the physical structure of these regions. In this paper we confine our attention to just one aspect of this area: the kinematic consequences of the interaction of the radio ejecta and the environment. Generally the EELR of Seyfert galaxies stretch many Kpc beyond their radio structure and show quiescent velocity fields implying that their ionization structure is dominated by the form of the nuclear radiation field. In contrast the EELR in radio galaxies are on the same physical scales as the radio structure and in this sense bear

---

*Affiliated with the Space Science Division of ESA at the Space Telescope Science Institute, 3700 San Martin Drive, Baltimore MD, USA and Nuffield Radio Astronomy Laboratory, University of Manchester, Jodrell Bank, Macclesfield, Cheshire, England

[†]Department of Astronomy, University of Manchester, Oxford road, Manchester, England.

[‡]Nuffield Radio Astronomy Laboratory, University of Manchester, Jodrell Bank, Macclesfield, Cheshire, England.

a close resemblance to the Seyfert Narrow Line region (NLR). Indeed in many cases the NLR themselves are also preferentially aligned with the radio structure [26].

In Section 2 we start by describing the role of radio ejecta in the formation of the NLR. In section 3 we then discuss the kinematic evidence that similar interactions are occuring around the jets in powerful radio galaxies and QSO's.

2. The Structure of the Narrow Line Region

The standard model of the NLR, as an assembly of emission line clouds confined by a hot medium, has as its observational basis the low filling factor of the NLR [15]. Many aspects of the model have been known to be unsatisfactory for a number of years. From a theoretical standpoint there are severe difficulties in accelerating and confining the clouds due to instabilities [8]. Observationally, two vital missing ingredients are that neither the geometry or the two-dimensional velocity field of the NLR are fully incorporated in the model. To a large extent our faith in the standard picture of the NLR has remained intact because of the success and elegance of the photoionization calculations in explaining the general form of the ionization structure of the NLR [5].

The discovery of the direct spatial and physical association between the radio structure of Seyferts and individual off-nuclear velocity components in the NLR ([22], [23], [6], [21], [12]) has demonstrated the crucial role of interactions between the radio ejecta and the ambient gas in both the formation of the NLR and in moulding our view of it. One of the best examples, NGC 5929 [22], is shown in Figure 1. The key feature common to all the objects is that their high velocity components are systematically closer to the nucleus than the radio blobs. This result is often not apparent in emission line images as generally they are heavily contaminated by emission from the ambient gas, but is supported strongly by recent HST images (e.g. Mkn 78 [4]).

In order to explain these results, we have developed a model for the formation and structure of the NLR in which ejected radio material drives shocks into the gas and both compresses and accelerates it ([18] , [19]). The high velocity components are formed by photoionization of the cooled post-shock material by the UV radiation from the nucleus. Since our primary intention was to model the emission line structure of the NLR, we deliberately side-stepped the issue of the production of radio emission in the interaction because the detailed physics of the processes is poorly understood (we note in passing that in NGC 1068 at least, there appears to be evidence that some radio emission can be produced in the bow shock itself [24]). Major distinctions between previous shock models [7] arise because the shock velocities involved are sufficently high that one must allow for the finite cooling length of the shocked gas and also for the ionizing effect of the nuclear source.

In plane parallel shocks, the cooling length is roughly $\propto V_s^4$, making the expected dis-

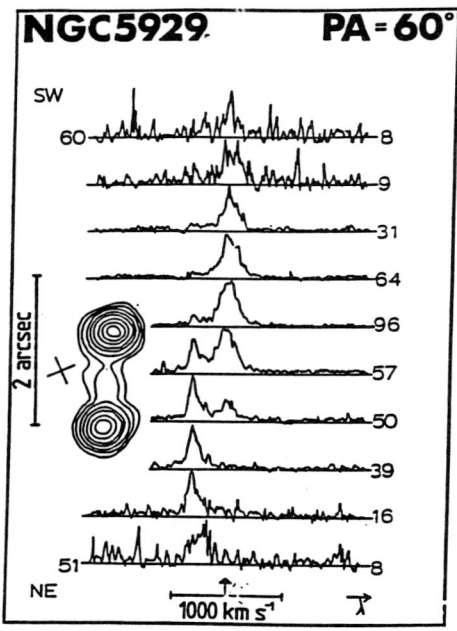

Figure 1. The observed [OIII] λ5007 line profiles of NGC 5929 superimposed on the radio structure. Note that the high velocity components associated with the radio blobs show opposite velocity shifts.

placement between the radio and optical emission very sensitive to the shock velocity, V_s. In the bowshock model, one must replace the concept of a single cooling length with an *effective cooling length* which represents the minimum point on the cooling curve which is a function of both the impact parameter and the shape of shock. The resulting line profiles then become suprisingly insensitive to V_s. For example with a quadratic bowshock shape, the effective cooling length is only $\propto V_s^{\frac{16}{13}}$ and, for a fixed density n_e and shock velocity, the viewing angle has the dominant effect on the profile structure. While our model makes a number of simplyfing assumptions (e.g. a steady state), it is nevertheless able to explain the major kinematic features of the NLR, including the displacement between the centres of the radio and optical emission, while avoiding fundamental physical difficulties inherent in previous models (e.g. cloud acceleration and stability). The other exciting aspect of the model is that we are able to explain the diversity of NLR profiles and velocity structures in terms of a simple two parameter family, namely, viewing-angle and plasmon/jet velocity (Figure 2). Clearly, in a more realistic model which includes instabilities (e.g. turbulence) and time-dependence, one might expect considerable microstructure. The overall appearance would then not be too dissimilar to that of Herbig-Haro objects and HII regions with numerous sheets and shells punctuated by bright clumps of

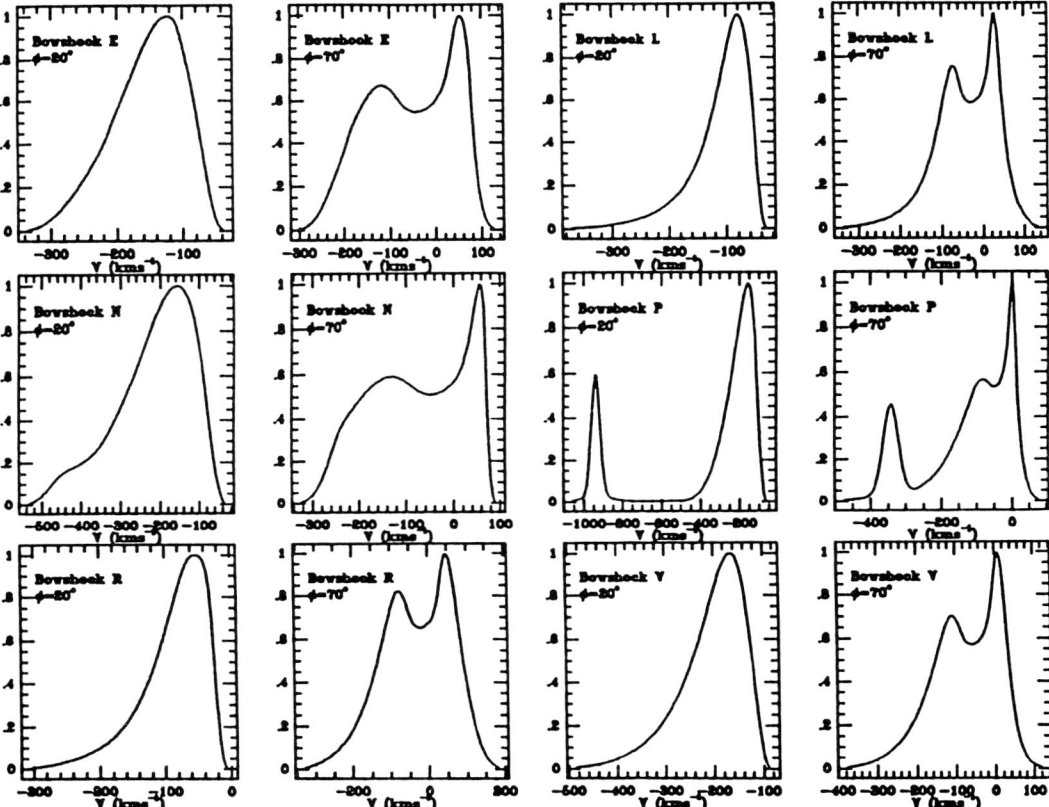

Figure 2. A montage of emission line profile predictions from the bowshock model. Six different models, labelled E, L, N, P, R and V, covering a range of shock velocities and densities are shown at two viewing angles 20^0 and 70^0. Note the diversity of profile shapes.

emission. Indeed this is exactly what one now seems to be seeing in HST images of the best resolved NLR (eg. NGC 1068, NGC 4151; [4]).

3. Jet-Cloud Interactions in Radio Galaxies

We already know from published emission line imagery that there is emission line gas close to many extragalactic radio jets. However much of this is likely to be quiescent material photoionized by the central AGN. Spectroscopic studies of a number of radio galaxies have shown broad lines and large velocity gradients in their EELR [2], [16]. Good examples are 3C 171 [9] and 3C 294 [11]. The literature, however, leaves the impression that these motions have no really coherent structure. An important development has therefore been the discovery of ordered motions which are a direct

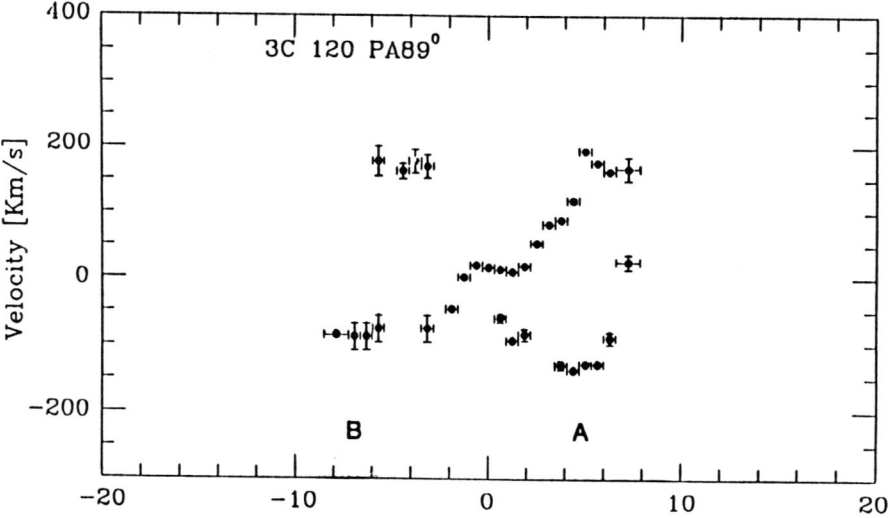

Figure 3. The observed velocity field along the jet of 3C120. Two velocity components are seen, separated by $\pm 300\,\mathrm{km\,s^{-1}}$ and localized to the jet.

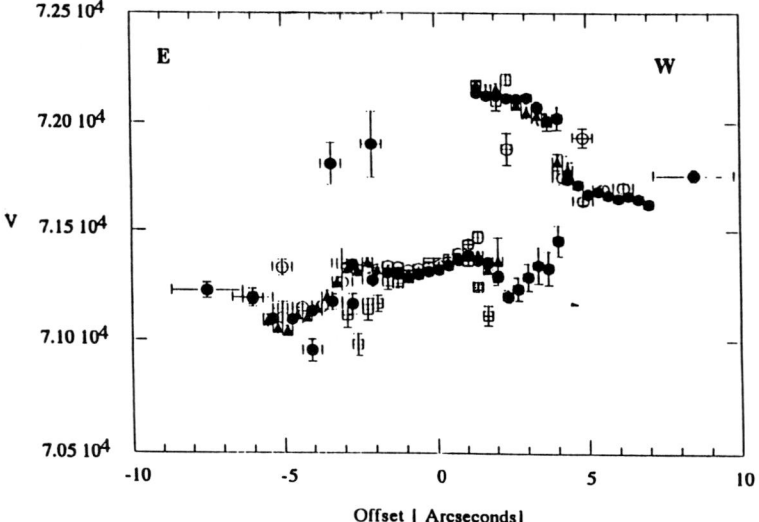

Figure 4. The observed velocity field along the radio axis of 3C 171 showing the split emission lines just behind the Eastern hot-spot.

consequence of the interaction of the jet with surrounding interstellar medium in a number of these systems, and which have close parallels with the origin of the motions we just described in the NLR. In 3C 120 [1], Figure 3, this interaction creates a spatially localized expanding cocoon of material around the jet. The size of the split line region transverse to the jet axis gives a direct measure of the cooling length of the gas and hence a dynamical time-scale.

In 3C 171, our new data (Figure 4) shows that, as in 3C120, the lines are split into two components in the Western lobe, separated by > 800 kms^{-1}, and that considerable kinematic disturbances are also present in the Eastern lobe. Similar kinematic structures are also found on the jets of 3C 405 and 3C 265, [16], and in these objects the velocity splitting is even larger $\sim 1600 - 1800$ kms^{-1}. As in the NLR the line emission of the disturbed gas systematically peaks behind the radio hot-spots and we believe that this is again due to the effect of long cooling times.

4. Are jets really two-sided?

Since the discovery of one-sided extragalactic radio jets, one of the most important issues has been whether such jets are truly one-sided or are in fact two-sided jets whose appearance is distorted by relativistic effects [14]. The two major problems are that at radio wavelengths, there is no means of measuring the velocities of the jets—except in superluminal cores—and that the expected surface brightness of any relativistic counter-jets can be dimmed by huge factors. It is important to realize that in two-sided relativistic jets, such cooling cocoons are also to be expected in the counter-jet direction, as the counter-jet is still highly supersonic. Although the radio continuum of the counter-jet is dimmed by relativistic aberration, the cocoon emission is not. Finding and measuring the dimensions of such cocoons is therefore of major significance as they provide a direct means of determining both the presence and location of any counter-jet. Recently we have discovered just such evidence. Figure 5 shows a localized region of disturbed gas in a region coincident with the faint radio extension on the counter-jet direction side of 3C 120. Probing the kinematic structure of gas in more distant superluminal sources with the HST now looks as it has the potential to provide a more universal proof for the beaming picture.

References

[1] Axon, D. J., Unger, S. W., Pedlar, A., Meurs, E. J. A, Whittle, D. M. and Ward, M. J., *Nature, Vol. 341 (1989) pp. 631-633.*

[2] Baum, S. A., Heckman, T. A. and van Bruegel, W. *Astrophys. J. , Vol. 389 (1990), pp. 208-222.*

[3] Baum, S. A., Heckman, T. A. and van Bruegel, W. *Astrophys. J. Suppl., Vol. 74 (1990), pp. 389-436.*

[4] Boksenberg, A., *in Science with the Hubble Space Telescope, ST-ECF/STScI workshop* ed P. Benvenutti and E. Schrier, *(1992) pp. 61-71.*

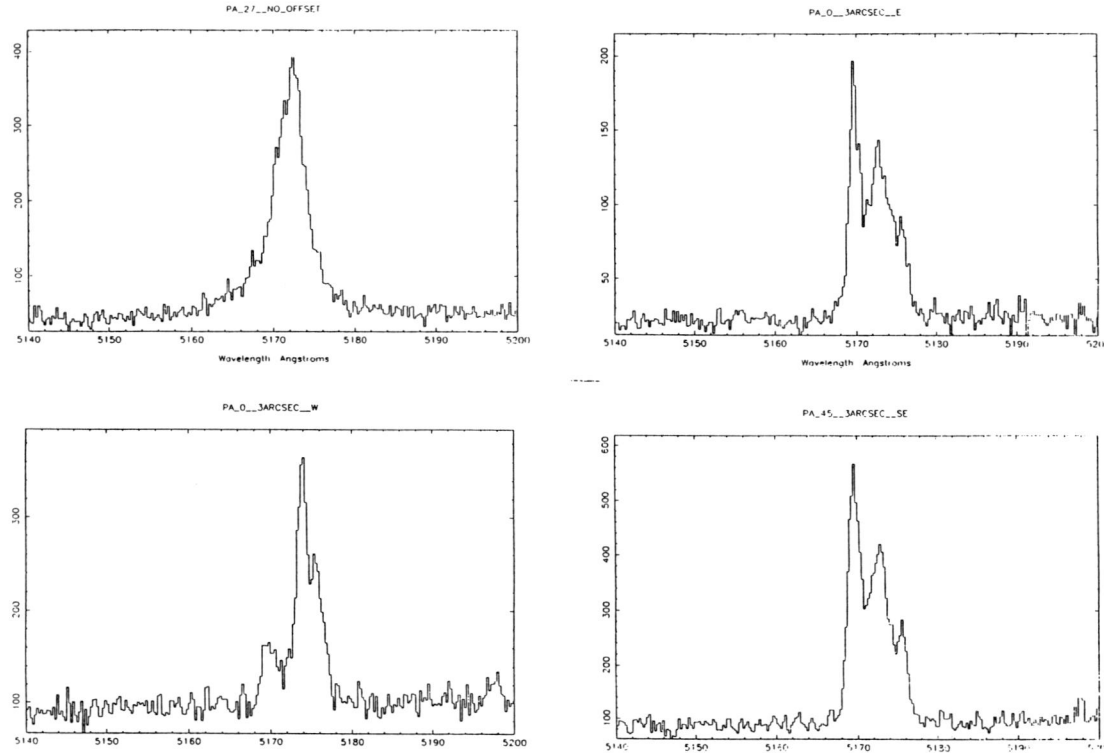

Figure 5. The observed emission line profiles in the counter-jet direction of 3C120. Top left the on-nucleus spectrum; top right PA 0 5 arcseconds East; bottom left PA 0 3 arcseconds West; bottom right PA 45 3 arcseconds South East.

[5] Ferland, G. J. and Osterbrock, D. E., *Astrophys. J.*, Vol. 300 (1986), pp. 658-668.
[6] Harrison, B. *PhD. Thesis University of Manchester (1988)*
[7] Hartigan, P. , Raymond, J. and Hartman, L. *Astrophys. J.*, Vol. 316 (1987), pp. 323-348.
[8] Hartquist, T. W., Dyson, J. E., Pettini, M. and Smith, L. J. *Mon. Not. R. astr. Soc.*, Vol. 221, (1986), pp. 715-726.
[9] Heckman, T. M., van Breugel, W. J. M. and Miley, G. K. *Astrophys. J.*, Vol. 286 (1984), pp. 509-516.
[10] McCarthy, P. J. , *ASP conference series*, Vol. 31, (1992) pp. 403-415.
[11] McCarthy, P. J. , Spinrad, H., van Bruegel, W., Dickinson, M., Djorgovski, S. and Eisenhardt, P. *Astophys. J.*, Vol. 365, (1990) pp. 487-501.
[12] Pedlar, A., Meaburn, J., Axon, D. J., Unger, S. W., Whittle, D. M., Ward, M. J. and Meurs,E. J.A., *Mon. Not. R. astr. Soc.*, Vol. 238, (1989), pp. 863-879.
[13] Robinson, A., Binette, L., Fosbury , R. A. E. and Tadhunter, C. N.) *Mon. Not. R. astr. Soc.*, Vol. 227, (1987), pp. 97-114.

[14] Scheuer, P. A. G. and Readhead, A. C. S. *Nature*, Vol. 277, (1979) pp. 182-185.
[15] Shields, G. A. and Oke, J. B. *Astrophys. J.*, Vol. 197 (1975), pp. 5-16.
[16] Tadhunter, C. N., Fosbury, R. A. E. and Quinn, P. J. *Mon. Not. R. astr. Soc.*, Vol. 240, (1989), pp. 225-254.
[17] Tadhunter, C. N. *Mon. Not. R. astr. Soc.*, .251 (1992), pp. 46p.
[18] Taylor, D., Dyson, J. E., Axon, D. J., and Pedlar, A. *Mon. Not. R. astr. Soc.*, Vol. 240, (1989), pp. 487-499.
[19] Taylor, D., Dyson, J. E. and Axon, D. J., *Mon. Not. R. astr. Soc.*, Vol. 255, (1992) pp. 351-368.
[20] Unger, S. W., Pedlar, A., Axon, D. J., Whittle, D. M., Meurs, E. J. A. and Ward M. J. *Mon. Not. R. astr. Soc.*, Vol. 228 (1987), pp. 671-679.
[21] Unger, S. W., Pedlar, A., Axon, D. J., Graham, D. A., Harrison, B., Saikia, D. J., Whittle, D. M., Meurs, E. J. A., Dyson, J. E.and Taylor, D., *Mon. Not. R. astr. Soc.*, Vol. 234 (1988), pp. 745-754.
[22] Whittle, D. M., Haniff, C. A., Ward, M. J., Meurs, E. J. A., Pedlar, A., Unger, S. W., Axon, D.J. and Harrison, B. A., *Mon. Not. R. astr. Soc.*, Vol. 222 (1986), pp. 189-200.
[23] Whittle, D. M., Ward, M. J., Meurs, E. J. A., Pedlar, A., Unger, S. W. and Axon, D. J., *Astrophys. J.*, Vol. 326 (1988), pp. 125-145.
[24] Wilson, A.S. and Ulvestad, J. S., *Astrophys. J.*, Vol. 319 ,(1987), pp. 105-117.
[25] Wilson, A.S., *Astrophysical Jets*, Vol 6, ST ScI Symposium Series ,1992. in press
[26] Wilson, A.S., Ward, M. J. and Hannif, C .A. *Astrophys. J.*, Vol. 334, (1988). pp. 104-120.

Knots in Extragalactic Radio Jets

E. Lüdke *

Abstract

The observational properties of knots in jets of two high-z quasars are presented and discussed. The knots along 3C9 and 3C309.1 jets have similar characteristics, suggesting that a common process may be responsible for knot formation. The knot's spectral indices tend to be flat at regions of disturbed flow and may be explained by particle re-acceleration due to small-scale shocks.

1. Introduction

Unresolved or partially resolved regions of enhanced synchrotron emission are often found in jets of extragalactic radio sources.

The VLA has provided maps of kiloparsec-scale jets with multiple knots of arcsecond dimensions in nearby FR I sources like M87 (z=0.0043) [1] and NGC6251 (z=0.023) [2] and provided valuable information about the physics of energy transports in jets.

A considerable amount of theoretical work has been done to explain the nature of such compact structures. However, most of the published data of high-z objects show unresolved knots along their jets. Spectral index information is difficult to obtain for sources with small angular sizes (< 30"), due to practical limitations. This is most unfortunate for distant sources since only knots are bright enough to be detected within low-brightness collimated flows.

This work is an attempt to give an interpretation of MERLIN, VLA and EVN observational data at 6 and 18 cm for two high-redshift quasars with bent and knotty jets, and to describe their radio properties. In calculating distances $H_o = 75\,km\,s^{-1}\,Mpc^{-1}$ and $q_o = 0.5$ are adopted.

2. Radio Observations

The MERLIN maps with knot identifications are shown in figure 1. Note the 3C9 jet curvature after K6 and the bow-shock shape of the 3C309.1 jet limited by K6, K7 and K8. Knot diameters $(2r)$, peak brightnesses (I_ν) and spectral indices (α) have been calculated. I fit scaling laws of the type $I_\nu \propto \nu^{-\alpha} r^{-\beta} D^\eta$ to the data, since most

---

*University of Manchester, Nuffield Radio Astronomy Laboratory, Jodrell Bank, Cheshire SK11 9DL, England

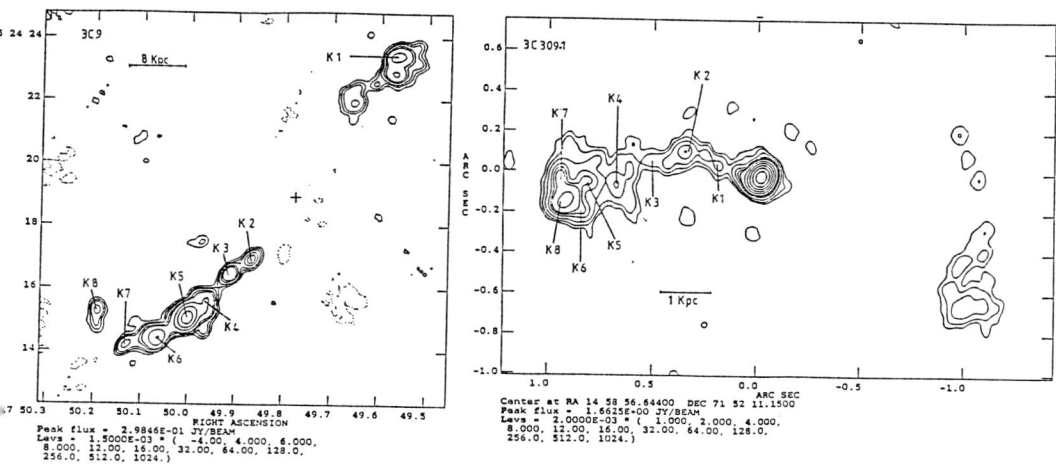

Figure 1. Knot identification in the MERLIN maps of 3C9 at 18 cm (0.35″ resolution) and 3C309.1 at 5 cm (60 mas). The cross in the 3C9 map is the core position) and 3C309.1 has a VLBI jet perpendicular to the source axis, probably disrupted by a strong plasma–ISM interaction [3].

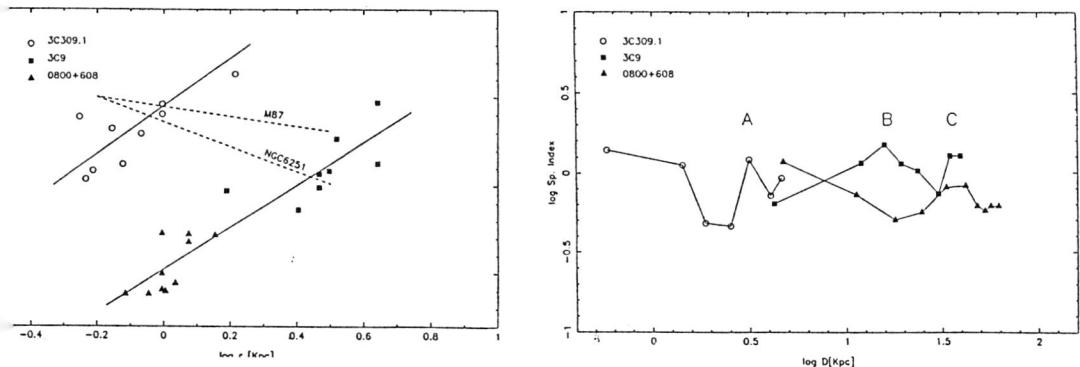

Figure 2. a) Peak brightness (mJy/beam) at 6 cm versus knot radii (kpc). The dashed lines represent the slope fit along the NGC6251 and M87 jets and full lines are the best fit to the observed data. Published data of 0800+608 are also shown [4]; b) spectral index variation along the jets. A,B and C are probably sites of particle re-acceleration (see text).

of theoretical models are able to define the exponents α, β and η. 0800+608 is a quasar with 11 knots along the jet and the published data [4] are also included.

Figure 2-a shows that there is a strong correlation between knot dimensions and peak brightness with ($\beta = -4.72 \pm 0.07$). Surprisingly, the slope of the best linear fit for each jet is the same and it is not typical of values found in jets of nearby radio galaxies ($\beta \approx 1.4$). A trend is also observed between brightness and distance from the core with $\eta = 0.76$ in the I_{18cm} equation for 3C309.1 while for 3C9, I_{6cm} has $\eta = 0.56$ between K2 and K5.

The spectral index variations along the jets of M87 [6] and NGC6251 are not significant but the knots along the quasar jets show irregular distributions (Fig. 2-b). However, if the particle population obeys a power-law energy distribution $N(E)dE \propto E^{-(2\alpha+1)}$ and if particle acceleration is occurring, then there should be a flattening of spectra at larger distances from the core. The 3C309.1 knots (K1 to K4) lose energy (probably by energy losses to the dense gaseous medium embedding the source) but K6, K7 and K8 are as flat as K2. 3C9 has two sites of maximum re-acceleration at K6 and K8, although the particles' average energy increases towards K5. The jet ends at K8 (which is actually two partially resolved knots) with particles having average energy similar to those in the regions between K4 and K5. The 0800+608 knots show similar behaviour and the jet also has more energetic particles in regions of greater disturbance and deviations from the original trajectory.

3. Alternative Explanations

Expressions for synchrotron radiation emitted by relativistic shock waves moving transversely to the jet magnetic field and adiabatic losses has been derived [5]. If the Doppler boosting factor follows $\delta \propto D^d$, the observed peak fluxes would obey $I_\nu \propto \delta^3$ and $d = 1.03$ along the 3C9 jet. This simple model is unable to explain the knot brightness distribution near regions where the jet decelerates, since d would not remain constant and the brightness could not be scaled with the core distance. Hence, the observed properties cannot be explained on basis of differential Doppler boosting only.

Are knots expanding plasmons? Adiabatically-expanding plasmons are also unable to explain the knot properties since the scaling laws typically have $\beta < 0$ ($\beta=7\alpha/3+1$) and $d = 3 + \alpha$ for plasmons expanding adiabatically in a free jet with uniform speed [6]). Plasmons could not increase their surface brightness as they expand.

The correlations can be understood if shock waves propagating along invisible jets (like M87) induce turbulence at the shear layer which increases both radio brightness and particle energy. If the jet suffers small-scale disruption due to strong interactions with the surrounding medium, energy transfer would be effective in creating turbulent instabilities at the boundaries. The amount of energy extracted from the jet relativistic flow will be dependent on the magnitude of the wave-medium interaction. This is seen in the present sample since knots tend to be brighter and have a flatter spectrum near bends and wiggles, rather than along the undisturbed flow.

4. Conclusions

The radio observations of 3C9 and 3C309.1 can be explained if high-z knots are unresolved working surfaces/shock waves on the boundary jet–intergalactic medium, probably due to energy and momentum transmission from the inner parts of the jet to the colder ISM. The magnitude of the jet perturbation would determine the energy transfer to the surfaces, which is consistent with the observations that knots contain particles of flatter energy distribution when the jet flow is more unstable. This is also supported by the similarities between observables in several radio sources with jets decelerated by surrounding material.

Finally, surveys for one-sided jets in quasars are needed to find more objects with large numbers of knots along the jet side to study their common properties. MERLIN observations are in progress to look for knots of galactic-scale jets in dense media and they often show knots with similar properties.

I would like to thank Dr. R.G. Conway and Dr. J.P. Leahy for useful comments, Dr. P.N. Wilkinson for his MERLIN+EVN map of 3C309.1 and the Brazilian agency CAPES for my PhD scholarship at NRAL.

References

1 Owen F.N., Hardee P.E. and Cornwell T.J., 1989, Ap.J., **340**, 698
2 Perley R.A., Bridle A.H. and Willis A.G., 1984, ApJSS, **54**, 291
3 Kus A.J. and Wilkinson P.N., *Parsec Scale Radio Jets*, Ed. Zensus J.A. and Pearson T.J., CUP (1990), p 161
4 Jackson N., Browne I.W.A., Shone D.L. and Lind K.R., 1990, MNRAS, **244**, 750
5 Marscher A.P, 1990, in *Parsec-scale Radio Jets*, Ed. Zensus J.A. and Pearson T.J., p.244, Cambridge University Press
6 Smith M.D. and Norman C.A., 1973, MNRAS, **164**, 243

Radio Emission and the Nature of Compact Objects in AGNs

Luis Colina [*]

Abstract

The radio properties of radio quiet active galaxies are revisited and considered under the starburst without black hole model. These radio properties are consistent with the luminosity, compactness and spectral index expected from a massive starburst process, where bright and compact radio supernovae and supernova remnants, i.e. *radio hypernovæ*, generate the radio emission.

1. Introduction

Since the discovery of quasars, theoretical and observational work has been done in order to characterize the variety of active galactic nuclei (AGNs), and to understand the physical mechanisms operating in these regions. The most popular scenario considers the presence of an accretion disc around a massive black hole. Under this scenario, different regimes of accretion and/or black hole masses plus some anisotropy in the radiation field could account for the whole variety of AGNs. Alternatively, a different model based on the evolution of a central compact star cluster, has been proposed and worked out in some detail (Terlevich, 1990 and references).

A natural way to ascertain the true nature of the compact objects in AGNs, is to look at the central regions of galaxies with the highest spatial resolution available, i.e. radio observations. High resolution VLBI and VLA observations with 5 and 250 mas resolution, respectively, allow the nuclei and cores of nearby active galaxies to be mapped with typical resolutions of a few parsecs. In this paper I briefly discuss the radio properties, absolute luminosity, spectral index, morphology and compactness of radio quiet active galaxies and compare these properties with the predictions of the starburst model.

A more detailed account of the ideas and results presented here can be found in Colina & Pérez-Olea (1993, in prep.) while a discussion on the radio emission expected from a starburst and its application to high luminosity and ultraluminous IRAS galaxies is presented in Colina & Pérez-Olea (1992).

---

[*]Departamento de Física Teórica, Universidad Autónoma de Madrid, Cantoblanco, 28049 Madrid, Spain.

2. Radio Properties of Radio Quiet AGNs

Radio quiet active galaxies represent most of all AGNs including LINERs, Seyfert galaxies, high luminosity and ultraluminous IRAS galaxies, and optical quasars. Their absolute radio luminosity at 1.49GHz ranges from a mean of 3 10^{20} W Hz^{-1} for starburst galaxies up to a mean of 2 10^{23} W Hz^{-1} for ultraluminous IRAS galaxies and quasars. Typical 1.49 GHz luminosities for Seyfert galaxies range from 10^{20} W Hz^{-1} to 10^{23} W Hz^{-1}.

Although there is a range of 4 orders of magnitude in radio luminosity among the radio quiet AGN sample, they share similar properties with regard to their spectral index, morphology and compactness. They have an almost universal steep radio spectrum characteristic of non-thermal synchrotron emission. The radio emission is mostly generated within the inner first kpc around the nucleus and the structure itself appears to show different morphologies. These range from diffuse emission without a compact radio core to compact radio cores with one or two less luminous peaks, or with a linear structure.

3. Radio Quiet AGNs and the Starburst Scenario

As already discussed in detail (Colina & Pérez-Olea 1992, 1993) the radio emission in nuclear or circumnuclear starbursts is mostly generated in radio supernova remnants. These remnants should be about 1000 times brighter than CasA, i.e. *radio hypernovæ*. The total radio luminosity produced by such supernova remnants is given by (Colina & Pérez-Olea, 1992)

$$L_{\rm NT}(\nu) = 1.77 \cdot 10^{22} \, (\nu/8.44 {\rm GHz})^{-0.74} \, \Upsilon_{\rm SNII} \quad {\rm W \, Hz^{-1}} \tag{1}$$

The supernovae rate ($\Upsilon_{\rm SNII}$) can be calculated by assuming that the measured radio luminosity is produced in supernovae and supernova remnants as given by expression (1). A comparison of this expression with the values of Table 1 shows that only a few of these luminous radio supernovae are needed in order to explain the radio luminosity of Seyferts and starburst galaxies. Also, since the radio emission of radio supernovae and their remnants have spectral indexes ($F_\nu \propto \nu^\alpha$) covering the range $\alpha = -0.5$ to -1.0 (Weiler et al. 1986), they mimic the radio spectral indexes measured in active galaxies, in general.

However, one of the criticisms and strong arguments against the starburst model for AGNs in general, is the fact that the radio structure of AGNs usually shows a very compact radio core (not resolved at the VLA scale) with a brightness temperature $T_b \geq 10^8$ K. On the contrary, the expected value for starbursts and HII regions would not be greater than 10^4 K.

This type of argument is no longer valid if compact and bright radio supernovae like those detected in some starburst galaxies, M82 (Kronberg et al. 1985) or NGC1808 (Saikia et al. 1990), are located in AGNs. These supernova remnants with brightness temperatures in excess of 10^8 K and diameters of the order of a few parsecs or less, mimic in luminosity and size the compact radio cores of AGNs. A comparison of the radio emission of some Seyferts and starbursts at the kpc and parsec scales is presented in Table 1. It is clear from these data that the compactness of both Seyferts and starbursts is the same, about 5—10 per cent of the radio flux produced within the first kpc around the nucleus comes from a compact source of no more than a few parsecs in size. This is also true for the archetypal starburst galaxy, M82, where the radio flux associated to the brightest radio supernovae, 41.9+58, corresponds to 4 per cent of the total radio flux produced in the starburst.

The existence of this type of supernovae, *radio hypernovæ*, in the nuclear regions of AGNs can be tested using VLA monitoring of a sample of nearby Seyferts.

Table 1. Radio luminosity of Seyferts and starburst galaxies

Galaxy	Type	logL(1.49GHz) W Hz^{-1}	Size kpc	logL(1.66GHz) W Hz^{-1}	Size pc	Reference
Mrk260	S2	21.75	0.5	20.73	0.5	1
Mrk928b	S2	22.81	2.7	21.42	1.8	1
Mrk171A	SB	22.57	1.3	20.54	0.9	1,2
NGC1808	SB/S2	21.78	2	20.78	< 25	3
Mrk297A	SB	22.67	2.0	21.39	1.3	1
M82	SB	21.95	0.5	20.49	< 2.3	4,5

1: Lonsdale et al. 1992; 2: Gehrz et al. 1983; 3: Saikia et al. 1990; 4: Hargrave 1974; 5: Kronberg et al. 1985.

4. Summary

It has been shown that the radio properties of nuclear or circumnuclear starbursts with radio hypernovæ, i.e., bright and compact radio supernovae 1000 times brighter than CasA, are consistent with the radio characteristics, absolute radio luminosity, spectral index, and compactness, observed in radio quiet AGNs. Therefore, before assuming a massive black hole model for all AGNs, one should consider more carefully the starburst model, at least for radio quiet AGNs that represent most of the AGNs.

Acknowledgements: This work has been supported by the Spanish Council for Research and Technology under grants PB87-0080 and PB90-0182.

References

Colina L., & Pérez-Olea D., *1992 M.N.R.A.S.*, , *in press*.
Colina L., & Pérez-Olea D., *1993, in preparation*.
Gehrz R., Sramek R., & Weedman D., *1983 Ap. J.*, **267**, *551*.
Hargrave P.J., *1974 M.N.R.A.S.*, **168**, *491*.
Kronberg P., Biermann P., & Schwab F., *1985 Ap. J.*, **291**, *693*.
Lonsdale C., Lonsdale C., & Smith E., *1992 Ap. J.*, **391**, *629*.
Terlevich R., *1990 in Windows on Galaxies, pp. 87*
Saikia D.J. et al., *1990 M.N.R.A.S.*, **245**, *397*.
Weiler K.W., et al., *1986 Ap. J.*, **301**, *790*.

The Radio Properties of Hidden Seyfert 1's: Implications for Unified Models

Edward C. Moran and Jules P. Halpern *

There are now ten published detections of Seyfert galaxies with hidden broad-line regions, indicated by the presence of broad permitted lines in their polarized-flux spectra (Antonucci & Miller 1985; Miller & Goodrich 1990; Tran et al. 1992; Ward 1992). As the figure below shows, the radio luminosities of all ten hidden Seyfert 1's are far out on the tail of the 20 cm luminosity function. Here we have plotted the combined Seyfert 1 and Seyfert 2 luminosity function for a complete distance-limited sample (Ulvestad & Wilson 1989). (The separate Seyfert 1 and Seyfert 2 luminosity functions do not differ significantly.) For comparison, we have indicated the radio luminosities of the hidden Seyfert 1's from the literature, although these are not all drawn from the same distance-limited sample. When measurements at 20 cm were not available, we extrapolated from 6 cm or 11 cm using the given spectral index, or by assuming $f_\nu \sim \nu^{-0.7}$ if the spectral index was not indicated. All ten of the hidden Seyfert 1's fall in the top 15% of the luminosity function, and eight of ten fall in the top 4%, a remarkably strong correlation. High radio luminosity must therefore be an intrinsic property of these objects.

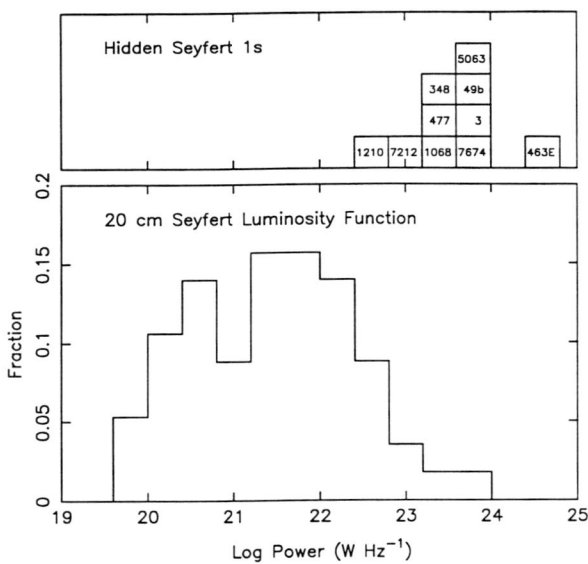

---

*Department of Astronomy, Columbia University, 538 W. 120th Street, New York, NY 10027

The radio luminosity in Seyferts is an isotropic property unless there is significant beaming of a core component. But in the "strong" unified Seyfert model, in which Seyfert classification is *purely* a viewing angle effect, the hidden Seyfert 1 nuclei cannot beam their radio emission toward the observer, so they should appear, if anything, *weaker* in the radio. The uniformly high radio luminosities of the hidden Seyfert 1's thus seem to rule out this "strong" unified Seyfert model.

Another version of the unified Seyfert model, described by Ward (1992), supposes that the *degree* to which the Seyfert nucleus is obscured is also an important parameter in determining the type of Seyfert we observe. Direct support for this model is provided by the Seyferts which show no evidence for broad Balmer emission, but for which broad Paschen-series lines are observed in the infrared spectrum. A particular geometrical picture is not mandatory to explain the spectral classification of Seyferts, but for some objects the degree of obscuration and the viewing angle must both be important. In Seyferts with linear radio features, very extended narrow-line emission is often observed along the radio axis; also, the radio axis tends to be aligned with optical "ionization cones" if they are present, and is perpendicular to polarization position angles in spectropolarimetrically discovered hidden Seyfert 1's. This evidence implies that for these cases there is less obscuration along the radio axis than along our line of sight to the nucleus. In terms of the degree-of-obscuration unifying scheme, the radio luminosity correlation for the hidden Seyfert 1's might then suggest that the radio-emission mechanism is actually responsible for "clearing out" any obscuring material in the path of the radio jet, and thus for creating some degree of cylindrical symmetry in the obscuring medium and the conditions right for scattering the non-stellar continuum and broad emission-line photons.

A challenge still exists for any unified Seyfert model: there are Seyfert nuclei which do not show broad Balmer lines in their total-flux spectra (implying obscuration), do have highly polarized non-stellar continua (implying reflection), yet show *no* broad emission in their polarized-light spectra (*e.g.*, four of eight objects in the Miller & Goodrich sample, and perhaps several of the galaxies observed as part of L. Kay's (1990) thesis). If further spectropolarimetric or infrared observations of these objects fail to turn up broad-line features, it must be concluded that they do not possess hidden broad-line regions, and that there are some "true" Seyfert 2's.

References

Antonucci, R. R. J., & Miller, J. S. 1985, *Ap. J.*, **297**, 621.
Kay, L. E. 1990, Ph. D. thesis, U. C. Santa Cruz.
Miller, J. S., & Goodrich, R. W. 1990, *Ap. J.*, **355**, 456.
Tran, H. D., Miller, J. S., & Kay, L. E. 1992, *Ap. J.*, **397**, 452.
Ulvestad, J. S., & Wilson, A. S. 1989, *Ap. J.*, **343**, 659.
Ward, M. J. 1992, in *Testing the AGN Paradigm*, ed. S. S. Holt, S. G. Neff, & C. M. Urry (AIP:New York), p. 500.

Anisotropic Optical Continuum Emission in Radio Quasars

J. C. Baker [*] R. W. Hunstead [*] V. K. Kapahi [†]
C. R. Subrahmanya [†]

Abstract

A preliminary analysis of radio and optical data for the low-frequency-selected Molonglo Quasar Sample provides new evidence that radio beaming, as inferred from core dominance of the radio emission, may be accompanied by an enhancement of the optical continuum.

1. Introduction

There is growing evidence that the optical continuum of radio quasars is not emitted isotropically (Tadhunter *et al.* 1987; Penston *et al.* 1990; Jackson & Browne 1991), implying that magnitude-limited samples could be seriously affected by an orientation bias (Kapahi & Shastri 1987). To minimise this bias we have defined a new *complete* sample from the 408 MHz Molonglo Reference Catalogue (MRC; Large *et al.* 1981), based on deep optical identifications from UK Schmidt plates. Quasars selected at low frequency are expected to be dominated by their unbeamed *extended* radio flux, rather than any relativistically beamed core component, and should therefore have their radio jet axes oriented randomly in the sky.

2. Sample Definition and Observations

Over 700 MRC radio sources, with peak flux density $S_{408} > 0.95$ Jy and falling in a 10° declination strip, $-20° > \delta > -30°$, were mapped at 843 MHz in 'snapshot' mode with the Molonglo Observatory Synthesis Telescope. Optical counterparts (complete to the plate limit, $B_J = 22.5$) were then identified from UK Schmidt survey plates. All 82 resulting QSO candidates in two RA regions (09^h–14^h and 20^h–06^h) were subsequently imaged with the VLA at 5 GHz; 31 of these had previously published redshifts.

For the 51 new QSO candidates, mostly fainter than $B_J \sim 19$, we obtained low resolution (10 Å blue, 25 Å red) optical spectra with the Anglo-Australian Telescope

---

[*] Department of Astrophysics, University of Sydney, Australia.
[†] T.I.F.R., Pune, India.

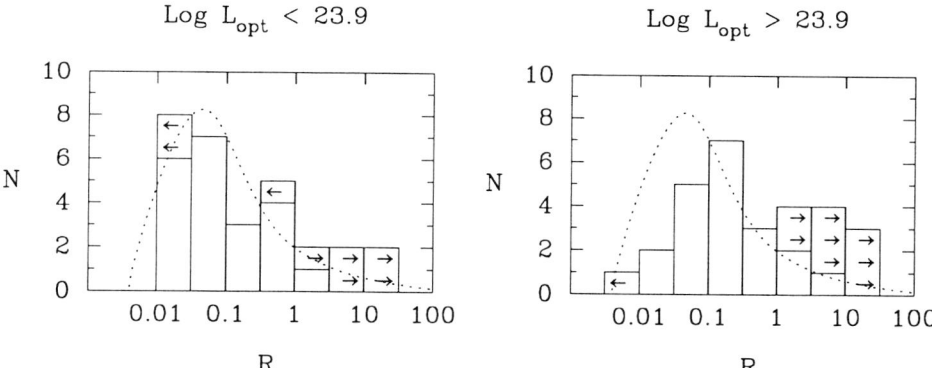

Figure 1. R-distribution for quasars falling below (*left*) and above (*right*) the median optical luminosity, $\log L \,[\mathrm{W\,Hz^{-1}}] = 23.9$ ($H_0 = 50\,\mathrm{km\,s^{-1}\,Mpc^{-1}}$; $q_0 = \frac{1}{2}$); only the 58 QSOs with R estimates are included here, with arrows indicating limits. Eleven compact steep spectrum sources are not included. The dotted line indicates the predicted distribution for a randomly oriented sample (Orr & Browne 1982) with parameters as for the 3CR sample.

covering the wavelength range 3000–10000 Å, and confirmed 38 to be quasars. The remaining 13 objects include some misidentifications and some for which the spectral classification is still uncertain. For the time being, this total of 69 QSOs comprises the Molonglo Quasar Sample; redshifts span the range 0.3–2.9 with median $z = 0.95$.

3. Preliminary Results

3.1 The R-Distribution

In common with previous studies, we use R, the ratio of core to extended radio flux (K-corrected to 10 GHz in the quasar rest frame), as an indicator of the orientation of the radio-jet axis with respect to the line of sight to the quasar (Hine & Scheuer 1980). In Figure 1 we show the R-distributions for quasars falling above and below the median optical luminosity; the R-values, or limits, have been derived from a preliminary deconvolution of the VLA images. The radio properties of the two subsamples are clearly different, with a significantly higher core dominance evident among the more optically luminous objects. Qualitatively, this is in agreement with the predictions of 'unified schemes' (Orr & Browne 1982, Kapahi & Saikia 1982) in which enhanced optical continuum emission accompanies a beamed radio component when the radio axis is observed at small angles to the line of sight.

The excess of high-R ($R > 5$) quasars in the optically luminous subsample could be a direct consequence of relativistic beaming (Orr & Browne 1982), bringing luminous but more distant quasars into our flux- and magnitude-limited sample as a result of extreme Doppler boosting. Note that for $\log L < 23.9$, $z_{\mathrm{median}} = 0.7$, whereas for $\log L > 23.9$, $z_{\mathrm{median}} = 1.5$. A further difference worth noting is that the median R

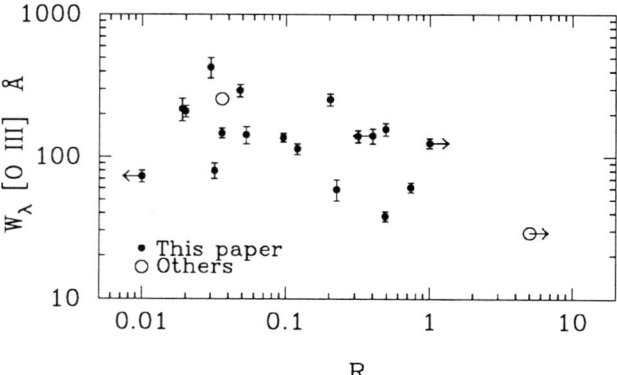

Figure 2. Equivalent width of the [O III] doublet vs R (core/extended radio flux). The strong anticorrelation seen is consistent with anisotropic radio and optical continuum emission.

for the more luminous quasars ($R \sim 1$) is higher than that for the low luminosity quasars ($R \sim 0.1$) by about one order of magnitude. This agrees with similar results based on apparent magnitudes (Kapahi et al. 1989). Since relativistic beaming is unimportant at large angles to the line of sight, we require some other orientation-dependent effect to dominate at small R. Such an effect could arise from thermal emission from a geometrically thin but optically thick accretion disk (Netzer 1985, 1987), or as a result of obscuration by dust in the form of a disk or torus (Antonucci & Miller 1985; Barthel 1989).

3.2 Emission-line Equivalent Widths

A widely used indicator of aspect dependence of the optical continuum has been the variation of emission line equivalent widths, W_λ, with R (eg., Wills & Browne 1986, Browne & Murphy 1987, Jackson & Browne 1991). If we assume that a particular line is emitted *isotropically*, but the continuum emission is anisotropic, we should expect an anticorrelation between W_λ and R. Traditionally, the [O III] $\lambda 5007$ emission line has been used for this test (Jackson & Browne 1989) because it is considered to be emitted isotropically, is relatively free from blending and has a well defined local continuum level.

In Figure 2 we show our preliminary plot of W_λ [O III] $\lambda\lambda 4959, 5007$ against R for the Molonglo Quasar Sample. We have chosen to measure the combined equivalent width of the [O III] doublet to avoid deblending uncertainties at the low resolution of our AAT spectra. A strong anticorrelation is evident (significant at the 99.5% level), even though the emission-line data for the brighter MQS quasars (generally those with previously published redshifts) are still incomplete, leading to an under-representation of high-R objects. Nevertheless, we confirm the trends reported in earlier studies, which in turn are likely to be biased in favour of high-R quasars.

The prominent upper envelope in Figure 2 arises from a paucity of high-R, high-W_λ

quasars, as expected in the beaming interpretation — objects with their radio axes directed towards our line of sight have strong optical continua and hence small emission line equivalent widths. Much of the scatter in W_λ at low R is probably intrinsic, but there may be additional contributions due to quasar variability or unrecognised line blending. The anticorrelation of W_λ with R is also seen for [O II] $\lambda\,3727$ but notably not for C IV $\lambda\,1550$ (which is produced much closer to the core), implying that C IV is probably emitted anisotropically.

4. Conclusions

We have presented evidence, from our preliminary analysis of the optical and radio properties of a new sample of 69 southern quasars, suggesting that radio beaming (inferred from R, the ratio of core to extended flux density) is accompanied by a corresponding enhancement of the optical continuum. Specifically, this is shown by the strong anticorrelation of [O III] equivalent width with R and, more generally, by the differing R-distributions for quasars of low and high optical luminosity.

References

Antonucci, R. R. J. & Miller, J. S., 1985, *Ap. J.*, **297**, 621.
Barthel, P. D., 1989, *Ap. J.*, **336**, 606.
Browne, I. W. A. & Murphy, D. W., 1987, *M.N.R.A.S.*, **226**, 601.
Hine, R. G. & Scheuer, P. A. G., 1980, *M.N.R.A.S.*, **193**, 285.
Jackson, N. & Browne, I. W. A., 1989, *Nature*, **338**, 485.
Jackson, N. & Browne, I. W. A., 1991, *M.N.R.A.S.*, **250**, 414 & 422.
Kapahi, V. K., Subrahmanya, C. R. & D'Silva, S., 1989, in *Active Galactic Nuclei*, IAU Symp. 134, p531.
Kapahi, V. K. & Saikia, D. J., 1982, *J. Astrophys. Astron.*, **3**, 465.
Kapahi, V. K. & Shastri, P., 1987, *M.N.R.A.S.*, **224**, 17p.
Large, M. I., Mills, B. Y., Little, A. G., Crawford, D. F. & Sutton, J. M., 1981, *M.N.R.A.S.*, **194**, 693.
Netzer, H., 1985, *M.N.R.A.S.*, **216**, 63.
Netzer, H., 1987, *M.N.R.A.S.*, **225**, 55.
Orr, M. J. L. & Browne, I. W. A., 1982, *M.N.R.A.S.*, **200**, 1067.
Penston, M. V. et al., 1990, *A. & A.*, **236**, 53.
Tadhunter, C. N., Fosbury, R. A. E., Binette, L., Danziger, I. J. & Robinson, A., 1987, *Nature*, **325**, 504.
Wills, B. J. & Browne, I. W. A., 1986, *Ap. J.*, **302**, 56.

The UV component in distant Radio Galaxies

A. Cimatti [*]   *S. di Serego Alighieri* [†]   *R.A.E. Fosbury* [‡]

Abstract

We analyse the properties of a sample of polarized radio galaxies with a wide range of redshifts. Our results suggest that a large fraction of UV light in distant radio galaxies is scattered anisotropic nuclear radiation.

1. Introduction

High redshift radio galaxies (HZRGs) have elongated morphologies (emission lines and continuum) aligned with the radio axis. The alignment of the continuum is strong in the ultraviolet and weak in the infrared around 1μ (rest frame) [2]. While the IR round component can be explained with an evolved stellar population, the nature of the UV aligned component is still a matter of debate. Two origins have been proposed for the UV light : **(a)** star formation induced in the ISM by the radio jet **(b)** dust/electron scattering of anisotropic nuclear light. We present new data which show that at least a considerable fraction of the UV light is not stellar.

2. The sample and the correlations

We have collected a sample of 41 radio galaxies with $0.1 \leq z \leq 1.2$ taken from our survey ([1]) and from the literature. We find that the degree of linear polarization P increases with redshift and with the total radio power at 178 MHz (P_r), while it decreases with the rest frame wavelength of the observation (λ_{rest}). The main problem in the interpretation of these correlations is that the quantities z–P_r–λ_{rest} are interdependent and it is not possible to clarify which correlations are intrinsic and which derived. Nevertheless some important clues can be deduced from these correlations. The main point is the presence of two groups: **(1)** radio galaxies with z>0.7 ($\lambda_{rest} \leq 3500$ Å) have high polarization and the plane of vibration of the electric vector perpendicular to the radio–optical axis (see the Figure). **(2)** Radio galaxies with z<0.5 have smaller polarizations and the orientation of the electric vector can be both perpendicular and parallel to the radio axis. High polarizations become dominant and homogeneous

---

[*]Dipartimento di Astronomia e Scienza dello Spazio, Largo E. Fermi 5, I–50125, Firenze, Italy.
[†]Osservatorio Astrofisico di Arcetri, Largo E. Fermi 5, I–50125, Firenze, Italy.
[‡]Space Telescope European Coordinating Facility, Karl-Scwarzschild Str. 2, D–8046, Garching bei München, Germany. Affiliated to the Astrophysics Division of the Space Science Dept., European Space Agency.

beyond the same redshift threshold of the alignment effect. It is clear that there must be a strong link between the two phenomena. Because scattering is the only mechanism able to explain all the observed properties at high z (see [1]), we believe that at least a considerable fraction of the aligned UV continuum is due to anisotropic nuclear radiation scattered by the medium (dust/electrons) toward the observer. This result is in favour of the scheme which explains the differences between radio quasars and radio galaxies in terms of different orientations with respect to the observer. Finally, if a large fraction of UV light is not stellar, this implies that the determination of the ages of distant radio galaxies with purely stellar models is not fully consistent with the observations.

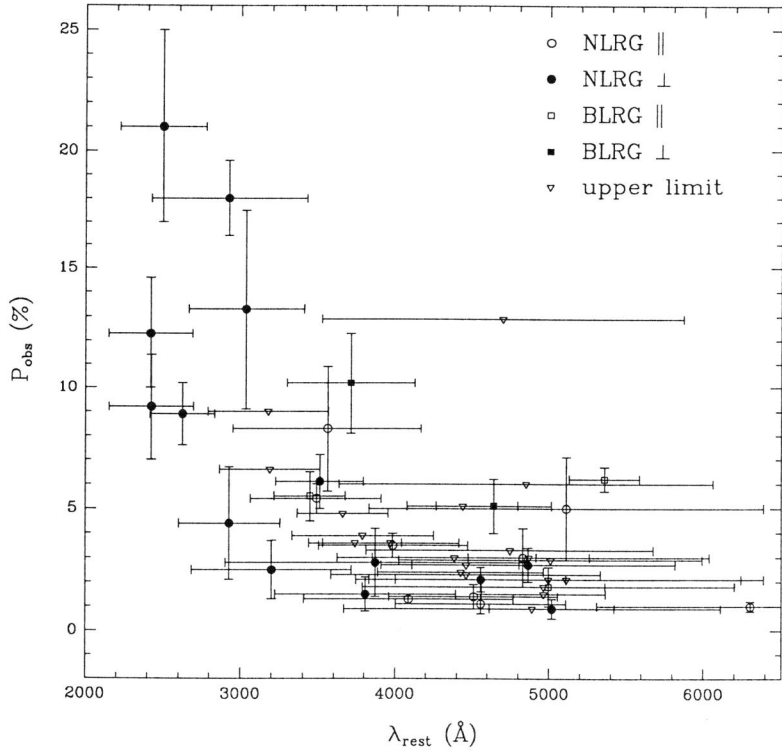

References

[1] di Serego Alighieri, S., Cimatti, A., & Fosbury, R.A.E., 1992. Ap.J., *in press*
[2] Rigler, M.A., Lilly, S.J., Stockton, A., Hammer, F., & Le Fevre, O. 1992. Ap.J., **385**, 61

A Connection between BL Lacertae Objects and Flat-Spectrum Radio Quasars?

Paolo Padovani [*]

Abstract

The properties of two complete samples of flat-spectrum radio quasars and radio-selected BL Lacs are analysed to look for any relationship between the two classes. It is shown that BL Lacs are not quasars with emission lines swamped by an enhanced optical continuum but their line luminosities are intrinsically weak. Moreover, an evolutionary connection between the two classes does not seem to be supported by the present data, while micro-lensing of quasars by stars in foreground galaxies can be ruled out as an explanation for the BL Lac phenomenon. BL Lacs and flat-spectrum radio quasars probably represent separate instances of relativistic beaming in low- and high-luminosity radio galaxies respectively.

1. Introduction

BL Lacertae objects are special types of active galactic nuclei, characterized by rapid variability, relatively high optical polarization, flat radio spectrum, radio core-dominance, superluminal motion, and weak or absent emission lines. Flat-spectrum radio quasars (FSRQs) are quasars with spectral index $\alpha \leq 0.5$ ($F_\nu \propto \nu^{-\alpha}$) at a few GHz, in many ways similar to BL Lacs (members of the two classes are often grouped together under the blazar category). The most striking difference between BL Lacs and FSRQs is the presence of strong emission lines in the latter objects. Although there are undoubtedly some borderline objects in which emission lines appear when the continuum is in a low state, a rest-frame equivalent width of 5 Å seems to separate quite well the two classes (see discussion in Stickel *et al.* 1991). The spectra of BL Lacs and FSRQs are thought to be dominated by emission from a relativistic jet closely aligned with the line of sight (Blandford & Rees 1978). This implies the existence of larger numbers of objects, the so called "parent populations", intrinsically identical to BL Lacs and FSRQs respectively but misdirected with respect to us. The luminosity functions of BL Lacs and FSRQs are in good agreement with the predictions of the beaming hypothesis (Padovani & Urry 1992; Urry, Padovani, & Stickel 1991) if Fanaroff-Riley type I and II (FR I and FR II) radio galaxies constitute the parent populations for BL Lacs and FSRQs respectively.

---

[*]European Southern Observatory, Karl-Schwarzschild-Str. 2, W-8046 Garching bei München, Germany. Present address: II Universitá di Roma, Via E. Carnevale, I-00173 Roma, Italy.

Two explanations have been proposed (amongst others) for the difference in the emission line properties: gravitational micro-lensing of the optical continuum, with consequent amplification relative to the line emission (hence the low equivalent widths: Ostriker & Vietri 1990); optical continuum dominated by a beamed component which swamps the emission lines, intrinsically similar to those of quasars (Blandford & Rees 1978). This has recently been taken up by Vagnetti *et al.* (1991), who suggested an evolutionary connection between FSRQs and BL Lacs, with strong-lined objects changing into weak-lined ones because of an increase with time of the Doppler factor. Here I discuss this second possibility using the 2 Jy FSRQs (Wall & Peacock 1985; Padovani & Urry 1992) and the 1 Jy BL Lacs (Stickel *et al.* 1991). As regards the first explanation, I refer the reader to Padovani (1992), where I show that micro-lensing cannot be the dominant phenomenon which determines the properties of BL Lacs. Two recent papers present other independent arguments against micro-lensing: Abraham *et al.* (1991) and Gabuzda *et al.* (1992). Throughout this paper the values $H_0 = 50$ km s^{-1} Mpc^{-1} and $q_0 = 0$ have been used.

2. BL Lacs as quasars with swamped emission lines?

The emission lines of BL Lacs do not seem to be intrinsically similar to those of quasars. [O III] luminosities are available for eight 1 Jy BL Lacs (Stickel *et al.* 1992), i.e. more than 70% of the objects for which [O III] was detectable. The mean value is $L_{\text{OIII}} \simeq 2 \times 10^{41}$ erg s^{-1} with a spread of less than an order of magnitude. [O III] luminosities were found for fourteen 2 Jy FSRQs in the literature, i.e. $\sim 70\%$ of the objects for which [O III] was detectable. The mean value in this case is $L_{\text{OIII}} \simeq 2 \times 10^{43}$ erg s^{-1}, that is two orders of magnitude larger than for BL Lacs, with a spread of more than two orders of magnitude, and no overlap between the two classes. The mean redshifts of the two subsamples are $\langle z(FSRQs) \rangle = 0.52$ and $\langle z(BL) \rangle = 0.18$ but this cannot be the explanation for this large difference: the ratio of the squares of the luminosity distance corresponding to the two mean redshifts would in fact imply a difference of only a factor of 10. The extended radio luminosities of FSRQs are also larger than those of BL Lacs and the two distributions (radio data on the extended emission are available for about 70% of the objects) are significantly different, with the former being similar to that of FR Is and the latter being similar to that of FR IIs: the parent populations of the two classes are then clearly different. Figure 1 plots the [O III] luminosity versus the extended radio luminosity for BL Lacs and FSRQs. It seems that a rough proportionality between L_{OIII} and L_{ext} exists and that the two classes fall in different regions of the plane. Although the small number of objects prevents one from drawing any other conclusion, the fact that the isotropic, unbeamed components in BL Lacs are smaller than those of FSRQs seems quite well established. The small equivalent widths of BL Lacs cannot be due to the swamping of the lines by the beamed optical continuum also because the Doppler factors, derived from SSC arguments, are actually *smaller* for BL Lacs than for FSRQs, contrary to what one

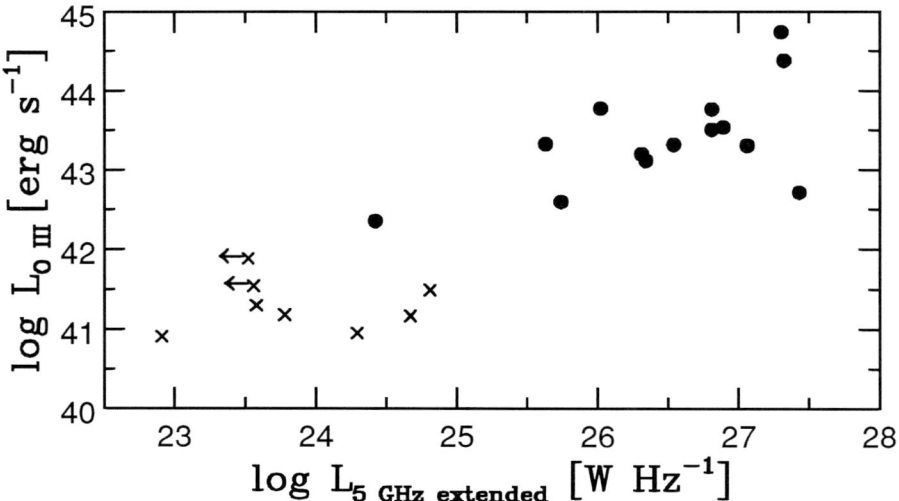

Figure 1. [O III] luminosity versus extended radio power at 5 GHz for radio-selected BL Lacs (crosses) and flat-spectrum radio quasars (filled points)

would expect if BL Lacs were more strongly beamed (Ghisellini et al. 1992). Note that even in the case of radio galaxies there is a relationship between radio power and emission line luminosity, with FR IIs having stronger lines than FR Is (Morganti et al. 1992).

The beaming hypothesis predicts that line luminosities should be, on average, similar for the beamed and the unbeamed population, as observed for the extended radio emission. Jackson & Browne (1990) have shown that radio quasar [O III] lines are typically stronger by a factor of about ten than those of FR IIs matched in redshift and extended radio power. This could be explained, for example, by dust extinction in the narrow line region of radio galaxies or by a contribution of the broad line region of quasars to the [O III] luminosity (see Tadhunter, this volume). The problem does not seem to be present for the BL Lacs - FR Is unification. A cursory look in the literature, in fact, shows that the mean [O III] luminosity for about a dozen FR I radio galaxies has the same value as for BL Lacs.

Turning to the possible evolutionary connection between FSRQs and BL Lacs, I have argued in Padovani (1992) that the supposed spectral continuity between the two classes can be explained by small statistics and BL Lac contamination of the quasar samples used by Vagnetti et al. (see Vagnetti & Spera, this volume, for a different opinion). The differences in the intrinsic properties of BL Lacs and FSRQs do not support this hypothesis, since they would imply a very strong luminosity evolution, much stronger in fact than any other evolution observed in AGN. Finally, the suggestion of Vagnetti et al. that strong-lined objects change into weak-lined ones because the lines are swamped by a beamed component does not apply to BL Lacs which have intrinsically weak lines, as shown above.

In conclusion, an evolutionary connection between BL Lacs and flat-spectrum radio quasars is not supported by the available data, although larger complete samples are needed for a definite test, while micro-lensing cannot be the dominant phenomenon which determines the properties of BL Lacs. BL Lacs are *not* quasars with emission lines swamped by the optical continuum: their small equivalent widths are due to the lines being intrinsically weak. The main conclusion is that the available observational data favour a scenario in which BL Lacs and flat-spectrum radio quasars are examples of similar relativistic phenomena in radio galaxies of different powers, with no direct connection between the two classes.

References

Abraham, R. G., McHardy, I. M. & Crawford, C. S., 1991, MNRAS, 252, 482
Blandford, R. D. & Rees, M. J., 1978, *Pittsburgh Conference on BL Lac Objects*, p. 328, ed. Wolfe, A., University of Pittsburgh Press, Pittsburgh.
Gabuzda, D. C., Cawthorne, T. V., Roberts, D. H. & Wardle, J. F. C., 1992, ApJ, 388, 40.
Ghisellini, G., Padovani, P., Celotti, A. & Maraschi, L., 1992, ApJ, in press.
Jackson, N. & Browne, I. W. A., 1990, Nature, 343, 43.
Morganti, R., Ulrich, M.-H. & Tadhunter, C. N. 1992, MNRAS, 254, 546.
Ostriker, J. P. & Vietri, M., 1990, Nature, 344, 45.
Padovani, P. 1992, MNRAS, 257, 404
Padovani, P. & Urry, C. M., 1992, ApJ, 387, 449.
Stickel, M., Fried, J. W. & Kühr, H., 1992, A&AS, in press.
Stickel, M., Padovani, P., Urry, C. M., Fried, J. W. & Kühr, H., 1991, ApJ, 374, 431.
Urry, C. M., Padovani, P. & Stickel, M., 1991, ApJ, 382, 501.
Vagnetti, F., Giallongo, E. & Cavaliere, A., 1991, ApJ, 368, 366.
Wall, J. V. & Peacock, J. A., 1985, MNRAS, 216, 173.

The Difference Between BL Lacs and QSOs

P. A. Hughes, M. F. Aller and H. D. Aller *   D. C. Gabuzda [†]

Abstract

We present evidence that BL Lac objects are a distinct class, rather than QSOs viewed close to, or within, the 'critical cone' of their collimated flow: both statistical analysis of the Stokes parameters and VLB polarimetry imply that the magnetic field structures in BL Lacs and QSOs are intrinsically different.

The low degree of polarization in the quiescent state of radio variable AGN, and the occurrence of occasional "rotation" events, have provided evidence for the presence of a highly tangled magnetic field. We would anticipate the Stokes parameters (Q, U) *versus* time plots of BL Lac objects and QSOs to show systematic differences if the relative strengths of tangled and ordered field components differ between classes. Fig. 1 illustrates typical Q-U plots for each class, derived from single dish observations, showing that the dispersion is large compared to the offset from zero for the BL Lac (interpreted as due to the presence of a weak mean field, so that evolution is dominated by the random walk of points in the Q-U plane as 'new' magnetic cells are advected into the window of observation), but small compared to the offset for the QSO (interpreted as due to a stronger mean field: the random walk is now about a point displaced from the origin by virtue of this stronger, axial field). A statistical analysis of 51 sources confirms this trend. Such a difference could arise if the integrated polarized flux were dominated by core emission (with randomly oriented PA) in BL Lacs but by jet emission in QSOs. VLB polarimetry shows that the core polarized flux is in fact dominant in of order 50% of BL Lacs, but this cannot explain the breadth of the trend noted above. VLB polarimetry also shows (Fig. 2; Gabuzda *et al.*, 1992, *Ap. J.*, **388**, 40; Cawthorne *et al.*, *Ap. J.*, submitted) that in BL Lac objects the electric vector PA of "knots" is nearly **parallel** to the VLBI structural axis, the orientation anticipated if these knots are associated with transverse shocks, which amplify the perpendicular magnetic field component. QSOs, however, display a weaker trend for the electric vector PA of their "knots" to be **perpendicular** to the structural axis, suggesting that the magnetic structure of QSO jets includes a longitudinal component of magnetic field. If QSO knots are also associated with shocks, the weaker correlation between "knot" PA and VLBI jet direction might then be due

---

*Astronomy Department, University of Michigan, Ann Arbor MI 48109-1090 USA. This work was supported in part by grant NSF AST-9120224.

[†]Department of Physics and Astronomy, University of Calgary, Calgary Alberta T2N 1N4 CANADA.

to a failure of the perpendicular magnetic field of the shock to dominate locally, particularly if the enhanced emissivity is dominated by particle acceleration (favoured by a parallel field) rather than compression of the turbulent magnetic field. Alternatively, QSOs may not develop the same plane shock structures as those inferred in BL Lac objects. Independent of specific interpretation, the observations highlight well-defined trends that imply an intrinsic physical difference between BL Lac objects and QSOs.

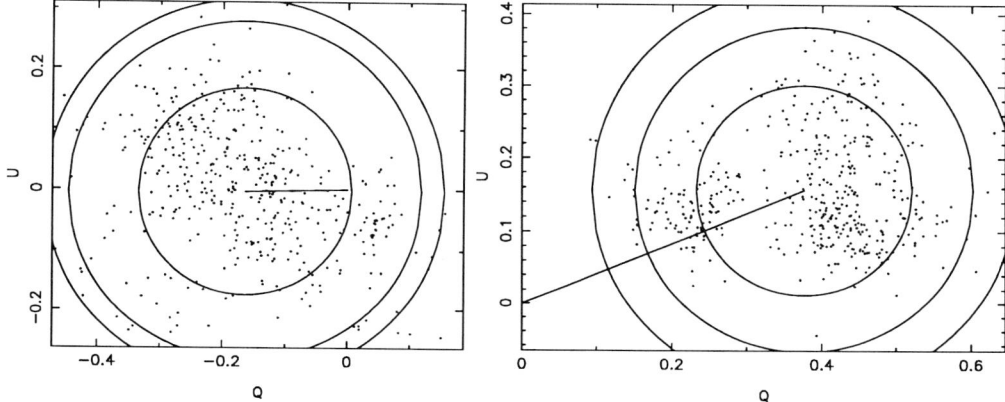

Fig. 1. Q-U diagrams for BL Lac object OJ 287 (left) and QSO 3C 454.3 (right) at 8.0 GHz. Each point is one observation between 1978.0 and 1991.0, and points populate the plane as the net magnetic field evolves. A line shows the offset of the centroid of points from the origin, and the circles enclose 66%, 95% and 99% of the sample.

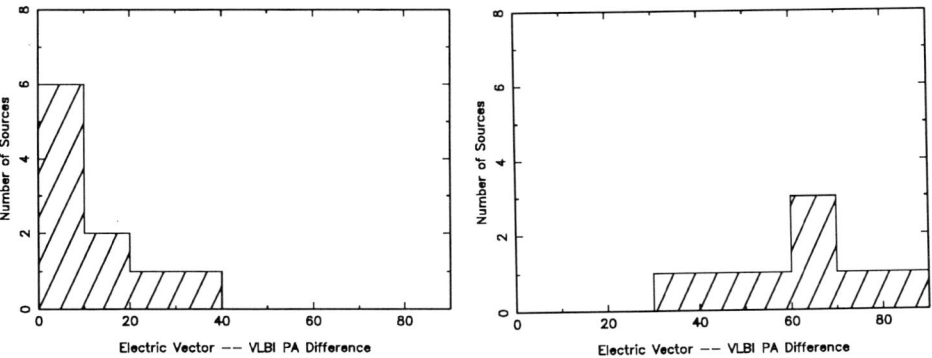

Fig. 2. Histograms of the difference between VLBI "knot" electric vector and jet direction in BL Lac objects (left) and QSOs (right).

The Evolutionary Unified Scheme and the $\theta - z$ Plane

F. Vagnetti & R. Spera *

In the *evolutionary unified scheme* (Vagnetti et al. 1991) a slower evolution in the optical band is found for flat spectrum quasars than for steep spectrum quasars and this is attributed to the presence of a slowly evolving relativistically beamed component $L_b = L_j[\Gamma(1 - \beta \cos \theta)]^{-(2+\alpha_{opt})}$, in addition to an isotropic component L_i. In turn, the slow evolution of the beam is attributed to a cosmological increase of Γ, rather than to an evolution of L_j/L_i. In the present work we consider the evolution in the optical band as a function of the viewing angle, and determine an evolution of Γ through a comparison with the average evolution of flat- and steep-spectrum quasars.

We assume pure exponential luminosity evolution in the look-back time T for the optical luminosity, and for both its components. For sources oriented at a given θ:

$$L_{opt}(\theta, z) = L_o e^{kT} = L_{io} e^{k_i T} + L_{jo} e^{k_j T}[\Gamma(1 - \beta \cos \theta)]^{-(2+\alpha_{opt})}. \quad (1)$$

Adopting subscripts s and f for steep- and flat-spectrum quasars respectively, their evolutionary ratio can be written as follows:

$$\frac{L_s(z)}{L_f(z)} = c e^{\Delta k T} = \frac{a e^{\Delta k_1 T} + \langle [\Gamma(1 - \beta \cos \theta)]^{-(2+\alpha_{opt})} \rangle_s}{a e^{\Delta k_1 T} + \langle [\Gamma(1 - \beta \cos \theta)]^{-(2+\alpha_{opt})} \rangle_f}. \quad (2)$$

Here $\Delta k = k_s - k_f$ is taken from Vagnetti et al. (1991); $\Delta k_1 = k_i - k_j$, $c = L_{os}/L_{of}$, and $a = L_{io}/L_{jo}$ are free parameters. The beamed terms in Eq. 2 have to be averaged over the angular intervals corresponding to the two populations, and an appropriate orientation probability must be used.

In fact, it has been shown by Cohen (1989) that, in a flux-limited sample, low luminosity sources can be observed more easily if their beamed component lies near the line of sight, due to Doppler amplification. The probability of observing sources at different angles is then altered. Cohen's formula holds for purely beamed sources. For the case of two radio components (core and lobes) we obtain:

$$p(\theta) \propto \{1 + f[\Gamma(1 - \beta \cos \theta)]^{-(2+\alpha_r)}\}^p \sin \theta, \quad (3)$$

p being the slope of the integral counts in the radio band, and f being the intrinsic fraction of power in the beamed component, which yields Cohen's case in the limit $f \to \infty$.

---

*Dipartimento di Fisica, Università di Roma "Tor Vergata", via della Ricerca Scientifica 1, I-00133 Roma, Italy. Work supported by CNR and MURST.

We assume, following Barthel (1989), that all quasars are oriented within an angle $\theta_{max} \simeq 45°$. We further assume that flat-spectrum quasars lie between 0 and a given angle θ^*, and steep-spectrum quasars between θ^* and θ_{max}. θ^* can be evaluated counting flat- and steep-spectrum quasars in the same samples already used for the radio-optical analysis Vagnetti et al. (1991). We then assign some appropriate values to the parameters and apply Eq. 2 at different redshifts to find $\Gamma(z)$. The result is an increasing Γ towards low redshift, in agreement with our previous findings Vagnetti et al. (1991).

Having determined $\Gamma(z)$, the quasar luminosity (Eq. 1) can be compared to the luminosity of the host galaxy. The latter can dominate the former at low redshift, but this depends on the viewing angle. Beyond a critical angle $\theta_G(z)$, low luminosity quasars are expected to be seen as radiogalaxies. This view agrees with the lack of steep spectrum quasars at $z \lesssim 0.3$ and with the observation of weak nuclear activity in some low redshift radiogalaxies (Yee & DeRobertis 1989). It is possible that many low redshift radiogalaxies are in fact steep-spectrum quasars of low luminosity, outshone by their host galaxies. We note that this effect would hold also for $\Gamma = $ const, in fact it is based only on two ingredients: (i) the presence of an anisotropic component, and (ii) the evolutionary nature of the quasar luminosity.

Following the suggestion of previous authors (Cohen 1989; Impey et al. 1991; Padovani & Urry 1992), we then assume a distribution of Lorentz factors of the form $n(\Gamma) \propto \Gamma^{-x}$. We choose $x = 2$ and $\Gamma_{min} = 1$, while Γ_{max} is determined by Eq. 2, after replacement of $p(\theta)$ (Eq. 3) with $p^*(\theta) \propto \int_1^{\Gamma_{max}} p(\theta)\Gamma^{-x}d\Gamma$. A critical angle $\theta_G(\Gamma, z)$ separating quasars from radiogalaxies is now found for each Γ in the distribution. We find that low redshift objects with $\Gamma \lesssim 5$ are observed as radiogalaxies for all the orientation angles. Complete results and discussion will be presented elsewhere.

References

Barthel, P.D. 1989, Ap.J. 336, 606
Cohen, M.H. 1989, in *BL Lac Objects*, L. Maraschi et al. Eds., p. 13
Impey, C.D., Lawrence, C.R., and Tapia, S. 1991, Ap.J. 375, 46
Padovani, P. and Urry, C.M. 1992, Ap.J. 387, 449
Vagnetti, F., Giallongo, E., and Cavaliere, A. 1991, Ap.J. 368, 366
Yee, H.K.C., and De Robertis, M.M. 1989, in *Active Galactic Nuclei*, D.E. Osterbrock and J.S. Miller Eds., p.457

Luminosity Functions and Continuum Energy Distributions

Radio Luminosity Functions of Active Galaxies

J.A. Peacock *

1. Introduction

1.1 Philosophy

It seems that every time a new population of extragalactic object is discovered, the first reaction of astronomers is to construct a luminosity function. Beyond sheer botany, this serves the useful purpose of giving a check on the completeness of surveys. The long-term motivation is Physics: the hope that the luminosity function and its change with redshift (the nearest we can get to an evolutionary track for a single object) will tell us something about how these spectacular sources operate. However, it has to be said that this objective remains in the far distance, despite nearly three decades of effort.

Why the radio waveband? Apart from the weight of history (radio astronomers take the blame for starting AGN research), the lack of foreground extinction and the lack of catalogue contamination by galactic objects are still very powerful advantages.

1.2 Notation

There are a few (arbitrary) conventions commonly adopted in the literature on this subject. The comoving density of objects per unit $\log_{10}$ power is denoted by ρ. The Hubble constant, where quoted explicitly, is given in the form $h = H_0/100$ kms^{-1}Mpc^{-1}. Unless otherwise specified, $\Omega = 1$ and $h = 0.5$ are assumed.

1.3 Radio astronomers' who's who

Radio astronomy takes the brutalist line of ordering the Universe according to output. Define $P \equiv \log_{10} L_{21\,\rm cm}/\rm WHz^{-1}sr^{-1}$; at $P \gtrsim 24$ we find the classical radio galaxies and quasars, conventionally divided roughly into Fanaroff-Riley (1974) FRII objects like Cygnus A and compact sources such as 3C273. At intermediate powers $23 \lesssim P \lesssim 24$, we find FRI sources: twin-jet objects such as 3C31, often lurking in clusters. At $P \lesssim 24$, we find all the rest of astronomy: 'radio-quiet' quasars, starburst galaxies and normal galaxies. This compressed scale has long been a disadvantage of the radio waveband, but this is now being overcome by VLA integrations into the μJy regime, allowing us to discuss the radio properties of these objects.

---

*Royal Observatory, Blackford Hill, Edinburgh EH9 3HJ, UK

2. Monovariate LFs

First consider what we know empirically about the abundance of radio AGN at high redshift, and what constraints this information may set on models of structure formation.

2.1 Results

No significant new datasets relevant to the luminosity function of powerful radio sources have been published since the study of the RLF published by James Dunlop and myself in 1990. This was based on nearly-complete redshift data on roughly 500 sources down to a limit of 100 mJy at 2.7 GHz, plus fainter number-count data and partial identification statistics.

The main conclusions of this study were firstly to affirm long-standing results that the RLF undergoes differential evolution: the highest luminosity sources change their comoving densities fastest. Nevertheless, because the RLF curves, the results can be described by a model of pure luminosity evolution for the high-power population, in close analogy with the situation for optically-selected quasars (Boyle et al. 1987; see also Boyle's contribution to this volume). It is not clear how much emphasis to place on this result; particularly, limited statistics make it uncertain just how well luminosity evolution is obeyed. For example, Goldschmidt et al. (1992) have produced evidence that the PG survey is very seriously incomplete at $z \lesssim 1$; if confirmed, this would imply that the evolution of quasars of the very highest luminosities is *less* than for those a few magnitudes weaker. Until we have samples of thousands of objects, the details of the LFs must be regarded as no better than provisional.

Sticking with PLE for the radio LF, the characteristic luminosity in this case increases by a factor $\simeq 20$ between the present and a redshift of 2. Similar behaviour applies for both steep-spectrum and flat-spectrum sources, which provides some comfort for those wedded to unified models for the AGN population. At higher redshifts, the uncertainties increase as the data thin out, but there is evidence that the luminosity function cannot stay at its $z = 2$ value at all higher redshifts. The form of this 'redshift cutoff' is uncertain: we cannot at present distinguish between possibilities such as a gradual decline for $z > 2$, or a constant RLF up to some critical redshift, followed by a more precipitous decline. We therefore present a 'straw man' model designed to concentrate the minds of observers, in which the luminosity evolution goes into reverse at $z \simeq 2$ and the characteristic luminosity retreats by a factor $\simeq 3$ by $z = 4$ (Figure 1).

This model predicts the following fraction of objects at $z > 3.5$ as a function of 1.4-GHz flux-density limit: 0.5% at 100 mJy; 3% at 1 mJy. Without some form of cutoff, these numbers would be about a factor of 5 higher. The reason for the increased ease of detecting a cutoff at low flux density is that the RLF is rather flat at low powers; for $\rho \propto P^{-\beta}$ and $S \propto \nu^{-\alpha}$, we expect $dN/dz \propto (1+z)^{-\beta(2+\alpha)-1/2}$. Steep spectra and a

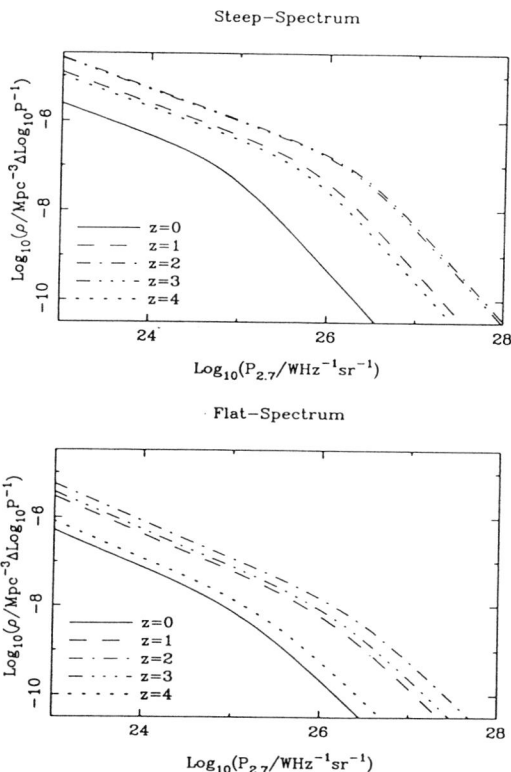

Figure 1. The evolving RLF, according to the pure luminosity evolution model of Dunlop & Peacock (1990). The main features are a break which moves to higher powers at high redshift, but which declines slightly at $z \gtrsim 2$. the strength of the break and the rate of evolution are comparable for both radio spectral classes.

steep RLF thus discriminate against high redshifts, but at low powers the flatter RLF helps us to see whatever high-z objects there are more easily. It should be relatively easy to test for the presence of a cutoff on the basis of these predictions.

2.2 Interpretation

Whether or not the redshift cutoff is real, we seem to have direct evidence that the characteristic comoving density of radio galaxies has not altered greatly between $z \simeq 4$ and the present. Integrating to 1 power of 10 below the break in the RLF, we find

$$\rho \simeq 10^{-6} h^3 \text{ Mpc}^{-3}.$$

Is this a surprising number? In models involving hierarchical collapse, the characteristic mass of bound objects is an increasing function of time. At high mass, the

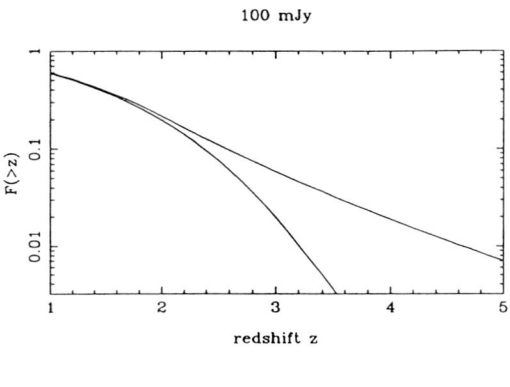

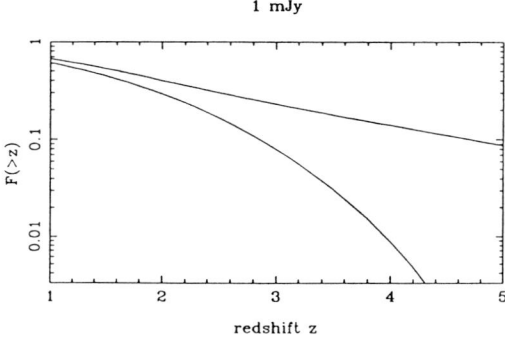

Figure 2. A plot of the integral redshift distributions predicted for two samples limited at 1.4-GHz flux densities of 100 mJy and 1 mJy. The upper line shows a prediction for a luminosity function which is held constant for $z \gtrsim 2$; the lower line shows the prediction of the 'negative luminosity evolution' model of Dunlop & Peacock (1990).

abundance of objects falls exponentially if the statistics of the density field are Gaussian. Clearly, a model such as CDM (which falls in this class) will be embarrassed if the density of massive objects stays high to indefinite redshifts. The analysis of this problem, using the Press-Schechter mass-function formalism (Press & Schechter 1974) was first given by Efstathiou & Rees (1988) for optically-selected quasars.

There are two degrees of freedom in the analysis: what mass of object is under study, and what are the parameters of the fluctuation power spectrum? For the first, Efstathiou & Rees had to construct a long and uncertain chain of inference leading from quasar energy output, to black-hole mass, to baryonic galaxy mass, to total halo mass. For radio galaxies, things are much simpler, because we can see the galaxy directly. Infrared observations imply that, certainly up to $z = 2$, the stellar mass of radio galaxies has not changed significantly. At low redshift, there is direct

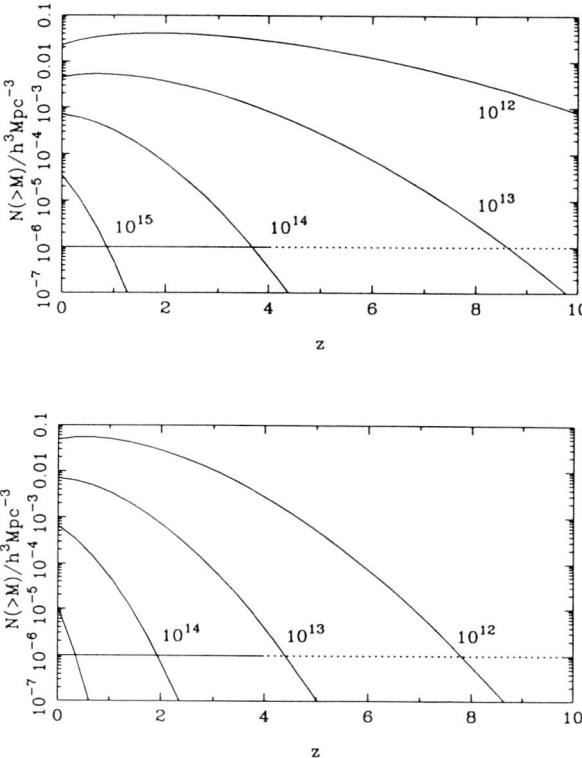

Figure 3. The epoch dependence of the integral mass function in CDM, calculated using the Press-Schechter formalism as in Efstathiou & Rees (1988). The normalization is to the COBE detection of CMB fluctuations. Results are shown for two Hubble constants: the 'standard' $h = 0.5$ (upper panel) and $h = 0.3$ (lower panel). The extra small-scale power in the former case means that many more massive hosts than the observed radio-galaxy number (horizontal line) are predicted, even at $z \gtrsim 10$.

evidence that the mass of radio galaxies exceeds $10^{12}\ M_\odot$, so it seems reasonable to adopt this value at higher redshift. Figure 3 shows the Press-Schechter predictions for two COBE-normalized CDM models. The low-h model which fits the shape of the galaxy-clustering power spectrum (Peacock 1991) intersects the observed number density at low-ish redshifts (7 – 8), whereas the 'standard' $h = 0.5$ model with its higher degree of small-scale power predicts many more objects. This is clearly only a suggestive coincidence at present, but it is clearly interesting that the model which most nearly describes large-scale structure also predicts that the formation of massive objects should occur near the point at which we infer a lack of high-z AGN.

3. Bivariate LFs

3.1 Radio-quiet quasars

Ever since it was discovered that radio-quiet quasars both existed and substantially outnumbered their radio-loud brethren, much argument has taken place over the relation between these two populations. A most influential paper in this regard was that of Schmidt (1970); he argued that the two populations should be regarded as extremes of a single continuous distribution of radio-to-optical luminosity ratio – the 'colour factorization' hypothesis. This held that the joint radio-optical LF could be factorized thus: $\rho = \phi(L_{\rm opt})\psi(L_{\rm rad}/L_{\rm opt})$. A correlation between the emission in the two bands is implied, so that the radio detection rate should not be a function of redshift (the high-z quasars in an optically selected sample have higher luminosities in both bands, compensating for the fact that more distant objects should be harder to detect to a given radio flux threshold).

Over the years, however, evidence accumulated that this was not the case. Peacock, Miller & Longair (1986) showed that there was a strong redshift dependence of the radio detection rate in the PG sample. They furthermore showed that there was a very strong redshift dependence in the radio-loud fraction even considering a fixed *luminosity* threshold: for quasars with $M_B \simeq -25$, the fraction with $P_{2.7} \gtrsim 10^{25}$ WHz^{-1}sr^{-1} falls from about 25% at $z \simeq 0.3$ to about 3% at $z \simeq 1.5$. A simpler picture which fits these observations is to say that there are two independent populations of quasar, for which the radio and optical outputs are not strongly correlated, and which have rather different degrees of cosmological evolution. What then is to be the criterion for separating these populations? This was provided by the observations of Miller, Peacock & Mead (1990), which showed that the quasar RLF was bimodal, with a clear 'zone of avoidance' centred around 5-GHz powers of about 10^{24} WHz^{-1}sr^{-1}. For the radio-loud class, there was indeed no correlation between radio and optical output. However, at low redshifts, a clear correlation existed for the radio-quiet class (see Figure 4), suggesting that they are to be interpreted as an extension of the Seyfert population (Edelson 1987). It seems clear that there are two physically unconnected populations of quasars, and that a simple criterion on radio power suffices to distinguish them. Speculation naturally turns to the role of the underlying galaxy in causing this dichotomy, with radio-loud quasars having elliptical hosts, and radio-quiet spiral. In any case, it is clear that radio-to-optical colour is a physically useful parameter only for the radio-quiet class, not the radio-loud quasars for which it was originally intended.

Figure 4 also demonstrates very clearly the trend of increasing radio-loud fraction with optical luminosity at a given redshift. This confirms the early finding by Wills & Lynds (1978) that radio-loud quasars had a much flatter optical LF ($\rho \propto L^{-1}$) than optically-selected quasars (*cf.* Boyle et al.'s bright-end slope of $L^{-2.5}$). This suggests a qualitative explanation for the lack of radio detections at low luminosities

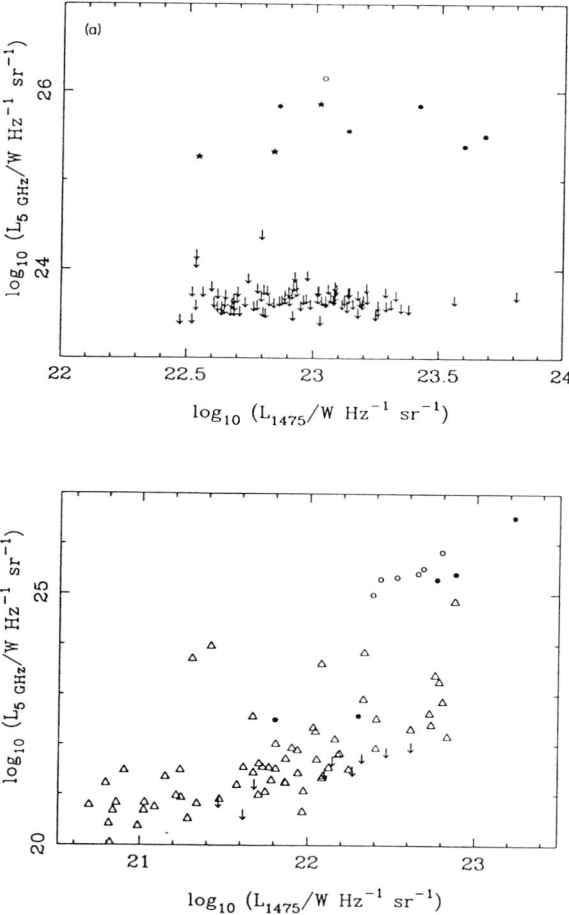

Figure 4. The distribution of radio and optical luminosities for (a) the $z \simeq 2$ CTIO sample of Miller, Peacock & Mead (1990); (b) the $z < 0.4$ PG quasars, taken from the same reference. Note (i) the lack of radio-loud quasars at low optical luminosities; (ii) the lack of quasars with $23 \lesssim \log_{10} L_5 \lesssim 25$; (iii) the correlation between radio and optical output for the radio-quiet class only.

and high redshifts: if the optical LFs for radio-loud and -quiet quasars both undergo comparable amounts of luminosity evolution, then the different slopes for the LFs will produce just such a trend. This possibility seems worth investigating in more detail.

One outstanding question posed by these observations is what causes the cutoff in quasar radio output around 5-GHz powers of 10^{25} WHz^{-1}sr^{-1}. Above this level, it seems that roughly 50% of steep-spectrum and 100% of flat-spectrum radio sources are quasars, independent of power and redshift – although there is still some uncertainty at the highest redshifts, where no $z > 4$ radio galaxies are yet known. For lower powers, however, the only significant quasar-like contribution is the $\sim 0.5\%$ of sources which are BL Lacs (at a given *extended* power). It may be tempting to identify the break here with the FR borderline, but this tends to happen at lower powers: for $24 \lesssim P \lesssim 25$ there are many FRII galaxies, but few quasars. This seems to form something of a puzzle for unified schemes in which on-axis FRII sources appear as quasars. A way out may however have been found in a very nice observation by Owen (1991); he has shown that the FR borderline is a function of galaxy optical luminosity, with the transition taking place at higher powers for the more luminous galaxies. If we speculate that quasars might be associated systematically with the more massive host galaxies, this could account for the lack of weak FRII quasars.

3.2 Starburst galaxies

A further important set of active galaxies which lurk in the same regime of radio power as radio-quiet quasars are the 'starburst' galaxies, distinguished by blue optical-UV continua and strong emission from dust which make them very bright in the IRAS 60-μm band. It has been increasingly clear since the work of Windhorst (1984) that such galaxies make up a substantial part of the radio-source population below $S \simeq 1\,\text{mJy}$. The fact that the radio counts have an inflection at this point has often been taken to indicate the possibility of evolution for this population. However, incomplete identification data at the faint radio level has tended to leave this issue somewhat inconclusive.

An important development has therefore been the study of evolution in the complete 'QDOT' sample of IRAS galaxies by Saunders *et al.* (1990). They found strong evolution, with either density evolution $\rho \propto (1 + z)^6$ or luminosity evolution $L \propto (1 + z)^3$ fitting the data. This has immediate implications for the radio waveband, because there exists an excellent correlation between output at 6 cm and 60 μm for these sources. Rowan-Robinson *et al.* (1992) have exploited this to investigate the implications of IRAS evolution for the faint radio counts. They find that, while the luminosity-evolution model gives a good match to the faint count data, the observed counts are grossly over-predicted by the density-evolution model.

4. Conclusions

There seems to be some evidence that pure luminosity evolution at about the $L \propto (1 + z)^3$ rate fits roughly the luminosity functions of radio galaxies, radio-loud and radio-quiet quasars and starburst galaxies. How is it possible for such different objects all to show such qualitatively similar evolution? Leaving aside the possibility of a grossly incorrect cosmological model (normal galaxies in the near-infrared show little evolution: Glazebrook 1991), one is tempted to look for an explanation which owes more to global changes in the Universe than in the detailed functioning of AGN. One obvious candidate, long suspected of playing a role in AGN, is galaxy mergers; Carlberg (1990) suggested that this mechanism could provide evolution at about the right rate. Why the evolution does not look like density evolution is still a major stumbling-block, but it seems that we should be looking at this area quite intensively, given that mergers have been implicated in both AGN and starbursts, and that there may be some evidence for their operation from the general galaxy population (Broadhurst, Ellis & Glazebrook 1992).

References

Boyle, B.J., Shanks, T. & Peterson, B.A., 1987. *Mon. Not. R. astr. Soc.*, **235**, 935.
Broadhurst, T.J., Ellis, R.S. & Glazebrook, K., 1992. *Nature*, **355**, 55.
Carlberg, R.G., 1990. *Astrophys. J.*, **350**, 505.
Dunlop, J.S. & Peacock, J.A., 1990. *Mon. Not. R. astr. Soc.*, **247**, 19.
Edelson, R.A., 1987. *Astrophys. J.*, **313**, 651.
Efstathiou, G. & Rees, M.J., 1988. *Mon. Not. R. astr. Soc.,*, **230**, 5P.
Fanaroff, B.L. & Riley, J.M., 1974. *Mon. Not. R. astr. Soc.,*, **167**, 318.
Glazebrook, K., 1991. PhD thesis, Univ. of Edinburgh.
Goldschmidt, P., Miller, L., La Franca, F. & Cristiani, S., 1992. *Mon. Not. R. astr. Soc.*, **256**, 65P.
Miller, L., Peacock, J.A. & Mead, A.R.G., 1990. *Mon. Not. R. astr. Soc.*, **244**, 207.
Owen, F.N., 1991. in proc. Ringberg jets meeting.
Peacock, J.A., Miller, L. & Longair, M.S., 1986. *Mon. Not. R. astr. Soc.*, **218**, 265.
Peacock, J.A., 1991. *Mon. Not. R. astr. Soc.*, **253**, 1P.
Press, W.H. & Schechter, P., 1974. *Astrophys. J.*, **187**, 425.
Rowan-Robinson, M., Benn, C.R., Broadhurst, T.J., Lawrence, A. & McMahon, R.G., 1992. *Mon. Not. R. astr. Soc.*, submitted.
Saunders, W., Rowan-Robinson, M., Lawrence, A., Efstathiou, G., Kaiser, N., Ellis, R.S. & Frenk, C.S., 1990. *Mon. Not. R. astr. Soc.*, **242**, 318.
Schmidt, M., 1970. *Astrophys. J.*, **162**, 371.
Schneider, D.P., van Gorkom. J.H., Schmidt, M. & Gunn, J.E., 1992. *Astr. J.*, **103**, 1451.
Visnovsky, K.L., Impey, C., Folz, C.B., Hewett, P.C., Weymann, R.J. & Morris, S.L., 1992. *Astrophys. J.*, **391**, 560.
Wills, D. & Lynds, R., 1978. *Astrophys. J. Suppl.*, **36**, 317.
Windhorst, R., 1984. PhD thesis, Univ. of Leiden.

The Quasar Luminosity Function

B.J.Boyle*

Abstract

Recent work on the quasar luminosity function at optical and X-ray wavelengths is reviewed. It is shown that the evolution of the quasar luminosity function in these regimes is marked by a strong and approximately similar power law increase in luminosity, $L \propto (1+z)^{3\pm 0.5}$, between the present epoch and $z \sim 2$. At $z > 2$, a slow-down in the rate of quasar evolution is witnessed in both regimes with possible evidence for a decrease in the space density of quasars being seen amongst optically faint ($M_B > -27$) QSOs at $z > 3.5$.

1. Introduction

The quasar luminosity function (LF) is one of the most fundamental statistics relating to the quasar population. Estimates of the quasar LF and its evolution with redshift are normally obtained from the statistical analysis of large unbiased quasar surveys with complete spectroscopic identification. As such, the rapid increase in the number of such surveys in recent years, particularly in the optical and X-ray regimes, has led to a dramatic improvement in our knowledge of the quasar LF and its evolution. The purpose of this review is to describe the current observational status of the quasar LF in the optical (4400Å) and X-ray ($\sim$ 2keV $\equiv$ 6.2Å) regimes.

2. The Optical Luminosity Function

As a result of the recent improvement in quasar statistics at $B > 20$, it has become increasingly clear (Koo 1983, Marshall 1987, Boyle et al. 1988, Koo & Kron 1988) that the low redshift ($z < 3$) quasar optical LF (OLF) exhibits a significant break in its power law slope at faint absolute magnitudes. The redshift dependence of this 'break' reveals that the OLF undergoes strong *luminosity* evolution between $z = 0$ and $z \sim 2$ with a significant cut-off occurring in this evolution at $z \gtrsim 2$. A qualitative picture of this result can be obtained from the binned estimate of the quasar OLF at $z < 2.9$ plotted in figure 1, derived by Boyle et al. (1991) from a composite catalogue containing over 700 quasars.

Based on the appearance of the OLF in figure 1, Boyle et al. (1991) fitted these data using a two power law function for the OLF, $\Phi(M_B, z)$:

---

*Institute of Astronomy, University of Cambridge, Madingley Road, Cambridge CB3 0HA, U.K.

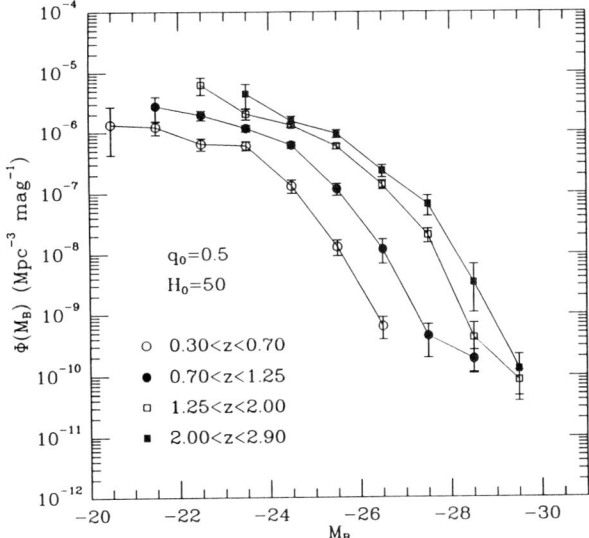

Figure 1. Binned estimate of the quasar OLF at $z < 2.9$ for a $q_0 = 0.5$ universe (Boyle et al. 1991).

$$\Phi(M_B, z)dM_B = \frac{\Phi^*}{10^{0.4[M_B-M_B(z)](\alpha+1)} + 10^{0.4[M_B-M_B(z)](\beta+1)}}dM_B$$

where α and β correspond to the bright and faint end slopes of the optical LF respectively. The evolution of the LF is uniquely specified by the redshift dependence of the 'break' magnitude, $M_B(z)$:

$$M_B^*(z) = M_B^* - 2.5k\log(1+z) \qquad z < z_{\max}$$
$$M_B^*(z) = M_B^* - 2.5k\log(1+z_{\max}) \qquad z > z_{\max}$$

corresponding to a power law evolution in optical luminosity, $L_{\text{opt}}^*(z) \propto (1+z)^k$, at low redshifts ($z < z_{\max}$) with an unevolving population at higher redshifts.

For $q_0 = 0.5$ and an assumed optical spectral index $\alpha_{opt} \sim 0.5$ ($f_\nu \propto \nu^{-\alpha_{opt}}$)[1], Boyle et al. (1991) obtained a satisfactory fit with values of $\alpha = -3.9 \pm 0.1$, $\beta = -1.5 \pm 0.15$, $M_B* = -22.4 \pm 0.2$, $k = 3.45 \pm 0.1$, $z_{\max} = 1.9 \pm 0.1$ and $\Phi^* = 6.5 \times 10^{-7}\text{mag}^{-1}\text{Mpc}^{-3}$.

At high redshifts, the derived fit to the OLF by Boyle et al. (1991) suggests that the strong luminosity evolution 'cuts-off' at $z = 1.9$ and that the distribution of quasars over the redshift range $1.9 < z < 2.9$ is consistent with an unevolving population. This result is qualitatively similar to that derived by Kron et al. (1991) based on a

---

[1] For power law luminosity evolution, any increase or decrease in the adopted value of the spectral index α_{opt} also increases or decreases the derived value of k by the same amount.

sample of 150 quasars with $B < 22.5$mag. Using a grism-selected sample of faint quasars, Schmidt et al. (1991) also find that the comoving space density of quasars with $M_B \lesssim -26$ remains approximately constant over the range $2 < z < 3$. Warren et al. (1991) find some evidence for continuing luminosity evolution beyond $z = 2$, with the integrated space density of quasars with $M_B < -26$ peaking at $z \sim 3$. Nevertheless, this evolution is much slower than that derived at lower redshifts. Thus, there is considerable agreement that the strong optical evolution of quasars witnessed at $z < 2$ dramatically slows down at $z \gtrsim 2$, with little or no evidence for any decline in the comoving space density of quasars over the redshift range $2 < z < 3$.

At $z > 3$, the picture that emerges is somewhat contradictory. At bright magnitudes, a number of surveys for high redshift quasars (Hazard et al. 1986, Mitchell et al. 1990, Irwin et al. 1991) have established that the space density of quasars with $M_B < -27$ remains unchanged between $z = 2$ and $z = 4$. In contrast, surveys for fainter quasars at high redshift (Warren et al. 1991, Schmidt et al. 1991) indicate that the space density of quasars with $-27 < M_B < -26$ is a factor of 3–10 lower at $z = 4$ than at $z = 2$. Schmidt et al. (1991) model this decline as an exponential decrease in the comoving space density of quasars with $M_B < -26$ beyond $z > 3$ by a factor of 3.2 per unit redshift. Although these two opposing results can be reconciled straightforwardly by a model which invokes a luminosity dependence in the density evolution of quasars at high redshifts (*i.e.* faint quasars undergo a rapid decline in their numbers at $z > 3$, while bright quasars exhibit little evolution), the steep OLFs derived from each individual survey are inconsistent with such a view (see discussion following Marshall 1991). However, samples of quasars at $z > 3$ are still small and the results derived from them are particularly sensitive to large corrections for incompleteness (see e.g. Warren et al. 1991). It is possible that a clearer picture of quasar evolution at high redshift will only emerge with larger optical surveys for quasars at $z > 3$, ideally selected at magnitudes $R \gtrsim 22$, thereby probing below the break in the OLF at these redshifts.

3. The X-Ray Luminosity Function

In common with the optical regime, the dramatic increase in the number of X-ray selected quasars identified in recent years has also led to a much improved understanding of the quasar X-ray ($\sim$ 2keV) luminosity function (XLF) and its evolution with redshift. Two recent surveys have played a particularly significant rôle: the Einstein Extended Medium Sensitivity Survey (EMSS, Stocke et al. 1991) containing 427 quasars with fluxes $S(0.3–3.5\text{keV}) \gtrsim 10^{-13}$ergs s^{-1}cm^{-2} and the ROSAT quasar survey (Shanks et al. 1991, Griffiths et al. 1993, in preparation) which contains 42 quasars identified to a much deeper flux limit, $S(0.5–2.0\text{keV}) \gtrsim 6 \times 10^{-15}$ergs s^{-1}cm^{-2}.

Using the EMSS quasars, Maccacaro et al. (1991) demonstrated that the low redshift ($z < 0.2$) XLF exhibited a pronounced break to a flatter slope at low luminosities

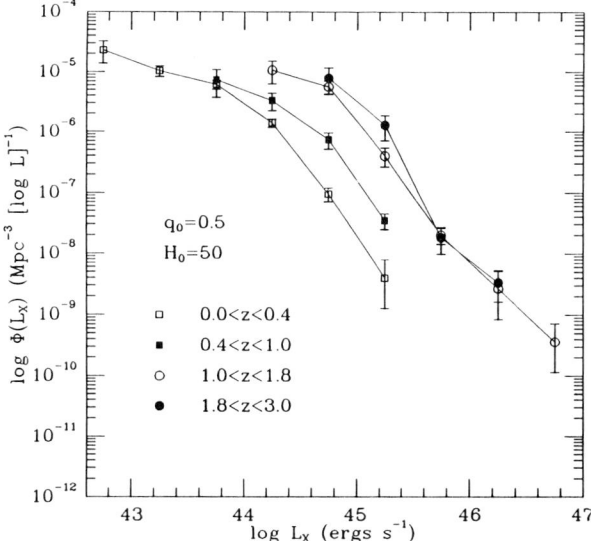

Figure 2. Binned representation of the XLF derived from the EMSS and ROSAT samples for a $q_0 = 0.5$ universe based on the analysis of Boyle et al. (1993).

and that the evolution of the XLF was consistent with luminosity evolution. By combining the ROSAT sample with the EMSS, Boyle et al. (1993) were able to demonstrate the existence of a break in the XLF at higher redshifts ($z > 0.2$) and confirm that luminosity evolution was a satisfactory fit to the XLF at $z \lesssim 2$. Both these results are clearly illustrated by the binned representation of the XLF plotted in figure 2 obtained from the analysis of the combined EMSS and ROSAT data-set by Boyle et al. (1992).

Based on the appearance of the binned XLF in figure 2, Boyle et al. (1993) modeled the $z = 0$ XLF as a two power law function of the form:

$$\Phi(L_X) = K_1 L_{X_{44}}^{-\gamma_1} \quad L_X < L_X^*(z=0)$$
$$\Phi(L_X) = K_2 L_{X_{44}}^{-\gamma_2} \quad L_X > L_X^*(z=0)$$

where $L_{X_{44}}$ is the 0.3-3.5keV X-ray luminosity expressed in units of 10^{44}ergs s^{-1}. The evolution of the XLF was parameterized as a power law in $(1+z)$:

$$L_x^*(z) = L_x^*(0)(1+z)^k$$

Assuming $q_0 = 0$ and an X-ray spectral index $\alpha_x = 1$, Boyle et al. (1993) derived an acceptable fit to the combined EMSS/ROSAT data-set with the following parameter values for the XLF: $\gamma_1 = 1.7 \pm 0.2$, $\gamma_2 = 3.4 \pm 0.1$, $\log L_X^* = 43.84 \pm 0.1$, $K_1 = K_2 L_{X_{44}}^{*(\gamma_1-\gamma_2)} = 0.57 \times 10^{-6}$Mpc^{-3}($10^{44}$ergs s^{-1})$^{-1}$ and $k = 2.75 \pm 0.1$.

Table 1. Summary of luminosity function parameters ($q_0 = 0.5$)

LF parameters	Optical (4400Å) (Boyle et al. 91)	X-ray (2keV) (Boyle et al. 92)	Radio (2.7GHz) (Dunlop & Peacock 90)
α	3.9 ± 0.1	3.4 ± 0.1	3.0 ± 0.1
β	1.5 ± 0.2	1.7 ± 0.2	1.8 ± 0.2
L^* (ergs s^{-1}Hz^{-1})	4.2×10^{29}	5.8×10^{25}	1.8×10^{33}
Φ^* (log L^{-1}Mpc^{-3})	1.6×10^{-6}	1.6×10^{-6}	7×10^{-9}

In common with Della Ceca & Maccacaro (1991), Boyle et al. (1993) also found evidence for a departure from the strong power law evolution at higher redshifts. Boyle et al. (1993) demonstrated that the power-law evolution model in a $q_0 = 0.5$ universe (marginally rejected at the 95% confidence level) gave a significantly better fit if a cut-off in the evolution similar to that required in the optical was introduced at $z = 2$. This cut-off is evident from the binned XLF in figure 2, with the XLF at $1.8 < z < 3$ exhibiting little evolution from its position at $1.0 < z < 1.8$. Nevertheless, there is still significant uncertainty over the high redshift evolution of the XLF and further samples of X-ray selected quasars at high redshift will be required before the precise nature of this cut-off is established.

4. Discussion

4.1 The Luminosity Function

From the preceding sections it is clear that the quasar LF and its evolution exhibits a number of striking similarities between the optical and X-ray regimes. Moreover, a similar form for the quasar radio LF (RLF) and its evolution has also been derived by Dunlop & Peacock (1990) based on a complete sample of 171 flat-spectrum quasars selected at 2.7GHz. In order to compare the different LFs, we adopt a standard parameterization:

$$\frac{dN}{d\log L\, dV} = \Phi(L,z) = \frac{\Phi^*}{[\frac{L}{L^*(z)}]^{\alpha+1} + [\frac{L}{L^*(z)}]^{\beta+1}}$$

where the comoving space density of quasars (dN/dV) is quoted per unit *logarithmic* luminosity interval in order to facilitate comparison between the values obtained for the normalization (Φ^*) of the LF in the different regimes.

Table 1 summarizes the best-fit parameter values obtained for the monochromatic OLF, XLF and RLF at 4400Å, 2keV and 2.7GHz respectively. The value of $L_X^*(0)$ in the broad band 0.3–3.5keV has been corrected to a monochromatic 2keV luminosity assuming $\alpha_x = 1$.

From table 1 we can see that both the OLF and XLF exhibit similar normalisations and slopes, in marked contrast to those obtained for the RLF. The difference in the normalisations for the OLF and the RLF demonstrates that at the characteristic 'break' luminosity L^*, almost 99% of the optically-selected quasar population is radio-quiet.

4.2 Evolution

Figure 3 summarizes the observed evolution of the OLF, XLF and RLF in a $q_0 = 0.5$ universe. As can been seen from this figure, the evolution of all the LFs at $z < 2$ can be represented by power law luminosity evolution with the X-ray ($k = 2.75 \pm 0.1$), radio ($k \sim 3$) and optical ($k = 3.45 \pm 0.1$) all exhibiting very similar rates of evolution. Since radio quasars only comprise $\sim 1\%$ of the optical/X-ray quasar population at L^* there is no *a priori* reason to expect them to exhibit a similar rate of evolution. This is particularly true given the evidence that the strong evolution with redshift in the environment of radio-loud quasars (Yee & Green 1987) is radically different to that seen seen in optically-selected quasars (Ellingson *et al.* 1991).

It should be noted that the differences between the rates of evolution obtained for the optical, radio and X-ray regimes are only significant at the level of the *statistical* errors which have been quoted here. It is likely that, with the ever-increasing numbers of quasars now used in the derivation of the quasar LF, systematic rather than the statistical errors will begin dominate the derivation of the LF parameters. For example, a decrease in the mean adopted optical spectral index of 0.2 (see Francis *et al.* 1991) coupled with a similar increase in the adopted X-ray spectral index would render the evolution derived in the optical and X-ray regimes statistically indistinguishable from each other. Other systematic effects such as incompleteness of existing optical surveys (Goldschmidt *et al.* 1992) or a spread in the optical spectral index (Giallongo and Neushafer 1992) may also reduce the rate of evolution derived in the optical to the point at which all three regimes exhibit nearly identical evolution. As such, the observation by Saunders *et al.* (1991) that IRAS galaxies also exhibit an evolution consistent with $L \propto (1+z)^{3\pm1}$ (albeit only derived for $z < 0.1$), makes this "one plus zee to the three" power law evolution seem remarkably (but as yet inexplicably) ubiquitous.

At $z \gtrsim 2$ all three regimes exhibit evidence for a cut-off in the strong luminosity evolution witnessed at lower redshifts. In the optical and X-ray the distribution of quasars is consistent with an unevolving population, $L^*(z) =$constant, while the radio population undergoes negative luminosity evolution at these redshifts.

Unfortunately, there is still considerable uncertainty over the nature of quasar evolution at $z > 3$. At these redshifts little or no data exists for the radio and X-ray regimes and the results in the optical regime lead to a somewhat ad-hoc model (as discussed above) which invokes markedly different rates of evolution for high and low luminosity quasars. Nevertheless, whatever the nature of quasar evolution at $z > 3$, it is still clear that quasars have a 'brief but brilliant' career during the first 2Gyr of

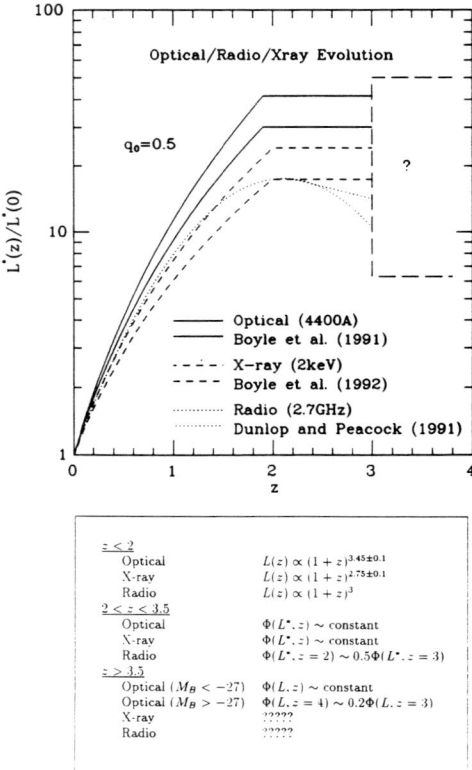

Figure 3. a) The evolution of L^*, normalized to the present epoch break luminosity, $L^*(0)$, plotted as function of redshift for the optical, X-ray and radio regimes. The statistical errors on the derived rates of evolution in the optical and X-ray regimes are indicated by the line pairs plotted.
b) Summary of the functional forms used to fit the evolution of the quasar LF.

the Universe, followed by an equally dramatic burn-out at subsequent epochs. Further surveys of quasars at high redshift should, in future, help complete the picture of quasar evolution.

Acknowledgements BJB was supported by a Royal Society University Research Fellowship during the course of this work.

References

Boyle,B.J.: 1992, in Texas-ESO/CERN symposium on Relativistic Astrophysics, Cosmology and Particle Physics, ed(s)., *J.Barrow, L.Mestel and P.Thomas*, Ann N.Y. Acad. of Sci. No 647, 14

Boyle,B.J., Shanks, T., and Peterson, B.A.: 1988, *MNRAS* **238**, 957

Boyle, B.J., Jones, L.R., Shanks, T., Marano, B., Zitelli, V. and Zamorani, G.: 1991, in The Space Distribution of Quasars, ASP Conference Series No. 21, ed(s)., *D. Crampton*, Provo: Brigham Young, 191

Boyle,B.J., Griffiths,R.E., Shanks,T., Stewart,G.C. and Georgantopoulos,I.: 1993, *MNRAS* in press

Della Ceca, R., Maccacaro,T., Gioia,I.M., Wolter,A. and Stocke,J.T.: 1992, *Ap.J.* **389**, 491

Della Ceca, R. and Maccacaro,T.: 1991, in The Space Distribution of Quasars, ASP Conference Series No. 21, ed(s)., *D. Crampton*, Provo: Brigham Young, 150

Dunlop,J.S. and Peacock,J.A.: 1990, *MNRAS* **247**, 19

Ellingson,E., Yee,H.K.C. and Green,R.F.: 1991, *Ap.J.* **371**, 41

Francis, P.J., Hewett, P.C., Foltz, C.B., Chaffee, F.H., Weymann, R.J. and Morris, S.L.: 1991, *Ap.J* **373**, 465

Giallongo,E. and Vagnetti,F.: 1992, *Ap.J.* in press

Goldschmidt,C.R., Miller,L., La Franca,F. and Cristiani,S.: 1992, *MNRAS* in press

Hazard,C., McMahon,R.G. and Sargent,W.: 1986, *Nature* **322**, 38

Irwin,M.J., McMahon,R.G. and Hazard,C.: 1991, in The Space Distribution of Quasars, ASP Conference Series No. 21, ed(s)., *D. Crampton*, Provo: Brigham Young, 117

Hewett,P.C., Foltz,C.B. and Chaffee,F.H.: 1993, *Ap.J.Lett.* in press

Koo,D.C.: 1983, *Quasars and Gravitational Lenses: 24th Liege Astrophysical Colloqium*, University of Liege, 240

Koo,D.C. and Kron,R.G.: 1988, *Ap.J.* **325**, 92

Kron,R.G., Bershady,M.A., Munn,J.A., Smetanka,J.J., Majewski,S. and Koo,D.C.: 1991, in The Space Distribution of Quasars, ASP Conference Series No. 21, ed(s)., *D. Crampton*, Provo: Brigham Young, 32

Maccacaro,T., Gioia,I.M.,Stocke,J.T.: 1984, *Ap.J.* **283**, 486

Maccacaro,T., Della Ceca,R., Gioia,I., Morris,S.L., Stocke,J.T. and Wolter,A.: 1991, *Ap.J.* **374**, 117

Marshall,H.L.: 1987, *A.J.* **94**, 628

Marshall,H.L.: 1991, in The Space Distribution of Quasars, ASP Conference Series No. 21, ed(s)., *D. Crampton*, Provo: Brigham Young, 184

Saunders,W., Rowan-Robinson,M., Lawrence,A., Efstathiou,G., Kaiser,N., Ellis,R.S. and Frenk,C.S.: 1990, *MNRAS.* **242**, 318

Schmidt,M., Schneider,D and Gunn,J.: 1991, in The Space Distribution of Quasars, ASP Conference Series No. 21, ed(s)., *D. Crampton*, Provo: Brigham Young, 109

Shanks,T., Georgantopoulos,I., Stewart,G.C., Pounds,K.A., Boyle,B.J. and Griffiths,R.E.: 1991, *Nature* **353**, 315

Stocke,J.T., Morris,S.L., Gioia,I.M., Maccacaro,T., Schild,R., Wolter,A., Fleming,T.A. and Henry,J.P.: 1991, *Ap.J.Supp.* **76**, 813

Warren,S.J. and Hewett,P.C. and Osmer,P.S.: 1991, in The Space Distribution of Quasars, ASP Conference Series No. 21, ed(s)., *D. Crampton*, Provo: Brigham Young, 139

Yee,H.K.C. and Green,R.F.: 1987, *Ap.J.* **319**, 28

UK ROSAT Deep & Extended Deep Surveys

L.R. Jones[1], I.M. McHardy[1], M. Merrifield[1]
G. Branduardi-Raymont[2], K.O. Mason[2], P.J. Smith[2]
M. Rowan-Robinson[3], M.H. Jones[3], A. Lawrence[3] G.P. Efstathiou[4]
T.J. Ponman[5], P. Willmore[5], D. Allan[5], A. Jeffries[5] J. Quenby[6],
H. Lehto[7], G. Luppino[8], J. Pye[9], C. Page[9], A. Pedlar[10]

Abstract

We present early results from the UK ROSAT Deep and Extended Deep Surveys. A total of 240 faint X-ray sources have been detected, most of which are expected to be QSOs and Seyfert galaxies at redshifts $z < 3$, although normal galaxies and starburst galaxies are also present. We will use these surveys, together with our parallel VLA 20cm & 6cm radio surveys and multicolour optical CCD surveys, to determine the evolution of the faint end of the X-ray and optical luminosity functions (LF) of QSOs, study the multiwaveband emission mechanisms of QSOs, map their distribution over a 'wedge' of high redshift sky, and investigate the X-ray evolution of distant clusters of galaxies.

The Multiwaveband Surveys.

The ROSAT survey was performed in a region of high-latitude sky of very low, and uniform, Galactic column density ($7\,10^{19}\,\mathrm{cm^{-2}}$), as determined by our 21cm and IRAS 100μm measurements. The deep survey reaches a limiting X-ray flux of $4\,10^{-15}\,\mathrm{erg\,cm^{-2}\,s^{-1}}$ (0.5–2 keV) over a 40 arcmin diameter region of sky and contains 96 faint X-ray sources. The extended survey stretches over a $4° \times 40$ arcmin strip starting from the position of the deep survey, with a limiting flux of $10^{-14}\,\mathrm{erg\,cm^{-2}\,s^{-1}}$ (0.5–2 keV).

Deep VLA radio maps at 20cm (and at 6cm in the deep survey area only) have been constructed to flux limits of 0.5 mJy on the deep survey field and 2 mJy on the extended survey. Of the 73 radio sources detected in the deep survey field, 11 (15%) are coincident with X-ray sources; the corresponding numbers in the extended survey are 16 out of 119 (13%). Initial CCD imaging of a small part of the survey has shown that the optical counterparts of the X-ray sources are very faint; 25% have $V > 22.5$ mag. Further wide-field multicolour (UBR) CCD imaging to R=24 mag is

---

[1]University of Southampton, [2] M.S.S.L., [3] Queen Mary & Westfield, [4] Oxford University, [5] Birmingham Univeristy, [6] Imperial College, [7] Turku University, [8] University of Hawaii, [9] Leicester University, [10] Jodrell Bank

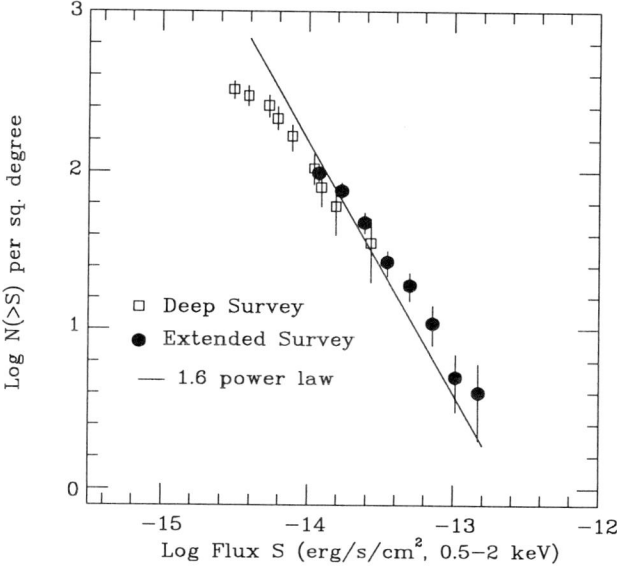

Figure 1.

scheduled. This will allow faint QSO selection by *optical* colours in addition to the X-ray selection, and allow use of upper limits on X-ray and optical luminosities.

Identifications.

We have used long-slit spectroscopy to classify 23 of the optically brightest counterparts (with $V < 19.5$). These sources cover a wide range of X-ray flux and comprise: 9 AGN, 2 normal galaxies, 1 cluster of galaxies and 11 stars. The AGN, at redshifts $0.02 < z < 0.27$, include 4 Seyferts, 2 radio galaxies and one possible starburst galaxy, and are mostly low luminosity, low activity objects. The normal galaxies have $L_X \sim 10^{40}\,\mathrm{erg\,s^{-1}}$ and are spirals. The relatively high proportion of stars and non-QSO AGN is almost certainly due to our current bright spectroscopic limit; many more QSOs are expected to appear in our sample as we identify the fainter objects. In the deep survey area, we have identified all X-ray sources with an optical magnitude $V < 19$. The number of AGN (> 4 at $z < 0.25$) may be considerably more than predicted from the X-ray LF of Boyle et al. (this volume). These low luminosity, low redshift AGN may contribute significantly to the X-ray background. The number of clusters of galaxies in our survey requires fewer high luminosity clusters in the past. The low redshift EMSS X-ray LF (& that of Piccinotti et al.) predicts > 8 clusters with $L_X > 10^{43}\,\mathrm{erg\,s^{-1}}$ in our survey; < 2 have been detected.

X-ray source counts.

A very preliminary integral LogN-LogS plot is shown in Fig. 1. Source searching was performed using the Cash algorithm, including the effects of a variable point-spread function and mirror vignetting. The counts from the two surveys are consistent with each other in the overlap region, and with the EMSS AGN slope of 1.6 at bright fluxes. Fainter than $\approx 10^{-14}\,\mathrm{erg\,cm^{-2}\,s^{-1}}$, the slope flattens, probably due to the increasing number of intrinsically faint QSOs selected from beyond the break in the X-ray luminosity function. Our classifications are amongst the few available at this time for objects at these extremely faint X-ray fluxes. Further work is in progress both on those objects already classified, and to extend the optical identification program using multi-object spectroscopy.

Luminosity Dependence of Optical Activity in Radio Galaxies

James S. Dunlop * *John A. Peacock* [†]

Abstract

We have previously reported evidence for radio-luminosity dependence of the strength of the near-infrared 'alignment effect' in radio galaxies (Dunlop & Peacock 1991). Here we present evidence for an associated radio-luminosity dependence in the level of optical activity found in radio galaxies at $z \simeq 1$. We find that this correlation, the strength of which is maximised by considering a combination of radio power and spectral index, is very similar and probably closely related to the correlation between radio-jet power and L_{NLR} found by Rawlings & Saunders (1991); the available evidence suggests that both correlations in fact arise from an underlying correlation with environment.

These correlations along with (i) the universal shape of the supposedly stellar UV-continuum in powerful high-redshift radio galaxies; (ii) the detections of significant optical/UV polarisation in several 3CR radio galaxies; (iii) the *inaccuracy* of the optical-radio alignments; (iv) the close spatial correspondence between the extended UV continuum and line emission; and (v) the correlation between radio-lobe depolarisation and extended optical emission, indicate that a large fraction of the optical/UV activity and the optical alignment effect in the 3CR sample is the result of Thomson scattering of a 'flat' ($f_\nu \propto \nu^{-0.2}$) quasar continuum emitted within a broad cone centred on the radio axis.

To investigate the relationship between the optical/UV activity in high-z radio galaxies and their radio properties, we have considered a composite sample of radio galaxies with $z \simeq 1$ which spans a decade in radio luminosity. The sample consists of 18 galaxies from the 3CR sample, 14 galaxies from the 1-Jy sample, and 14 galaxies from the Parkes Selected Regions (see Dunlop & Peacock (1992) for further details). The level of optical activity in these sources was quantified via the f_{5000} parameter introduced by Lilly (1989). A two-component spectral model consisting of an 'old' red component (produced by a 1-Gyr 'Burst model' of galaxy evolution at an age of 10 Gyr) plus a 'flat' power-law spectrum ($f_\nu \propto \nu^{-0.2}$) was fitted to the multi-waveband photometry of each object by varying the relative contributions of the two components;

---

*Chemical & Physical Sciences, Liverpool John Moores University, Byrom Street, Liverpool L3 3AF, England.

[†]Royal Observatory, Blackford Hill, Edinburgh EH9 3HJ, Scotland.

f_{5000} is simply the fractional contribution made by the blue flat-spectrum component at a rest wavelength of 5000Å.

In contrast to the result of Lilly (1989), we find that the radio parameter with which f_{5000} correlates most strongly is radio luminosity P ($p = 0.0004$) rather than radio spectral index α ($p = 0.72$). Nevertheless, some of the bluest sources do have extremely steep radio spectra and, as a result, the best radio-based predictor X of f_{5000} is given by $log_{10} X = 0.72 log_{10} P + 0.42\alpha - 20.37$. The correlation between X and f_{5000} is shown in Fig. 1. Interestingly, analysis of the data used by Rawlings & Saunders (1991) to deduce the $Q - L_{NLR}$ correlation shows that $log_{10} Q = 0.74 log_{10} P + 0.45\alpha + 0.22 log_{10} D + const$ (where D = radio source diameter), which suggests that the two correlations are closely related. The implications of this result are discussed in Dunlop & Peacock (1992).

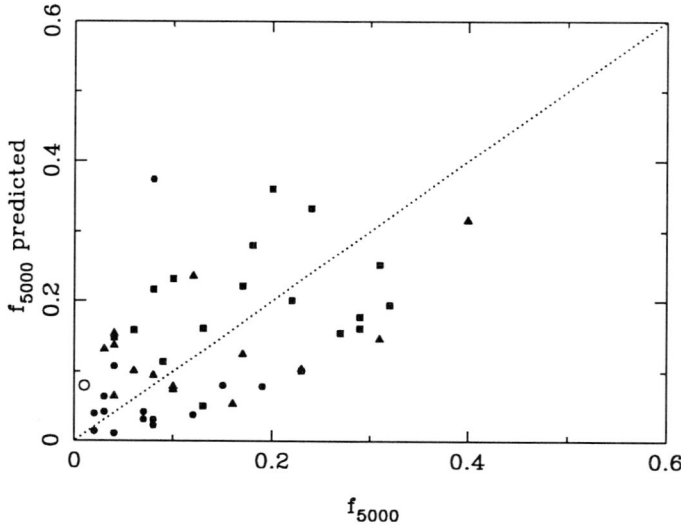

Figure 1. f_{5000} predicted versus f_{5000} observed for the combined 46-source 3CR/1-Jy/PSR sample (squares = 3CR sources, triangles = 1-Jy sources, circles = PSR sources). The extremely red radio galaxy 3C65 is indicated by the open circle.

References

Dunlop, J.S. & Peacock, J.A., 1991. In: *'Physical Cosmology'*, p.542, eds. Blanchard, A., *et al.*, Editions Frontieres.
Dunlop, J.S. & Peacock, J.A., 1992. *Mon. Not. R. astr. Soc.*, in press.
Lilly, S.J., 1989. *Astrophys. J.*, **340**, 77.
Rawlings, S. & Saunders, R., 1991. *Nature*, **349**, 138.

Modelling the Quasar Luminosity Function in Hierarchical Models for Structure Formation

Martin Haehnelt *

Abstract

The strong evolution of the host object population postulated in hierarchical models for structure formation is invoked to explain the observed strong evolution of the space density of quasars. The quasar activity is interpreted as marking the advent of a new step in the hierarchic build-up of bigger and bigger dark matter halos. The Press–Schechter formalism within the CDM scenario is used to estimate the number of newly forming dark matter halos. Pronounced peaks are found in the number density of newly forming massive black holes, capable of explaining the short time scale of the evolution of the quasar population. A gratifying fit to the observed luminosity function is obtained.

1. Quasar evolution in the CDM scenario

Soon after the discovery of the first quasars it was noticed that quasars are a strongly evolving population of objects. With the increasing number of known intermediate redshift quasars it became possible to determine the time evolution of the luminosity function of quasars [1]. The main feature of the luminosity function is a characteristic break luminosity which decreases with time. The quasar luminosity function is most naturally interpreted as a superposition of many generations of short-lived quasars with a life time $\sim 10^8$yr and a characteristic mass that decreases with time as $\sim (1+z)^3$.

In hierarchical models for structure formation, such as the CDM scenario, larger and larger structures build up by merging of smaller structures and the smaller structures are at least partially erased. The natural way to interpret the quasar phenomenon in a hierarchical model would be in terms of the formation of new black holes within each new hierarchy of objects that results from merging of structures on smaller scales. Quasars would resemble the final phase of the core collapse of a newly forming dark matter halo [2]. Life times of quasars would be rather short and several generation of quasars would occur naturally.

---

*Institute of Astronomy, Madingley Road, Cambridge CB3 0HA. The author was supported by an Isaac-Newton-Studentship and by the Gottlieb Daimler- and Karl Benz-Foundation.

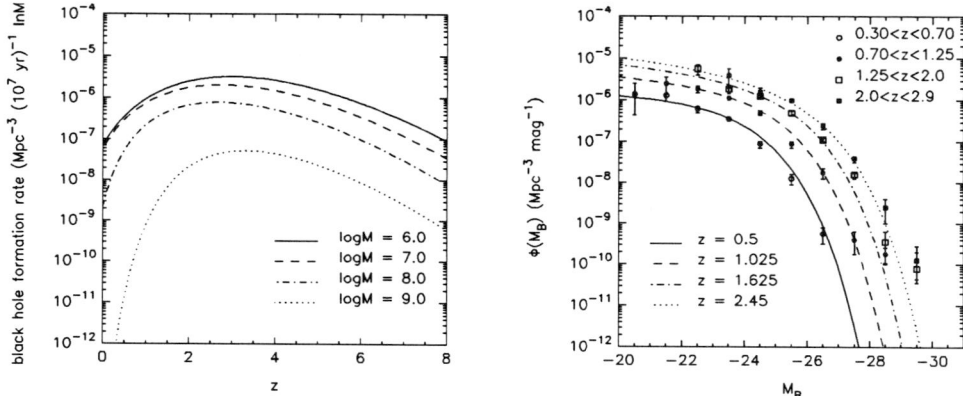

Figure 1. Black hole formation rate as a function of redshift for a CDM spectrum with bias parameter $b = 2.5$ (left) and modelled quasar luminosity function superimposed on the observed luminosity function (right).

2. The number density of newly forming black holes and the quasar luminosity function

The Press-Schechter formalism [3] was used to estimate the number density of newly forming dark matter halos. A simple parameterization reflecting the changing conditions for forming a black hole was used to relate the mass of a newly forming black hole to the mass of its host object. The resulting formation rate is shown in Fig. 1a. It is seen that the model can produce a sharp rise and fall in the number of newly formed black holes. The evolution time scale for smaller mass black holes is much longer than for the more massive ones and the peak of the formation rate shifts to higher redshifts for the bigger masses. These are the key ingredients necessary to model the observed quasar luminosity function successfully. In Fig. 1b the obtained black hole formation rate is convolved with a model lightcurve for an individual quasar. Superimposed is the observed luminosity function [1].

References

[1] Boyle, B.J., Jones, L.R., Shanks, T., Marano, B., Zitelli, V. Zamorani, G., 1991. in: *The Space Distribution of Quasars*, ASP Conference Series No. 21, p.191, ed. Crampton, D., Astronomical Society of the Pacific, San Francisco

[2] Rees, M.J., 1992. in:*First Light in the Universe: Stars or QSO's*, eds. Rocca-Volmerange, B., Dennefeld, M., Guiderdoni, B., Tran Thanh Van, J., in preparation

[3] Press, W.H., Schechter, P.L., 1974. *Astroph. J.*, **187**, 425

Active Galactic Nuclei in Clusters of Galaxies

Alastair C. Edge *

Abstract

A cross correlation of the active galactic nuclei (AGN) in the Véron-Cetty-Véron catalogue with the Abell cluster catalogues reveals 84 AGN that are associated with clusters. The Véron-Cetty AGN appear to be distributed, spatially and in velocity, within clusters in the same way as the member galaxies with a core radius of 500 kpc and a velocity dispersion of 1090^{+98}_{-90} km s^{-1}. The rapid growth in redshift studies of clusters and in high quality X-ray data will provide a much more detailed picture of the inter-relation of AGN and clusters in the near future.

1. Introduction

Active galactic nuclei are found in a wide variety of host galaxies and environments. The high densities in clusters provide a powerful test of the effect of environment on AGN. Despite the importance of such studies, little has been published on AGN in clusters since the general trend for fewer AGN in clusters was noted by Osterbrock (1960). The best quantitative work has been by Dressler *et al.* (1985) who analysed 1268 spectra in 14 clusters and found a lower incidence of AGN and emission line galaxies in clusters (at 1 %) than in the field (at 5 %) and a large fraction of Seyfert 2s. However, the percentage of AGN in field galaxies in Dressler *et al.* (1985) is substantially higher than the value of 1 % derived from the CfA redshift survey by Huchra & Burg (1992). This may be related to the fact that the Dressler *et al.* (1985) field sample contains more high luminosity galaxies compared with the cluster sample as it is a magnitude (rather than volume) limited sample and that the percentage of AGN is a strong function of absolute magnitude (Huchra & Burg 1992). Also, a number of studies of AGN specifically in clusters have been made. A study of Markarian galaxies in Zwicky clusters by Petrosian & Turatto (1986) shows that Markarian galaxies are apparently drawn at random from the cluster population, both spatially and in velocity. In this paper I have expanded on that work and examined Véron AGN in Abell clusters. A full presentation of the results is given in Edge (1992).

---

*Institute of Astronomy, Madingley Road, Cambridge CB3 0HA.

2. Analysis

The fifth edition Véron-Cetty & Véron (1991) catalogue was cross-correlated with the northern, southern and supplementary Abell catalogues (Abell, Corwin & Olowin 1989). For all clusters, coincidences with AGN with measured redshifts of less than 0.3 within 90 arcmin were used. These values were chosen to match the redshift distribution of clusters (there are virtually no Abell clusters with redshifts above 0.3) and the expected clustering scale of galaxies around clusters (less than 4 Mpc). These coincidences were taken and the cumulative distribution of offset angles was tested against the expected parabolic curve. A Hubble constant of 50 km s^{-1} Mpc^{-1} and q_0 of 0.0 are used throughout.

Both the Abell and Véron-Cetty samples are known to be incomplete at some level. These deficiencies need to be borne in mind when using either catalogue. However, the goal of the analysis, to identify cases of coincidence, is independent of any incompleteness problems as long as either sample is not biased against objects in the other. These effects can be tested after the 'real' coincidences have been removed. The lack of a complete sample only prevents a conclusive quantitative analysis of how many clusters or cluster galaxies have AGN.

3. Results

The cumulative histogram of offsets from optical cluster centre for the AGN in the sample was compared with the expected parabola. There was a clear excess above a parabola for offsets less than 40'. A K-S test for these curves gave a consistency with a parabola at less than the 10^{-4} per cent level. The excess is consistent with between 70 and 100 sources being physically associated with the clusters on a scale of 3–4 Mpc. As redshifts are known for all the AGN and most (80 %) of the clusters, it is possible to split the coincidences into cases where the AGN are background, foreground and close in redshift. The foreground and background cases are consistent with a parabola at the 30–40% level but the other case is not.

Although the overall sample shows no significant bias, the different selection techniques should be examined individually. Most nearby, optically selected AGN are colour selected from large numbers of plates and radio, infrared and X-ray slew survey selection are relatively uniform over the sky (on the few degree scale). However, there is a bias in the X-ray selection from X-ray images, where the field of view is relatively small (40×40'), and many of the *Einstein* and *EXOSAT* pointings were centred on Abell clusters. An analysis of the distribution of the X-ray imaging selected AGN shows an overdensity of a factor of 50–70% which shows that any Einstein observation was more likely to be of an Abell cluster than a 'random' area of sky. This effect is one against background sources rather than for any particular sources within cluster so the premise that the catalogues are not biased against each other is still true.

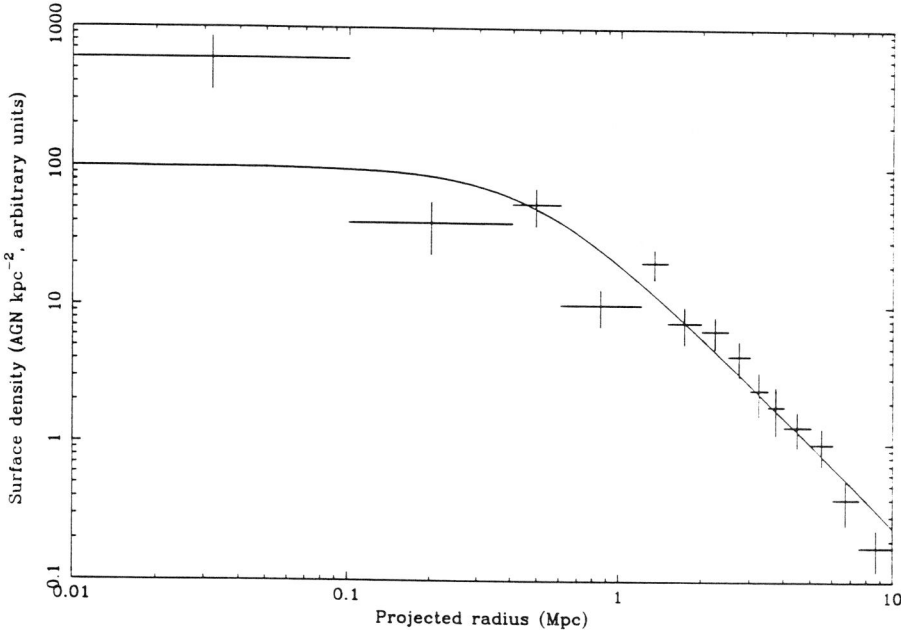

Figure 1. The surface density of Véron-Cetty AGN around Abell clusters. The solid line is a King model with a core radius of 500 kpc.

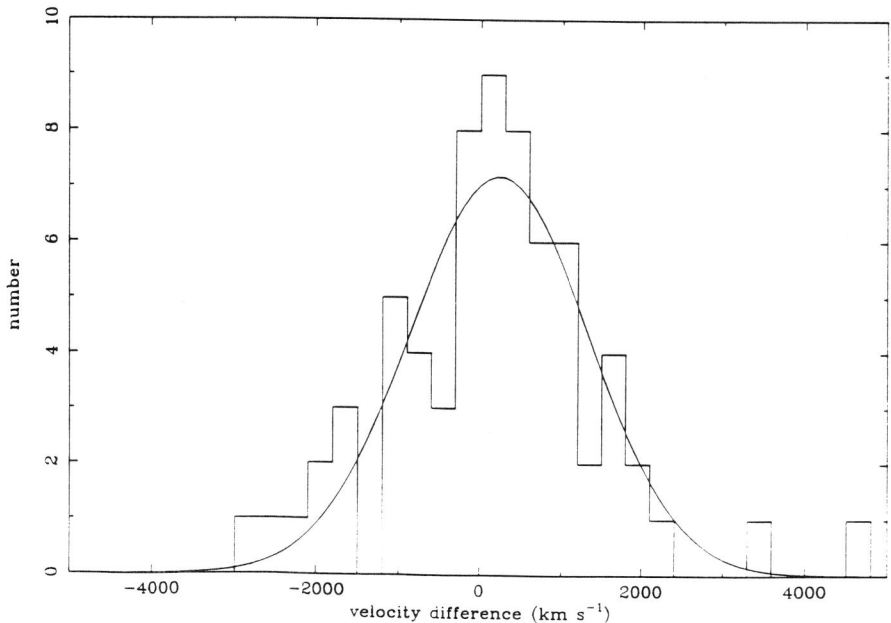

Figure 2. The histogram of the differences in velocity between the AGN (of all types) and clusters. The solid curve is a gaussian with a sigma of 1090 km s^{-1}.

There are 84 AGN with a projected offset less than 4 Mpc and a difference in redshift of less than 0.017. The AGN are predominately Seyfert 1s and are selected from a wide variety of catalogues (Markarian galaxies, radio, IRAS and X-ray sources). For the sixteen cases where the cluster redshift is unknown, the AGN redshift was compared with the value of m_{10} for the cluster. Plotting the surface density of AGN with respect to the cluster centre (Fig. 1), shows that AGN more than 100 kpc from the centre are consistent with a King model with a core radius of 500 to 1000 kpc. Within 100 kpc there is an excess of five AGN (six are found but only one is expected). Five of the central six AGN are in the brightest galaxy in the cluster and as such may not be representative of the overall AGN population. The differences in the redshift of the cluster and AGN show a strong peak at zero (Fig. 2). The 70 coincidences within 4 Mpc result in a velocity dispersion of 1090^{+98}_{-90} km s^{-1} (68 % errors). This value is 40 % higher than the median velocity dispersion of 744 km s^{-1} found in a large sample of clusters (Zabludoff, Huchra & Geller 1990) and the value of 696 km s^{-1} found for Markarian galaxies in Zwicky clusters (Petrosian & Turatto 1986). This is, at least in part, due to the rounding of the AGN redshifts in the Véron-Cetty catalogue (quoted to the third decimal place) and the errors in the mean cluster velocities which are based on less than 6 galaxies in 70 % of the clusters.

4. Conclusions

From these data, it appears that the Véron-Cetty AGN are drawn randomly from the cluster population, both spatially and in velocity. Petrosian & Turatto (1986) draw a similar conclusion for Markarian galaxies (which are predominantly starbursts). Therefore there is a strong indication that AGN and starburst activity are not 'triggered' at any particular radius from, or relative velocity to, the cluster by interaction with the ICM or galaxy-galaxy mergers.

References

Abell, G.O., Corwin, H.G., Olowin, R.P., 1989, *Astrophys. J. Suppl.*, **70**, 1.
Dressler, A., Thompson, I.B., Shectman, S.A., 1985, *Astrophys. J.*, **288**, 481.
Edge, A.C., 1992, *Mon. Not. R. astr. Soc.*, submitted.
Huchra, J., Burg, R., 1992, *Astrophys. J.*, **393**, 90.
Osterbrock, D.E., 1960, *Astrophys. J.*, **132**, 235.
Petrosian, A.R., Turatto, M., 1986, *Astr. & Astrophys.*, **163**, 26.
Véron-Cetty, M.-P., Véron, P., 1991, ESO Scientific Report No. 10
Zabludoff, A., Huchra, J.P., Geller, M., 1990, *Astrophys. J. Suppl.*, **74**, 1.

Clustering Properties of AGNs and their Contribution to the X–ray Background

Luigi Toffolatti [*]

1. Introduction

The AutoCorrelation Function (ACF) of the intensity fluctuations of the X–ray background (XRB) is an integrated view of the clustering properties of the source populations contributing to the XRB, so studies of the ACF provide information on the origin of the XRB as well as on clustering evolution of the underlying sources.

Recently *ROSAT* Deep Surveys have shown that about 40% of the soft XRB ($E \leq 3\,\mathrm{keV}$, henceforth SXRB) is contributed by sources brighter than $S(0.5\text{--}2\,\mathrm{keV}) \geq 7 \times 10^{-15}\,\mathrm{erg\,s^{-1}\,cm^{-2}}$ (Hasinger et al. 1991; Shanks et al. 1991). On the other hand only a few percent of the hard XRB ($3 < E < 60\,\mathrm{keV}$, henceforth HXRB) has already been resolved into sources. *ROSAT* observations also show that the SXRB is actually very smooth on arcminute scales with a current upper limit on the ACF in the 0.9–2.4 keV band $W(9') \leq 2 \times 10^{-3}$ (Hasinger 1992). As regards the HXRB, the results of the ACF analyses on sub-degree angular scales have been used to put significant constraints on clustering of AGNs and galaxy clusters (Martín-Mirones et al. 1991; Carrera et al. 1992). Here we will discuss data on ACFs to obtain limits on clustering, clustering evolution and volume emissivity of X–ray sources contributing to the SXRB and HXRB.

2. Constraints on the clustering scale and the clustering evolution of AGNs and on their contribution to the XRB

Recent analyses of large samples of optically selected QSOs have produced consistent values of the clustering scale, $12 \lesssim r_0 \lesssim 20$ Mpc ($H_0 = 50\,\mathrm{km\,s^{-1}\,Mpc^{-1}}$), and of the evolution of the correlation function with $\epsilon \geq -1.2$, a constant comoving clustering scale being slightly favoured (Boyle 1991; Andreani & Cristiani 1992). We find limits compatible with this optical results. It is interesting to notice that clustering with $r_0 = 20\,\mathrm{Mpc}$ and $\epsilon = -1.2$ would imply that soft spectrum AGNs produce less than 40% of the SXRB and this limit shifts to 50% if $r_0 = 12\,\mathrm{Mpc}$. Whatever the origin of the remaining background, its sources must cluster rather weakly.

Galaxy clusters and Active Star Forming (ASF) galaxies are obvious candidates to produce the remaining ~50% of the SXRB. Galaxy clusters are known to cluster on large scales. As a consequence their contribution to the background is bound to be

---
[*] Osservatorio Astronomico di Padova, vicolo dell'Osservatorio 5, 35122 PADOVA (ITALY)

small ($f \lesssim 15\%$), unless they are very extended and of low surface brightness. On the other hand ASF galaxies could yield a significant fraction of the SXRB, because they probably cluster on a scale comparable (or smaller) to that of normal galaxies.

It is worth noticing that the constraints on both AGNs and ASF galaxies have been obtained under the hypothesis that any other contribution to the background is smoothly distributed in the sky. Therefore constraints on clustering are expected to be even more stringent than those derived above. For instance let us consider the case in which AGNs and ASF galaxies together produce a fraction f$\simeq$80% of the SXRB (leaving room for contributions of galaxy clusters and galactic stars). Let us also assume that AGNs cluster with $r_0 = 12$ Mpc and $\epsilon = -1.2$ and give about 50% of the background; as a consequence they would saturate the autocorrelation level allowed by ROSAT limits. Then ASF galaxies giving the residual fraction $f \simeq 30\%$ are bound to have $r_0 \lesssim 10$ Mpc even in the case $\epsilon = 0$ (cf. Danese et al. 1992).

As for the HXRB, Mushotzky & Jahoda (1992) claimed a possible detection of positive autocorrelations on large scales, $\theta \geq 6°$. This signal could well be due to nearby, hard spectrum AGNs clustered with $r_0 \approx 20$ Mpc. Moreover, if the clustering evolution is confined to $\epsilon \geq -1.2$, AGNs could also produce $\sim 60\%$ of the HXRB, without violating the presently available limits on autocorrelations on scales of a few degrees (Martín-Mirones et al. 1991; Carrera et al. 1992). An alternative explanation of the possible detection of large scale autocorrelations requires that the local volume emissivity of the low luminosity sources is $\rho_x(2 - 10 keV) \approx 5 \times 10^{38}$ erg s^{-1} Mpc^{-3}, twice that associated with galaxies (Jahoda et al. 1991) and of the same order of the upper limit derived by Boldt (1992) from the dipole moment of the distribution of the X-ray flux, but much higher than that derived from direct observations.

Eventually, the limits on clustering derived from SXRB and HXRB correlations suggest that soft, highly luminous AGNs and hard, less luminous AGNs have similar clustering properties, with scales slightly larger than that of the galaxies, and evolution at least as steep as implied by the comoving clustering model ($\epsilon \geq -1.2$).

References

Boyle, B.J., 1991. In *Texas-ESO/CERN Conference on Relativistic Astrophysisc*
Andreani & Cristiani, 1992. ApJ, in press
Boldt 1992. In *The X-ray Background*, eds. X. Barcons & A.C. Fabian
Carrera et al. 1992. MNRAS, in press
Danese et al. 1992. ApJ, submitted
Hasinger 1992. In *The X-ray Background*, eds. X. Barcons & A.C. Fabian
Hasinger, Schmidt & Trümper, 1991. A&A, **246**, L2
Jahoda et al. 1991. ApJ, **378**, L37
Martín-Mirones et al. 1991. ApJ, **379**, 507
Mushotzky & Jahoda 1992. In *The X-ray Background*, eds. X. Barcons & A.C. Fabian)
Shanks, T. et al. 1991. Nature, **353**, 315

Energy Distributions of AGN

Chris Impey *

Abstract

AGN emit energy across the electromagnetic spectrum from radio waves to gamma rays. Observations in any single waveband give a very incomplete view of the relevant physical processes; higher energies generally probe smaller scales near the nucleus. AGN with compact radio cores show nonthermal emission from relativistic particles at many wavelengths. Many observed properties are altered by beamed emission from a relativistic jet. Low luminosity AGN without strong radio cores show thermal emission from cool dust in the far infrared, and from hot gas in the ultraviolet. Anisotropic obscuration leads to observed properties which are a function of orientation. A new window on AGN has been opened with the detection of high energy gamma rays.

1. Introduction

The greatest challenge in research on Active Galactic Nuclei (AGN) is to understand the physical mechanisms behind the prodigious energy output of these distant sources. Progress has been slow because the continuum emission extends over at least *eighteen* decades in frequency (from 10^8 to 10^{26} Hz). Unfortunately, we operate our narrow bandwidth detectors in a broad bandwidth universe. Until the past decade, most of our information had come from two limited windows at radio and optical frequencies. Space missions and new detector technologies have opened up a variety of new wavebands, for example in the millimetre, far infrared, X-ray and γ-ray parts of the spectrum. This review of multifrequency spectra of AGN will concentrate on large bandwidth features (due to space limitations, the bibliography is selective and brief). In other words, the emphasis will be on thermal ($\Delta\nu/\nu \sim h\nu/kT \sim 1$) and nonthermal ($\Delta\nu/\nu \sim (1-v^2/c^2)^{-3/2} >> 1$) processes, rather than narrow ($\Delta\nu/\nu \sim 1/c(2kT/m)^{1/2} << 1$) atomic and molecular processes. It is presumed that a full energy budget of AGN will be required to answer fundamental questions about the power source.

---

*Steward Observatory, University of Arizona, Tucson, Arizona 85721, USA. The author was partially supported by NASA grants NAS7-918 and NAG5-1235, and by NSF grant AST-901181.

2. AGN Across the Electromagnetic Spectrum

This broad perspective is relevant to all aspects of AGN : theoretical, instrumental, and observational. Figure 1 shows the theoretical prejudice as to the main constituents of AGN, taken from [1]. Activity is found over a wide range of scales, with a tendency for higher energy radiation to emerge from smaller scales. This is illustrated by the following sequence : extended Mpc jets (radio), kpc-sized starburst nuclei (far infrared), broad line regions on pc scales (optical), thermal accretion disk on mpc scales (ultraviolet), and a putative μpc black hole magnetosphere (X-rays). The theoretical paradigm is strikingly simple, the global properties of the AGN are determined only by the mass, spin and fuelling rate of a supermassive black hole. However, there are two major complications. First, there is considerable *reprocessing* of the primary radiation of AGN. This can involve reprocessing to lower energies, such as thermal radiation by dust, or to higher energies, such as inverse Compton scattering of synchrotron radiation. Second, most AGN emit *anisotropically*. Examples include relativistically beamed emission from compact radio sources, anisotropic obscuration from gas and dust tori, and the effects of the host galaxy itself. It is clearly unrealistic to expect a simple observational diagnostic of any of the elements of Figure 1.

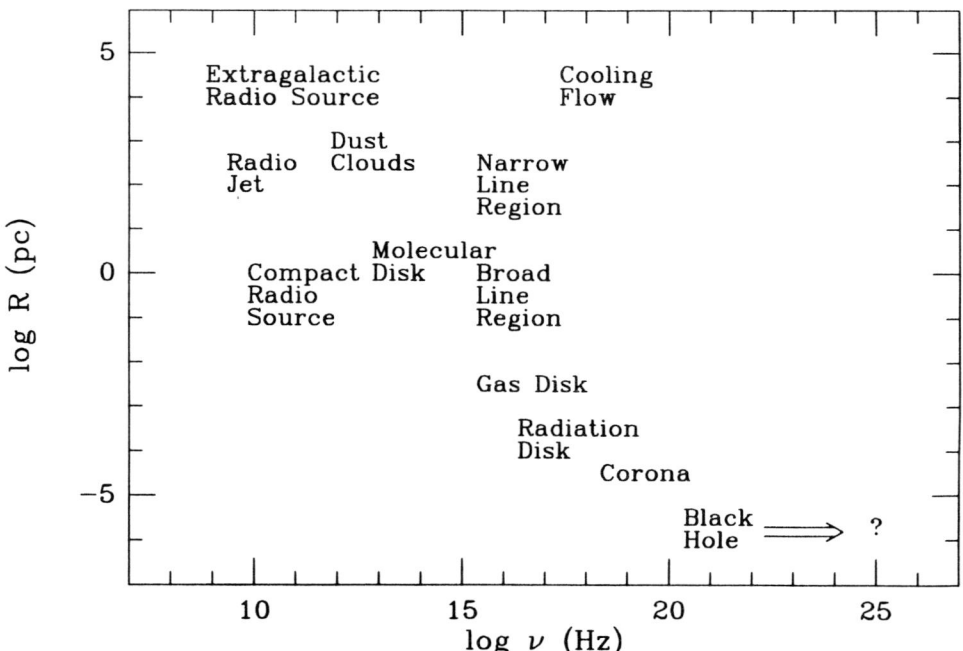

Figure 1. Schematic diagram of the active components associated with the nuclei of galaxies, showing the general inverse correlation between energy and scale.

The instrumental situation is improving rapidly. The ideal dataset would be an energy distribution from radio waves to γ-rays. However, more than 95% of the nearly 10,000 AGN catalogued by [2] and [3] were selected at either visual or radio frequencies, and only a small fraction have been detected at any frequency outside those two windows. Figure 2 shows the current and projected capabilities for detecting AGN across the electromagnetic spectrum. The sensitivity limits are plotted for surveys that cover $\geq \pi$ steradians, deeper surveys are possible over much smaller areas. At low energies, the MIT-Green Bank survey reaches to $\sim$ 50 mJy at 5 GHz. A proposed VLA snapshot survey would improve this limit by a factor of 50. The ISO mission will reach two orders of magnitude deeper than IRAS in the far infrared, but only over small areas of sky. In the near infrared, the 2MASS survey is expected to detect a few times 10^4 new AGN. The planned Digital Sky Survey with the Sloan Telescope will yield $\sim$ 100,000 optically selected QSOs and AGN. The ROSAT survey is expected to increase the number of X-ray selected AGN by more than an order of magnitude, to about 20,000. Finally, the GRO mission and the Whipple Observatory have allowed the study of AGN for the first time in high energy γ-rays.

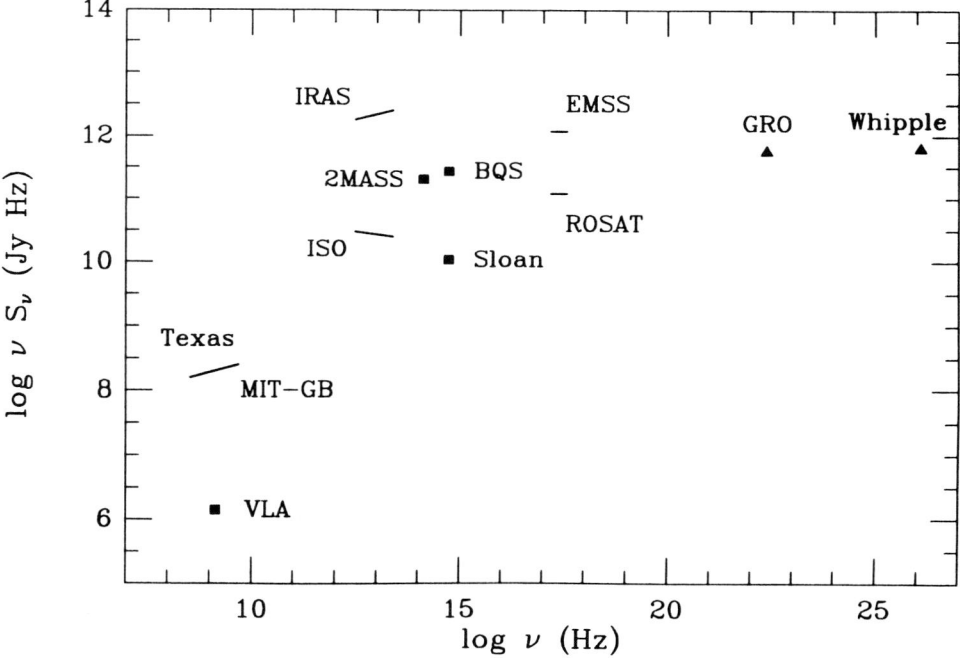

Figure 2. Detection capabilities for AGN surveys across the electromagnetic spectrum, including current and projected capabilities from ground and space-based facilities.

A complex observational path is used to classify AGN. The hope is that the taxonomy will illuminate the fundamental connections between different types of AGN [4]. Unfortunately, the classification of AGN rarely depends on simple or fundamental physics. For example, the terms "radio-quiet" and "infrared-luminous" have a lot to do with the sensitivity of the instruments in those two spectral regimes. The distinction between resolved and unresolved AGN depends on the host galaxy luminosity and its contrast with the nonstellar continuum, but also on the quality of the imaging. Key optical AGN diagnostics such as polarization (Blazar vs. Quasar), broad lines (Seyfert 1 vs. Seyfert 2), and line-free continua (BL Lac vs. Quasar) depend on spectral resolution and signal-to-noise. Figure 3 contains a schematic plot of the major continuum components of AGN, ignoring the finer details of emission line classification. One component is a "thermal" AGN spectrum, typified by the energy distribution of the PG quasars [5]. The second component is "nonthermal" at all frequencies, representative of the energy distribution of blazars [6]. One contribution not plotted is the composite spectrum of stars in the host galaxy. In the absence of extremely hot stars and large amount of dust, the bolometric luminosity of the galaxy is dominated by optical and infrared radiation. Varying combinations of these components can reproduce the energy distributions of most AGN.

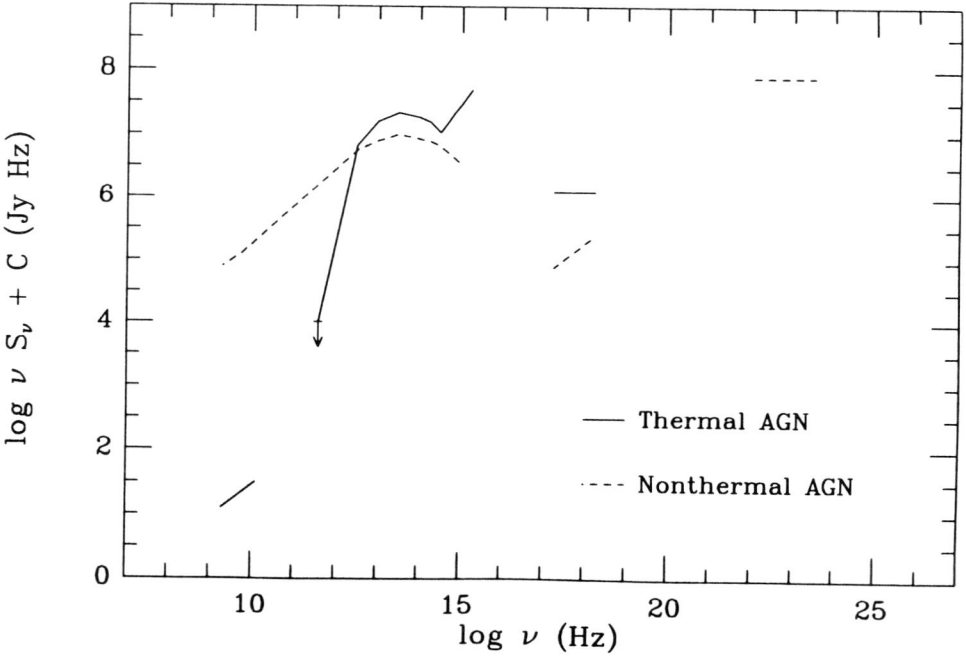

Figure 3. Average energy distributions from radio to gamma rays for the two main contributors to the energy budget : nonthermal emission (dashed line) and thermal emission (solid line).

Comparison of Figures 2 and 3 illustrates why AGN have only recently been accessible across the electromagnetic spectrum. Radio surveys have historically been sensitive to populations whose bolometric luminosity is dominated by other spectral regimes. The poor sensitivity of infrared and X-ray surveys (in terms of energy per unit bandwidth) biases them to luminous and/or nearby AGN. At present, any AGN that can be detected in γ-rays will have its energy budget dominated by high energy photons. As examples of how misleading single band classification can be, consider the following. Optically selected quasars often have their energy budget dominated by ultraviolet radiation and soft X-rays [5]. Radio galaxies are selected by their MHz emission but have energy output that peaks in the far infrared [7]. Most dramatically, blazars are selected by their GHz emission, but have prodigious γ-ray outputs [8].

3. Thermal AGN

There is general agreement that most AGN without a strong radio core emit thermal radiation at *most* wavelengths from $0.1\mu m$ to 1mm. Given the low space density of strong radio sources, this statement applies predominantly to low redshift and low luminosity AGN, and it represents a major change from a decade ago, when it was assumed that the continuum emission of quasars and Seyfert galaxies had a nonthermal origin. At optical and infrared wavelengths, the evidence for quasars rests on (a) the lack of variability and high polarization, (b) the fact that the energy minimum near $1\mu m$ in the rest frame corresponds to the maximum temperature for dust near an AGN, and (c) the millimetre spectra that fall too steeply in some cases to be due to synchrotron emission. In the ultraviolet, spectral decomposition indicates a few times 10^4 K black body component, and this component sometimes shows up as a soft X-ray turn-up also. Extensive references, and a good summary of the issues can be found in [9].

4. Nonthermal AGN

AGN with compact radio emission show the signatures of emission from relativistic particles at *most* wavelengths from $0.1\mu m$ to 1mm. This applies regardless of whether the AGN was selected by radio, optical, or X-ray techniques. This class has become known as "blazars", although there are significant differences between the properties of the two main categories, BL Lac objects and highly polarized quasars. Nonthermal AGN exhibit a correlated set of properties including (a) smooth synchrotron spectra from 10^{12} to 10^{16} Hz, (b) rapid variability and high linear polarization, and (c) one-sided radio jets. Most blazars show apparent superluminal motion of the VLBI radio components, which is strong evidence for relativistic source motion close to the line-of-sight. Further evidence that the synchrotron emission is not being emitted isotropically comes from the lack of expected synchrotron-self-Compton X-rays, and the brightness temperatures which can be as high as 10^{12} K. Inevitably, some transition objects like 3C 273 mix thermal and nonthermal components [10].

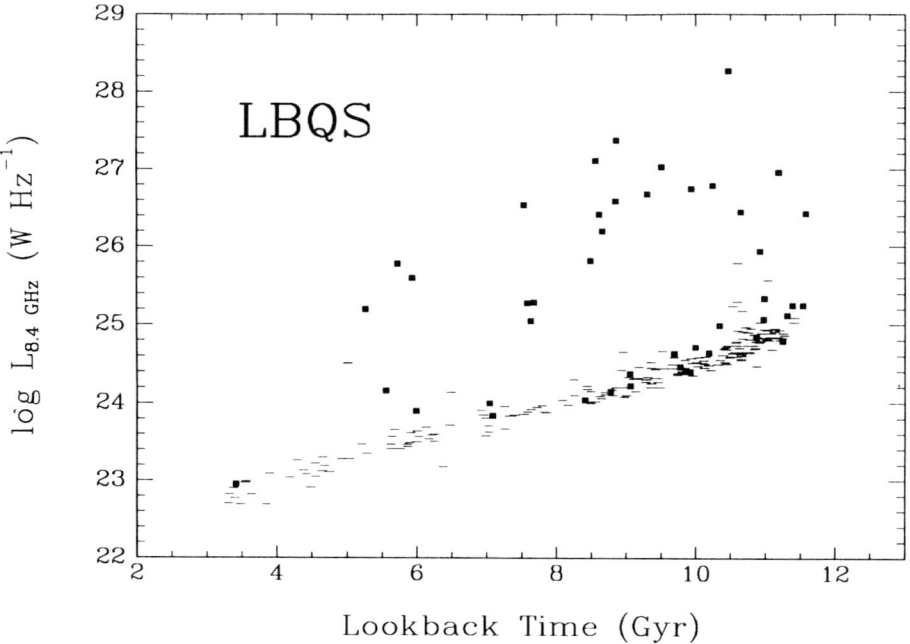

Figure 4. Radio luminosity as a function of lookback time ($H_0 = 50$, $q_0 = 0.5$) for 250 optically selected quasars from the Large Bright Quasar Survey.

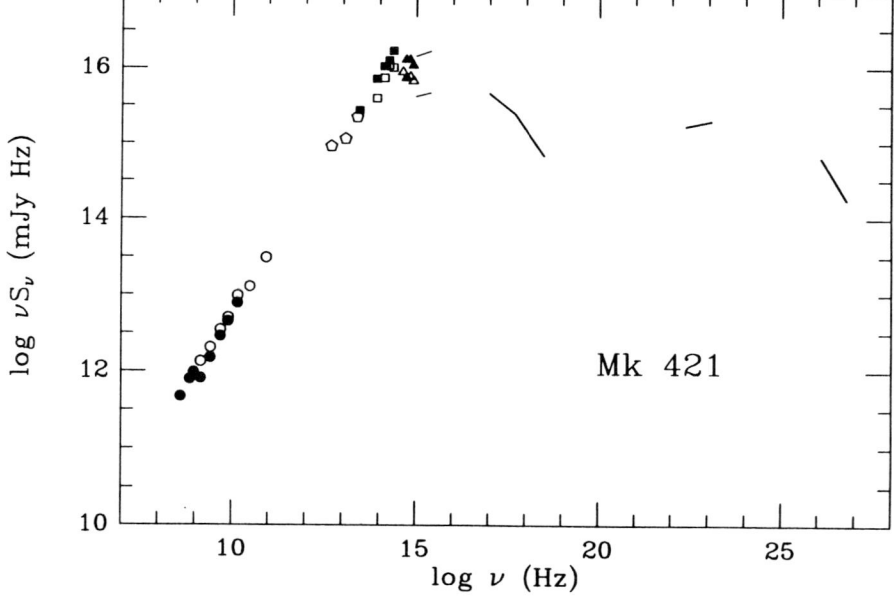

Figure 5. Energy distribution of the BL Lac object Mk 421 from radio to gamma rays, including recent results from GRO and the Whipple Observatory.

5. Unification

Many AGN are affected by anisotropic emission or obscuration. A number of unification schemes are motivated by the fact that the observed properties will therefore be dependent on orientation. Unification should help to simplify the complex taxonomy of AGN; the nuances of this growing field are discussed in [11]. The proposed agent of anisotropic obscuration is an opaque, dusty torus (opening angle $\phi \sim 60\,\mathrm{deg}$). The torus is mostly opaque to the broad line region and the radio core, and mostly transparent to the narrow line region (due to large size) and hard X-rays (due to low opacity). Unification models were given a great boost by the demonstration that the Seyfert 2 galaxy NGC 1068 has an embedded, obscured Seyfert 1 nucleus [12]. Broad and narrow line radio galaxies have now been unified in a similar way. The proposed agent of anisotropic emission is a relativistic jet (opening angle $\phi \sim 10 - 20\,\mathrm{deg}$). Statistical evidence indicates that high luminosity FRII radio galaxies are quasars with strong radio cores (including highly polarized quasars from the "blazar" class) seen at large angles to the jet axis [13]. Similarly, low luminosity FRI radio galaxies are proposed as the misaligned population of BL Lac objects [14].

The most dramatic aspect of AGN continua is the fact that the ratio of the radio to optical powers of 10% of quasars is $10^4 - 10^5$ times higher than that of the remaining 90% [15]. The distinction between radio-loud and radio-quiet quasars has so far resisted unification in terms of an orientation effect. Recent radio work on optically selected quasars shows that the radio-loud fraction is independent of redshift, but that strong radio emission only switches on above $M_B = -24$ [16] (corresponding to $z = 0.5$ for the LBQS in Figure 4). Another strong result is that broad absorption line quasars are almost exclusively radio-quiet [17]. The implication is that quasars either create a relativistic, highly collimated jet, leading to luminous radio emission, *or* a subrelativistic, uncollimated wind, leading to accelerated broad line clouds. The fact that the radio-loud fraction is independent of lookback time indicates that distinction between these populations is established in the collimation zone very close to the central power source.

6. The Highest Energies

A new window on AGN has been opened with the launch of the Gamma Ray Observatory. Sixteen AGN have been detected at GeV energies; all have compact radio cores and are probably beamed. The two most plausible explanations for the gamma rays are (a) self-Compton scattering of millimetre photons from the radio core, and (b) Compton "reflection" of photons from the accretion disk. The self-Compton scattering model predicts that gamma ray variations should be amplified versions of the compact radio core variations. All explanations are challenged by the recent detection of TeV cosmic rays from the BL Lac object Mk 421 [18] (each of the photons has an energy of about an erg!). The global energy distribution is shown in Figure 5.

References

[1]   Blandford R.D., *SAAS-Fee Advanced Course 20 : Active Galactic Nuclei. Springer-Verlag, 1990.*
[2]   Véron-Cetty M.-P., Véron P., *A Catalog of Quasars and Active Galactic (5th Edition). European Southern Observatory, 1991.*
[3]   Hewitt A., Burbidge G., 1989, ApJS, 69, 1
[4]   Lawrence A., 1987, PASP, 99, 30
[5]   Sanders D.B., et al., 1989, ApJ, 247, 29
[6]   Impey C.D., Neugebauer G., 1988, AJ, 95, 307
[7]   Golombek D., Miley G.K., Neugebauer G. 1988, AJ, 95, 26
[8]   Dermer C.D., 1992, Science, in press
[9]   Bregman J.N., 1991, Astronomy and Astrophysics Review, 2, 125
[10] Impey C.D., Malkan M.A., and Tapia S., 1989, ApJ, 347, 145
[11] Antonucci R.R.J., 1993, ARAA, in press
[12] Antonucci R.R.J., Miller J.S., 1985, ApJ, 297, 621
[13] Barthel P.D., 1989, ApJ, 336, 606
[14] Browne I.W.A., 1983, MNRAS, 204, 23P
[15] Kellerman K, et al., 1989, AJ, 98, 1195
[16] Hooper E., Impey C., Foltz C., Hewitt P., 1992, in preparation
[17] Stocke J.S., et al., 1992, ApJ, 396, 487
[18] Punch M., et al., 1992, Nature, 358, 477

Absorption in the ROSAT X-ray Spectra of Quasars

Belinda J. Wilkes, Martin Elvis, Fabrizio Fiore
Jonathan McDowell [*]

Abstract

The first ROSAT X-ray spectra of two high-redshift quasars reveal unexpectedly strong absorption when compared with similar luminosity objects at low-redshift. A third quasar shows none. A fourth, low-redshift, radio-loud quasar (3C351) with extended radio structure, shows absorption possibly due to a warm absorber with a strong OVII absorption edge.

1. Introduction

X-ray spectral observations of quasars have been confined to low redshift objects (z≤0.5) whose proximity makes them bright enough to study and also to those with relatively bright X-ray flux ($\alpha_{ox} \lesssim 1.5$). ROSAT, with its high sensitivity, enables us to observe the spectra of high redshift (z>2) and large α_{ox} quasars for the first time. We have begun a ROSAT observing program to study the X-ray spectra of quasars selected to cover the full range of continuum properties. In particular this sample includes objects at high redshift, with relatively faint X-ray flux and with a full range of radio properties: strong, weak, extended and compact. We are also carrying out a follow-up observing program to obtain multi-wavelength (infrared – ultra-violet) data for all our ROSAT-observed quasars.

2. Sampling the full quasar population with ROSAT

To date we have received and analysed data for > 25 quasars. Their spectra are generally steeper than those seen at higher (*e.g.* Einstein IPC) energies, as observed in general with ROSAT [1]. Our current sample includes 4 high-redshift (z>2.8) quasars with sufficient counts (> 350) to obtain spectral information (Table 1). Given the high redshift, the rest frame energy range is similar to the EXOSAT ME and *Ginga* energy ranges for low-redshift quasars (~ 1 − 10 keV) allowing us to study any change in slope with redshift and luminosity respectively. A comparison of the 1.7-17.36 keV (rest frame) energy indices derived for radio-loud quasars from *Ginga* [15] and those

---

[*]Center of Astrophysics, 60 Garden St., Cambridge, MA 02138, USA.

observed with ROSAT as a function of redshift or luminosity shows no change in X-ray slope with either quantity (15% and 30% probability of a correlation [4]). The mean slope for the ROSAT observed radio-loud quasars is 0.88±0.12, well within the error bars of the *Ginga* slopes. The flattening seen in the *Ginga* data is due to the difference in X-ray slope between radio-loud and radio-quiet quasars [15, 12] and is likely the cause of a similar trend in the ROSAT survey data [7].

Table 1. ROSAT X-ray observations

Quasar	z	V	Date	$\log(L_x)^a$	N_H^b	Counts	exp(s)
S5 0014+81	3.384	16.5	3/91		14.4^c	398	5951
Q0420−388	3.123	16.9	2/91	46.82	2.07	360	15611
PKS0438−436	2.852	18.8	2/91	46.98	1.40	595	10725
3C351	0.371	15.28	10/91	44.93	2.26^d	1420	13068
PKS2126−158	3.275	17.3	5/91	47.83	4.95	572	3424

a: *Einstein* value in erg s^{-1}, [14]; b: 10^{20}cm^{-2} [10]; c: [11]; d: [3]

3. Absorption in high-redshift quasars

Two of the high-redshift quasars in our sample show strong absorption in excess of the measured Galactic column density: PKS0438-436, [13] and PKS2126-158 [4]. Q0420-388 shows no significant absorption while the Galactic column density towards S5 0014+81 is too high. Results for a single power law plus absorption fits are given in Table 2 along with the column density of the excess absorption assuming it is intrinsic to the quasar. The ROSAT/PSPC spectral resolution cannot distinguish between excess absorption along the line-of-sight or intrinsic to the quasar so both possibilities are considered.

Table 2. ROSAT X-ray spectral results for all the quasars

Quasar	α_E	3rd parameter	$N_H(\text{fit})^a$	$\chi^2(\text{dof})$
S5 0014+81	$1.1^{+1.2}_{-0.7}$	-	17^{+23}_{-8}	1.13(18)
Q0420−388	1.09±0.65	-	2.7±1.5	1.25(15)
PKS0438−436	$0.7^{+0.4}_{-0.3}$	$N_H(\text{int})^b=1.0^{+0.7}_{-0.4}$	$7.0^{+6.1}_{-2.3}$	1.03(17)
PKS2126−158	0.6±0.6	$N_H(\text{int})^b=2.1^{+1.7}_{-0.7}$	11^{+10}_{-4}	1.32(20)
3C351[5]:				
Single power law	0.47±0.16		0.39±0.25	2.18(28)
Double power law	$-0.04^{+0.34}_{-0.25}$	$\alpha > 3.1$	4.8±3.0	1.03(26)
Partial covering	$2.1^{+0.4}_{-0.3}$	$F_C = 0.93^{+0.04}_{-0.05}$	200^{+70}_{-40}	0.97(27)
Warm absorber	0.5 FIXED	$U = 0.12^{+0.006}_{-0.02}$	200^{+20}_{-90}	1.04(28)

a: 10^{20}cm^{-2}; b: Intrinsic $N_H/10^{22}$cm^{-2} with Galactic value at z=0.

Absorption along the line-of-sight to high redshift quasars is a well known and heavily studied phenomenon with "Lyman α forest" and metal line systems being the dominant sources. Given the X-ray column densities observed ($\sim 10^{21-22}$cm^{-2}), the only possible explanations are damped Lyman α systems at the high end of the observed range or highly-ionized metal line systems [13]. For PKS0438-436, the only published optical spectrum shows no obvious strong line-of-sight absorption features [6]. Additional observations are necessary. PKS2126-158 has > 10 metal line systems [9] which may explain the absorption.

Most low-redshift quasars of comparable luminosity do not show intrinsic absorption. Possible reasons for this difference are high redshift, high luminosity, importance of beaming or a selection effect since strong absorption significantly weakens the observed X-ray flux. Both absorbed quasars may well be highly beamed similar to the BL Lac object PKS2155-304 which has strong OVIII Lyα absorption [2, 13].

4. Absorption in low-redshift quasar 3C351

3C351 is a lobe-dominated quasar (z=0.371) and is among the most X-ray quiet of radio-loud quasars (α_{ox}=1.6). A single power-law fit to the X-ray spectrum of 3C351 yields a flat slope, N$_H$ significantly below the Galactic value (Table 1) and a high χ^2(Table 2, Figure 1). The strong deficit between 0.6 and 0.9 keV is significant at $\sim 10\sigma$. Clearly a more complex model is required.

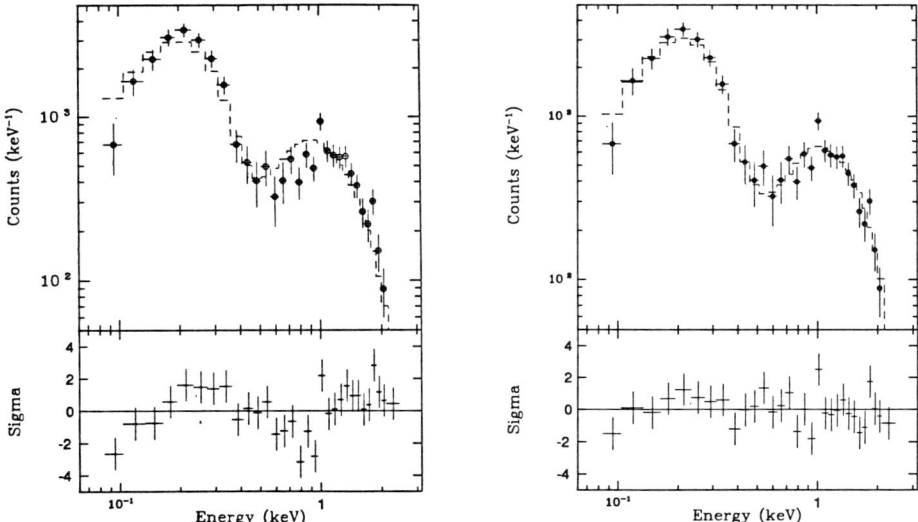

Figure 1. a) Single power law fit for 3C351 with residuals beneath showing strong deviations; b) best fit warm absorber model.

Three models were attempted: a double power law; partial covering; and a warm absorber. All three are acceptable (Table 2) although only the warm absorber model succeeds with typical values for the high energy power law slope. The parameters for the warm absorber model are not unique, equally acceptable fits can be found with a range of high energy slopes and ionization parameters [5]. The strongest absorption edge feature lies in the range 0.58−0.76 keV (1σ, one interesting parameter), implying OIV−OVII as the most likely absorbing ions. The ionization parameter and column density of the absorber are well constrained to $0.1 - 0.2$ and $0.5 - 2 \times 10^{22}$ atoms cm^{-2} respectively.

Whichever model is correct, this observation of 3C351 limits the possible causes for 'X-ray quietness'. Quasi-simultaneous X-ray, optical and ultraviolet (HST) observations of 3C351 rule out variability as the cause of its steep α_{OX}. Spectral fits that allow for intervening absorption increase the intrinsic emitted X-ray flux of 3C351 by only a minor part of the difference in α_{OX}. If a warm absorber model applies then the α_{OX} of 3C351 originates in weak X-ray emission relative to the optical while in the other models steep or hard continua contribute to the X-ray quietness of 3C351.

Acknowledgements. This work was supported by NASA grants NAG5-1536, NAG5-1724, NAG5-1872, NAG5-1883 (ROSAT), NAS5-30934 (RSDC), NAGW-2201 (LT-SARP)

References

[1] Brinkmann, W. 1992 *MPE Report #235 1992 p. 195*
[2] Canizares, C. R. and Kruper, J. 1984 *ApJL* **278**,L99
[3] Elvis, M., Lockman, F. J. and Wilkes, B. J. 1989 *AJ* **97**, 777
[4] Elvis, M, Fiore, F., Wilkes, B., McDowell, J. and Bechtold J. 1992 *in preparation*
[5] Fiore, F., Elvis, M., Mathur, S., Wilkes, B. and McDowell, J. 1992 *in preparation*
[6] Morton, D. C., Savage, A. and Bolton, J. G. 1978 *MNRAS* **185**, 735
[7] Schartel N., Fink H., Brinkmann W. and Trümper J. *MPE Report #235 1992 p. 195*
[8] Shastri, P., Wilkes, B. J.,Elvis, M. and McDowell, J. C. 1992 *ApJ submitted*
[9] Sargent, W. L. W, Steidel, C. and Boksenberg, A. 1990 *ApJ* **351**,364.
[10] Stark, A.A., Gammie, C.F., Wilson, R.W., Bally, J., Linke, R.A., Heiles, C. and Hurwitz, M. 1992, *Ap. J. Suppl.*, **79**, 77
[11] Veron-Cetty, M.-P. and Veron, P. 1991 ESO Scientific Report No. 10
[12] Wilkes, B. J. and Elvis, M. 1987 *ApJ* **323**, 243
[13] Wilkes, B. J., Elvis, M., Tananbaum, H., McDowell, J. C. and Lawrence, A. 1992a *ApJL* **393**, L1
[14] Wilkes, B. J. Tananbaum, H., Worrall, D. M., Avni, Y., Oey, M. S. and Flanagan, J. 1992b *ApJS in preparation*
[15] Williams, O.R., et al. 1992, *ApJ*, 389, 157

Dust in AGNs

Ari Laor * *Bruce T. Draine* [†]

Abstract

We use the lack of a significant silicate or silicon carbide emission feature in bright AGNs at $\lambda \sim 10$ μm to constrain the properties of the dust. We first calculate the optical properties of grains in the $0.005 - 10$ μm size range, over the 1000 μm$-$ 1Å wavelength range. We use these grain models to calculate the emission of optically thin and of optically thick dust, incorporating both absorption and scattering in the radiative transfer. A galactic dust composition in any configuration which is optically thin at 10 μm produces a very strong silicate emission feature and is clearly ruled out. We list what grain compositions, grain size distributions, and dust optical depths are consistent with the absence of a 10 μm feature. Independent dynamical arguments lead to a very similar set of constraints on the dust properties. We finally list the implications of these constraints on dust reddening, and line emission from AGNs.

1. Introduction

About a third of the bolometric luminosity of quasars and bright Seyfert 1 galaxies, is emitted in the $1-100$ μm range (Sanders *et al.* 1989). The emission mechanism is not yet established, but an attractive possibility is thermal reprocessing of the optical to far UV emission from the central continuum source by nearby dust. This hypothesis is supported by the IR spectral shape, its low amplitude variability, lack of significant polarization, and the near IR response to UV variations (e.g. Clavel, Wamsteker & Glass 1989). Recent calculations have shown that the observed IR continuum can be fitted with optically thin dust in a spherical distribution (Barvainis 1987), with dust in a highly warped disk (Phinney 1989), and possibly also with optically thick dust in a torus (Pier & Krolik 1992).

IR spectroscopy of most bright AGNs, however, does not reveal features which can be associated with dust. In particular, the 9.7 μm silicate feature is clearly seen in absorption in many starburst galaxies, but is generally undetectable in bright AGNs down to a level of 20-50% (e.g. Roche *et al.* 1991). The objective of this work is to obtain constraints on the composition, size, and optical depth of dust in the central parts of AGNs, given the apparent absence of the 9.7 μm silicate feature.

---

*Institute for Advanced Study, Princeton, NJ 08540, USA. This author was supported by NSF grant PHY91-06210.

[†]Princeton University Observatory, Peyton Hall, Princeton, NJ 08544, USA. This author was supported by NSF grant AST - 9017082 and NASA grant NAGW-1973.

2. The calculations

We consider three possible materials for dust in AGNs: amorphous silicate with a composition like that of olivine; crystalline graphite; and SiC. We calculated the dielectric functions of these materials (e.g Draine & Lee 1984) over the wavelength range 1000 μm–1Å. We then used these functions to calculate the optical properties of the grains using Mie theory, the Rayleigh-Gans approximation and geometric optics, for grains in the 0.005 − 10 μm size range. We used two grain mixture: a graphite+silicate and a graphite+silicon carbide mixture. We constructed various dust models assuming different grain size distributions (e.g. Mathis Rumple & Nordsieck 1977, MRN). The opacity of various graphite+silicate dust models is displayed in Figure 1.

A detailed radiative transfer code including both absorption and scattering, and a range of grain sizes at each grid point, was used to calculate the emission of an illuminated slab. Figure 2 displays some of our results. We used these results to obtain constraints on the possible properties of dust in AGNs in either "optically thin" or optically thick configurations.

3. Conclusions

1. Warm graphite + silicate or graphite + SiC dust (T>200K), with the MRN size distribution, and a column density $N_H \lesssim 10^{23}$ cm^{-2}, produces a strong silicate or SiC emission feature. Such dust cannot be responsible for the observed infrared emission from AGNs.
2. If the observed 10 μm emission comes from dust which is optically thin at 10 μm, then the dust must be either depleted in silicates by at least a factor of five, or have a grain size distribution extending to $a_{max} = 10$ μm, with a power- law size distribution with $\beta = d\ln n/d\ln a \gtrsim -2.5$. Qualitatively similar constraints apply to graphite + SiC dust.
3. MRN dust with a high optical depth at 10 μm produces an emission feature with an amplitude of about 57%, in excess of the typical observational limit for most objects.
4. If the observed 10 μm emission comes from dust which is optically thick at 10 μm, then the dust might either be depleted in silicates by a factor of two or more, have a grain size distribution extending to 10 μm with $\beta > -3.5$, be heated at large optical depth by a source other than the incident optical-UV flux, or some combination of the above. Qualitatively similar constraints apply to a graphite + SiC grain mixture.
5. Reddening of the broad emission lines and continuum is unlikely to be common. MRN dust can have a significant reddening effect only if its covering factor is about 0.1. MRN-like dust depleted in silicates can have a covering factor of about 0.3, and therefore is more likely to intersect our line of sight, but it will also produce a noticeable 2175Å absorption feature. Dust with a covering factor of about one can

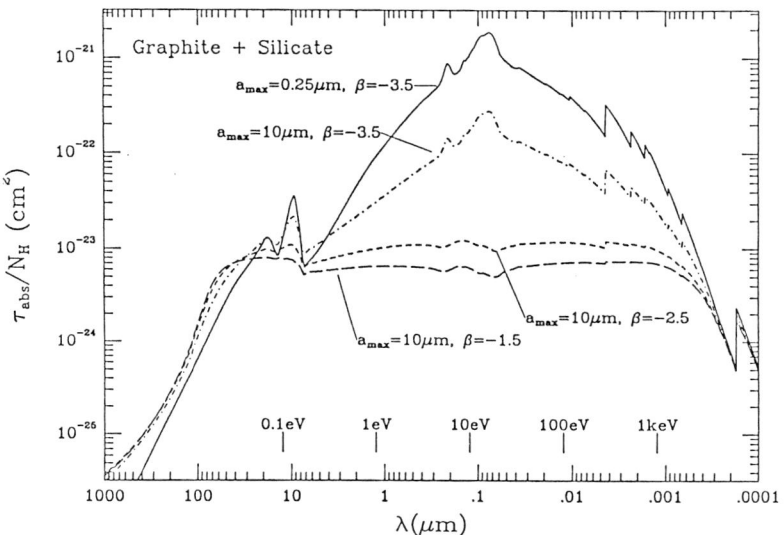

Figure 1. Absorption cross sections per H nucleus calculated for the MRN mixture of graphite and silicate spheres with a dust to H mass of 100. The size distribution extends from $a_{\min} = 0.005$ μm to $a_{\max} = 0.25$ μm in the upper curve, and to $a_{\max} = 10$ μm in the lower three curves, the slopes of the power law size distributions are indicated in the figure.

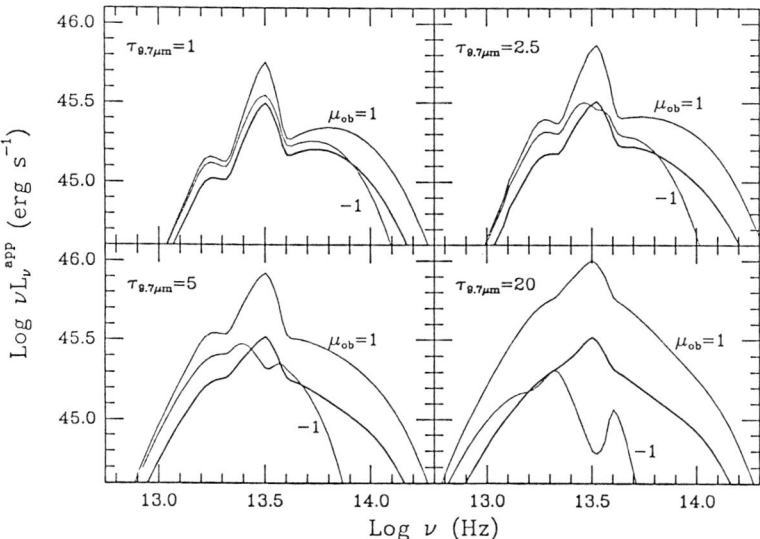

Figure 2. The emission of a slab composed of "MRN dust" illuminated by a source at $\mu_{\rm in} = 0.5$, where μ is cosine angle from slab normal, for different values of $\tau_{9.7\,\mu m}$. The upper dashed curve in each panel is for $\mu_{\rm ob} = 1$, and the lower curve is for $\mu_{\rm ob} = -1$ (from the back side of the slab). The solid curve represents the emission averaged over 4π.

produce only little reddening $[E(B-V) \leq 0.033]$, unless $L_{\rm IR}/L_{\rm bol} \Rightarrow 1$.

6. Dynamical arguments also indicate that high column density dust, and low column density dust dominated by large grains, both of which do not produce a significant $\sim 10\mu$m emission feature or reddening, are more likely to survive the effects of the incident radiation pressure.

7. Dust cannot reside in the BLR clouds if their distance from the continuum source is smaller than $0.2L_{46}^{1/2}$pc. The innermost radius at which even the largest grains can survive is probably just outside the BLR.

8. Dust can reside in the NLR, and it will have a significant effect on the gas phase emission when $U \gtrsim 0.01$. If the observed IR emission originates in clouds with $U \gtrsim 0.1$ and $N_{\rm H} \gtrsim 3 \times 10^{23}{\rm cm}^{-2}$ then the photoionized cloud face will be a source for narrow line emission, possibly similar to that observed.

A complete description of the calculation of the grain properties, the various dust models, and the solution of the radiative transfer is given by Laor & Draine (1993). We also discuss there constraints on the properties of the associated gas phase, dynamical constraints, and processes which can modify the grain size distribution and composition. We finally list a number of specific observational predictions which will allow further constraints on the possible solutions suggested above for the absence of a $\lambda \sim 10$ μm feature in bright AGNs. Further discussions of the effects of dust on line emission in AGNs is given by Netzer & Laor (1993)

References

Barvainis, R. 1987, ApJ, 320, 537
Clavel, J., Wamsteker, W., & Glass, I. 1989, ApJ, 337, 236
Draine, B. T., & Lee, H. M. 1984, ApJ, 285, 89
Laor A., & Draine, B. T. 1993, ApJ, 402, in press
Mathis, J. S., Rumpl, W., & Nordsieck, K. H. 1977, ApJ, 215, 425
Netzer, H., & Laor, A. 1993, ApJL, in press
Phinney, E. S. 1989, in Theory of Accretion Disks, eds. F. Meyer, *et al.*, Kluwer
Pier, E. A. & Krolik, J. H. 1992, ApJ, in press
Roche, P. F., Aitken, D. K., Smith, C. H., & Ward, M. J. 1991, MNRAS, 248, 606
Sanders, D. B., Phinney, E. S., Neugebauer, G., Soifer, B. T. & Mathews, K. 1989, ApJ, 347, 29

First Simultaneous UBVRI Photopolarimetric Observations of a Sample of Normal Quasars

L.O. Takalo and A. Sillanpää [*]
M.R. Kidger and J.A. de Diego [†]

Abstract

We present the results from truly simultaneous UBVRI photopolarimetric observations on "normal" quasars at the Nordic Optical Telescope. Some of the observed quasars show no measurable polarization, this being consistent with previous measurements of these objects. On the other hand in PG 1008+133 and PG 1351+64 we detected low (1.5%) polarization. In PG1351+64 this polarization shows also wavelength dependence and in PG 1008+133 the position angle show wavelength dependence. These are the first such observations in these quasars.

1. Observations

We have selected quasars from the Veron-Cetty & Veron (1991) catalogue. The only selection criteria being: the object is bright enough to be observed in reasonable integration time. In practice this means that Vmag< 16. And that the object can be observed from La Palma, using the Nordic Optical Telescope (Declination $> -30°$).

The observations were made at the Nordic Optical Telescope during February 1991 and July 1992. We used the Turku photopolarimeter. This instrument gives truly simultaneous results in five (UBVRI) colour bands. A detailed description of the instrument and the observing method can be found from Takalo et al. (1992, and references therein). The total integration time depended on the object brightness, ranging from one to four hours.

2. Results

The main results are listed in Table 1., where we list the observed objects according to the measured polarization. The criteria for detecting polarization was; at least 3 sigma in more than three filters. We have not corrected for the interstellar polarization, since all the observed quasars are higher than 20 degrees from the galactic

---

[*]University of Turku, Tuorla Observatory, Tuorla, SF-21500 Piikkiö, Finland
[†]Instituto de Astroficica de Canarias, 38200 La Laguna, Tenerife, Spain

Table 1. The observed quasars, with the V band measurements and radio and X-ray data collected from Veron-Cetty and Veron (1991).

Object	V	Pol	radio	X-ray
Detected: (3σ in 3 filters)				
Markarian 335	13.82(0.04)	0.73(0.15)	RQ	X
I ZW 1	14.24(0.06)	0.50(0.18)	RQ	X
PG 1008+133	15.29(0.03)	1.12(0.33)	RQ	-
PG 1351+64	14.28(0.01)	0.88(0.18)	RQ	X
	15.09(0.08)	0.96(0.44)		
KUV 18217+64	14.27(0.04)	0.74(0.16)	RQ	X
Possible detections: (3σ in 2 filters)				
PG 1634+706	14.53(0.07)	0.42(0.19)	RQ	X
PG 1718+481	14.48(0.04)	0.62(0.16)	RQ	-
3C 351	15.10(0.05)	1.04(0.52)	RL	X
II ZW 136	14.54(0.03)	0.20(0.13)	RQ	X
Not-detected:				
NAB 0205+02	15.02(0.02)	0.75(0.37)	RQ	-
PKS 0405-12	14.68(0.05)	0.15(0.20)	RL	-
B2 0742+31	15.63(0.02)	0.24(0.31)	RL	-
3C 232	16.07(0.07)	0.41(0.55)	RL	-
Markarian 205	14.97(0.02)	0.41(0.26)	RQ	X
S4 1435+63	16.57(0.10)	0.17(0.67)	RL	-
3C 273	12.70(0.08)	0.19(0.27)	RL	X

plane. There the interstellar polarization is thought to be small (e.g. Serkowski, Mathewson, and Ford 1975).

The highest measured polarization were in PG 1008+133 and PG 1351+164. These two quasars were also the only ones showing wavelength dependent polarization (PG 1351+164) and wavelength dependent position angle (PG 1008+133).

PG 1351+164 was observed during both observing runs. The observed polarization and its wavelength dependence was the same in both observations. Photometrically we detected a 0.8 magnitude variability in the V-band, with similar variations in the other bands.

As can be seen from Table 1, most (8/9) of the detected or possible detected quasars are radio quiet and (7/9) of these have also been detected in X-rays. On the other hand only two out of the seven non-detected ones are radio quiet or have been detected in X-rays. We do not know if this is due to a selection effect, or if there is a clear physical reason. There are too few observed objects for any statistical analysis.

References

Serkowski, K., Mathewson, D.S., and Ford, V.L.: 1975, *ApJ*, *196*, *261*
Takalo, L.O., Sillanpää, A., Nilsson, K., Kidger, M., de Diego, J.A., and Piirola, V.: *1992*, *A&AS*, *94*, *37*
Veron-Cetty, M.P., and Veron, P.: 1991, *ESO Scientific Report No.10*.

Intermediate Resolution Spectropolarimetry of Three Quasars

J.A. de Diego, E. Pérez and M.R. Kidger [*]   L.O. Takalo [†]

Abstract

We present spectropolarimetric observations of the quasars 3C273, CTA102 and 3C345. A synchrotron origin is the most likely explanation for the polarization of 3C273. The quasar CTA102 may have strongly polarized Fe II lines and, along with 3C345, some minima in the polarization are alongside the red wing of the MgII and CIII] lines.

1. Introduction

An often debated problem is that of the variability and polarization of the emission lines in AGNs. Despite the work done on Seyfert-2 and radio galaxies (e.g., McLean et al. 1983; Antonucci & Miller 1985; Antonucci 1992; Jackson in this meeting) and on quasars (Goodrich & Miller 1988), which demonstrates that the lines are usually polarized, the lack of appropriate instruments has severely limited the number of research groups dedicated to this important task.

The aims of our observations were, first, to get a spectropolarimetric sample of quasars. The objects are chosen for a) their brightness, brighter than 17 mag, b) their polarization, larger than 3%, and c) their known variability. Besides these, some other objects, like 3C273, of which polarization has never been reported to be larger than 2.5%, were also included in the sample. The second aim is to see if the emission lines are polarized and search for possible variability. Comparing lines and their continuum polarization and variability we should find new and powerful parameters for studying the actual models for quasars and for the origin of their polarized radiation.

Here we present the first results of our observations taken with the William Herschel Telescope (4.2 m) sited in the Observatorio del Roque de Los Muchachos on the island of La Palma, on 1991, June 11-12. We used the Faint Object Spectrograph, which has a resolution ranging from 4Å in the blue to 8Å in the red parts of the spectrum. A description of the instrument and reduction procedure can be found in Rutten & Dhillon (1992) and De Diego et al. (1992). Another set of observations has been taken on 1992, July 2-3, still being reduced. From here on, all the wavelengths will be given in the object's rest frame.

---

[*]Instituto de Astrofísica de Canarias, 38200 La Laguna (Tenerife), Spain
[†]Turku University Observatory, SF-21500 Piikkiö, Finland

2. 3C 273

Our observations cover the range from 3200 to 8000Å, and we have observed Hα, Hβ, Hγ, Hδ, [O III] (λ5007), He I (λ5876) and Fe II – [Ne III] (λ3869) lines. Fe II blends are important on the whole spectrum, particularly from 4000 to 5300Å.

The spectropolarimetry of this object has already been discussed elsewhere (De Diego et al. 1992).

3. CTA 102

The observations range from 1900 to 4500Å. We can distinguish the following lines: C III] (λ1909), Fe II blends between 2200 and 2700Å, Mg II (λ2798), [Ne V] (λ3426), [O II] (λ3727), [Ne III] (λ3869-3965), Hδ and Hγ.

The PD, which is rather low for this object, has a strong wavelength dependence (fig. 1). A detailed analysis shows a curious similarity between the PD and the theoretical Fe II spectrum (cf. figure 15 from Netzer 1990). In some AGNs it has been reported that the peaks of some polarized lines are shifted with respect to the centre of the line (cf. McLean et al. 1983; Antonucci & Miller 1985); if this is the case for the features near C III] (λ1909) and Mg II (λ2798), this red-shift is about 290Å.

There are random variations in the PA around 160°. The dispersion of the angle is greater than the estimated measurement errors. On the other hand, the polarized flux closely follows the PD, unlike in 3C273 and 3C345, and has a net drop from the blue extreme to the red one.

4. 3C 345

This object has been observed in the range from 2400 to 6000Å. The spectrum shows Fe II blends in the ranges 2300-2700Å and 5100-5300Å. The most prominent lines are Mg II (λ2798) and Hβ. Other lines are [Ne V] (λ3426), [Ne III] (λ3869), Hγ, Hδ and [O III] (λ5007).

The typical PD of this source ranges between 2 and 12%, usually with no wavelength dependence (from observations with the Nordic Optical Telescope, NOT). Our observations show a deep minimum (fig. 2), red-shifted nearly 80Å from the position of the Mg II line; another minimum is also visible that coincides with the Hβ line. Furthermore, the PD increases about 2% from 2400 to 4500Å. Then it stays constant at a value of 5.5% till the extreme blue, except for the drop on the Hβ line.

The NOT data show either no or complex wavelength dependent PA. The former is also the case for the 2400 to 3000Å data obtained in these observations (fig. 3). But, from 3000 to 4000Å, the PA presents a clear increase of around 8° and possible small variations ($\sim$ 1°) between 4000 and 5600Å.

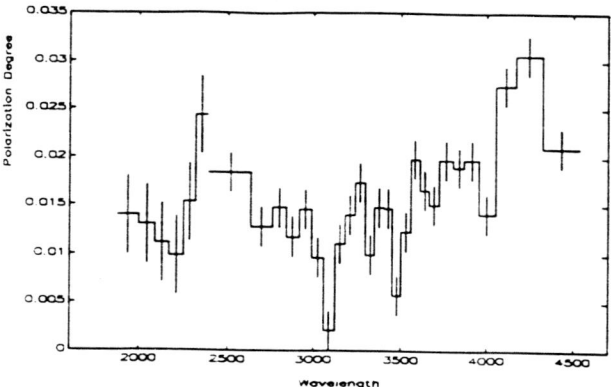

Figure 1. Wavelength dependence of the polarization degree in CTA 102.

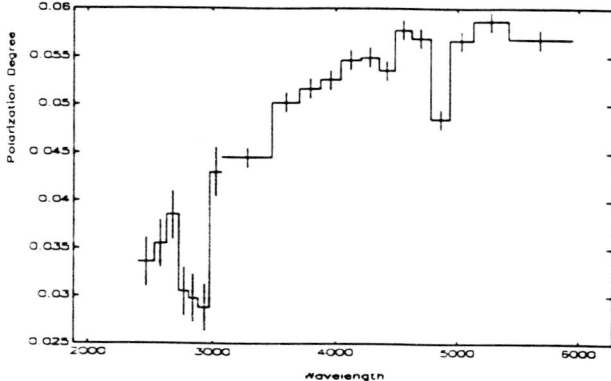

Figure 2. Wavelength dependence of the polarization degree in 3C 345.

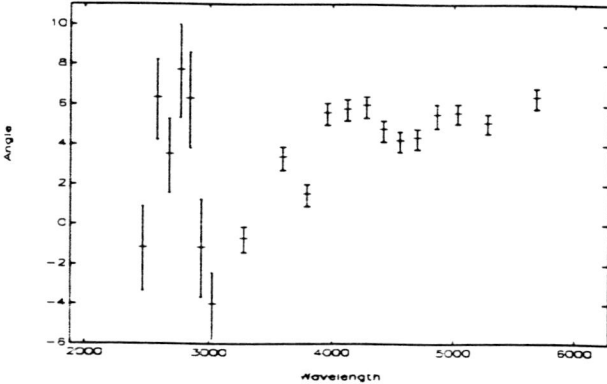

Figure 3. Wavelength dependence of the polarization position angle in 3C 345.

The polarized flux decreases continuously from the blue to the red extremes. Its appearance corresponds to the flux spectrum itself, but this is steeper than the polarized flux one. There are two minima, a large one at the position of Mg II (red shifted) and a much smaller one coincident with Hβ.

5. Conclusions

We have shown spectropolarimetric observations of a sample of three quasars that show large differences among all the objects.

The Hα and Hβ lines do not share the polarization of the continuum in 3C 273. There is a sharp decrease at the position of the lines in the PD spectrum. In spite of this, the PD spectrum of 3C 345 presents a less polarized Mg II line with the minimum redshifted 80Å with respect to the centre of the line, whilst the minimum on the Hβ line is not displaced. In CTA 102, this may be also the case for the C III] and Mg II lines, but in that case the shift would be about 290Å. In CTA 102, the Fe II might be strongly polarized: the increases in the PD spectrum agree with the theoretical models for Fe II lines. In the case of 3C 273, the most likely explanation for the polarization is a synchrotron origin (De Diego et al. 1992).

References

Antonucci R.R.J., Miller J.S. 1985 *ApJ, 297, 621*
Antonucci R. 1992 in *Testing the AGN Paradigm*, ed. S. Holt, S. Neff and M. Urry (New York: American Institute of Physics
De Diego J.A., Pérez E., Kidger M.R., Takalo L.O. 1992 *ApJ, 396, L19*
Goodrich R.W, Miller J.S. 1988 *ApJ, 331, 332*
McLean I.S., Aspin C., Heatcote S.R., McCaughrean J. 1983 *Nature, 304, 609*
Netzer H. 1990 in *Active Galactic Nuclei*, ed. T.J.-L. Courvoisier and M. Mayor (Berlin: Springer-Verlag
Rutten R.G.M., Dhillon V.S. 1992 *A&A, in press*

Active Galactic Nuclei Which Emit Strongly at 25 μm

Ramon D Wolstencroft [*]   *Carol J Lonsdale* [†]
Quentin A Parker [‡]

A galaxy which radiates strongly at 25 μm is likely to have an active nucleus. For Seyfert galaxies R=F (25 μm)/F(60 μm) is typically 0.2 to 0.5 (Miley & Neugebauer, 1985; De Grijp et al., 1985) and for quasars first detected by IRAS R ranges from 0.2 to 1.1 (Clowes, Leggett & Savage, 1991); whereas for IRAS galaxies for which star formation is the principal source of emission R is usually in the range 0.05 to 0.2. It appears that a 'high' value of R ($\gg 0.2$) strongly suggests the presence of an active nucleus while a 'low' value (R $\simeq 0.2$) does not exclude this. Hill, Becklin & Williams (1988) note that galaxies with high R tend to be compact at 10 μm which supports the idea that high R may be associated with nuclear activity.

The origin of the strong 25 μm emission from galaxies with active nuclei is unclear. Thermal emission from hot dust surrounding and heated by a power law source is an obvious possibility: however spherically symmetric models do not always provide good fits to the spectral energy distribution of Seyfert galaxies (Rowan-Robinson & Crawford 1989) and it appears that a disc geometry, perhaps combined with a high optical depth, may be needed in some cases. Other factors that may influence the models include the possible destruction of the very small grain component close to the AGN and the clumpiness of the dust distribution. Sanders et al. (1988) have suggested that warm (R > 0.2) ultraluminous IRAS galaxies may represent a phase in the evolution from quasars 'buried' in dust to 'classical' optical quasars that are almost dust-free: if this picture of quasar evolution has merit then we might expect high R values to indicate an early stage in this evolution when the dust is dense and close to the nucleus.

In order to develop these ideas we are assembling for further study a sample of 100 extragalactic sources which emit strongly at 25 μm. The sample is being taken from the IRAS Faint Source Database. It comprises a complete but very small subset of the $\sim$ 107,000 25 μm only sources in the database as well as some sources detected also at 60 μm which have high R values. Apart from the arguments given above, a major motivation for studying this sample is related to our discovery (Wolstencroft et al., 1991) that the 25 μm only sources are distributed essentially isotropically over the

---

[*] Royal Observatory, Edinburgh EH9 3HJ, Scotland
[†] Infrared Processing and Analysis Center, California Institute of Technology, Pasadena, USA
[‡] Anglo-Australian Observatory, Coonabarabran, NSW, Australia

sky, suggesting that they are principally either (a) very distant (i.e. galaxies) or (b) very close (e.g. very cool M dwarfs) or (c) a combination of both. In the former case this could mean that the 25 μm only source population contains a large number of warm (R > 0.2) galaxies likely to be predominantly AGNs. So far we have carried out optical spectroscopy (with FLAIR on the UK Schmidt Telescope and on the SAAO 1.9m) of approximately 100 sources with optical counterparts having 7.0 <B < 18.5 and which have a range of galactic latitude: we have found as anticipated that 90% of the faint counterparts (B> 17) are galaxies and 93% of the bright counterparts (B < 15) are stars, many of them being late type dwarfs. Based on our preliminary results it seems likely that the isotropy of the 25 μm only sources may be explained at least in part by (c) above, namely by a combination of nearby M dwarfs plus galaxies with faint optical counterparts (B > 17) — including those with optical counterparts fainter than the limits of our current spectroscopic survey (i.e. B > 8.5). Overall 30% of the sample observed so far are galaxies (13 < B < 18.5), and based on standard line ratios essentially all these galaxies are AGNs, including quasars. They have values of $R_{min} = F(25\mu m)/(60\mu m)_{min}$ in the range 0.1 to 1.6 (median 0.45). This success rate in discovering new AGN suggests that our identification procedure is reliable (for details see e.g. Wolstencroft et al., 1991) and that the 25 μm only sources will provide an abundant supply of new AGNs.

References

Clowes, R.G., Leggett, S.K. & Savage, A., 1991. *Mon Not.R. ast.Soc. 250, 597*
De Grijp, M.H.K., Miley, G.K., Lub, J. & De Jong, T., 1985. *Nature 314, 240*
Hill, G.J., Becklin, E.E. & Wynn-Williams, C.G., 1988. *Astrophys. J. 330, 737*
Miley, G.K., Neugebauer, G. & Soifer, B.T., 1985. *Astrophys. J. 193, L11*
Rowan-Robinson, M. & Crawford, J., 1989. *Mon. Not.R.astr.Soc 238, 523*
Sanders, D.B., Soifer, B.T., Elias, J.H., Neugebauer, G & Matthews, K., 1988. *Astrophys. J. 328, L35*
Wolstencroft, R.D. et al., 1991. *'Digitised Optical Sky Surveys', ed. H.T. MacGillivray & E.B. Thomson, Kluwer, Dordrecht, p.471.*

The Broad Line Region: Variablility and Structure

Emission-Line and Continuum Variability in Active Galactic Nuclei

Bradley M. Peterson *

Abstract

Large-scale multiwavelength spectroscopic monitoring campaigns are producing new information about the central regions of AGNs. Reverberation mapping experiments are now being undertaken, and these are providing useful constraints on the structure and kinematics of the broad-line region and leading to reinvestigation of the physics of the line-emitting gas. In this contribution, the fundamental assumptions of reverberation mapping, some of the principal results of recent monitoring campaigns, and questions that have arisen from recent work are briefly reviewed.

1. Introduction

The broad emission lines in the spectra of AGNs vary in response to changes in the luminosity of the central ionizing source with a time delay due to light travel-time effects within the broad-line region (BLR). It is in principle possible to make use of these light travel-time effects to map out the geometry and kinematics of the BLR through detailed comparison of the continuum and emission-line variability (e.g., [1]). This technique, known as "reverberation" or "echo" mapping, requires large amounts of high-quality data. Recent campaigns to measure continuum and emission-line variations in AGNs are for the first time providing data suitable for this purpose. These new data are leading to important new inferences about the nature of the BLR and the central source. In this contribution, I will discuss progress made in application of reverberation mapping techniques and mention some of the areas where further progress can be made in the near future. A more complete review of the subject will be provided elsewhere [2].

2. Fundamental Assumptions

Suppose the central continuum source undergoes an instantaneous outburst, a δ-function in time. The photons generated in this outburst propagate outward in a thin

---

*Department of Astronomy, The Ohio State University, 174 West 18th Avenue, Columbus, OH 43210 USA. The author is grateful for support of this work by the US National Science Foundation through grant AST-9117086.

spherical shell that expands at the speed of light. Some of these photons are absorbed by clouds of gas in the BLR which reprocess the absorbed energy into emission lines. To a distant external observer, the emission-line response to the continuum outburst at any time τ relative to detection of the continuum pulse is due to clouds on a surface of constant time delay, or isodelay surface, which is an expanding paraboloid. By measuring the emission-line response $\Psi(\tau)$ as a function of time delay, the BLR gas distribution can be determined. For real AGNs, the continuum luminosity is a complicated function of time $C(t)$, and the observed emission-line response $L(t)$ is usually assumed to be given by

$$L(t) = \int_{-\infty}^{\infty} \Psi(\tau) C(t-\tau) d\tau, \qquad (1)$$

where $\Psi(\tau)$ is the "transfer function" that describes the response of a particular emission line as a function of the BLR geometry and its orientation relative to the observer. For example, if the BLR gas clouds are confined to a thin spherical shell of radius r, it is simple to show that the transfer function has a rectangular form and is non-zero between $\tau = 0$ (when the gas along the line of sight responds) and $\tau = 2r/c$ (when the gas on the far side of the BLR responds). The aim of reverberation mapping is to use the observables $C(t)$ and $L(t)$ to solve eq. (1) for $\Psi(\tau)$. Reconstruction of $\Psi(\tau)$ from the observables requires large amounts of high-quality data. With limited data bases, such as those produced in AGN monitoring experiments, it is not possible to obtain a unique solution for $\Psi(\tau)$. However, it is often possible to reject certain solutions as unlikely, and conversely identify other solutions as being more probable [3].

There are several implicit assumptions underlying the reverberation mapping technique:

1. It is assumed that the continuum emission originates in a single, central source that radiates isotropically.

2. It is assumed that there is a simple, although not necessarily linear, relationship between the observable continuum and the ionizing continuum that is driving the emission lines.

3. It is assumed that the light-travel time (which is of order days to weeks) is the most important time scale.

The third assumption requires that emission-line response is virtually instantaneous. This depends on the recombination time and the resonance-line diffusion time; estimates of these time scales are in both cases about an hour or less, so these do not present a problem. The third assumption also requires that the BLR crossing time for individual gas clouds is long compared to the light-travel time. The typical cloud velocities can be estimated from the emission-line widths, and these give crossing times of order a few years, which is much longer than the light-travel time.

With sparse data sets, direct solution of eq. (1) is not possible. However, by cross-correlation of the continuum and emission-line light curves, $C(t)$ and $L(t)$, it is under some conditions possible to obtain a estimate of the physical scale of the BLR. For linear line response to the continuum, the cross-correlation function and $\Psi(\tau)$ are related in a simple fashion. In this case, the cross-correlation function is given by convolution of $\Psi(\tau)$ and the continuum autocorrelation function [4]. The location of the maximum in the cross-correlation function, known as the "lag", is the most frequently quoted result of AGN emission-line variability studies. It is important to realize that both determination and interpretation of the lag is fraught with difficulty because of its dependence on the continuum behaviour (as reflected in the continuum autocorrelation function), the tendency of cross-correlation to pick out the smallest physical scales (because these have the most coherent response), and technical difficulties related to the non-regular intervals between astronomical observations (e.g., [5], [6], [7], [8]).

3. Results of Observational Campaigns

The massive amounts of data required for reverberation mapping have necessitated extraordinary cooperative efforts among observers. Cooperation is necessary not only to ensure adequate temporal sampling of $C(t)$ and $L(t)$, but also to obtain the multi-wavelength observations necessary to address the problem in the most complete way through observation of a wide range of continuum energies and as many emission lines as possible. Several consortia of observers have been formed in an effort to obtain data sets suitable for reverberation mapping. Two of the larger collaborations are LAG (described in this volume by van Groningen) and the International AGN Watch, a group of over 100 astronomers who have carried out multiwavelength spectroscopic monitoring programs on two Seyfert galaxies, NGC 5548 (1988 – present) and NGC 3783 (1991 – 92). In the remainder of this review, I will concentrate on the results obtained by the AGN Watch on NGC 5548 as this program has been the most comprehensive to date and illustrates the potential of these studies.

Between 1988 December and 1989 August, the ultraviolet spectrum of the Seyfert 1 galaxy NGC 5548 was observed with *IUE* once every four days as part of an interagency (NASA/ESA/SERC) program, thus providing 60 observations over a 240-day period [9]. This program was complemented by a concurrent ground-based effort that involved more than 20 telescopes and astronomers from more than a dozen countries; between 1988 December and 1989 October, 246 optical spectra, 52 photometric observations, and numerous broad-band CCD images were obtained by participants in the AGN Watch study [10], [11], [12], thus providing contemporaneous optical continuum, Balmer-line, and helium-line light curves. Also, a light curve for the "small

Table 1

NGC 5548

Cross-Correlation Results

Feature (1)	τ_{peak} (Days) (2)	F_{var} (3)
UV cont. a	...	0.32
optical cont. b	2	0.13
N v λ1240 a	2	0.40
He II λ1640 a	2	0.36
"Small bump" c	6	0.11
He II λ4686 d	7	> 0.17
He I λ5876 d	9	0.08
Lyα λ1215 a	10	0.18
C IV λ1549 a	10	0.14
Hγ λ4340 d	13	0.11
Hα λ6563 d	17	0.06
Hβ λ4861 b, d	20	0.09
C III] λ1909 a	22	0.15
Mg II λ2798 a	34 – 72::	0.07

a Clavel et al. 1991
b Peterson et al. 1991
c Maoz et al. 1993
d Dietrich et al. 1993

blue bump", a strong, broad blend of Fe II lines and Balmer continuum in the range 2000 – 4000 Å, has been obtained by combining the *IUE* and optical spectra [13]. The ground-based monitoring program is still continuing [14].

During the 1988 – 89 campaign, the continuum went through three major outbursts, or "events", of 50 – 100 days duration. The broad emission lines show the same pattern of variability, with a short lag relative to the continuum variations. Cross-correlation analysis has been used to determine the lags relative to the UV continuum for various features, and these are summarized in Table 1. The lag quoted here, τ_{peak}, is the location of the maximum of the cross-correlation function. Also given is a measure of the amplitude of variability, F_{var}, which is the rms fractional variation of the feature during the campaign.

The principal results of the AGN Watch study of NGC 5548 are:

1. No significant phase lag is seen between the UV and optical continuum variations.

2. The UV continuum varies with greater amplitude than the optical continuum, even after accounting for the dilution of the optical continuum by a constant starlight contribution.

3. The lags for the various emission lines are all quite small, and show a pattern of increasing lag with decreasing ionization level, thus providing evidence for radial ionization stratification of the BLR.

4. The highest ionization lines vary with larger amplitude than do the low ionization lines.

The BLR size inferred from the lags is about an order of magnitude smaller than predicted by standard photoionization equilibrium models, which means that the density of ionizing radiation striking the C IV/Lyα zone is about two orders of magnitude higher than previously supposed. Revised models based on the new data [15] suggest that the most significant modification to the previous models is the particle density, which is now thought to be $n_H \approx 10^{11}$ cm^{-3}, nearly two orders of magnitude higher than in earlier models.

The light curves obtained in the 1988 – 89 program are sufficiently well sampled that it is possible solve eq. (1) directly for $\Psi(\tau)$ [16], [17]. Even though these are the most complete AGN light curves ever obtained, the amount of data is still too small for the transfer functions to be well constrained. For example, for some of the lines with larger lags, the transfer functions have peaks that are well displaced from zero delay, i.e., they show a deficit of response at small delays relative to a spherical shell. However, because of the limited amount of data available, it is not possible to make definitive statements about whether or not this apparent deficit at small delays is real. If it is real, it might indicate that we see the BLR in NGC 5548 as a face-on disk. The problem with this interpretation is that the broad-line profiles require untenable axial velocities. On the other hand, the lack of response at small delays might indicate that the emission lines are emitted anisotropically by the BLR clouds, with most of the radiation directed back towards the continuum source. Indeed, photoionization models indicate that for the parameters now thought to characterize the BLR, much of the line radiation escapes from the same cloud face through which the ionizing photons enter [15].

There is some evidence that the BLR is geometrically thick, i.e., the emission lines are produced over some range in distance from the central source — it would be unthinkable, of course, to suppose that Lyα and the Balmer lines are produced in completely different regions rather than having overlapping emissivity distributions that peak at different distances. Compelling evidence that a thick geometry is appropriate comes from the fact that the cross-correlation lags depend on the continuum behaviour [18],

[19]; the various emission lines show more rapid response for the shorter-duration outbursts. Moreover, there is also evidence from the relative amplitude of response for different lines that the shape of the ionizing continuum changes from one outburst to the next [2].

4. New Questions

One of the most important results of the NGC 5548 campaign is the demonstration of the simultaneity of the UV and optical continuum variations, which places strong constraints on the origin of the continuum emission. In particular, such close correspondence between the variations in these two continuum bands cannot be accounted for in standard thin accretion disk models. Recent efforts to account for AGN continuum radiation (e.g., [20]) thus have been directed towards reprocessing of a primary spectrum, perhaps by an accretion disk, into lower energy (UV/optical) photons. Part of the attraction of such models is that there is independent evidence for reprocessing in the X-ray spectra of some AGNs [21]. The relationship between the X-rays and the UV/optical continuum is far from clear, and extensive simultaneous monitoring of X-rays and the longer wavelength continuum will probably be needed to establish how they are related.

The infrared emission from AGNs also appears to be due to reprocessing, in this case by dust [22]. In Fairall 9 and other sources, the IR continuum variations lag behind those of the UV/optical. At the distance inferred from the optical/IR lag, graphite grains would be heated by the central source to close to their sublimation temperature. The IR continuum may thus be attributable to hot dust that is as close to the central source as it can be and still survive. Observations of additional sources should establish whether reprocessed emission from dust is a general characteristic of AGN spectra, or whether the IR continuum is an extension of the power-law continuum observed at X-ray energies.

Reverberation mapping experiments have not yet made use of the velocity-dependent information available in the resolved line profiles of AGNs. In principle, eq. (1) can be solved as a function of line-of-sight velocity and thus distinguish between different BLR geometries and velocity fields that have similar one-dimensional transfer functions $\Psi(\tau)$ (e.g., [23], [24]). This has not yet been done successfully, as existing data sets are too noisy, too inhomogeneous, or not well resolved in time.

The nature and origin of emission-line profile variability is not yet understood. The emission-line profiles in AGNs show pronounced variations, but on time scales longer than the BLR light-travel time; they are clearly not attributable only to excitation inhomogeneities. The time scales for profile variations are of the same order as the cloud crossing times for the BLR, and may indicate real changes in the BLR geometry on time scales of many months to years.

The amplitude of the response of various emission lines also provides an important diagnostic of the unobservable ionizing continuum [16], [25]. The response of the emission lines to continuum variations is apparently non-linear for virtually all of the important lines, including Lyα (see Shields & Ferland, this volume).

Similar monitoring campaigns on additional sources are of great importance to test the generality of the NGC 5548 results. One of the important requirements for new programs is to improve the temporal resolution as well as to obtain higher signal-to-noise spectra. It is important to determine (a) whether the apparent deficit of line response at zero delay is in fact real and if it is ubiquitous, or only occurs in some sources, and (b) the response times for the highest ionization level lines, such as He II and N V, which appear to respond almost instantaneously to continuum variations. In 1993, the AGN Watch has a limited, high sampling-rate campaign on NGC 5548 planned with *Hubble Space Telescope* and *IUE*, and if it is successful, this program should address some of the important issues related to the most rapid variations.

Reverberation mapping experiments on additional sources will reveal how the physics of the BLR changes with luminosity; for example, if the BLR cloud particle densities, ionization parameter, and shape of the ionizing spectrum are the same in all AGNs, the BLR radius should scale like $L^{1/2}$. The few existing data are consistent with such a relationship, but extension of the range of L over which this can be tested is of great importance, even though similar campaigns on quasars present a number of additional technical difficulties.

5. Conclusions

Reverberation mapping can in principle provide spatial resolution at the microarcsecond level in AGNs. Suitable data bases are now allowing us for the first time to probe directly the inner geometry of AGNs, although the information that we can extract with confidence is limited by the quality of the data and the temporal resolution of the observations. Nevertheless, important results already have been obtained. For example, the radiation density in the BLR is now known to be 100 times higher than thought previously, and many other old assumptions also are found to be incorrect. Reverberation mapping results are fundamentally changing our view of the inner regions of these objects.

References

[1] Blandford, R.D., & McKee, C.F. 1982, ApJ, 255, 419
[2] Peterson, B.M. 1993, PASP, 105, in press
[3] Horne, K. 1993, in preparation
[4] Penston, M.V. 1991, in Variability of Active Galactic Nuclei, ed. H.R. Miller & P.J. Wiita (Cambridge, Cambridge Univ. Press), p. 343
[5] Gaskell, C.M., & Peterson, B.M. 1987, ApJS, 65, 1

[6] Edelson, R.A, & Krolik, J.H. 1988, ApJ, 333, 646
[7] Maoz, D., & Netzer, H. 1989, MNRAS, 236, 21
[8] Robinson, A., & Pérez, E. 1990, MNRAS, 244, 138
[9] Clavel, J., et al. 1991, ApJ, 366, 64
[10] Peterson, B.M., et al. 1991, ApJ, 368, 119
[11] Dietrich, M., et al. 1993, preprint
[12] Romanishin, W., et al. 1993, in preparation
[13] Maoz, D., et al. 1993, ApJ, in press
[14] Peterson, B.M., et al. 1992, ApJ, 392, 470
[15] Ferland, G.J., Peterson, B.M., Horne, K., Welsh, W.F., & Nahar, S.N. 1992, ApJ, 387, 95
[16] Krolik, J.H., Horne, K., Kallman, T.R., Malkan, M.A., Edelson, R.A., & Kriss, G.A. 1991, ApJ, 371, 541
[17] Horne, K., Welsh, W.F., & Peterson, B.M. 1991, ApJ, 367, L5
[18] Netzer, H., & Maoz, D. 1990, ApJ, 365, L5
[19] Clavel, J., et al. 1992, ApJ, 393, 113
[20] Collin-Souffrin, S. 1991, A&A, 249, 344
[21] Nandra, K., Pounds, K.A., Stewart, G.C., George, I.M., Hayashida, K., Makino, F., & Ohasi, T. 1991, MNRAS, 248, 760
[22] Clavel, J., Wamsteker, W., & Glass, I.S. 1989, ApJ, 337, 236
[23] Welsh, W.F., & Horne, K. 1991, ApJ, 379, 586
[24] Pérez, E., Robinson, A., & de la Fuente, L. 1992, MNRAS, 256, 103
[25] Pogge, R.W., & Peterson, B.M. 1992, AJ, 103, 1084

Results of the LAG Monitoring Campaign

Ernst van Groningen [*]   *Ignaz Wanders* [*]

Abstract

We present evidence for changes in the transfer function of the broad line region (BLR) in NGC 3516 on time scales of a couple of months. If this occurs in most AGN, it would mean a serious complicating factor for mapping the BLR by reverberation methods. Furthermore, the Hβ profile in this object shows a time-variable dip at the same velocity shift as the strong absorption in C IV $\lambda 1550$. A corresponding feature is *not* present in Hα. The question addressed is whether the Hβ dip is absorption or a dip between two emission peaks. These results are part of the LAG[1] project: during the first five months in 1990, a sample of 6 Seyfert-1 galaxies and 2 QSO's were monitored spectroscopically and photometrically. Significant line-profile changes were found in the lower luminosity objects in the sample. The higher luminosity objects displayed continuum variations only.

1. The time variable transfer function in NGC 3516

In this section we will show that the transfer function of the BLR in NGC 3516 changed significantly over a time scale of several months. This observation could have a profound impact on the use of reverberation mapping as a means to map the structure of the BLR, because it is one of the basic assumptions of reverberation mapping that the transfer function does not change during the experiment.

Both cross-correlation analysis and total-flux reverberation mapping of the BLR in NGC 3516 show that the Hα emission lags the continuum variations by ~ 14 days, indicating that the BLR in this object is ~ 14 light days across under the assumptions of the standard BLR model. However, we first notice that the total flux transfer function (TF) is not constant in time. Figure 2 shows that the response of Hβ to continuum variations has a smaller amplitude towards the end of the time series. A similar behaviour was found in NGC 5548 (Clavel et al. 1991). This could be due to nonlinear line-response effects, to spectral-index variations in the ionizing continuum, or to physical changes in the BLR.

Perry et al. (1993) point out that, again under the assumptions of the standard BLR model, emission lines observed at different epochs should be of equal shape if the

---

[*]Astronomiska observatoriet, Box 515, S-751 20 Uppsala, Sweden
[1]Lovers of Active Galaxies, a collaboration between >50, mainly European, astronomers

continuum light curve preceding the observations has equal history over at least the light travel time through the BLR. They show for several objects that this is not the case, implying that at least one of the assumptions of the standard model is in error. In the remainder of this section we will show that NGC 3516 displays changes in the line profiles that are not correlated with changes in the continuum.

To study the evolution of the actual *shape* of the broad-line profiles, a number of corrections have to be applied before profiles at different epochs can be compared – *i)*: non-BLR features (underlying galaxy, narrow lines) have to be removed. *ii)*: the pure broad line profiles have to be *normalised* so that a useful comparison of the shape of the broad emission line at different epochs can be done. The difference between two normalised profiles at two epochs will show *relative changes in the emissivity at different projected velocities*. On the other hand, if the line intensity changes proportional to the ionizing continuum, then the pre-normalised profiles are just scaled versions of each other, and the difference spectrum between the normalised ones will be zero.

We applied this procedure to the 22 Hα spectra and 21 Hβ spectra, taken during the LAG monitoring campaign (see Section 3.). The integrated area between -5500 and +5500 km/s was scaled to a common value. We then compared these normalised profiles with the normalised *mean* Hα and Hβ profiles and produced a series of difference spectra. Some representative examples are shown in Figure 1. These 'residual-shape spectra' were used to measure the residual flux in three broad bands in the profile – the blue wing ($-5000 < \Delta v < -1700$ km s^{-1}), the line core ($-1700 < \Delta v < +1700$ km s^{-1}) and the red wing ($+1700 < \Delta v < +5000$ km s^{-1}). In this way three light curves are constructed representing *the relative importance of the emissivity in a particular part of the emission line at a given epoch relative to the mean*. These emissivity time series are presented in Figure 2 together with the measured continuum and total line-flux light curves from Wanders et al. (1993). It can be seen that both the Hα and the Hβ emission-line *shapes* behave almost identically, and that both are independent of the continuum and total line-flux light curves.

This suggests that the continuum variations are not the sole driving mechanism behind profile variations and that real physical changes in the BLR occur on a time scale of a couple of months.

These three observations (variable TF, line profiles not the same at epochs where they should be, and line shape variations independent of continuum or total light flux variations) all point into the direction of real physical changes in the BLR in the sense that emissivity must be redistributed. Because the BLR crossing time is large compared to the time scales over which the changes are seen, this would favour models of the BLR in which clouds are constantly being produced and destroyed.

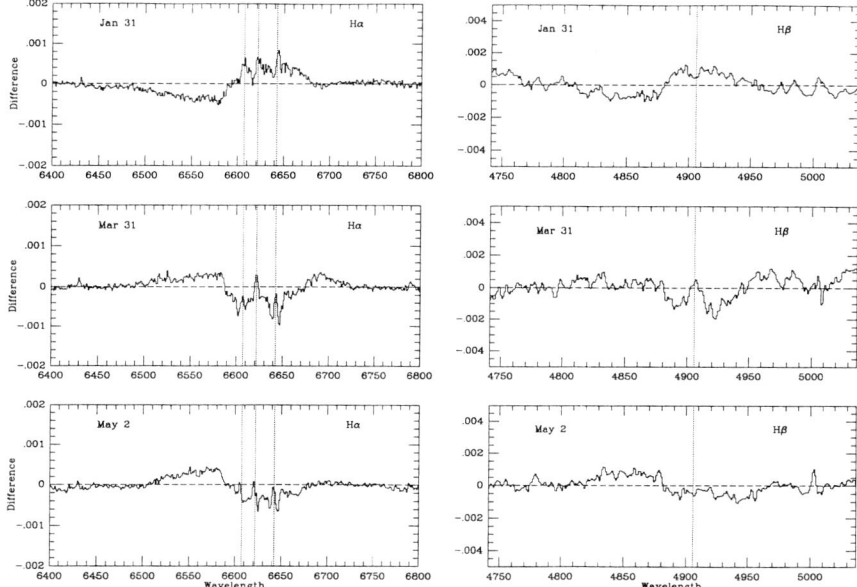

Figure 1. The difference between a set of normalised Hα and Hβ profiles and the normalised mean profiles of these lines, for a number of representative epochs. Note the similarity between the two lines. See text for a full explanation.

2. Hβ absorption in the broad line region of NGC 3516

During the first half year in 1990 the Hβ line in NGC 3516 developed a weak spectral feature best described as a dip at a blue shift of about 1300 km/s relative to the line peak. This corresponds to the velocity shift of the strong absorption observed in C IV λ1550. The presence of a dip of comparable strength in Hα can be excluded. The line profiles of both lines at all epochs are displayed in Figure 3

– the spectra are on a common velocity scale and have the same resolution (~ 200 km/s). The narrow lines and the underlying continuum have been subtracted and the velocity zero-point corresponds to the peaks of the narrow Hα and Hβ lines. The location of the dip is marked with the dashed line. The dip emerges on the Hβ profile around the end of February 1990; it is most pronounced around mid-April and then slowly fades towards the end of the time series (June 1990). There is no sign of a similar dip on the Hα profile.

The fact that the dip varies slowly in time and the fact that the spectra were taken with different instruments on different telescopes (see Section 3.) excludes the possibility that instrumental or reduction errors cause the dip. Absorption in the underlying galaxy continuum can also be excluded as the cause of the dip – no strong features are present at this wavelength in the spectra of other galaxy bulges.

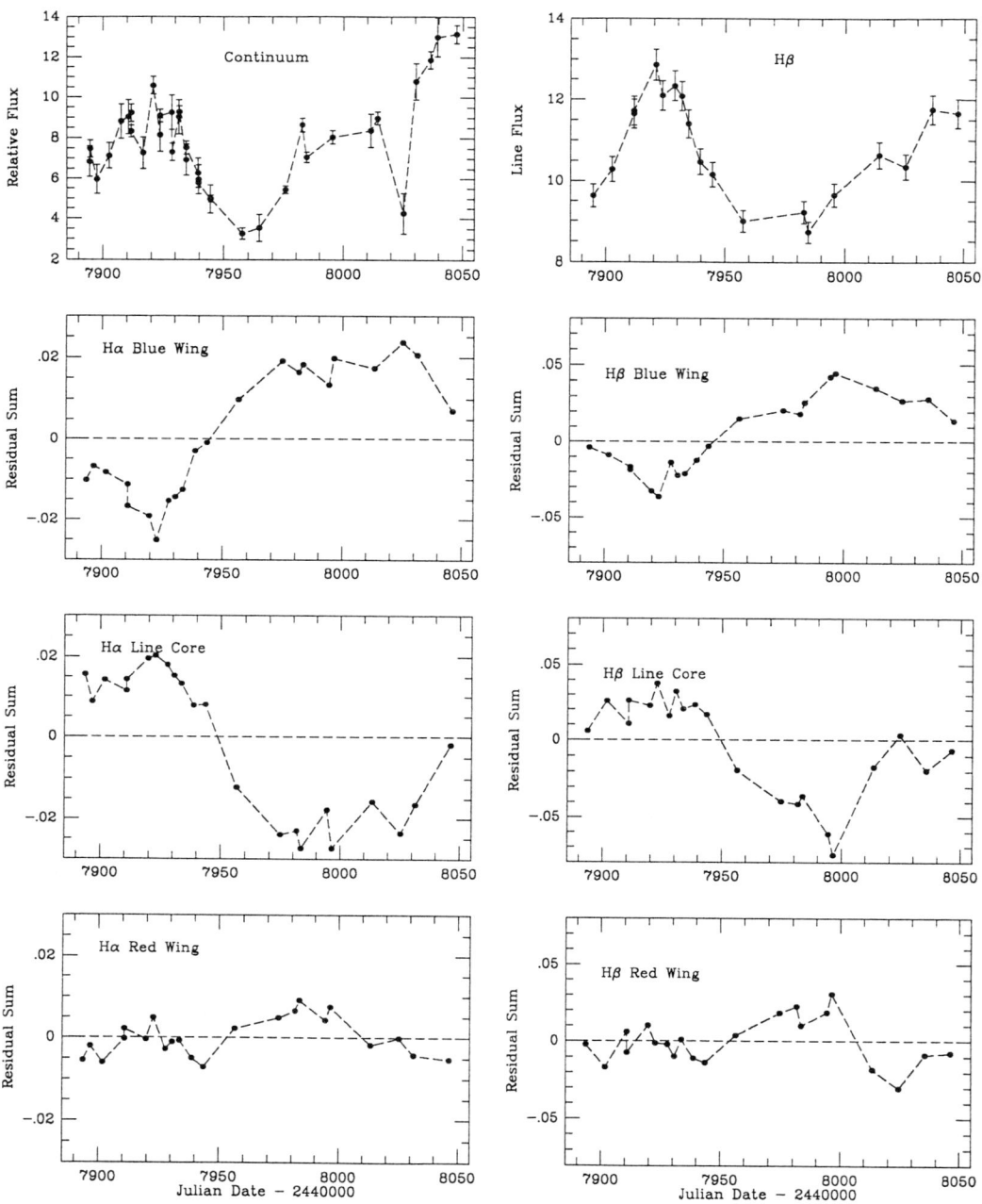

Figure 2. Residual line fluxes in three parts of the broad line profiles for Hα and Hβ. This figure shows that there are common global changes in the profile shape of these lines. Comparison with the measured light curves of the optical continuum and the Hβ flux, shows no clear correlation

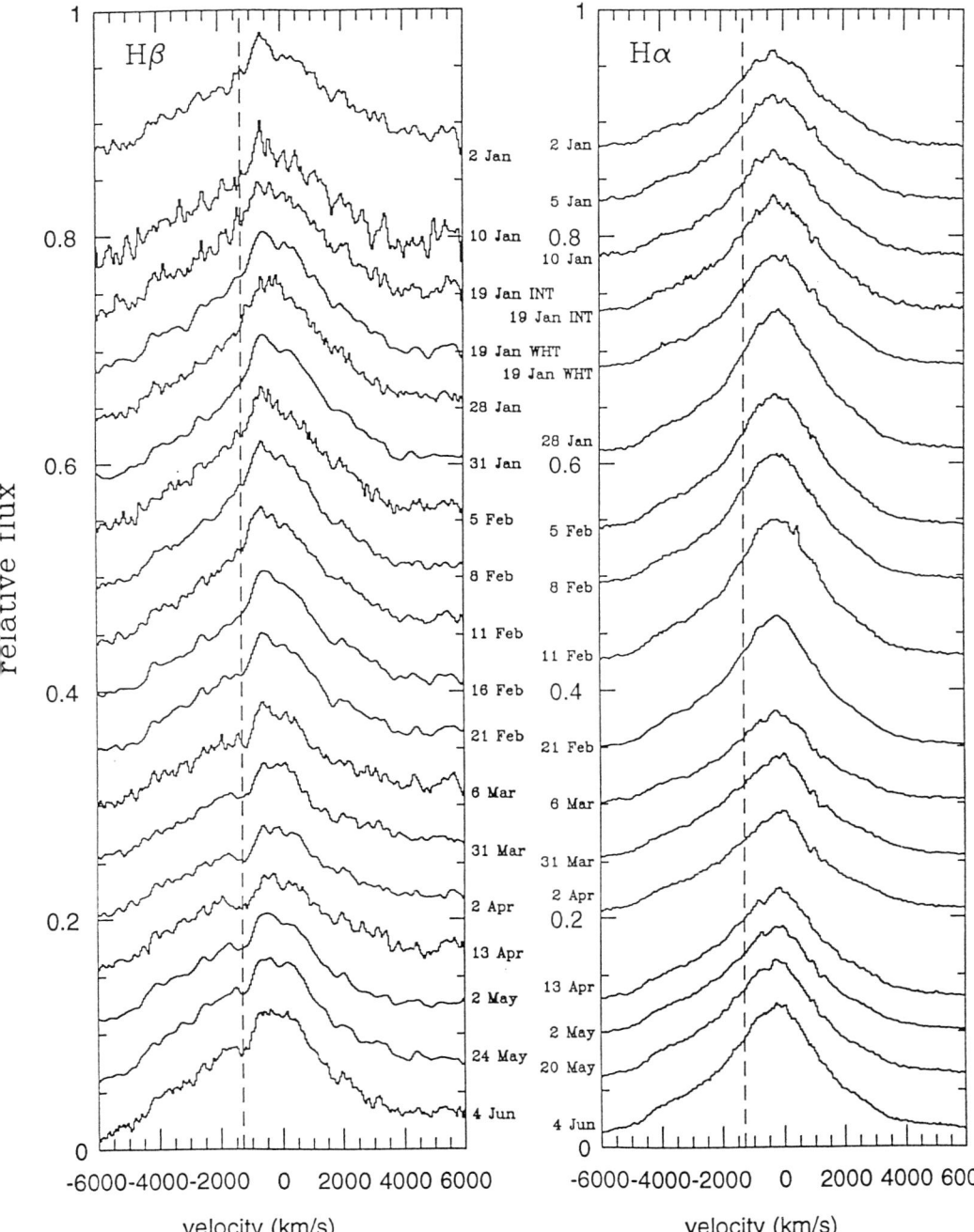

Figure 3. The profiles of Hα and Hβ in 1990. Narrow lines and underlying galaxy emission have been removed. Dashed line marks the position of the dip in Hβ, which begin to develop in the middle of February. Also note the absence of a corresponding dip in Hα.

172   THE BROAD LINE REGION: VARIABILITY AND STRUCTURE

The exact correspondence in projected velocity between the Hβ dip and the strong absorption in C IV λ1550 is demonstrated in Figure 4. Here, an IUE spectrum of June 1981 is displayed together with the Hβ profile, shifted to a common velocity scale. The strength of the C IV λ1550 absorption is variable (Walter et al. 1990), however, it remained at the same velocity between 1980 and 1990.

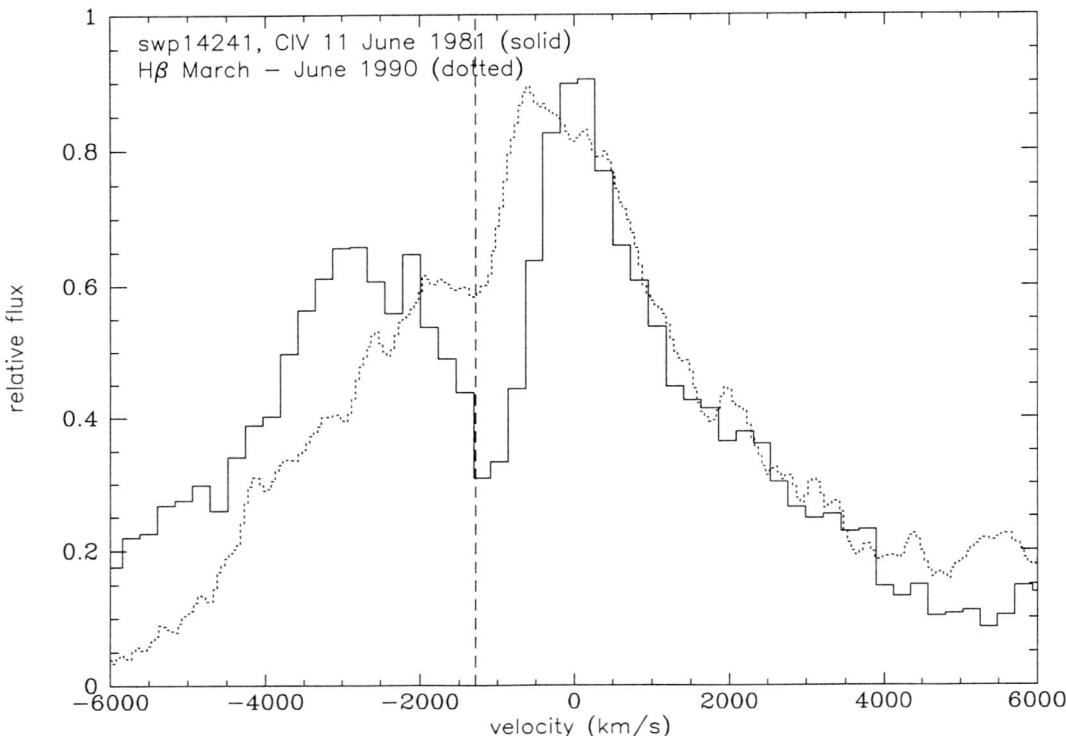

Figure 4. The profile C IV λ1550 in NGC 3516 as observed by IUE in June 1981 (solid line), and the Hβ mean profile in the period March to June 1990. The position of the dip in Hβ is marked and corresponds almost exactly to the velocity of the C IV absorption.

Is the dip caused by Hβ absorption of the non-stellar continuum or underlying emission line, or does it represent a real dip between two emission peaks? At this stage we cannot decide what is the cause of the dip, however, arguments can be used for and against both possibilities: *A dip between two peaks:* It is not unusual for Seyfert-1 galaxies to have a secondary peak in their Balmer line profiles that is more pronounced in Hβ than in Hα (Akn 120 is a good example, e.g. Korista, 1992). This interpretation would then imply that we observed a genuine profile change in Hβ and a much weaker corresponding change in Hα. In fact, the Balmer increment in the "dip emitting region" must have changed by at least 25% on a time scale of a couple of weeks. *Balmer absorption:* Here the problem is how to get significant Hβ absorption ($EW \sim 0.8$Å) without the corresponding Hα absorption. This is only possible if the absorbing material itself also emits Balmer line radiation – in this case, the brightness temperature in Hα of the intervening material should be comparable to that of the absorbed continuum, while the Hβ brightness temperature must be significantly lower. A good candidate for the absorbing material would then be the broad line clouds.

3. The variability sample

Observations: LAG undertook to monitor the BLRs of 6 Seyfert-1 galaxies and 2 QSOs during the first 5 months of 1990. We obtained spectroscopy of either Hα or Hβ for most objects while both lines were observed in NGC 3516 and NGC 4151. The excellent weather in January and February allowed a high sampling rate in this period (on average ~ 1.4/week), however, this became considerably worse in the later part of the campaign.

Results: Significant line and continuum changes were observed in most of the objects. Some of the results are presented elsewhere in this proceedings. At the time of writing the results could be summarized as follows.

PKS 1302-102 and PKS 1217+023. The latter object was included in the LAG sample because it was suspected to display rapidly variable Lyα emission (Gondhalekar 1990). However, during the LAG campaign both quasars only showed some marginal continuum variations, while the line fluxes remained constant. As a consequence, the spectra could be used to obtain very high S/N profiles of the Hβ lines in these objects, as discussed in Jackson et al. (1992).

NGC 3227. This object showed continuum and line variations on a level of 20–30 %. Cross-correlating the continuum and the Hα time series yields a delay of 17 ± 7 days (Salamanca et al. 1993).

NGC 3516. The most important results for this object were discussed in Sections 1. and 2.. A full account of the data reduction is given in Wanders et al. (1993).

NGC 4151. About 2 dozen high S/N spectra of both Hα and Hβ were obtained for this object, but are not yet fully analyzed.

NGC 4593. Dietrich et al. (1993) find a very compact BLR in this source. Cross-correlation analysis indicates a luminosity weighted radius of 5 ± 0.5 ($H\alpha$) and 6 ± 1 ($H\beta$) lightdays.

Mkn 279. The results for this object are presented by Stirpe (1993). Mkn 279 displays line profile variations that are correlated with the continuum flux, in the sense that the Hα line becomes redder when the continuum becomes brighter. Stirpe combined the LAG results with data from her earlier variability campaigns to show that this behaviour persists over at least 5 years. Hence, it seems that the transfer function in this object is illumination-dependent.

Mkn 876. Some continuum-flux variations were observed but the Hα line profiles and total flux did not change significantly during the campaign. On the other hand, major changes in the profiles are observed over longer time spans, when comparing the LAG results with earlier spectra.

4. LAG: a short project description

LAG was founded in 1988 by the late Michael Penston in order to apply for the large chunk of observing time at the La Palma Observatory known as the International Time or CCI Time[2]. The collaboration obtained 5% of the observing time at all La Palma telescopes, in the period March 1989 to August 1990. Currently, the collaboration consists of 54 members from 26 institutions in 8 European countries and 1 in the USA. The list of members is subject to evolution but a fair impression of it can be obtained from the co-author list and acknowledgements of the first LAG publication (Penston et al. 1990).

The goal of the collaboration is to study several aspects of AGNs that require considerable amounts of observing time or special scheduling that are not normally feasible. Some of these aspects for which results have been published are:

- The extended narrow line region in AGN – studies the effect of the nuclear radiation field on the circum-nuclear environment (Penston et al. 1990; Vila-Vilaró et al. 1993).

- The spectrophotometric monitoring reported above.

- Spectropolarimetry of Seyfert 2 galaxies – to look for hidden Seyfert-1 nuclei (Jackson et al. 1993)

References

Clavel, J., et al.: 1991, Astrophys. J. **366**, 64
Dietrich, M., Kollatschny, W.: 1993, these proceedings
Gondhalekar, P.: 1990, MNRAS, **243**, 443
Jackson, N., et al.: 1992, Astronomy & Astrophysics, **262**, 17
Jackson, N., et al.: 1993, in preparation
Korista, K.: 1992, Astrophys. J. Suppl. **79**, 285
Penston, M.V., Robinson, A., Alloin, D., Appenzeller, I., Aretxaga, I., Axon, D.J., Baribaud, T., Barthel, P., Baum, S.A., Boisson, C., de Bruyn, A.G., Jablonka, P., Jackson, N., Kollatschny, W., Laurikainen, E., Lawrence, A., Masegosa, J., McHardy, I., Meurs,

---
[2]Comité Científico Internacional

E., Miley, G., Moles, M., O'Brian, P.T., O'Dea, C., del Olmo, A., Pedlar, A., Perea, J., Pérez, E., Pérez-Fournon, I., Perry, J., Pilbratt, G., Rees, M., Robson, I., Rodríguez-Pascual, P., Rodríguez Espinosa, J.M., Santos-Lleo, M., Schilizzi, R., Stasínska, G., Stirpe, G.M., Tadhunter, T., Terlevich, E., Terlevich, R., Unger, S., Vila-Vilaró, V., Vílchez, J., Wagner, S., Ward, M.J., Yates, G.J.: 1990, Astrononomy & Astrophysics, **236**, 53

Perry J.J., et al.: 1993, in preparation

Salamanca, I., et al.: 1993, in preparation

Stirpe G.M.: 1993, these proceedings

Vila-Vilaró B. et al.: 1993, in preparation

Wanders I., et al.: 1993, Astronomy & Astrophysics, in press

Walter, R., Ulrich, M.-H., Courvoisier, T., Buson, L.,: 1990, Astronomy & Astrophysics, **233**, 53

A Relation between the Profiles and Intensities of Broad Emission Lines

Giovanna M. Stirpe [*]

Abstract

Two Seyfert 1 galaxies of comparable intrinsic luminosity, NGC 5548 and Markarian 279, have varied by a factor 2 in the optical band within a period of 5 years (1985–1990) during which both have been frequently observed. The large amplitude of the long-term variations reveals evidence that the profile of the broad emission lines depends on the luminosity of the line itself, in the sense that the asymmetry of the lines is stronger when the objects are brighter. This indicates that the structure of the broad line region in these two sources is stable on time scales of 5 years, and that their transfer functions are illumination-dependent in at least part of the radial velocity space. The effect, however, is absent in other, lower luminosity Seyfert 1 galaxies.

1. Introduction

The technique of reverberation mapping described by Blandford & McKee (1982) has been applied successfully in recent years to derive the transfer function (TF) of the broad line region (BLR) of well-monitored Seyfert 1 galaxies (e.g. Krolik et al. 1991). Until now the common assumption has been that the shape of the TF does not depend on the illumination of the BLR: this implies that if the continuum of a source were constant, the resulting broad lines would have the same profile irrespective of their intensity. The data presented in this paper, however, indicate that the broad lines of two Seyfert 1 galaxies, Markarian (Mkn) 279 and NGC 5548, display different characteristic profiles at different intensities, and therefore that the respective TFs have illumination-dependent shapes in at least part of the radial velocity range.

2. Line profile variations

Mkn 279. The Lovers of Active Galaxies (LAG) collaboration monitored 8 active galactic nuclei (AGN) on short time scales during the first five months of 1990. The sample included Mkn 279. The amplitude of its variation during this period was not very high (30% continuum under Hα, 20% broad Hα flux), but the source was about a factor 2 brighter than observed in previous years (Stirpe & de Bruyn 1991, Stirpe

---

[*] Osservatorio Astronomico di Bologna, Via Zamboni 33, 40126 Bologna, Italy.

1991). Profile variations took place during the entire period 1985–1990. In particular, the red side of Hα, which was almost straight in 1987 and 1988, was convex during the LAG campaign, and became even more convex when the source went through a shallow maximum in February 1990 (Stirpe et al., 1993). The LAG spectra show that the profile variations were not entirely caused by light travel time effects, but were more closely related to the luminosity of the line. This can be seen in Fig.1, in which a rough indicator of the asymmetry of the broad line is plotted against the total broad Hα flux. Clearly, the line tends to become more asymmetric when its flux is higher. Notice that the profile change cannot be due to a slow evolution of the structure of the BLR, because the time arrow changes direction several times in Fig.1. For instance, when the LAG campaign started the object was close to the upper right corner of the diagram, after which it increased slightly in both luminosity and asymmetry, and finally it descended along both axes towards the centre of the diagram. Figure 2 shows how the long-term profile variation occurs entirely in the red wing, while the blue wing remains almost unchanged on a time scale of several years. The profile varies between $\sim$ 500 and $\sim$ 4000 km s^{-1}, with a relative maximum at $\sim$ 2500 km s^{-1} (the discrepancy beyond 4500 km s^{-1} is not significant because of contamination by the atmospheric O$_2$ B-band and by the subtracted [S II] lines). An analysis of the other available spectra yields the same results, with differing amounts of excess on the red side. With the line in its highest state, the 'red excess' with respect to the scaled minimum profile is $\sim$ 6% of the broad line flux.

NGC 5548. The profile of Hα in NGC 5548 was analyzed in the same way as that of Mkn 279, using spectra obtained between June 1985 and January 1990. The resulting diagram (Fig.3) shows that a similar but *opposite* relation exists between the profile and intensity of the line. Rosenblatt & Malkan (1990) similarly found that Hβ has a characteristic profile in its low state, and a different characteristic profile in its high state. Both sides of Hα change profile on long time scales, contrary to the case of Mkn 279. Notice that the broad lines in NGC 5548 are blue-asymmetric with respect to their rest wavelengths, and that also in this case the line asymmetry is enhanced when the illumination of the BLR is stronger.

Other AGN. No relation of the kind described above has yet been found in other Seyfert 1 galaxies: of the other variable objects monitored by LAG (NGC 3227, NGC 3516, NGC 4151, NGC 4593) none displayed a similar phenomenon, though all underwent some profile variability. All are intrinsically weaker than Mkn 279 and NGC 5548 by at least a factor 2. The higher luminosity objects did not vary enough to allow the appropriate analysis to take place.

3. Conclusions

Within the reverberation scenario, the profile/luminosity relation leads to several conclusions. Firstly, the different characteristic shapes associated with a given line

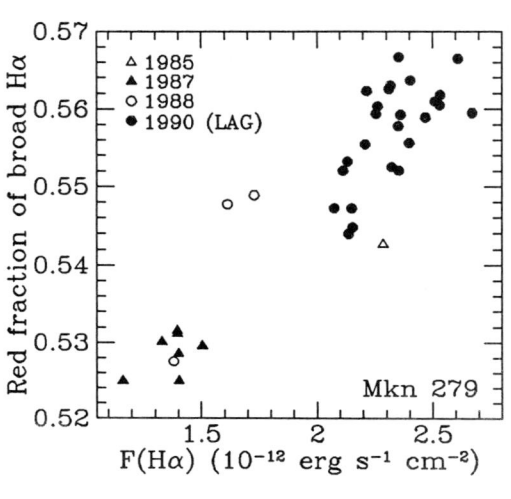

Figure 1. The fraction of broad Hα flux at positive radial velocities, plotted against the total broad Hα flux in Mkn 279. The spectra used for the diagram were obtained in different observing seasons, as indicated by the symbols.

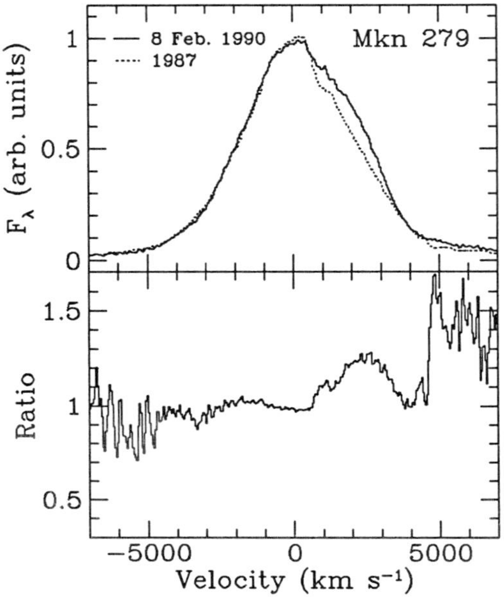

Figure 2. Top: the broad component profiles of Hα in the highest and lowest observed states. The latter is a mean of the spectra obtained in 1987, scaled so that the blue wings coincide. Bottom: the ratio between the two profiles.

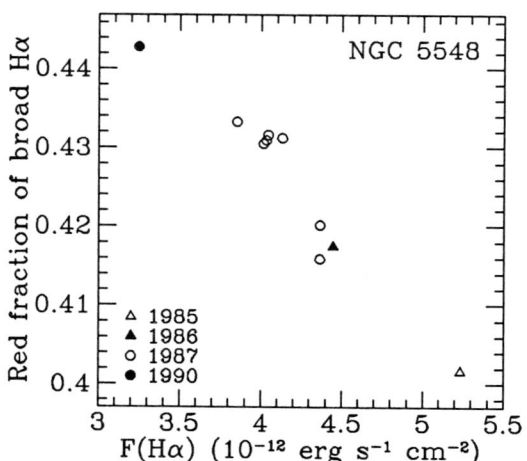

Figure 3. Same as Fig.1, for NGC 5548

luminosity imply that the TFs in Mkn 279 and NGC 5548 are illumination-dependent, and that this dependence varies throughout the radial velocity space. Secondly, the fact that a line 'remembers' what it has to look like at a given luminosity, even after a few years, implies that the BLRs of Mkn 279 and NGC 5548 are stable on time scales of at least 5 years; this is comparable with the dynamical crossing time of these regions as derived by the width of the lines, under the assumption that the BLR is gravitationally bound. Thirdly, the opposite behaviours of the relation in Mkn 279 and NGC 5548 imply either an intrinsic difference in the kinematic structures of their BLRs, or that their structures are strongly anisotropic, and therefore that the line profiles are orientation-dependent. Fourthly, the absence of the effect in lower luminosity AGN may mean that their BLRs are less stable, or that light-travel time effects prevail in their line profiles.

A search for similar effects in other objects is highly desirable to establish how widespread they are, and whether they really affect only a particular kind of AGN. Also, further observations of Mkn 279 and NGC 5548 would help to determine whether their BLRs are stable on time scales significantly longer than the dynamical crossing times. Neither of these goals requires intensive monitoring campaigns, but only occasional observations of the same objects, and large amplitude variability of the sources on long time scales.

References

Blandford R.D., McKee C.F., 1982, ApJ 255, 419
Krolik J.H., Horne K., Kallman T.R., Malkan M.A., Edelson R.A., Kriss G.A., 1991, ApJ 371, 541
Rosenblatt E.I., Malkan M.A., 1990, ApJ 350, 132
Stirpe G.M., 1991, Proc. Heidelberg Workshop on *Variability of Active Galaxies* (Duschl W.J., Wagner S.J., Camenzind M. eds.), Springer-Verlag, Berlin, p.71
Stirpe G.M, de Bruyn A.G., 1991, A&A 245, 355
Stirpe G.M., et al., 1993, in preparation

Broad-Line Profile Variability in NGC4593

M. Dietrich and W. Kollatschny *

Abstract

Balmer emission line variations of the integrated flux as well as profile variations for the Seyfert galaxy NGC4593 are presented. This active galaxy was observed in an international monitoring campaign. The broad emission line profiles consist of at least three components varying in an independent way. The FWZI of the broad line profiles of the Hα line remained constant during the intensity variation indicating a turbulent velocity field of the BLR.

1. Introduction

It has been shown by simple kinematical model calculations that the response of the broad line profiles to an outburst of the central continuum source of an active galactic nucleus (AGN) is different for radial, rotational, and turbulent motions (Pérez, Robinson, de la Fuente 1992, Welsh, Horne 1991). The detailed study of variations of the broad emission line profiles provides a powerful tool to investigate the structure and the kinematics of the central broad-line region (BLR), which cannot be directly resolved. Therefore, the investigation of the temporal evolution of broad emission line profiles with respect to continuum variations yields information for distinguishing between these possible types of the kinematics of the BLR gas. The analysis of the variability of emission line profiles requires spectra with high spectral resolution and high signal-to-noise ratio monitored over at least several months with a sampling rate of a few days.

2. Observations

The Seyfert galaxy NGC4593 was part of a sample of AGN which were monitored by the LAG (Lovers of Active Galaxies) collaboration at La Palma in 1990 (e.g. Dietrich, Kollatschny et al.,1993). LAG is an European collaboration including astronomers from England, France, Germany, Italy, Spain, and Sweden, among others.

Optical spectra of this galaxy were taken with the 4.2m William Herschel Telescope (WHT) and the 2.5m Isaac Newton Telescope (INT) at La Palma on the Canary Islands. From January until June, observations were obtained at 18 epochs for the Hα and 9 epochs for the Hβ emission line. The spectral resolution is 3 Å.

---

*Universitätssternwarte, Geismarlandstr.11,D-3400 Göttingen, F.R.G.

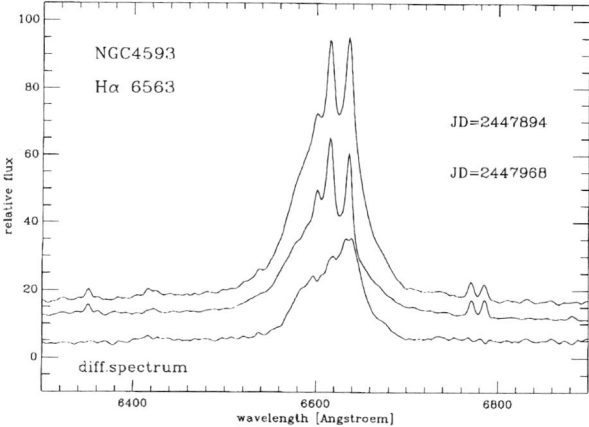

Figure 1. Hα spectra for two epochs of NGC4593 and the corresponding difference spectrum at the bottom (JD=2447894 and JD=2447968)

To investigate the broad line flux variations, the spectra have to be scaled to a uniform flux standard. For this purpose the narrow forbidden emission lines, which are constant on time scales of at least a couple of decades are used. The variations of the Hα spectral region in NGC4593 for two different epochs are shown in Fig. 1. At the bottom the corresponding Hα difference line profile for the two epochs is shown. The narrow lines vanish in the difference and only the variation of the broad emission line flux is visible.

3. Line Profile Variations in NGC4593

The variations of the Hα emission line are shown in Fig. 2 from the beginning of the campaign (Jan. 2nd) until the end (Jun. 6th). The integrated line flux declined within 70 days by nearly 60%. Furthermore, a strong short time scale variation occurred during this period. At the end of January (JD=2447916) the Hα flux diminished by 30% within 5 days.

The temporal resolution of the spectra was insufficient to resolve the broad-line region spatially. The auto-correlation function (ACF) of the sampling window and of the continuum are comparable. From the ACF of the Hα variability an upper limit of 25 light days was estimated for the size of the BLR of NGC4593.

NGC4593 showed strong profile variations during this monitoring campaign. The temporal evolution of the Hα line profile is presented in Fig. 3. The Hα spectra were scaled to the spectrum of epoch JD=2447968, since it represented the minimum state of the Hα line flux during this observing period. At least three components can be found in the broad line profile; besides the strong central component, blue and

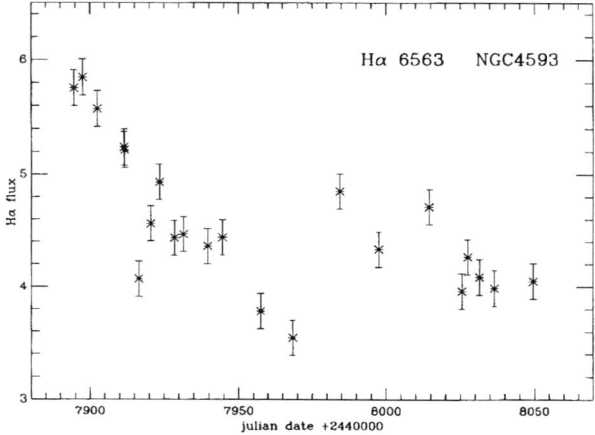

Figure 2. Variations of the integrated Hα line flux of NGC4593 from the beginning of January (JD=2447894) until the beginning of June (JD=2448049)

red shifted components are clearly visible. Furthermore, the blue component does not vary with the same amplitude as the red one. The full width at zero intensity (FWZI) of the broad Hα line profile remained constant while shape and intensity varied significantly. Fig. 3 shows the evolution of the Hα line profile for the first half of the declining phase at the beginning of the campaign and the variations of the profile for the declining flux during the second half of the monitoring campaign of NGC4593. In addition to the inner blue and red components, two further outer components with relative velocities of $\sim \pm 3000 \mathrm{kms}^{-1}$ can be seen (Dietrich, Kollatschny, et al.,1993). These outer components do not vary in phase with the inner components. Furthermore, the inner blue component does not vary in the same way as the red one.

4. Summary

Preliminary results of the LAG monitoring campaign for the Seyfert galaxy NGC4593 include the following:

- over time intervals of months the FWZI of the difference spectra remains constant; therefore, turbulent motion seems to be the dominant component of the velocity field;
- the broad-line profiles of NGC4593 consist of at least 3 components;
- these components are nearly symmetric with respect to the central component;
- single components have narrow, steep profiles; typical width is ~ 1000 km s^{-1};

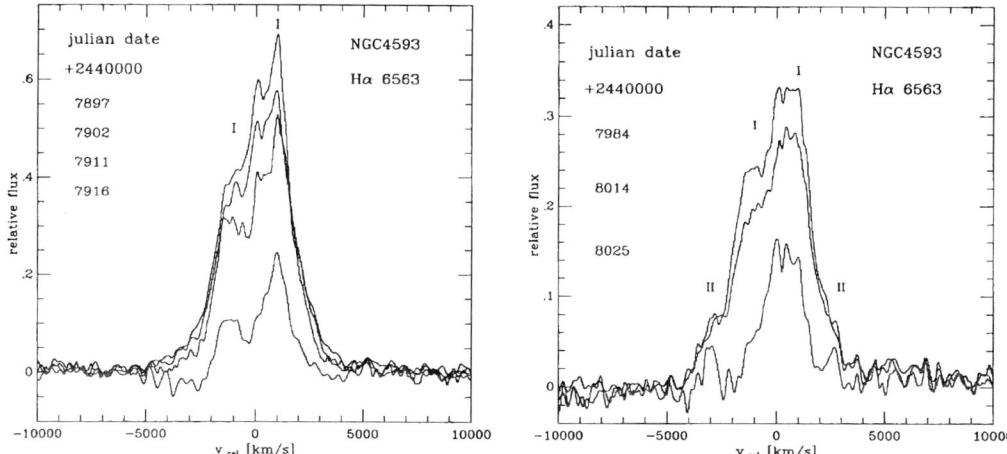

Figure 3. Hα difference profiles relative to the minimum state (JD=2447968) for 4 epochs at the beginning of the monitoring campaign of NGC4593 (left) and for three epochs during the second half of the campaign (right)

- the relative velocities of the components remain the same over several periods of increasing and decreasing continuum and line emission;
- the components do not vary with the same amplitude or phase; no clear correlation of the individual components with the continuum flux is apparent;
- because of the intensity variations of single components, simple accretion disk models to explain the line profile variations can be ruled out.
- the individual components show different and variable Balmer decrements; therefore, they have to originate in physically different regions

Acknowledgements. This work has been supported by BMFT grants Verbundforschung Astronomie FKZ05 5GO42A1, FKZ 50 OR 90045 and DFG grant Ko 857/13-1

References

Dietrich, M., Kollatschny, W., et al., Astron.&Astrophys., in press (1993)
Perez, E., Robinson, A., de la Fuente, L., MNRAS, **256**, 103 (1992)
Welsh, W.F., Horne, K., ApJ., **379**, 586 (1991)

Deconvolution of Variable Seyfert 1 Profiles

D. Michael Crenshaw [*]

Abstract

A technique is presented to deconvolve variable Seyfert 1 profiles that consist of components that differ in radial velocity coverage and time scale of variability.

1. Introduction

Variable structure in the emission-line profiles of AGN can in principle be used to map out the kinematics of the broad-line region (BLR). However, profile variations can also occur if the profile consists of more than one component and each component is variable in flux on a different time scale (Peterson et al. 1990). In order to use profile variations as a probe of the BLR, it is important to devise techniques to deconvolve components that may represent physically distinct emission-line regions. One possible technique will be outlined using data obtained from the 1989 campaign to monitor the Seyfert 1 galaxy NGC 5548 with the *International Ultraviolet Explorer* (IUE), as well as IUE spectra obtained by Webb and Crenshaw in April, 1990, when NGC 5548 was in a particularly low state.

Figure 1 shows that the profile of C IV $\lambda 1550$ is much narrower in the low state, suggesting that there is a strong narrow component that has not varied as much as the rest of the profile. In addition, it is clear that excess emission in the blue wing, which Peterson et al. (1990) claim is a distinct component that varies on a longer timescale than the majority of the broad-line flux, has nearly disappeared. A relatively simple technique has been developed to accurately deconvolve these components. This technique works if each component has a constant width and position, and varies only in intensity. In the case of NGC 5548, this assumption can be tested by comparing the results obtained from three different "events" during the 1989 campaign; during each event, the continuum and emission-line light curves experienced a large increase and decrease over a period of 50 to 100 days (Clavel et al. 1991).

---

[*] Computer Sciences Corporation, NASA/Goddard Space Flight Center, Code 681, Greenbelt, MD 20771.

2. Deconvolution

The deconvolution technique, as performed on the C IV profile of NGC 5548, is outlined below:

– Given three major components (a,b,c), where the time scale of variability (τ) in response to a continuum pulse is such that:

$$\tau_a \ll \tau_b \ll \tau_c$$

– For data obtained over short time intervals ($< \tau_b$) the profile flux at each wavelength can be separated into variable and constant parts:

$$f(\lambda, t) = a(t)f_a(\lambda) + f_{bc}(\lambda)$$

– The only component that contributes to the red wing is "a". If the integrated flux from 1560 to 1580 Å is $F_r(t)$ for each observation and F_{avg} for the average profile for the event, then

$$a(t) = F_r(t)/F_{avg}$$

– The variable (a) and constant (bc) components as a function of time can be determined by a linear regression fit to:

$$f(\lambda, t) = a(t)f_a(\lambda) + f_{bc}(\lambda)$$

– $f_a(\lambda)$ is the slope and $f_{bc}(\lambda)$ is the intercept of the fit.

– Fits to data from each of the three separate events yield "a" components that are very similar. An average "a" component can now be scaled to the red wing of each C IV profile and subtracted from each C IV profile in the data set.

– Only component "b" contributes to the blue wing at 1526 - 1540 Å. A linear regression fit is now performed for all of the IUE archive data:

$$f_{bc}(\lambda, t) = b(t)f_b(\lambda) + f_c(\lambda)$$

3. Results

The final components for the average C IV profile are given in Figure 2. Component "a" is very broad (9000 km sec^{-1}) and is symmetric around zero km sec^{-1}. It responds on a time scale of 8 - 16 days to sharp continuum changes (Clavel et al. 1991). Recent studies on the size, geometry, physical conditions, and kinematics of the BLR in NGC 5548 are characterizing this component. Component "b" is quite broad and is roughly symmetric, but its centroid is at about -2000 km sec^{-1} as compared to the narrow optical emission lines. A new result is that the blue component responds to continuum changes on a time scale of 100 - 300 days. Component "c" is very narrow and is centered on zero km sec^{-1}; it is constant over a 10 year period, and is apparently associated with the narrow components of the optical emission lines.

4. References

Clavel, J., et. al. 1991, ApJ, 366, 64

Crenshaw, D.M., and Blackwell, J.H., 1990, ApJ, 358, L37

Peterson, B.M., Reichert, G.A., Korista, K.T., and Wagner, R.M. 1990, ApJ, 352, 68

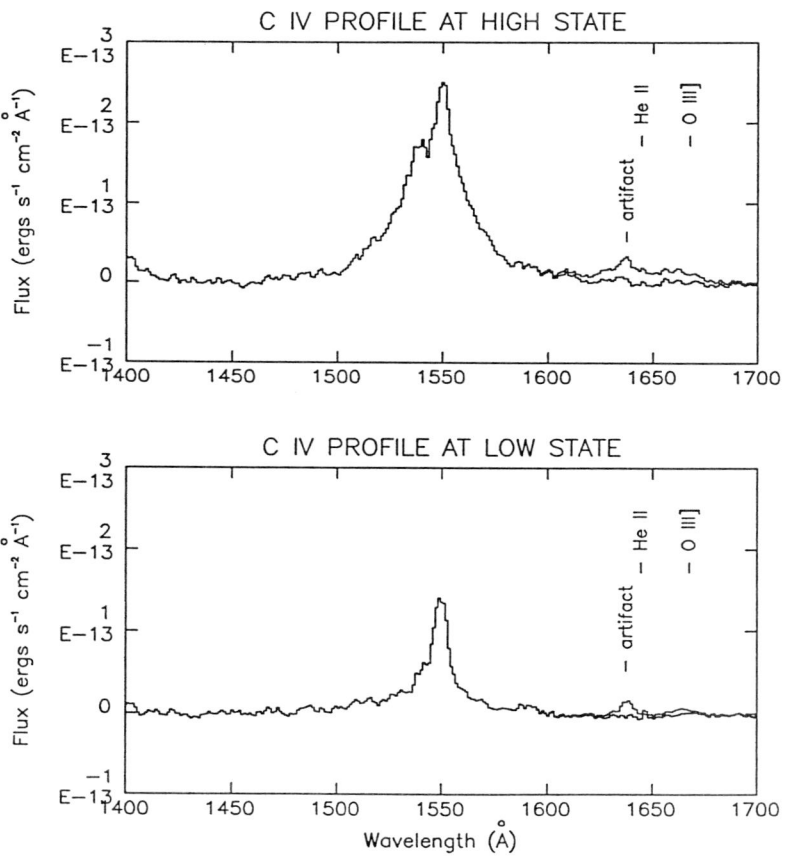

Figure 1. C IV profiles (original and deblended)

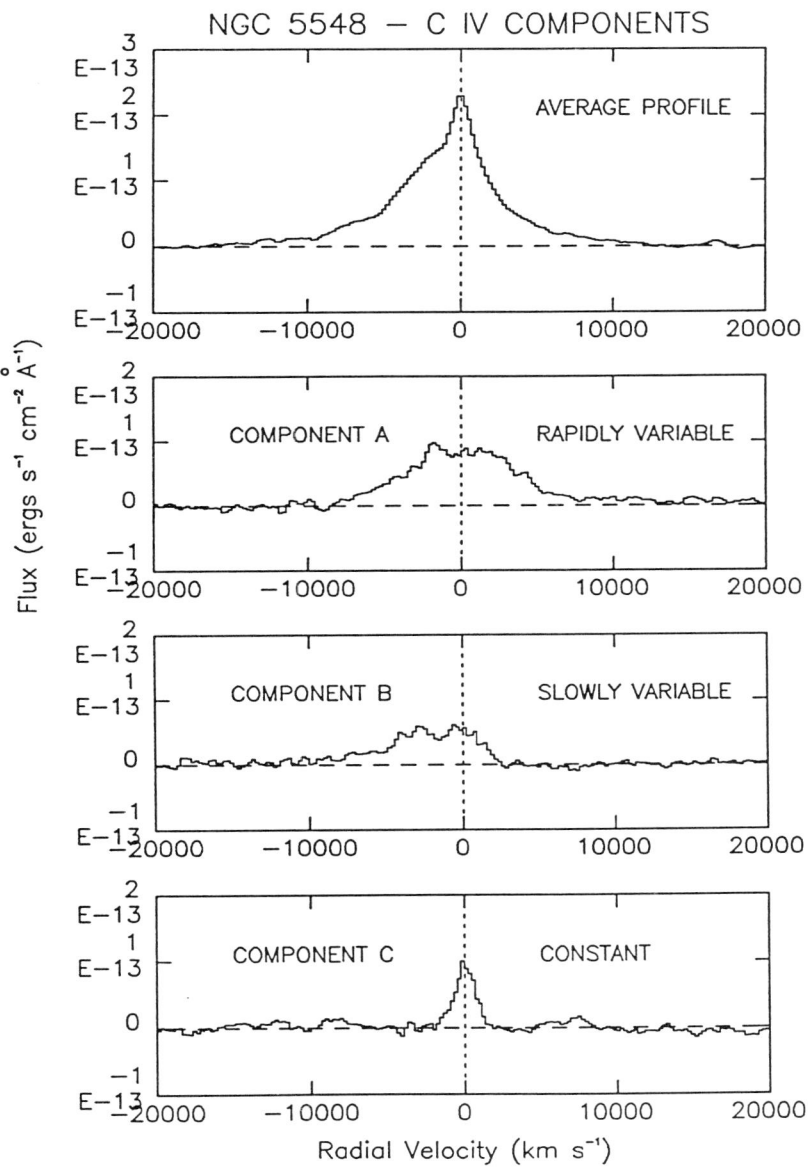

Figure 2. Final components for C IV

Ultraviolet Variability of AGN

Pedro M. Rodríguez–Pascual [*]

Abstract

IUE mean and standard deviation spectra are presented for three Seyfert 1 galaxies. This simple computational technique proves very useful to study the structure of broad emission lines profiles.

1. Introduction

Variability was shown to be an important property of most AGN in earlier studies of these objects. Later on, it was demonstrated it might be a useful tool to understand the physical conditions and geometry of the line emitting gas. Results presented here will be based on simple computations on three data–sets that fulfil some basic requirements: (a) homogeneity in the data, (b) flux variations much larger than typical measurement errors and (c) large number of measurements. IUE data fulfil these basic conditions for a number of AGN. We have selected three Seyfert 1 for which at least one "event", i.e. a turn–down or turn–up in the light curve, is observed with acceptable sampling. For these objects, a simple analysis has been performed on their IUE spectra. This analysis consists in computing the average and the standard deviation of the flux at each wavelength.

Mean spectrum: $f_\lambda = \frac{\sum_{i=1}^{N} f^i_\lambda}{N}$; σ spectrum: $\sigma_\lambda^2 = \frac{\sum_{i=1}^{N}(f^i_\lambda - f_\lambda)^2}{N}$

where f^i_λ is the flux of i-th spectrum at wavelength λ and N is the total number of spectra.

2. Results

2.1 NGC 4151

Although this Seyfert 1 galaxy has been monitored with IUE since early after launch, only the last run of observations in 1991 has been included in the analysis (Rodríguez–Pascual et al., 1993). The data quality is exceptionally high. In each half–shift at least two spectra in the short wavelength range (1150–1950Å) and one in the long wavelength (1950–3200Å) were taken. For each pair of spectra in the same range, one of them is optimally exposed in the continuum and faint lines (strong lines saturated) and the other is optimally exposed in the strong emission lines (continuum and faint lines underexposed). Combining both spectra results in a single spectrum with high signal-to-noise ratio in the continuum *and* the emission lines.

[*]ESA IUE Observatory, Villafranca del Castillo, Madrid, Spain

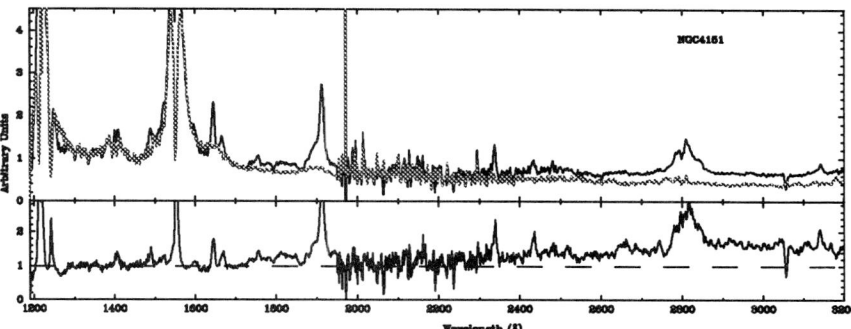

Figure 1. *Top*: Mean (*solid*) and σ_λ (*dotted*) spectra of NGC4151. *Bottom*: f_λ/σ_λ

A total of 22 data–points form the light curve, which spans 36 days. The duration of the "event" (defined as the time between a maximum and a minimum) is 20 days. The amplitude of the variations, characterized by $F_{var} = \frac{F_{max}-F_{min}}{F_{max}+F_{min}}$, is 0.54.

Figure 1 shows the mean and σ spectra normalized to their value at 1450Å (*top*). First result is that σ_λ is steeper than f_λ. This may be due to either a real change in the spectral shape of the continuum or the presence of a feature whose flux increases with wavelength in the IUE range and that varies much less than the continuum. Such a feature could be blended FeII lines and Balmer continuum emission, that have been proven to contribute significantly to the observed flux of some Seyfert 1 galaxies.

Line profiles in the σ spectrum are fairly symmetric. In CIV, HeII and SiIV the whole profile, except a narrow emission feature, varies with the same amplitude as the continuum. This is more clearly seen at the bottom of Fig.1, where the ratio f_λ/σ_λ is shown, again normalized to its value at 1450Å. Only a small fraction of the CIII] line varies. Moreover, the feature in the σ spectrum lies at a wavelength which might indicate it is SiIIIλ1892. MgII remains nearly constant, even though its profile is broad.

2.2 NGC 5548

During 1989, a campaign was carried out to monitor the variability of NGC5548. IUE spectra were taken every 4 days and three events occurred in the 240 days of campaign (Clavel et al., 1991). Although the amplitude and duration was different for each event, $F_{var} = 0.60$ and 60 days may be adopted as typical values.

As in the case of NGC4151, the σ spectrum is steeper than f_λ (Fig. 2). Previous studies of this galaxy showed an important contribution of blended FeII lines and Balmer continuum emission in the long wavelength range of IUE (Wamsteker et al., 1990).

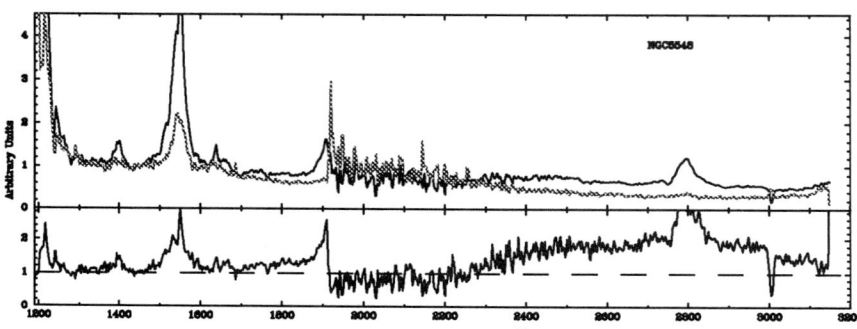

Figure 2. *Top*: Mean (*solid*) and σ_λ (*dotted*) spectra of NGC5548. *Bottom*: f_λ/σ_λ

The σ profile of CIVλ1550 is almost symmetric and blue shifted with respect to the peak of the mean profile. The f_λ/σ_λ ratio shows a narrow core and the far wings of the line vary much less than the continuum. These results suggest the presence of, at least, three components in the CIVλ1550 line: a narrow core, a broad variable component and a very broad constant component. Both broad features are blue shifted with respect to the narrow one.

In the rest of the emission lines, the variations are rather small. Only HeII and Lyα show indications of variability, although the later one is strongly contaminated by geocoronal emission.

2.3 Fairall 9

IUE observations of Fairall 9 are described by Recondo-González *et al.* (1992). The UV light curve of this Seyfert 1 galaxy shows a decrease of a factor 30 in 10 years, then an increase over 3 years, followed by a "plateau" region, with some evidence of faster variations. For the whole data-set, $F_{var}=0.92$ and $N=80$.

As in the previous cases, the σ_λ spectrum is steeper than f_λ. However, the difference is smaller, suggesting either the contribution of FeII lines and Balmer continuum is small, or they vary with large amplitude.

The broad component of emission lines varies with the same amplitude as the continuum. The only exception is CIVλ1550, where the blue wing seems to vary with a smaller amplitude. Once again, the f_λ/σ_λ ratio shows mostly narrow line profiles.

3. Discussion

First result is the hardening of the far UV spectrum as the continuum increases, as is shown by a σ spectrum steeper than f_λ. It is suggested that the contribution of blended FeII lines and Balmer continuum emission in the long wavelength range of

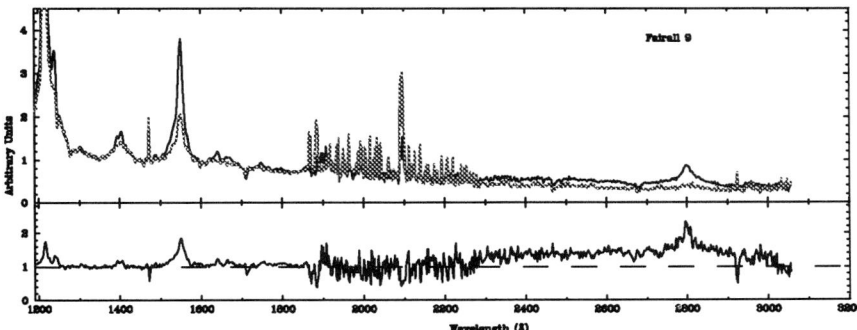

Figure 3. *Top*: Mean (*solid*) and σ_λ (*dotted*) spectra of Fairall 9. *Bottom*: f_λ/σ_λ

IUE causes the observed change in the continuum, but an intrinsic hardening cannot be ruled out.

In contrast to the continuum, the emission lines behave differently from object to object. The only common result is that the σ_λ line spectrum is different to the f_λ line spectrum; the σ_λ profiles do not match the f_λ profiles and the lines ratios change dramatically. Indeed, some strong broad lines in f_λ may disappear in σ_λ. The f_λ/σ_λ ratio appears as a good indicator of non–variable features. It always shows narrow prominent components in all the broad lines. In some cases, broad substructure is also seen, indicating the presence of several components in the broad lines.

Line profile variations in a single line plus different variations in different lines could be explained in terms of several broad line sub–regions with different kinematics and physical conditions. However, more complex scenarios than gas regions photoionized by a single central continuum source must be explored.

References

Clavel, J. et al., 1991, *Ap.J.*, **336**, 64.
Recondo–González M.C., Wamsteker W., Cheng F., Clavel J., 1992, *these Proceedings*
Rodríguez–Pascual P.M., et al., 1992, in preparation
Wamsteker W., Rodríguez P.M., Wills B.J., Netzer H., Wills D., Gilmozzi R., Barylak M., Talavera A., Maoz D., Barr P., Heck A., 1990, *Ap.J.*, **354**, 446.

Broad-Line Variations in NGC5548

W. Kollatschny and M. Dietrich *

The Seyfert 1 galaxy NGC5548 has been monitored between 1988 December and 1989 October in the optical by an international collaboration (Peterson et al., 1991). Parallel to the optical monitoring, this galaxy was observed every 4 days with the IUE-satellite from Dec. to Aug. (Clavel et al., 1991). The internal calibration of the spectra was done by scaling with respect to narrow forbidden lines.

The optical emission lines Hα, Hβ, HeI5876, and HeII4686 show the same variability pattern as the UV continuum, the Lyα, and CIV1548 lines, but with different delays for the various lines (Fig. 1a — d) (Dietrich, Kollatschny, et al., 1992). We estimated the extent of the broad line region for different lines by cross-correlating the UV-continuum and the emission lines. For the Hα and Hβ lines we got lags of 17 and 19 days respectively. The helium lines originate closer to the central continuum source, as indicated by delays of 7 days (HeII4686) and 11 days (HeI5876).

In Figure 2a,b we have plotted the temporal evolution of the difference spectra of Hα and Hβ for several epochs with respect to that epoch when the galaxy was in the minimum state. The full width at zero intensity (FWZI) remained constant. This indicates a turbulent velocity field in the broad-line region of NGC5548. Furthermore, a blue component ($v_{blue} \approx -2000 kms^{-1}$) is visible in addition to the central component. This blue component varies independently and with a different amplitude than the central component (Kollatschny and Dietrich, 1991). Therefore, it has to originate in a physically different region.

Acknowledgements. This work has been partly supported by BMFT grant Verbundforschung Astronomie FKZ 05 5GO42A1 and DFG grant Ko 857/13-1.

References

Clavel, J., et al., ApJ., **366**, 64 (1991)
Dietrich, M., Kollatschny, W., Peterson, B.M., et al., ApJ., in press (1992)
Peterson, B.M., et al., ApJ., **368**, 119, (1991)
Kollatschny, W. and Dietrich, M., Proc. of the Workshop 'Variability of Active Galaxies', Duschl et al., eds. Springer Verlag, p.57, (1991)

---

*Universitätssternwarte, Geismarlandstr. 11, D-3400 Göttingen, F.R.G.

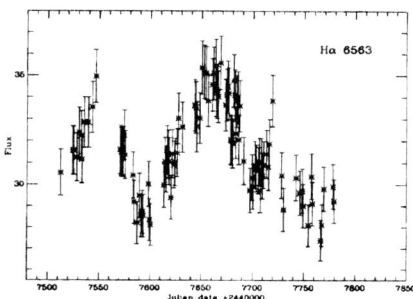

Fig. 1a Hα light curve in units of [10^{-13} erg s^{-1} cm^{-2}]

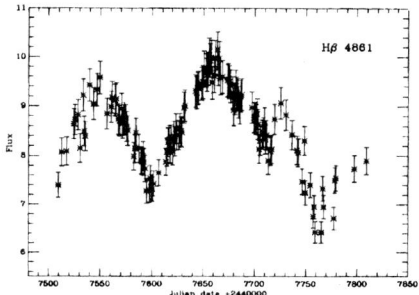

Fig. 1b Hβ light curve in units of [10^{-13} erg s^{-1} cm^{-2}]

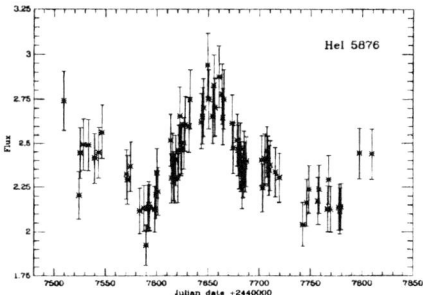

Fig. 1c HeI5876 light curve in units of [10^{-13} erg s^{-1} cm^{-2}]

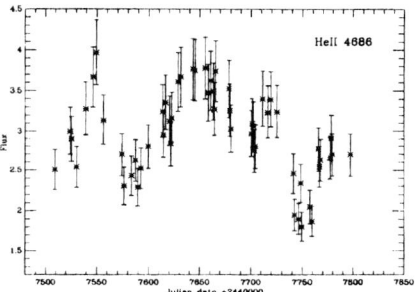

Fig. 1d HeII4686 light curve in units of [10^{-13} erg s^{-1} cm^{-2}]

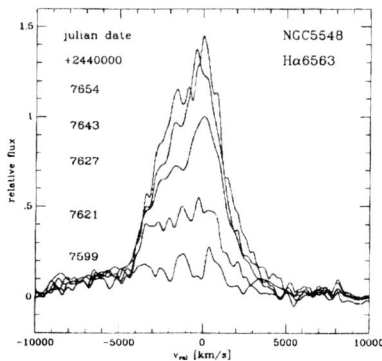

Fig. 2a Temporal evolution of Hα difference spectra with respect to epoch JD=2447749

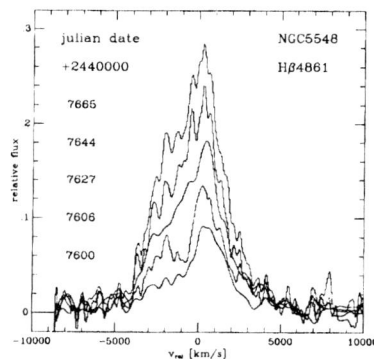

Fig. 2b Temporal evolution of Hβ difference spectra with respect to epoch JD=2447749

NGC 4593: A Low Luminosity Compact Seyfert 1 Nucleus

María Santos-Lleó [*][†]   Jean Clavel & Paul Barr [‡]
Ian S. Glass [§]

Abstract

We present the results of a monitoring campaign of the low luminosity Seyfert 1 NGC 4593, at X-ray, ultraviolet, optical and near IR frequencies. The rapid and large amplitude fluctuations detected set interesting upper limits to the sizes of the various emitting regions. We investigate the properties of the nuclear energy distribution, its pattern of variability and the relationship between the emission processes at different frequencies. We also study the variations of the broad emission lines in relation to those of the continuum. The implications of our data are discussed. It is found that they are consistent with a very hot accretion disk illuminated by the hard X-ray source.

1. Observations

Repeated observations of the nucleus of NGC 4593 have been performed from the near IR to X-rays. Though the temporal density of our observations is poor, they have the advantage of a wide wavelength coverage. Besides, NGC 4593 turned out to be highly variable at all frequencies.

2. Results and implications

The strong and rapid variations detected in all wavebands imply that the continuum source is unusually compact: ≤ 1.1 lt-hr in the hard X-rays, ≤ 3 lt-hrs in the soft X-rays, 70 lt-hrs at 1450 Å, and 37 lt-days at 2.5 microns. The latter corresponds roughly to the dust evaporation radius (~ 33 lt-d, Barvainis, 1987) in such a low luminosity source. Such an IR behaviour is reminiscent to that of Fairall 9, a more luminous Seyfert 1, (Clavel et al., 1989). After correcting for reddening and careful removal of "parasitic" emission (mainly stellar, plus blended FeII lines and BLR

---

[*] ESA IUE Observatory, P.O. Box 50727, 28080 – Madrid, Spain.
[†] Observatoire de Paris, 92195 Meudon Principal Cedex, France
[‡] ISO Observatory, Code SAI, ESTEC, Postbus 299, 2200 AG Noordwijk, The Netherlands. Affiliated to the Astrophysics Division Space Sciences Department.
[§] South African Astronomical Observatory, P.O. Box 9, Observatory, 7935 South Africa.

Balmer continuum), the energy distribution from 1.2 μ to 1330 Å obeys a power-law whose index ($\alpha = -0.85 \pm 0.05$) is significantly steeper than that of the average Seyfert or QSO, while both the IR-to-X-ray and the hard X-ray slopes are "normal". This implies that the "big-bump" in NGC 4593 is either absent or shifted to shorter wavelengths. The hard X-rays, UV and optical, are all linearly correlated (Fig. 1). Moreover, any time delay between the UV and optical band is less than 2 days. These results cannot be accommodated in the framework of the standard geometrically thin accretion disk model (Courvoisier and Clavel, 1991; Clavel et al., 1992).

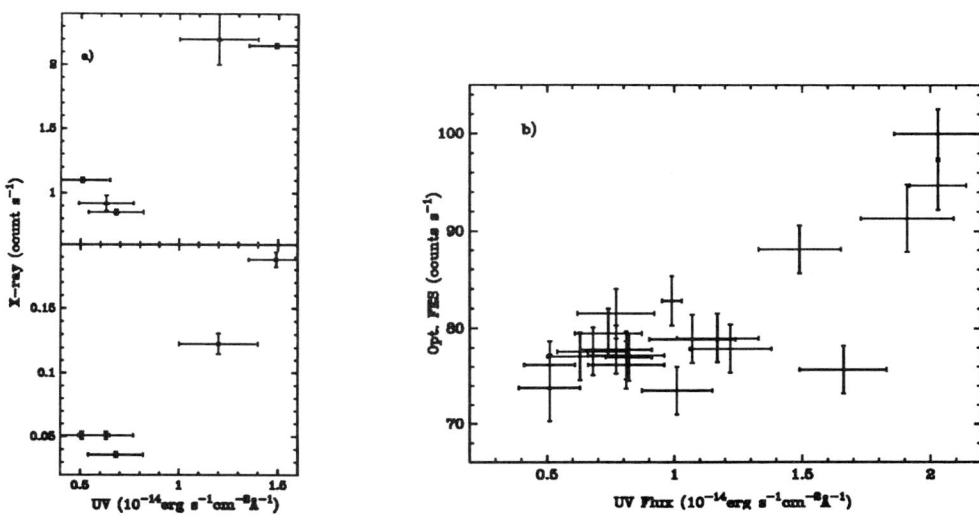

Figure 1. a) X-rays vs UV flux at $\lambda_{obs} = 1447$ Å (top hard X-rays, bottom soft X-rays); b) Optical FES counts, from the IUE satellite, vs UV flux.

Most of the emission lines are variable and their variations are correlated with those of the ultraviolet continuum (Fig. 2). The higher ionization lines are broader, and vary more rapidly than features from lower ionization species. The wings of both CIVλ1549 and Ly$\alpha\lambda$1216 vary more rapidly and with a larger amplitude than the line cores. These results are consistent with a stratified picture of the BLR, where the higher velocity gas is ionized by and lies closer to the central continuum source.

The absence of a "big bump" in the far UV together with the detection of a soft X-ray excess can be explained in the accretion disk scenario, given the properties of the nuclear emission in NGC 4593. The relatively narrow and rapidly variable emission lines imply a small nuclear mass ($2\,10^6 M_\odot$), consistent with its low bolometric luminosity ($5\,10^{43}\,erg\,s^{-1}$). This small central mass allows the existence of a close by and therefore very hot accretion disk emitting in the soft X-rays. The UV-optical behaviour suggests that most of this flux is reprocessed hard X-rays radiated by a very compact source that illuminates the disk (Zdziarski et al., 1990). The near IR is probably thermal reprocessing of the UV flux by hot dust grains.

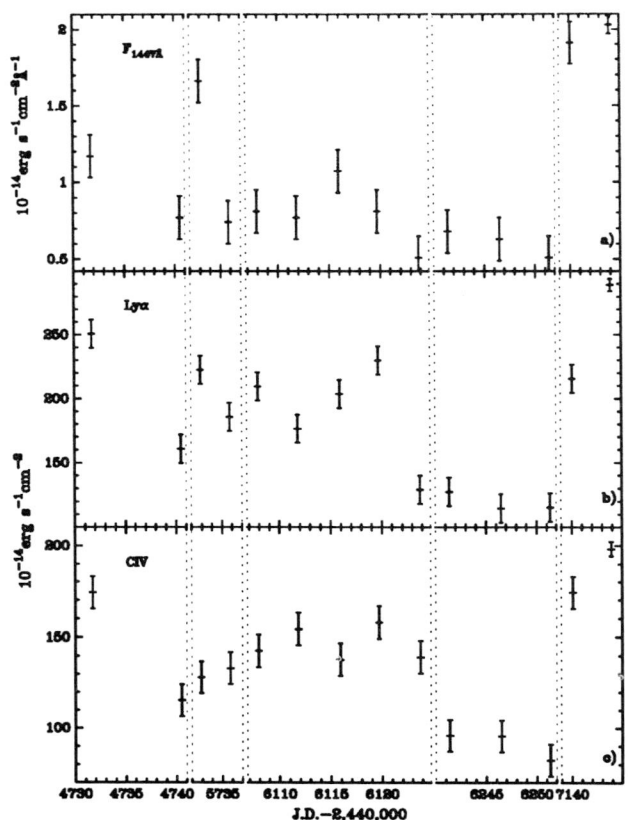

Figure 2. UV continuum (a), Ly$\alpha\lambda$1216 (b) and CIVλ1549 (c) light curves. Only those epochs when consecutive observations are taken within a few days are shown.

References

Barvainis, R. 1987, *Astrophys.J.*, **320**, 537.
Clavel, J., Wamsteker, W., Glass, I.S. 1989, *Astrophys.J.*, **337**, 236.
Clavel, J., Nandra, K., Makino, F., Pounds, K., Reichert, G.A., Urry, C.M., Wamsteker, W., Peracaula-Bosch, M., Stewart, G.C., Otani, C., 1992, *Astrophys.J.*, **393**, 113.
Courvoisier, T.J.-L. and Clavel, J. 1991, *Astron. Astrophys.*, **248**, 389.
Zdziarski, A.A., Ghisellini, G., George, I.M., Svensson, R., Fabian, A.C., Done, C., 1990, *Astrophys.J.*, **363**, L1.

UV Continuum Origin and BLR Structure in F-9

M.C. Recondo-González *   W. Wamsteker *   F. Cheng [†]
J. Clavel [‡]

Abstract

Thirteen years of IUE observations of the Seyfert 1 galaxy Fairall-9 have shown a large variability in the UV continuum and UV emission lines, $Ly\alpha\lambda 1216$ and $CIV\lambda 1550$. The relation between UV continuum and X-ray (no delay) is similar to that found for lower luminosity objects and suggests that X-ray reprocessing causes the UV continuum. We also use the line variability to support and refine a gaussian decomposition of the profiles of $Ly\alpha$ and CIV, based on a previous study of $H\beta$. The decomposition of $Ly\alpha$ and CIV shows a well identified component structure valid for both lines. The components respond very differently to the changes in the ionizing continuum brightness, confirming the different physical nature of the material associated with them.

1. The UV-X ray continuum relation

We show in figure 1 (1a and 1b) the UV and X-ray light curves of F-9, while fig. 1c and 1d show relations between these two quantities. Although the time resolution in the CCF (1d) is of course limited by the sparse X-ray sampling no evidence is found for delays of the size suggested by the IR (400 days). We see here also that at high UV brightness the rather tight correlation between the 2-10 keV flux and F(1338Å) breaks down. Although the details of this behaviour are currently not fully understood it has been suggested that at low levels the UV continuum is the result of reprocessing of X-rays emitted above the disk while the huge UV variations could be associated with major accretion events.

2. The UV lines: model and lags

The decomposition of UV lines into gaussian components is based on the results of $H\beta$ and on the difference method. The lines show a well identified component structure which is valid for the variable profiles of both $Ly\alpha$ and CIV at all epochs.

---

*ESA IUE Observatory, Apartado 50727, 28080 Madrid, Spain
[†]Center for Astrophysics (USTC), Hefei, Peoples Rep. China.
[‡]ISO Observatory, Code SAI, ESTEC, Postbus 299, 2200 AG Noordwijk, The Netherlands.

This structure is: I. The narrow line with 1700 kms^{-1} (FWHM) and II. The broad line composed of three components:

Red   : 5800 $km\,s^{-1}$ (FWHM) at $\Delta v \sim$ 3100 $km\,s^{-1}$
Central : 3600 $km\,s^{-1}$ (FWHM) at $\Delta v \sim$ 0 $km\,s^{-1}$
Blue   : 6200 $km\,s^{-1}$ (FWHM) at $\Delta v \sim$ -3900 $km\,s^{-1}$

The light curves of the broad line components (central, red and blue), their strong dependence on the UV continuum and the various CFs with the continuum, confirm that the gas is photoionized by the UV continuum, but the physical nature of the material associated with each component is different (for example, the blue variations are smaller than the other ones and the ratio Lyα/CIV for the red component is higher than the other ones). The Correlation Functions (fig. 2 for Lyα) give delays of $\sim$450, 250 and 0 light days for the central, blue and red components, respectively. The delay of the central component suggests a close relation with the dust found at 400 light days by Clavel et al. (1989).

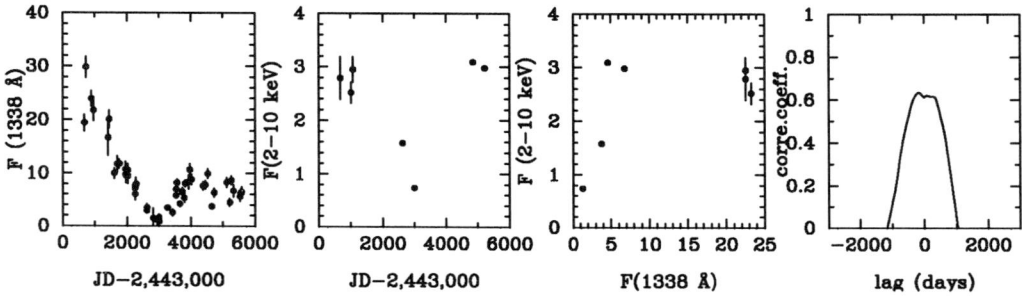

Figure 1. (a) UV and (b) X-ray variability, (c) X rays-UV correlation, (d) CCF.

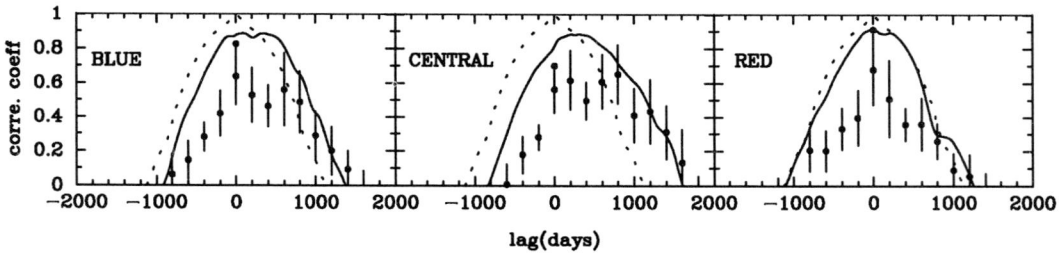

Figure 2. CCF and DCF of the line components with the UV continuum for Lyα. The dotted line shows the continuum ACF for reference.

References

Clavel J., Wamsteker W., Glass I.S., *1989, Ap. J., 337*, p.236.

UV Emission Line Intensities and Variability: A Self-Consistent Model for Broad-Line-Emitting Gas in NGC 3783

Anuradha Koratkar [*]   *Gordon M. MacAlpine* [†]

Abstract

Short-wavelength IUE archival data for NGC 3783 were re-extracted and analysed to constrain numerical modelling parameters in detailed photoionization analyses. The He IIλ1640, C IVλ1549, and C III]λ1909 line intensities and trends can be reasonably well reproduced by a two cloud component model. In order to produce satisfactory line intensities or trends for other higher ionization lines and lower ionization lines, still more gas components are necessary with gas density ranging from roughly 10^{11} cm^{-3} to 10^9 cm^{-3} or less. In going from the inner to outer clouds, the optical depth increases and the gas density decreases approximately as r^{-2}.

The amount of dust obscuration along the line of sight, as required by the models, is consistent with reddening estimates from HeII line ratios, and CNO abundance ratios derived from intercombination line intensities suggest abundances of carbon, nitrogen and oxygen lower by a factor of about 2 relative to solar. The HeII line rest equivalent widths from the models suggest a gas covering factor of order 0.25.

1. Introduction

Although sophisticated numerical photoionization models have been employed for two decades in investigations of the broad–line–emitting regions (BLRs) of Seyfert 1 galaxies, some of the most fundamental BLR properties such as the intrinsic ionizing radiation field, the ionization parameter, emitting gas density, chemical abundances, and the "covering factor" (the fraction of solid angle occupied by optically thick gas around the ionizing radiation source) are still open to question. In the present work, we use recent estimates for the geometry of line-emitting clouds (Koratkar and Gaskell 1991) and new UV line intensity measurements in efforts to constrain photoionization models for NGC 3783.

For NGC 3783, 29 spectra were found to be suitable for measuring the weak lines of He IIλ1640 and O III]λ1663, along with C III]λ1909 and C IVλ1549 in the IUE

---

[*]Space Telescope Science Institute, 3700 San Martin Drive, Baltimore, MD 21218, U.S.A
[†]Department of Astronomy, University of Michigan, Ann Arbor, MI 41809, U.S.A.

archives. Unfortunately, NGC 3783 has a relatively low redshift so Ly$\alpha\lambda$1216 is significantly blended with geocoronal emission and could not be measured. To investigate variability of weaker lines while retaining reasonable S/N, the spectra were coadded into different groups. In the figures the individual measurements are represented by their quality (Q) or as coadded measurements.

2. Variability Trends

From the data, we were able to examine important trends which constrain BLR gas and ionizing radiation field characteristics in NGC 3783. These trends are discussed below and illustrated in Figures 1-5.

Fig. 1 – HeIIλ1640 rest frame equivalent widths: The measured HeII rest equivalent widths decrease somewhat with increasing λ1335 continuum flux suggesting that, as the continuum at λ1640 is increasing, the ionizing radiation flux in the clouds increases somewhat more below 4 Ryd than it does above this He$^+$ ionization threshold.

Fig. 2 – HeIIλ1640 and CIVλ1549: The near simultaneous increase in both HeII and CIV emission with rising continuum flux, suggests that the ionizing continuum does not undergo dramatic changes in spectral shape as it varies. Large changes in energetics of the ionizing radiation field would be expected to produce different effects on recombination and collisionally excited emission.

Fig. 3 – Ly$\alpha\lambda$1216/CIVλ1549: Since Lyα in NGC 3783 was not measured, Lyα/CIV data for other Seyfert 1 galaxies (see Koratkar and Gaskell 1991) were used to investigate this trend. As illustrated, globally this ratio is flat or slightly increasing with increasing continuum flux at λ1335; but for each individual galaxy the trend can be anywhere from flat to slightly steeper than the global slope, or even marginally decreasing. For a fixed or only moderately changing ionizing continuum shape, an increase in luminosity implies an increase in U, the ionization parameter. For U near the canonical value of 10^{-2}, photoionization models predict that the Lyα/CIV ratio should decrease in going to higher luminosity. The observed flat or increasing trends may indicate larger effective values for U in some cases.

Fig. 4 – HeIIλ1640/CIVλ1549 : The HeII/CIV ratio in NGC 3783 exhibits a mild increase with rising λ1335 continuum flux. This is another indication of U significantly higher than the canonical value.

Fig. 5 – CIII]λ1909/CIVλ1549: This ratio is characterized by a marginal decrease with increasing continuum at λ1335. Again, for a fixed or only moderately changing continuum spectral shape and U of roughly 10^{-2}, the models predict a more marked line ratio decrease for increasing continuum luminosity. The lack of strong correlation in our data suggests a value for U near but somewhat above 10^{-2}.

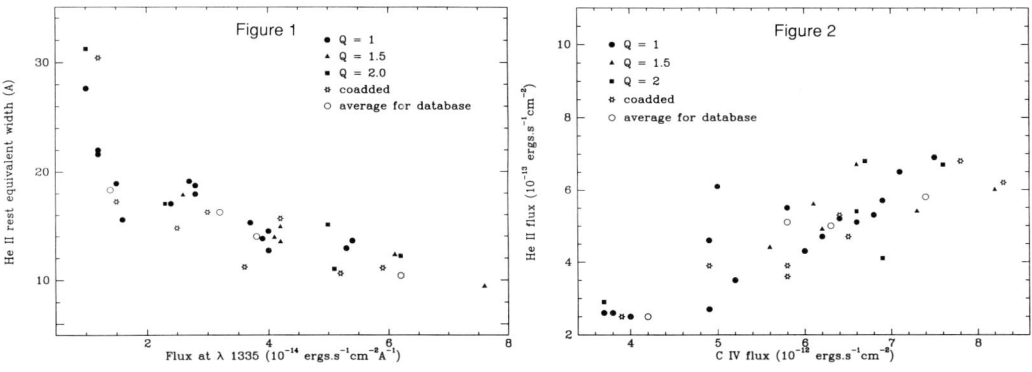

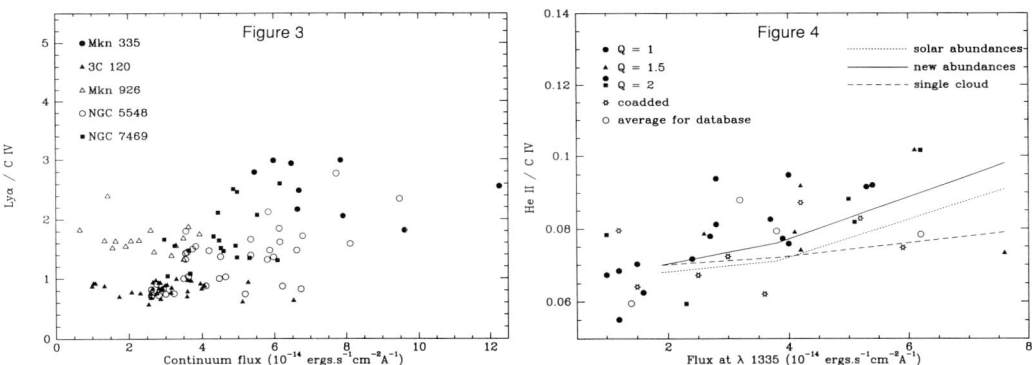

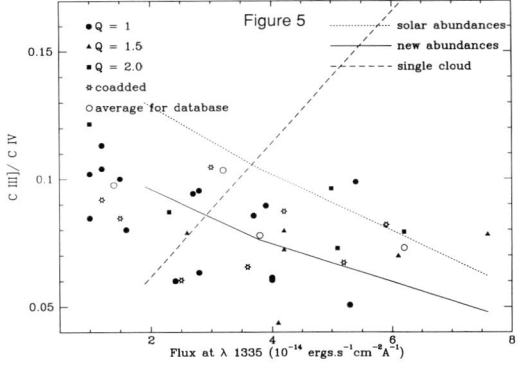

3. Results

The above trends and the measured line intensities were used to develop a self-consistent model for the BLR in NGC 3783. For further details see Koratkar and MacAlpine 1992. A single-component model with clouds at 24 lt. days produced a fair fit to most measured line intensities in an averaged spectrum, but could not reproduce the line variability *trends* with varying ionizing radiation flux (see Figs. 4 and 5). All the data for He IIλ1640, C IVλ1549, and C III]λ1909 can be reasonably well reproduced by two cloud components. One component has a source-cloud distance of 24 light days, gas density around 3×10^{10} cm^{-3}, ionization parameter range (as the continuum varies) of 4×10^{-2} to 2×10^{-1}, and cloud thickness such that $\approx 50\%$ of the carbon is doubly ionized and $\approx 50\%$ is triply ionized. The other component is located approximately 96 light days from the source, is shielded from the source by the inner cloud, has a density around 3×10^9 cm^{-3}, is characterized by an ionization parameter range 10^{-3} to 3×10^{-2}, and the cloud thickness was such that $\approx 45\%$ carbon is doubly ionized and $\approx 55\%$ is singly ionized.

In order to produce satisfactory line intensities or trends for O VIλ1034, N Vλ1240, Mg IIλ2800, and possibly Ly$\alpha\lambda$1216, still more gas components are necessary, located at both smaller and larger distances from the ionizing radiation source. The overall picture that emerges from this investigation involves clouds distributed over a range of distances from a central radiation source, from probably 10 light days or less to more than 100 light days. In going from the inner to outer clouds, the optical depth increases and the gas density decreases approximately as r^{-2} from roughly 10^{11} to 10^9 cm^{-3} or less. Effective cloud ionization parameters cover a large range from at least several times 10^{-1} to several times 10^{-4} (including allowance for continuum variability).

The amount of dust obscuration along the line of sight, as required by the models, is consistent with reddening estimates from HeII line ratios, and CNO abundances derived from intercombination line intensities are low, but within a factor of two of their solar values. The HeII line rest equivalent widths from the models suggest a gas covering factor of order 0.25.

Acknowledgements. This work was supported by NASA grant NAG5-1198 to the University of Michigan.

References

Koratkar, A. P., & Gaskell, C. M., 1991, ApJS, 75, 719
Koratkar, A. P., & MacAlpine, G. M., 1992, ApJ, 410, 000

Non-linear Anisotropic BLR Models

P.T. O'Brien & M.R. Goad *

Abstract

We have devised a BLR modelling code, PROSYN, which computes the non-linear anisotropic line response to a variable ionizing continuum. PROSYN computes the line response for a spatially extended BLR taking into account the variation in the physical properties of individual clouds with distance from the central ionizing continuum source. We discuss the result of applying a particular constant density model to the Seyfert 1 galaxy NGC5548, and compare the model and observed light-curves of Lyα and MgIIλ2798. A non-linear anisotropic model fits the observations significantly better than a linear isotropic model.

1. Non-linear line response

The variability of the broad emission lines in AGN is believed to be due to the response of the line emitting gas to variations in the ionizing continuum flux. It is convenient to think of the line variations $L(v,t)$ in terms of a convolution of the continuum variations, $C(t)$, with a 'response function', $\Psi(v,\tau)$,

$$L(v,t) = \int_0^\infty C(t-\tau)\Psi(v,\tau)d\tau, \qquad (1)$$

where v represents velocity, t time and τ time-delay (Blandford & McKee 1982). If we assume that the cloud density does not change significantly during a continuum 'event', then the variations in continuum level result only in changes in ionization parameter (U). For optically thick clouds over a small range in variations in U it is a reasonable approximation to say that the line emissivity *changes* linearly with *changes* in the continuum level. In this approximation the response function is then weighted by the *'responsivity'* of the gas rather than the emissivity. This linear approximation works best for recombination lines whose strength is basically proportional to the flux of ionizing photons. However, if U changes by a large enough factor the linear approximation will break down. For example, at large U some lines may saturate, whilst others may decrease in intensity due to their parent ionic species being depleted as a result of ionization or the removal of the 'partially ionized' zone. These non-linear effects, together with anisotropic line emission, must be taken into account when comparing BLR models with observations (Netzer 1990; Sparke 1993; Goad, O'Brien & Gondhalekar 1993).

---

*Department of Physics and Astronomy, UCL, Gower Street, London, WC1E 6BT

2. NGC5548

Adopting the continuum luminosity and density for NGC5548 given by Ferland *et al.* (1992), the spherical constant density BLR model discussed in Goad & O'Brien (this volume) scales to the correct radial scale for NGC5548 within a factor of a few. Thus as a test case we used this model to compute Lyα and MgII light-curves for comparison with the observations of NGC5548 given by Clavel *et al.* (1991). Two different models were computed with different assumptions about how the gas responds to continuum variations:

model 1: The clouds radiate isotropically and respond linearly to changes in the ionizing continuum. For a given change in continuum level all the clouds are assumed to vary their line emission by *exactly* the same factor and in the same sense. Thus the response functions for this model are *emissivity weighted* in accordance with the initial steady-state conditions, and this weighting is presumed time-invariant. The centroids of the response functions for Lyα and MgII in this model are 8.7 and 20.0 light-days respectively.

model 2: The clouds radiate anisotropically and respond non-linearly to changes in the ionizing continuum. The time-dependent line emissivity and anisotropy are computed explicitly based on the local physical conditions using a photoionization code. The adopted form of the line anisotropy is given by:

$$\varepsilon_{obs} = \frac{\varepsilon_{in}}{2}(1 - cos\theta) + \frac{\varepsilon_{out}}{2}(1 + cos\theta), \qquad (2)$$

where ε_{obs} is the observed emissivity, ε_{in} is the emissivity of the inward cloud face (towards the ionizing continuum source), ε_{out} is the outward emissivity (away from the ionizing continuum source), and θ is the angle between the cloud radius vector and the observers line of sight. The variation in $\varepsilon_{totl}(=\varepsilon_{in} + \varepsilon_{out})$ as a function of U, and $\varepsilon_{in}/\varepsilon_{totl}$ as a function of radius, r, for Lyα and MgII for this model are shown in Goad & O'Brien.

In Fig 1(a,b) we show the observed continuum light-curve for NGC5548 (solid lines), the observed line light-curve (dashed lines), and the linear, isotropic model 1 light-curve (dotted lines) for Lyα and MgII respectively. The model light-curves were computed by convolving the emissivity weighted response functions with the continuum light curve using equation 1. The light-curves are only shown for epochs beyond twice the light crossing time of the outer radius of the BLR model ($R_{out} = 56.2$ light-days), where day 1 corresponds to the start of the observed UV continuum light-curve (JD = 2,447510.139). The response timescale is reasonably well-fitted for Lyα, but for both lines the model 1 light-curves have too large a response amplitude compared to the observations.

As noted by Krolik *et al.* (1991), the lower amplitude of the Lyα response compared to that of the UV continuum at 1300 Å could be due to a lower continuum variability

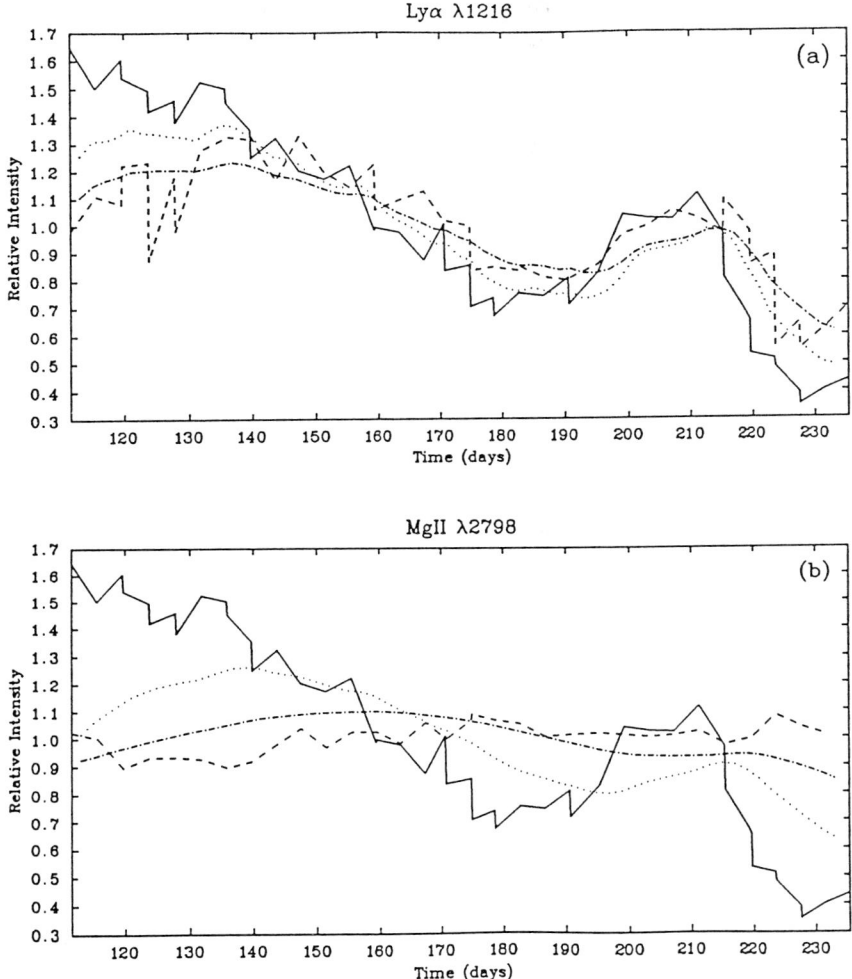

Figure 1. (a) The observed continuum light-curve of NGC5548 (solid line), the observed Lyα light-curve (dashed line), the linear isotropic model 1 Lyα light-curve (dotted line), and the non-linear anisotropic model 2 light-curve (dot-dashed line); (b) Same for MgIIλ2798.

amplitude shortward of the Lyman limit. This seems unlikely as the observed continuum is only a few hundred angstroms longward of the Lyman limit. If anything the observed hardening of the UV continuum with increasing continuum level implies that the shorter-wavelength continuum variability amplitude may be underestimated. An alternative explanation is that not all of the broad line gas responds in the same manner when the continuum changes. To allow for such effects we must abandon the assumption of linear response used in model 1.

In Fig 1(a,b) we show the non-linear, anisotropic model 2 light-curves (dot-dashed lines), computed directly from the observed line emissivities. (*i.e.* equation 1 was *not* used). These light-curves fit the observations significantly better than the model 1 light-curves for both lines in terms of time delay and response amplitude. The improved fit is due in part to allowance for anisotropic line emission, but is mainly due to the non-linear response. The innermost BLR clouds ($r < 1$ light-day) are optically thin to the Lyman continuum in this model (Goad & O'Brien). As the continuum level increases (larger U) these clouds saturate in Lyα, reducing the overall increase in line flux relative to that in model 1. The reduction in line response amplitude is even stronger for MgII, as the emissivity of the innermost clouds actually decreases for this line as the continuum level increases.

Although the gas responds non-linearly to continuum changes in model 2, at each epoch we can define an *'instantaneous' responsivity weighted* response function, where the weighting is based on the derivative of the line emissivity with respect to the continuum level at that epoch. Comparison of these functions illustrates the variation in responsivity of the gas as a function of continuum level. The optically thin gas in the innermost BLR causes the instantaneous response functions for both lines to have smaller amplitudes and larger centroids when the continuum level is high. The amplitude is actually negative for MgII at some epochs for very small time delays. This is possible for lines for which the emissivity can decrease with increasing continuum flux, and must be allowed for when recovering response functions from observations.

Although a better fit than model 1, the model 2 response amplitude is still too high compared to the observations. This may be because the NGC5548 BLR responds more non-linearly than model 2 (*e.g.* a larger fraction of the gas may be optically thin), or some of the line emission may come from another gas component. Another possibility is that the assumption made here of a constant continuum shape may be incorrect. To discriminate between these possibilities requires further modelling, including more lines.

References

Blandford, R.D. & McKee, C.F., 1982, Ap.J., 255, 419
Clavel, J. *et al.*, 1991, Ap.J., 366, 64
Ferland *et al.*, 1992, Ap.J., 387, 95
Goad, M.R., O'Brien, P.T. & Gondhalekar, P.M., 1993, MNRAS, in press
Krolik, J. *et al.* 1991, Ap.J., 371, 541
Netzer, H., 1990, *Lecture Notes in Physics*, 377
Sparke, L.S., 1993, Ap.J., in press

Anisotropic Line Emission from Extended BLR's

M.R. Goad & P.T. O'Brien *

Due to the combination of high radiation densities in the BLR and the high column densities of clouds which are optically thick to the Lyman continuum, many lines emitted from the BLR are likely to be optically thick, and hence will be emitted anisotropically. Ferland *et al.* (1992) discussed the case where the amount of anisotropy is constant for all clouds, but in reality the anisotropy will vary for a spatially extended BLR, depending on the local physical conditions. To study this effect, we have computed several models for spherical BLR's populated by an ensemble of spherical clouds in pressure balance with an intercloud medium, using our BLR modelling code, PROSYN (Goad, O'Brien & Gondhalekar 1993), which utilizes the photoionization code CLOUDY (Ferland 1991). Here we show some of the results for a constant pressure BLR model, where the cloud density (N) and column density (N_{col}) are constant with radius (r), and hence the ionization parameter (U) varies as r^{-2}. The model was normalized to have $N = 10^{10}$ cm^{-3}, $N_{col} = 10^{23.75}$ cm^{-2} and $U = 10^{-2}$ at a radius of 10 light-days. The inner and outer radii were set at 0.3 and 56.2 light-days respectively, giving a range in log U from 1.0 to -3.5.

The total line emissivity $\varepsilon_{totl} = \varepsilon_{in} + \varepsilon_{out}$, where ε_{in} is the emissivity of the inward cloud face (towards the ionizing continuum source) and ε_{out} is the outward emissivity (away from the ionizing continuum source). The distribution of ε_{totl} as a function of U is shown in Fig 1(a,b) for Lyα and MgII λ2798 respectively. For Lyα, $\varepsilon_{totl} \propto U^{0.86}$ from log $U = -3.5$ to ≈ 0.25, but saturates at larger U due to the clouds becoming optically thin to the Lyman continuum. For MgII the emissivity distribution is a complex function of U, peaking at log $U \approx 0.0$, and then falling rapidly at larger U due to the removal of the partially ionized zone.

The ratio of ε_{in} to ε_{totl} is shown in Fig 1(c,d) as a function of r. For both lines the ratio is ≈ 1 for $r > 1$ light day, but for the reasons discussed above, the lines become optically thin in the inner-most BLR. Clearly assuming a constant amount of anisotropy is a poor approximation for these lines in this model. This is a common property of many models, and hence to construct a physically realistic BLR model the local anisotropy must be computed. Finally, we note that although the anisotropy patterns shown in Fig 1(c,d) are similar, the dissimilar emissivity distributions result in a different non-linear response of these lines to a varying continuum source (O'Brien & Goad, this volume).

---

*Department of Physics and Astronomy, UCL, Gower Street, London, WC1E 6BT.

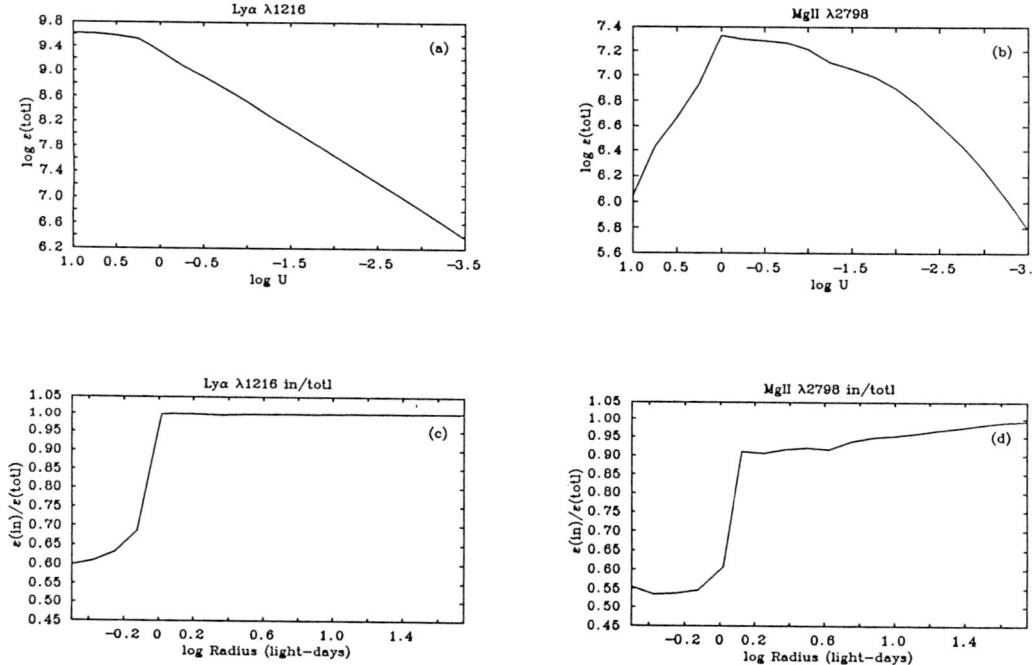

Figure 1. The distribution of ε_{totl} as a function of U for (a) Lyα and (b) MgII $\lambda 2798$. The ratio of ε_{in} to ε_{totl} as a function of r for (c) Lyα and (d) MgII $\lambda 2798$.

References

Ferland, G.J., 1991, OSU Internal Report 91-01)
Ferland *et al.*, 1992, Ap.J., 387, 95
Goad, M.R., O'Brien, P.T. & Gondhalekar, P.M., 1993, MNRAS, in press

Active Galactic Nuclei and Nuclear Starbursts

R. J. Terlevich [*]   G. Tenorio-Tagle [†]   J. Franco [‡]
B. J. Boyle [§]

Abstract

The Starburst model for radio-quiet Active Galactic Nuclei (AGN) postulates that the activity seen in most AGN is powered solely by young stars and compact supernova remnants (cSNR) in a burst of star formation at the time when the metal rich core of the spheroid of normal early type galaxies was formed. In this model, the broad permitted lines characteristic of the Broad Line Region (BLR) and their variability are originated in these cSNR. Combined analytic and numerical hydrodynamic simulations, with static photoionization computations have shown that cSNR can reproduce most of the basic properties of the BLR in low luminosity AGN.

We have explored the hypothesis that QSOs are the young metal rich cores of massive elliptical galaxies forming at z $\gtrsim$ 2.0. Only a small fraction ($\sim 5\%$) of the total mass of a normal spheroid, the core mass, is needed to participate in a burst to explain the observed luminosities and luminosity function of Quasars at z $\gtrsim$ 2.0. We predict that the progenitors of QSOs should look as dusty starbursts and about 4 times more luminous than QSOs themselves.

1. Introduction

The hypothesis that a Starburst can power the most extreme forms of nuclear activity has been proposed several times in the past (Shklovskii 1960, Field 1964, McCrea 1976), but was not favoured mainly because it failed to explain satisfactorily the observed large luminosity and variability of quasars, their radio emission, unresolved images, the presence of extremely broad permitted emission lines in the spectrum and their observed intensity ratios.

In the past few years we have started a systematic study of properties of high metallicity starbursts including the latest developments of the knowledge of massive star evolution and formation. A model was developed where the observed nuclear activity of QSOs and Seyfert galaxies is the *direct* consequence of the evolution of a massive

---

[*]Royal Greenwich Observatory, Madingley Road, Cambridge, CB3 0EZ, U.K.
[†]Instituto de Astrofísica de Canarias, 38200 La Laguna, Tenerife, Spain
[‡]Instituto de Astronomía UNAM, Apartado Postal 70-264, 04510 México D. F., México.
[§]Institute of Astronomy, Madingley Road, Cambridge, CB3 0HA, U.K.

young cluster of coeval stars in the high metal abundance environment of the nuclear region of early type galaxies (see Terlevich et al. 1992 hereafter TTFM 92, and Terlevich and Boyle, 1993).

Terlevich and Melnick (1985) computed the changes that a population of hot Wolf-Rayet stars (WARMERS) introduces in the UV spectrum of a young stellar cluster, and found that at about 3 Myr of evolution, when the most massive stars ignite the Helium in their core, the ionizing spectrum of the cluster is fundamentally modified by the appearance of a luminous and hot component, the *Warmer* component. Consequently the emission line spectrum is transformed from that of a typical low excitation HII region into a high excitation Seyfert 2. After 5 Myr, as the ionizing flux decreases, the ionization parameter also decreases, and the Seyfert 2 nucleus becomes a LINER (Cid-Fernandes et al. 1992). These results suggest the possibility of a direct relation between star formation and stellar evolution with some forms of nuclear activity. Following the evolution still further, when the cluster enters the supernova phase the Starburst can display properties similar to those observed in the BLR of AGN (TTFM 92). During this supernova or QSO phase, most of the luminosity is emitted by the young stars, while the broad permitted emission lines and their variability, i.e. the BLR luminosity, is due to cSNR activity (TTFM 92) and can be up to 50 percent of the total luminosity. Terlevich and Boyle (1993) have explored the hypothesis that QSOs are the young cores of massive ellipticals forming most of the dominant metal rich population in a short Starburst. They showed that the Starburst model can account for the numbers of high luminosity QSOs observed.

Here we present an overview of the model and describe some of the recent developments.

2. The evolution of luminous metal rich star clusters

The early evolution of a massive metal rich star cluster presents four different phases with well defined transitions (Terlevich, Melnick and Moles 1987). The first phase is dominated by HII regions ending with the appearance of the first extreme Wolf-Rayet or *warmers* which marks the beginning of the type 2 Seyfert phase. The explosion of the first SN of type Ib corresponds to the onset of non-thermal radio emission while the first explosions of type II SN lead to the formation of the BLR and the start of the QSO phase.

Recent observations show that the spectrum of at least some luminous SN exploding in HII regions have a striking resemblance to that of the BLR of Seyfert galaxies (Filippenko, 1989) and, conversely, that the flares of some Seyfert galaxies have the luminosity, lifetime and spectral signatures of type II SN (Terlevich and Melnick, 1988). The fundamental difference between "Seyfert-like" and normal type II SN can be understood if the former are associated with shocks that, after leaving the envelope of the star, expand into a region of *high circumstellar gas densities*. All theoretical computations of the evolution of SNR in dense environments show that after sweeping up a small amount of gas these remnants become radiative and deposit most of their energy in very short time scales thus reaching very high luminosities. Because of the large shock velocities, most of the energy is radiated in the extreme UV and X-ray region of the spectrum (Shull, 1980; Wheeler et al. , 1980; TTFM 92).

3. The light curve and variability of cSNR

The initial interaction of the SN ejecta with the ambient medium generates a hot shocked region enclosed by two shock waves: a leading shock moving outwards, and an inward "reverse" shock. The leading shock ($V \sim 10^4$ km s^{-1}) raises the gas temperature to $\sim 10^9$ K. The slower reverse shock ($V \sim 10^3$ km s^{-1}), thermalizes the supernova ejecta to temperatures of about $\sim 10^7$ K. The onset of the radiative phase in a dense medium occurs very early in the evolution and all evolutionary phases are substantially speeded up (TTFM 92). Given the strength of the radiation produced upon cooling, this phase plays a key role in the model. Such a strong cooling imply, as in the colliding cloud model (Daltabuit, MacAlpine & Cox, 1978), that a large flux of ionizing photons will emerge from the shocked gas. The wide range of gas temperatures in the cooling region results in a power-law-like spectrum at UV and X-ray frequencies (TTFM 92).

One of the most interesting properties of cSNR, is their rapid burst of radiative energy which makes them reach luminosities well in excess of 10^9 L$_\odot$. This happens in a few years (≤ 10 yr) through which more than 90% of the initial energy of the explosion (10^{51} erg) is radiated away, as the remnants acquire dimensions of the order of a few times 10^{16} cm. As most of the energy is radiated away before the kinetic energy of the ejecta is fully thermalized, cSNR do not enter the so called "Sedov" or quasi-adiabatic evolutionary phase. In a luminosity (L) vs time (t) diagram, remnants increase at first their luminosity as t$^{0.8}$ to reach within a few years (≤ 3 yr) an $L \sim 10^9$ L$_\odot$. This is followed by a slower decay (t$^{-11/7}$) leading to factors of 4 to 10 lower luminosities for the various runs. Our hydro models present four mayor energy bursts superimposed on the secular decay of the remnant luminosity. These events are all very short, typically of only a few months duration, and involving about 1-10% of the initial explosion energy. These four bursts of luminosity are produced by: 1) The onset of strong radiative cooling, within the swept up circumstellar matter. This leads to the absolute maximum luminosity (several times 10^9 L$_\odot$ for $n_0 = 10^7$ cm^{-3}) of the remnant. 2) The completion of thin shell formation at the edge of the rapidly expanding (≥ 2000 km s^{-1}) remnant. 3) The onset of strong radiative cooling behind the reverse shock, also leading to the condensation of the shocked ejected matter into a secondary inner shell, and 4) the merging, or collision, of both shells which leads to fragmentation and dispersal of the remnant. All these events occur prior to full thermalization, and, in all cases, more than 90% of the initial energy of the explosion is radiated away by the time the last burst is over.

We have compared the behaviour described above with the detailed spectroscopic observations of NGC 1566 which span over several years. The nucleus showed four major periods of activity, lasting about 1300 days, and releasing about 10^{51} erg each (Alloin etal, 1986). These we attribute to supernova explosions and their associated cSNR. The last of which was observed with high temporal resolution and showed 3, maybe 4, short energy bursts. Each burst presented a steep rise time ($\sim$ 20-30 days) and a longer decay of about 300-400 days. This behaviour is indeed very similar to that present in our numerical solution. Furthermore, a least square fit to the slope of the secular decay of NGC 1566 (ie, after removing the 3 or 4 bursts of energy from the light curve) gives -1.57, while the numerics gives $-11/7 = -1.57$. The observed luminosities are also in very good agreement with our predictions.

The general expected behaviour of the optical variability is supported by most studies which show that high luminosity QSO are less variable than low luminosity ones, in particular Seyfert galaxies. The analysis of the best data sets of photometry of QSOs with a long time base line is consistent, if not suggestive, of the variability being originated in supernova size events (Terlevich, 1990a, 1990b; Aretxaga & Terlevich, 1993).

4. The QSO luminosity function

One of the central objections raised against the Starburst model for AGN is that it cannot account for the numbers of high luminosity QSOs observed. The Starburst model requires star formation at high metallicity, and the obvious places with high metallicity are the cores of luminous ellipticals. Thus, a strong test of the objection would be to compare the predicted luminosity function (LF) for cores of elliptical galaxies at the time in which the "Starburst" activity is taking place with the QSO LF at high redshifts ($2 < z < 3$) where it is now generally accepted that QSO activity was at its peak (Schmidt, Schneider and Gunn 1991). The QSO LF in the blue (rest) pass band at these redshifts has recently been determined by Boyle (1991) using the large number of faint QSO catalogues which have been compiled in the last few years. In order to obtain the predicted LF for the cores of elliptical galaxies at high redshift to compare with the QSO LF, we can simply scale the present day luminosity function for elliptical galaxies in both luminosity and space density using *observed* or established properties of elliptical galaxies and their cores.

We have shown that the predicted young core luminosity function is an excellent match to the observed luminosity function for Quasars in the redshift range from 2.0 to 2.9 (Boyle, 1991; Terlevich, 1992; Terlevich & Boyle, 1993). Thus, **the young cores of ellipticals containing only 5 % of the total galactic mass are capable of producing the luminosity of even the most luminous Quasars.**

5. IRAS galaxies as progenitors of QSOs

One important prediction of the Starburst model regards the properties of the progenitors of QSOs. As discussed in the second section of this paper, the HII and Seyfert 2 phases are associated with large amounts of dust. This is due not only to the dust present in any star formation region but also to the large amount of dust synthesized by the most massive stars during the η-Carinae phase before becoming WR stars. During this phase up to 1 $M_\odot$ of dust per evolved massive star may be injected into the core ISM, enshrouding most of the massive stars. The emitted luminosity will be therefore, dominated by the far infra-red (FIR) luminosity and these systems will presumably be present in large numbers in the IRAS sample of Starbursts and Seyfert 2.

The HII and Seyfert 2 phase that precedes the QSO phase, lasts 8×10^6 years, or about 1/10 of the duration of the QSO phase. During this early phase, the cluster luminosity is at its maximum, and on average, about 4 times higher than during the QSO phase. The QSO progenitor should therefore be a luminous and short lived (short compared with the QSO life-time) FIR source. Comparative studies of the luminosity function of IRAS galaxies and QSOs should show if IRAS galaxies are, at a given epoch, about 4 times more luminous in bolometric units and about 10 times less frequent than QSO. This simple prediction applies only to the case of a coeval population; systems where the star formation time scale is longer will show a mixture of all four phases at all times and will therefore probably look like a QSO for most of their bright phase.

6. Conclusions

Recent results on evolving young luminous compact SNR can explain a large number of the observed properties of the BLR of AGN with a minimum number of free parameters. If our hypothesis is correct it should be possible to extend this simple model to explain the observed variability of Seyfert nuclei in the optical, UV and X-ray spectral range.

We suggest that perhaps most of the high redshift unresolved emission line objects found in optical surveys and classified as QSO due to the presence of broad permitted emission lines in their spectrum, are in fact the young cores of elliptical galaxies forming in a single starburst most of their metal rich stellar population.

Future work will explore important problems for the simple Starburst model like the BLR line profiles and the origin of the rapid X-ray variability.

Although more work is needed, the research done so far indicates that the Starburst model, based in simple assumptions, is potentially able to reproduce the main observed properties of radio quiet Active Galactic Nuclei.

References

Alloin, D., Pelat, D., Phillips, M.M., Fosbury, R.A.E. & Freeman, K., 1986, *Ap. J.*, **308**, 23.

Aretxaga, I. & Terlevich, R., 1993 in *The Nearest active Galaxies* **Ap. & Sp. Sc.**, in press

Boyle, B. J. 1991, in *Texas ESO/CERN Meeting on Relativistic Astrophysics*, Brighton, U.K.

Cid Fernandes Jr,. R., Dottori, H.A., & Gruenwald, R. B., 1992, *M.N.R.A.S.*, **255**, 165.

Daltabuit, E., MacAlpine, G. & Cox, D., 1978, *Ap. J.*, **219**, 372.

Field, G.B., 1964, *Ap. J.*, **140**, 1434.

Filippenko, A.V., 1989, *A. J.*, **97**, 726.

McCrea, W.H., 1976 in *The galaxy and the local group*, eds Dickens, R.J. & Perry, J.E. (RGO Bull 182).

Schmidt, M. Schneider, D.P. & Gunn, J.E., 1991, in *The Space Distribution of Quasars*, PASP conference series No. 21, ed. D.Crampton, (Brigham Young: Provo), p109.

Shull, J. M., 1980, *Ap. J.*, **237**, 769.

Shklovskii, I.S., 1960, *Soviet Astronomy*, **4**, 885.

Terlevich, R. 1990a, in *Windows on Galaxies*, ed. G. Fabbiano, J. Gallagher & A. Renzini, Kluwer: Dordrecht.

Terlevich, R. 1990b, in *Structure & Dynamics of the Interstellar Medium*, ed. Tenorio-Tagle, G., Moles, M. & Melnick, J., Springer-Verlag: Berlin.

Terlevich, R. 1992, in *Relationship Between AGN and Starburst Galaxies*, ed. A. Filippenko.

Terlevich, R. & Melnick, J. 1985, *M.N.R.A.S.*, **213**, 841.

Terlevich, R., Melnick, J. & Moles, M. 1987, in *Observational Evidence for Activity in Galaxies*, ed. E. Khachikyan, K. Fricke & J. Melnick, Reidel:Dordrecht.

Terlevich, R. & Melnick, J. 1988, *Nature*, **333**, 239.

Terlevich, R. Tenorio-Tagle, G., Franco, J., & Melnick, J., 1992,*M.N.R.A.S.*, **255**, 713. TTFM 92,

Terlevich, R. & Boyle, B. J. 1993, *M.N.R.A.S.*, **262**, 491.

Wheeler, J. C., Mazurek, T. J. & Sivaramakrishnan, A. 1980, *Ap. J.*, **237**, 781.

Rapidly Evolving Compact SNRs and the Nature of the Lag in AGNs

G. Tenorio-Tagle [*]   R. J. Terlevich [†]   M. Rozyczka [‡]
J. Franco [§]

Abstract

This is a short summary of several detailed calculations of strong radiative cooling behind supernova shock waves evolving in a high density medium. These lead to definite predictions about the *lag*, the observed delay between sudden changes in the continuum ionizing radiation followed, after some time, by changes in the intensity of the emmision lines from the broad line region of AGNs. A full description of these results is due to appear soon in a journal.

1. Introduction

In the starburst model of AGNs, sometimes viewed as exotic and/or unconventional, the applied physics are in fact most conventional, as it uses the little, or the lot, that we know about real events: the physics of stars and stellar evolution and their interaction with the surrounding gas, and with these predictions are made. In this model, the observed broad emission lines and their variability, are generated by "compact", strongly radiative supernova remnants which are expected to occur in the central regions of early type galaxies undergoing a violent nuclear burst of star formation. The activity of all stars and the frecuency of supernova explosions soon establishes a high pressure region around the cluster. This high pressure, as it acts upon the slow winds from the massive stars ($M \geq 8 \, M_\odot$), leads to the development of a high density circumstellar medium ($n_0 \geq 10^7$ cm^{-3}) around each of the potential supernova stars. It is in this high density circumstellar medium that the energy of each supernova is released, causing the most obvious property of "compact" supernova remnants: a burst of radiative energy which makes them reach luminosities well in excess of $10^9 \, L_\odot$. This happens during a few years (≤ 10) during which more than 90% of the initial energy of the explosion (10^{51} erg) is radiated away, while the remnants acquire dimensions of only a few times 10^{16} cm. The high circumstellar densities thus provoke the rapid onset of strong radiative cooling and with it the various hydrodynamical

---

[*]Instituto de Astrofísica de Canarias, 38200 La Laguna, Tenerife, Spain
[†]Royal Greenwich Observatory, Madingley Road, Cambridge, CB3 0EZ, U.K.
[‡]University of Warsaw Observatory, Al. Ujazdowskie 4, Pl-00478 Warszawa, Poland.
[§]Instituto de Astronomía UNAM, Apartado Postal 70-264, 04510 México D. F., México.

events that lead to the broad line region, to the continuum and line variability and also to the *lag* typical of these sources.

A summary of our research method and some of our recent findings are to be found in our paper I (Terlevich *et al.* 1992). Here however, we wish to concentrate on the *lag*. This delay, if interpreted as a result of the geometry of the source, implies a measure of the distance at which the broad line region sits away from the central ionizing source. Instead in the starburst model, the *lag* results from time dependent changes in the ionization parameter within the layer of gas encountered by the supernova blast wave, as matter adjusts itself to the drop in pressure suddenly promoted by strong radiative cooling. The *lag* is scale independent and its value has little or nothing to do with the size of the emitting source!

2. The nature of the *lag*

Detailed calculations of compact SNRs have shown in great detail the sequence of events that take place as remnants approach and reach maximum luminosity. The matter involved in the process, suddenly has to readjust to the large pressure inbalance promoted within the shocked gas by the onset of strong radiative cooling. The final outcome is the formation of a thin shell at the edge of the remnant, several orders of magnitude denser than the original background medium. Gas condensation however, does not happen immediately, as it requires of the passage of secondary shocks through the cool region for this to acquire the appropriate density and thus pressure. The secondary shocks emmanate from the hottest section of the remnant interior, overtaken earlier during the evolution when the shock speed was larger, and that take much longer to cool. They follow the blast wave in an attempt to communicate the interior pressure. However, given the increasingly larger densities behind them and thus the correspondingly shorter cooling distances, they also become rapidly radiative. Meanwhile cooling proceeds as a front, ahead of the secondary shocks, into gas more recently overtaken by the progressively slower blast wave to eventually catch up with it. The blast wave slows down for two reasons: because of geometrical dilution as the remnant grows and because it has now suddenly lost its piston pressure due to strong cooling behind it. The gas steadily overtaken by the cooling front is a source of ionizing continuum radiation, to be observed as a variation in the continuum of the AGN. This radiation is immediately absorbed by the reshocked matter, by the cool layer of gas continuously changing density after the passage of secondary shocks. The combination of a steadily denser layer constantly irradiated, as the cooling wave progresses through the layer of shocked gas, leads to a continuous decrease in the effective ionization parameter and results in a rapidly changing ionization structure of the photoionized gas.

The numerical calculations have shown that the width of the photoionized shocked region, traversed by the cooling front and continuously swept by secondary shocks to condense it into a cool thin shell, is only about 10^{13} cm (or about 10^3 secs $\times$ c), and yet, *lags* of up to several days, weeks, and even months, are generated for different lines. In general, the calculations predict shorter delays for high ionization lines than for low ionization ones. A detailed comparison with the results from the NGC5548 extensive monitoring campaign, agree both with the time delays for different lines, and with the intensity values reached by the various lines. Our calculations thus show that the compact supernova remnant model is capable of giving an accurate and detailed description of the temporal behaviour of the broad line region as well as accounting for all of its intrinsic parameters with a minimum of free parameters.

3. Coda.

A quote, and an addition, to J. D. Borrow and its world within the world: "An acid test of any theory is its ability to stand the experimental test of its predictions. This is because as scientists, we have come to regard predictability as an essential ingredient of any useful scientific law". This is only possible, if there is a prediction, at all, to be made within the "world", or scheme contemplated by the theory.

We expect many more events to be eventually resolved and carefully studied to elucidate their role in the starburst model of AGNs. We are aware of the existence of relativistic jets in a few (less than 10%) of the AGN sources. We also know about the X-ray variability and micro-variability detected in some Seyfert nuclei; all of these events, as long as the energies involved remain within the range expected from a violent stellar burst and their supernova output, deserve to be considered for an explanation within the starburst model for AGNs. The preliminary results are promising, given the detailed agreement with the large majority of sources (see paper I), and yet one is still to explore a wide number of possibilities. These, to name a few, may have to do with the environment of the nuclear starburst, or the effects of the increasingly larger number of stellar remnants and/or the detailed physics of remnants driven by fragmented ejecta.

References

Terlevich, R. Tenorio-Tagle, G., Franco, J., & Melnick, J., 1992,*MNRAS,255,713*. Paper I,

Supernova Explosions in QSO's? – II.

Itziar Aretxaga *   *Roberto Cid Fernandes* [†]   *Roberto Terlevich* [‡]

The observed anticorrelation between the variability amplitude and luminosity in QSOs can be naturally explained in the framework of the Starburst model as the multiple superposition of supernovae (SNe) and their associated compact remnants. Figure 1 shows the comparison between our models [1] and the QSO sample of Hook et al. [2], where med(σ) is the median value of the standard deviation from the mean magnitude of the sources.

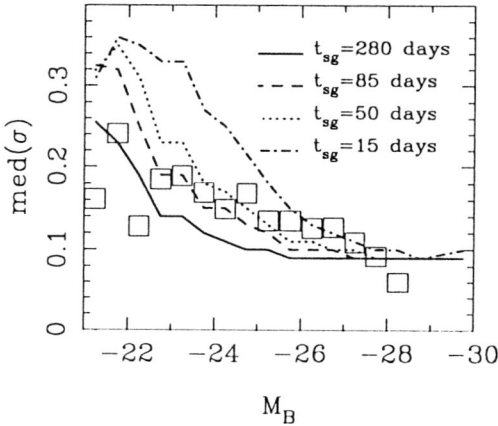

Figure 1. Variability in the QSO sample [2] (squares) and assigned models [1] (lines). The typical scatter about med(σ) is $\pm 0.1 - 0.05$ mag.

The curves correspond to different values of t_{sg}, the time for the onset of the radiative phase of the remnant associated to each SN outburst [3]. Due to cooling effects, t_{sg} decreases as the density and metallicity of the circumstellar medium increase. Since the density is related to the progenitor's mass loss rate and the latter should increase with Z, we may regard t_{sg} as a metallicity indicator. Figure 1 shows a *marginal* tendency for brighter sources to have shorter t_{sg} values, which could be interpreted as a luminosity-metallicity (or a mass-metallicity) relation in QSOs.

In the Starburst model, distant QSOs are thought to be the progenitors of the cores of present day elliptical galaxies [5]. Present day ellipticals do exhibit a mass-metallicity

---

*Dpto. Física Teórica, C-XI. Univ. Autónoma de Madrid. Cantoblanco. E - 28049 Madrid. Spain. Supported by the Basque government (BFI88.009) and Spanish DGYCIT (PB900182)

[†]Institute of Astronomy. Madingley Road, CB3 0HA Cambridge. U.K. Supported by the Brazilian institution CAPES

[‡]Royal Greenwich Observatory. Madingley Road, CB3 0EZ Cambridge. U.K.

relation $Z \propto M_E^\alpha$, with $\alpha \approx 0.25$–0.5. Taking into account the mass loss rate dependence on Z and the effects of metal cooling on t_{sg} we may relate t_{sg} and Z by $t_{sg} \propto Z^{-\beta}$, with $\beta \approx 0.75 - 1.1$. This means that for two sources of different luminosities their t_{sg} values are roughly related by

$$\log\left(\frac{t_{sg}^1}{t_{sg}^2}\right) \approx \log\left(\frac{Z_1}{Z_2}\right)^{-\beta} \approx \frac{\alpha\beta}{2.5}(M_B^1 - M_B^2). \tag{1}$$

If one considers NGC 4151 and NGC 5548 as typical $M_B \approx -21$ mag sources, where $t_{sg} \approx 260 - 280$ days [4], one gets the following t_{sg} values for QSOs in the sample:

M_B(mag)	-24	-28
t_{sg}(days)	62 – 170	8 – 88

where the range in t_{sg} is worked out from the limit cases of α and β. These values are completely consistent with the models in Figure 1.

The interpretation that Z increases with the QSO luminosity, in analogy with the mass-metallicity relation for elliptical galaxies, seems to provide a consistent explanation for the tendency observed in Figure 1 of luminous sources to have short values of t_{sg}. To confirm this effect better observations, i.e., precise photometry and a better sampling of the light curves, would be required. In any case, it is interesting to note that high metallicity in luminous QSOs has also been proposed on the basis of their emission line spectra [6].

References

[1] Aretxaga I. et al. 1992. *"First Light in the Universe: Stars or QSOs"*. Eds: B. Rocca-Volmerange et al., Editions Frontières, *in press*.
[2] Hook I.M. et al. 1992. *"Space Distribution of Quasars"*. Ed. D. Crampton, A.S.P. Conference Proceedings, vol 21.
[3] Terlevich R. et al. 1992. Mon. Not. R. astr. Soc., **255**, 713.
[4] Aretxaga I. & Terlevich R. 1992. *" The Nearest Active Galaxies"*. Astrophys. Space Sci., *in press*.
[5] Terlevich R. 1992. *" Relationships between Starburst and Active Galaxies"*. Ed. A. Filippenko, A.S.P. Conference Proceedings, *in press*.
[6] Hamann, F. & Ferland, G. 1992. Astrophys. J. Lett., **391**, L53.

High Metallicities in QSOs

Gary Ferland [*]   *Fred Hamann* [†]

Abstract

We describe constraints on the metallicity of quasar broad line region gas. The overall emission line spectrum is surprisingly insensitive to order of magnitude changes in the global metallicity Z. Indirect methods, employing photoionization models and explicit stellar chemical evolution models of the selective enrichment of the elements, must be used to infer the metal enrichment. Two line ratios, both involving NV λ1240, are developed to measure Z. The first is the ratio of NV to the collisionally excited line CIV λ1549. The second is the ratio of NV to the recombination line HeII λ1640. Both indicate nitrogen enhancements exceeding an order of magnitude above solar. These results imply high metallicities in high redshift quasars, a property they have in common with the cores of massive galaxies.

1. Introduction

One of the longest standing goals of AGN emission line research has been to use these lines to measure the composition of the emitting gas. This could then test models of both the quasar phenomenon and stellar nucleosynthesis (Davidson and Netzer 1979; Shields 1976). Fundamental uncertainties concerning the nature and geometry of the BLR have made this work difficult.

Variability studies have rejuvenated interest in the BLR by providing methods to directly measure quantities such as the distance between continuum source and emitting clouds. These have shown that the clouds are both denser and exposed to a far more intense radiation field than had been inferred (Peterson 1992; Ferland and Persson 1989; Ferland et al. 1992). These measurements remove several sources of uncertainty in photoionization modelling. Energized by this new knowledge, we have embarked on an effort to rethink what is known about the metallicities of the BLR.

2. BLR Abundances: An Old Problem, New Parameters

Here we first summarize the reasons that the measurement of abundances using BLR emission lines is so difficult, go on to outline the recent revision of many BLR parameters, and finally outline a scheme for measuring the metallicity Z by first determining

---

[*]Department of Physics and Astronomy, The University of Kentucky, Lexington, KY 40506, USA. This author is supported by grants from the NSF and NASA.

[†]Department of Astronomy, The Ohio State University, 174 W 18th Ave., Columbus OH 43210-1106, USA. This author is supported by a Columbus Fellowship.

the nitrogen abundance relative to He or C. The relationship between Z and the N abundance is then estimated by reference to stellar evolution calculations (see Hamann and Ferland, this volume).

2.1 Abundances from Emission Lines

Unlike the case of absorption lines, which can continue to grow ever stronger as the column density of absorbers increases, emission line intensities are not sensitive to the overall metallicity of the emitting gas. In equilibrium, heating (by photoionization) and cooling (primarily by collisionally excited emission lines) match, and the intensity of a strong collisionally excited emission line is set by the photoelectric heat input. Davidson (1977) discusses this further.

Figure 1 illustrates the difficulties. A series of photoionization model calculations were computed using the code described by Ferland (1993). Fig. 1 shows the intensity of the collisionally excited line CIV $\lambda 1549$ relative to the recombination line Lyα. A scale factor multiplies the solar abundances of the 13 heavy elements included in the calculation so that all are varied away from solar by the factor indicated along the x-axis. The abundance of helium is not changed.

The ratio of the intensities of the carbon and hydrogen lines changes by less than a factor of two despite a four order of magnitude variation in the C/H ratio. The CIV line is a major coolant in BLR clouds, and the intensity of the line simply cannot vary by much if energy is to be conserved. In fact, what happens is that the electron temperature varies inversely with the metallicity so that for high abundances the CIV line is able to carry the required cooling at a lower electron temperature. The line does in fact continue to decrease for metallicities smaller than that shown. In that limit carbon is very scarce and the cooling is increasingly carried by hydrogen free-free and free-bound continua. The lesson of Fig. 1 is that we could actually be observing a very wide range of metallicities amongst the quasar population, and not notice it from the relative intensities of the strongest emission lines.

It is surprising that so few attempts at measuring the abundances of quasar clouds have been made. Most work has centred on the use of weaker intercombination lines (Shields 1976). These have the advantage that they are less sensitive to the global energy balance of the cloud, but the twin disadvantages that they are difficult to detect at high redshift and the deduced abundance is sensitive to the assumed cloud density, which is essentially unknown. Shields notes that nitrogen may be overabundant in some objects, while more evidence for a large nitrogen to carbon ratio in one object is presented by Osmer (1980). The high ionization NV line used below has not been developed as an abundance indicator, and in fact most photoionization analyses studies have under-predicted this line by an order of magnitude (see the review by Davidson and Netzer 1979).

Despite the overall insensitivity of the collisionally excited lines to the global metallicity, it is still possible to measure *relative* abundances of the heavy elements. As

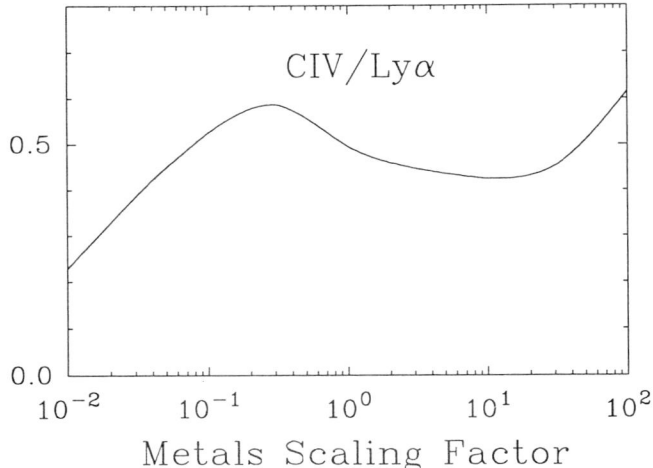

Figure 1. The CIV $\lambda 1549$/Lyα Ratio vs. Metals Scaling Factor. The density is $10^{10}\,\mathrm{cm}^{-3}$, the flux of hydrogen ionizing photons was $10^{20}\,\mathrm{cm}^{-2}\,\mathrm{s}^{-1}$, and the continuum is that described by Mathews and Ferland (1987). The CIV/Lyα ratio changes by less than a factor of two despite a 4 decade change in the C/H ratio.

shown in the accompanying paper (Hamann and Ferland, this volume) the abundance of nitrogen increases more rapidly than that of the other elements in typical models of galactic chemical evolution. For the simplest nuclear evolution models the N/He ratio scales as the square of the metallicity (the abundance of He changes little) while the N/C ratio increases roughly linearly (Hamann and Ferland 1993). By reference to these evolutionary models it is possible to infer the overall metallicity of the emitting gas if the abundance of nitrogen relative to He or the heavy elements can be measured.

2.2 Variability Studies; a Revolution in Cloud Parameters

It was long thought that cloud parameters (the density, and the ionization parameter U, a ratio of ionizing photon density to hydrogen density) were constrained by the relative intensities of the stronger emission lines, especially CIV $\lambda 1549$ relative to CIII] $\lambda 1909$ (Davidson 1977; Baldwin and Netzer 1978; Davidson and Netzer 1979). It was argued that this ratio set a limit to the density of less than $10^{9.5}\,\mathrm{cm}^{-3}$ and an ionization parameter in the neighbourhood of U$\sim 10^{-2}$.

Given the assumed density, the deduced ionization parameter can be recast as a measure of the flux of ionizing photons. The source-cloud separation can then be inferred if the continuum shape is known. This separation can also be measured from line-continuum reverberation studies. Such studies (reviewed by Peterson 1992) show that the source-cloud separation is much smaller, and that the flux of ionizing radiation is one to two dex larger, than had been deduced. Ferland and Persson

(1989) showed that Strömgren ionization fronts occur for the heavy elements at high ionization parameters, so that for an optically thick cloud lower ionization lines such as CIII] $\lambda1909$ are quite strong. The ratio of CIII] $\lambda1909$ to CIV $\lambda1549$ depends mostly on the continuum shape rather than U. Variability studies have also shown that the CIII] does not originate in the same region as the CIV line, removing it as a density indicator for the high-ionization region, and perhaps suggesting that the high ionization region is not optically thick to the ionizing continuum.

Ferland et al. (1992) have re-examined the case of the well-studied Seyfert galaxy NGC 5548. Reverberation studies have shown the flux of ionizing photons for the CIV region to be in the neighbourhood of $10^{20} - 10^{21}\,\mathrm{cm}^{-2}\,\mathrm{s}^{-1}$. The absence of strong CIII] emission at the CIV radius either sets a lower limit to the density of the CIV clouds, of roughly $10^{11}\,\mathrm{cm}^{-3}$, *or* suggests that this region is optically thin to the ionizing continuum. A density somewhere in the neighbourhood of 10^{10}–$10^{11}\,\mathrm{cm}^{-3}$ is suggested. We adopt $n\sim 10^{10}\,\mathrm{cm}^{-3}$ and $\phi(\mathrm{H})\sim 10^{20}\,\mathrm{cm}^{-2}\,\mathrm{s}^{-1}$ as typical of the high ionization region of BLR clouds, with an uncertainty of roughly one order of magnitude. This corresponds to an ionization parameter of roughly 1/3.

3. Nitrogen Abundance Indicators

Figure 2 shows the ionization structure of He, C, and N, computed for a BLR cloud, with the parameters just given and cosmic abundances (Grevesse and Anders 1989). The cloud naturally divides into He^+ and He^{++} zones. The CIV 1549 line-producing region fills much of the He^+ zone, while the NV 1240 line-producing region occurs within the He^{++} zone, where the HeII $\lambda 1640$ recombination line originates. The relative proportions of these various zones depends mostly on the shape of the ionizing continuum rather than the density or flux of photons, as long as the ionization parameter is large (Ferland and Persson 1989).

Figure 3 shows the relative intensities of the three strongest high-ionization lines as a function of ionization parameter, along with a few other line ratios. The high ionization ratios are insensitive to U as long as U is large (it is roughly $\log(U) = -0.5$ for the standard high-ionization line region parameters given above). This insensitivity to various parameters sets the stage for the use of these lines as a nitrogen abundance indicator. Notice that both NV/CIV and NV/HeII are predicted to be roughly an order of magnitude smaller than observed in luminous high redshift quasars Baldwin et al. 1993, and Hamann and Ferland this volume).

High abundances of nitrogen are indicated by the contrast between the predictions shown in Fig. 3 and actual observations. The NV/HeII ratio is typically 5 while the NV/CIV ratio nearer unity in intermediate redshift quasars. The nitrogen line is observed to be an order of magnitude stronger than predicted assuming solar abundances. It is not correct to simply scale the predicted ratio in Fig. 3 by the observed line ratio to deduce the nitrogen abundance for two reasons. First, the lines are optically thick so that radiative transfer effects are important, and second,

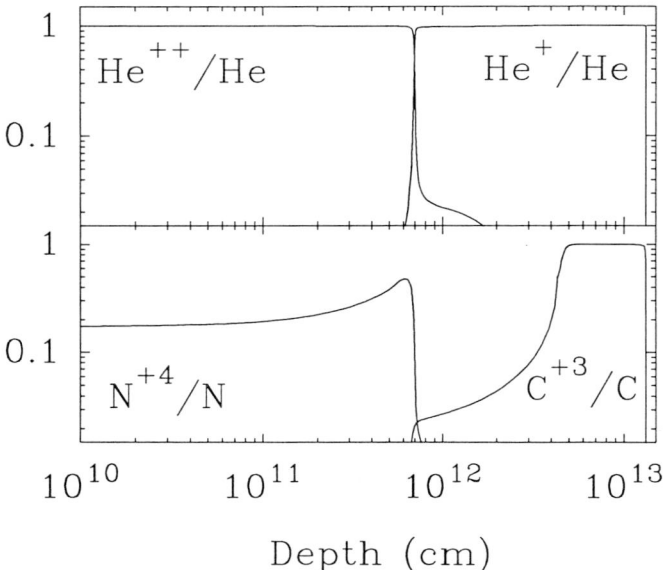

Figure 2. The He, C, and N ionization structure of a cloud with a density of 10^{10} cm^{-3}, a flux of ionizing photons of 10^{20} cm^{-2} s^{-1}, cosmic abundances, and the Mathews and Ferland continuum, is shown. The helium ionization structure is the upper panel, while the N^{+4} and C^{+3} creation zones are shown in the lower panel.

the lines shown are major coolants, so energy conservation controls their intensity. Explicit models, in which the abundances of the heavy elements are derived in a self consistent manner, are needed. These models are described further below.

It has long been recognized that the NV λ1240 line is far stronger than expected for simple photoionization of a gas with solar abundances (see the review by Davidson and Netzer 1979). In the past this has been attributed so an optically thin high ionization region which is not reproduced by conventional BLR calculations. We show below that such an explanation is ruled out, and that high nitrogen abundances are the more likely explanation.

The first of the two ratios we develop is NV λ1240/CIV λ1549. This is sensitive to the N/C abundance, which is (in the simplest case) linearly proportional to the metallicity of the gas. This ratio could be affected by a particularly adverse geometry. As Fig. 2 shows, N^{+4} and C^{+3} do not occur in the same region of a single cloud. If instead of the optically thick geometry shown in the figure, the BLR geometry is one in which the interior N^{+4} region is not associated with an adjoining C^{+3} region then it is possible to imagine situations where NV emission is not accompanied by CIV emission. (For Fig. 2 to be valid the NV and CIV creation zones do not need to be adjacent, but they should occur along the same line of sight from the continuum source). Indeed, such a geometry (in which NV regions were not accompanied by

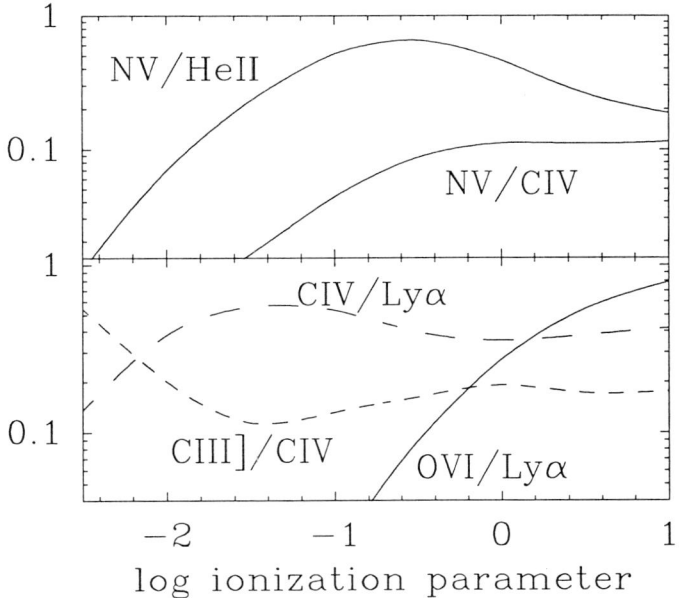

Figure 3. The line ratios NV/CIV and NV/HeII are shown as a function of the ionization parameter U. Several other ratios are shown in the lower panel. For large values of U the line ratios are not strongly sensitive to U.

CIV or Lyα regions) was the initial explanation for the great strength of NV $\lambda 1240$ relative to lower ionization lines. This geometry would cause the NV/CIV line ratio to *overestimate* the N/C ratio.

Although the suggestion that there exists an optically thin region which emits predominantly NV but little CIV might explain the NV/CIV ratio, this cannot explain the NV/HeII ratio. As Fig. 2 shows, any region producing NV emission *must also* make the HeII recombination line. Although it is impossible for a region producing NV emission to not also make HeII emission, it *is* possible for a HeII-emitting region to have nitrogen in stages of ionization other than N^{+4}. For such optically thin geometries the NV/HeII indicated by Fig. 3 would then *underestimate* the N/He ratio. It is simply impossible to produce NV without also making HeII, although the opposite is possible.

The fact that the two ratios are affected by an adverse geometry in the opposite manner, but yet both suggest that nitrogen is enhanced by an order of magnitude, suggests that nitrogen is indeed overabundant in luminous quasars. High N is needed for *any* geometry.

4. Implications for Quasars

Photoionization models with solar abundances under-predict the relative intensity of NV $\lambda 1240$ in high redshift quasars by more than an order of magnitude. The fact that both emission line ratios described above indicate an order of magnitude enhancement of N despite having opposite sensitivities to possible inhomogeneities suggests that N is indeed strongly enhanced relative to the other elements.

This inferred enhancement of N relative to C or He can be converted to a metallicity Z by reference to a complete simulation of the stellar nuclear evolution environment. This is discussed in the adjoining paper (Hamann and Ferland this volume). The implication is that the metallicity is indeed high. The fact that this occurs when the Universe is quite young suggests quasars are associated with an epoch of very rapid star formation. This inference is similar to events associated with the rapid buildup of heavy elements in the cores of massive galaxies, and suggests that quasars are indeed galaxies in the process of formation.

References

Baldwin, J. A., *et al.* 1993, *in prep.*
Baldwin, J.A., and Netzer., H., 1978, Ap.J. 226, 1.
Davidson, K., 1977, ApJ, 218, 20
Davidson, K., and Netzer, H., 1979, Rep Prog in Physics 51, 715.
Ferland, G.J., 1993, *Hazy, a Brief Introduction to Cloudy 84*, University of Kentucky CCS Internal Report.
Ferland, G.J., Peterson, B.M., Horne, K., Welsh, W.F., and Nahar, S.N., 1992, ApJ, 387, 95
Ferland, G.J. and Persson, S.E., 1989, ApJ 347, 656
Grevesse, N., and Anders, E., 1989, in *Cosmic Abundances of Matter*, AIP Conference Proceedings 183, Ed. C.J. Waddington, (New York: AIP)
Hamann, F., and Ferland, G. 1992, ApJL, 391, L53
Hamann, F., and Ferland, G. 1993, ApJ, *submitted*
Mathews, W.G., and Ferland, G.J., 1987, ApJ, 323, 456
Osmer, P., 1980, Ap.J. 237, 666.
Peterson, B.M., 1992, PASP in press
Shields, G. A. 1976, ApJ, 204, 330
Wheeler, J.C., Sneden, C., and Truran, J.W., 1989, Ann. Rev. Ast. Ap., 27, 279

The Chemical Evolution of QSOs

Fred Hamann [*]   *Gary Ferland* [†]

Abstract

One zone chemical evolution models are developed for application to QSO broad emission line regions. The elemental abundances derived from the broad line ratios imply that the gas is highly evolved, with metallicities ranging from ~ 1 to $\gtrsim 10$ times solar. The short timescales (i.e. $\lesssim 1$ Gyr if $q_o \approx 1/2$) and relatively flat initial mass functions (compared to the solar neighbourhood) needed to fit most of the high redshift line ratios are almost identical to the parameters used in one zone models of elliptical galaxies. We conclude that the QSO phenomenon generally follows an episode of rapid star formation, exactly like that expected in massive, young galactic nuclei.

An observed trend in the emission line data suggesting higher metallicities at high redshifts could result from a mass–metallicity–redshift relation among the QSOs. Thus the highest mass QSOs (and/or host galaxies) might form only at high redshifts (e.g. $z > 2$).

1. Introduction

The broad emission line spectra of QSOs show that heavy elements are present at redshifts up to nearly $z \sim 5$. Therefore some amount of star formation must have occurred before the QSOs 'turned on'. Unfortunately, the line strengths are not indicative of the overall metal content of the gas, but some of the line ratios *are* sensitive to the *relative* abundances (see Ferland & Hamann this volume). The relative abundances can in turn be used to constrain both the metallicity and the past evolution because the elements form by different processes and on different timescales; cf. [2]. By assuming the enrichment processes are understood, and including them in a simple one zone chemical evolution model, we use the observed N Vλ1240 Å to C IVλ1550 Å (hereafter NV/CIV) and NV to He IIλ1640 Å (NV/HeII) line ratios to constrain the chemistry and evolution of QSO environments. We implicitly assume that the broad emission lines form in well mixed interstellar gas. Our early results based on NV/CIV only are discussed in [3]. The evolution models and the new analysis based on NV/HeII are discussed in detail in [4].

---

[*]Department of Astronomy, The Ohio State University, 174 W 18th Ave., Columbus, OH 43210-1106, USA. This author is supported by a Columbus Fellowship.

[†]Department of Physics and Astronomy, The University of Kentucky, Lexington, KY 40506, USA.

2. Chemical Evolution Models

The evolution models describe a closed system assembled by the infall of primordial gas. Stars form over the mass range $0.087 \leq M_* \leq 100$ M$_\odot$, as specified by power law initial mass functions (IMFs) of the form, $\Phi \propto M_*^{-x}$. Chemical enrichment occurs as the dying stars eject processed gas back into the interstellar medium. The time delays between the birth and death of the stars are included.

In this contribution we contrast the abundance evolution in two models; one using 'standard' solar neighbourhood parameters and another using parameters like those derived (by others) for giant elliptical galaxies. In the solar neighbourhood model the IMF has a slope of $x = 1.1$ for $M_* \leq 1$ M$_\odot$ and $x = 1.6$ for $M_* > 1$ M$_\odot$. The star formation rate is set so that the fraction of the total mass in gas is 15% at the present epoch (13 Gyr). In the QSO model the IMF has a slope of $x = 1.1$ for all M_*, and the stellar birthrate is much faster, producing a gas fraction of 15% at 0.5 Gyr.

Figures 1 and 2 compare the evolution of the gas phase metallicities and relative abundances in the two models. In the solar neighbourhood model the abundances are solar at the time of the sun's formation (8.5 Gyr). In the QSO/giant elliptical model the metallicities reach ~ 10 Z$_\odot$ in ~ 1 Gyr and the abundance ratios are quite different from solar. In particular, note the high nitrogen abundance in the QSO model is due to secondary CNO processing in stellar envelopes (see [4]).

The abundances derived in the evolution models are fed directly into the spectral synthesis code CLOUDY to predict the QSO line fluxes. The sensitivity of the line strengths to the spectral synthesis parameters is discussed by Ferland & Hamann in this volume (also [1], [3], and [4]). Here we adopt a hydrogen (HI+HII) density of 10^{10} cm^{-3}, an ionizing photon flux of 10^{20} cm^{-2} s^{-1}, and a continuum shape described in [4] with $\alpha_{ox} \approx 1.24$.

3. QSO Evolution and Galaxy Formation

Figure 3 compares the predicted NV/CIV and NV/HeII line ratios to the QSO observations [4]. The solar neighbourhood model clearly does not fit most of the data. The timescales are too long and the N/C and N/He abundances are too low. The only way to fit the large line ratios at high redshift is to *both* shorten the timescales *and* use an IMF that favours high mass stars. These criteria are met by the QSO/giant elliptical model. The full range of high redshift data can be fit by simply changing the slope of the IMF. Flatter IMFs favour higher metallicities and larger line ratios.

The tendency for larger observed line ratios (Fig. 3) at higher redshifts suggests that the high redshift sources ($z > 2$) often have higher metallicities. This result is uncertain because the line ratios also depend on the physical conditions in the emitting regions (see [4] and Ferland & Hamann this volume). However, if the trend in the line ratios does result from a metallicity dependence, then the trend might imply a mass–metallicity–redshift relation in QSOs that is perhaps analogous to the well known mass–metallicity relation in nearby ellipticals. Thus, the high mass QSOs (and/or host galaxies) form preferentially at high redshifts. Some support for a mass–metallicity–redshift relation in QSOs comes from the positive correlation between the line ratios (i.e. metallicities) and the source luminosities (i.e. masses; see [4]).

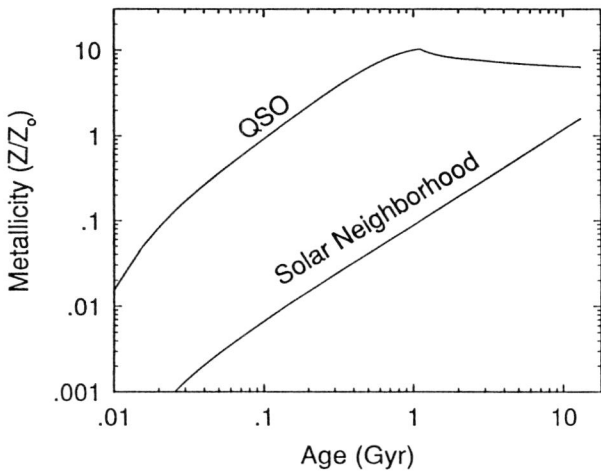

Figure 1. Metallicities normalized to solar are plotted for the solar neighborhood and QSO/giant elliptical models.

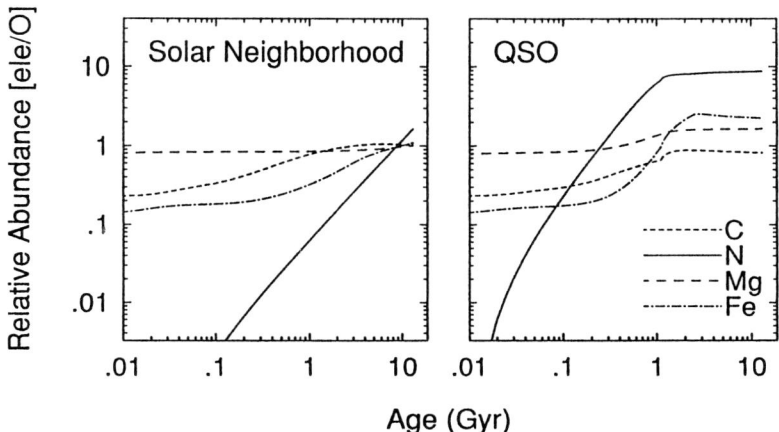

Figure 2. Abundances relative to oxygen and normalized to solar are plotted for both models.

References

[1] Baldwin, J. A., *et al.* 1993, *in prep.*
[2] Wheeler, J.C., Sneden, C., and Truran, J.W., 1989, Ann Rev Ast Ap, 27, 279
[3] Hamann, F., and Ferland, G. 1992, ApJL, 391, L53
[4] Hamann, F., and Ferland, G. 1993, ApJ, *submitted*

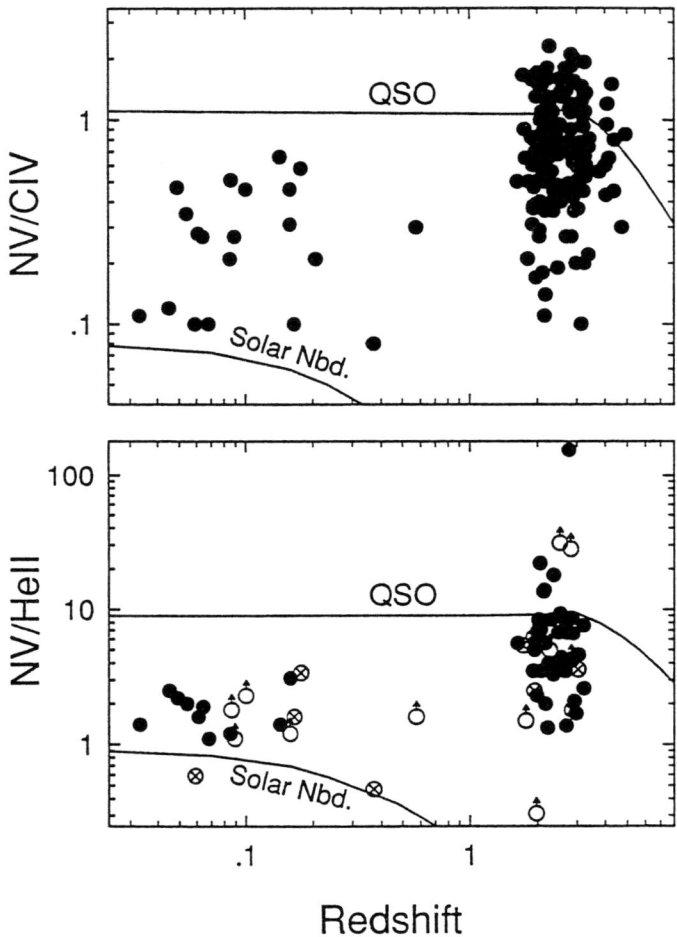

Figure 3. Theoretical line ratios are compared to the QSO observations for a cosmology with $H_o = 75 \, \text{km s}^{-1} \, \text{Mpc}^{-1}$ and $q_o = 1/2$. The curves are drawn assuming the evolution begins with the big bang. The data set is described in [4]. The open circles in the NV/HeII plot are lower limits because HeII is absent. The crossed circles *might* be lower limits because the He IIλ1640 Å measurements probably include some contribution from O III]λ1663 Å.

Nonlinearity of Lyα Response in Variable AGNs

Joseph C. Shields and Gary J. Ferland *

Abstract

Nonlinear relationships between Lyman α and continuum emission are observed in variable Seyfert nuclei as well as in ensembles of QSOs. We report the results of photoionization calculations that consider the relationship between Lyα emissivity and nebular conditions. The nonlinear response observed in variable Seyferts can be explained by increased destruction of the Lyα line at higher ionizing flux levels. The observed slope of Lyα versus continuum brightness can be employed as an independent diagnostic of the ionizing flux incident on the broad-line region (BLR) clouds. Observed Lyα equivalent widths provide additional constraints on the hardness of the ionizing continuum and the BLR covering factor.

1. Introduction

Under simple nebular conditions, hydrogen recombination lines can be treated as tracers of the flux of ionizing photons. For homologous continuum shapes, the flux in a recombination line is simply proportional to the local continuum, implying that the equivalent width of the line remains constant if the continuum (and line) flux levels vary in amplitude. Variable Seyfert nuclei tend to fail this expectation for Lyα, exhibiting a relationship that can be described as a power law $F_{Ly\alpha} \propto F_{contin}^{\beta}$, where $F_{Ly\alpha}$ and F_{contin} are measured fluxes in Lyα and the adjacent continuum, respectively, and $\beta < 1$ (typically $0.5 - 0.7$). Expressed in terms of Lyα equivalent width, this relationship implies $EW(Ly\alpha) \propto F_{contin}^{\beta-1}$, such that equivalent width is inversely related to continuum level.

Several possible explanations exist for nonlinear Lyα response in variable AGNs. One possibility is that our assumption of a luminosity-independent continuum shape is in error, such that the relative proportion of continuum photons at 1216 Å and shortward of 912 Å is a systematic function of luminosity. The ultraviolet continuum

---
*Department of Astronomy, Ohio State University, Columbus, OH 43210, USA. This work was supported by NSF grant AST 90-19692 and NASA grant NAGS-1366.

shape longward of 912 Å is indeed known to vary systematically with continuum amplitude, in the sense that the continuum gets harder as the source brightens [1]. However, the sense of this trend would naively lead us to expect $EW(\text{Ly}\alpha)$ to increase with continuum luminosity, in disagreement with the observations. The continuum region that specifies the Lyα flux level is, of course, that at ionizing energies, so this contradiction could reflect an error in extrapolation from the observable ultraviolet interval. If the observed equivalent width behavior arises from changes in the continuum spectral energy distribution, the observed pattern implies that the optical through extreme-ultraviolet (EUV) continuum softens as it brightens.

An alternate origin of nonlinear Lyα response is nonvariable line-emitting components included in the line flux measurement. Such contributions could arise from optically thin, fully ionized clouds, and also nebular components distributed on spatial scales for which the light-crossing time exceeds characteristic timescales for continuum variation. A mix of clouds responding linearly to the ionizing flux and nonvariable nebular components will result in a line-continuum relation with $\beta < 1$ when expressed as a power law. Unambiguous identification of a nonvariable Lyα component amounts to measuring an intercept line flux when the continuum flux approaches zero. Confident extrapolation to zero continuum levels, or alternatively, detection of curvature on a linear-linear plot, relies heavily on measurements of a variable source when it is in its faintest state. This task is observationally challenging since any given source may never evolve through a particularly faint state, and if it does, the data are correspondingly subject to increased measurement error.

Measurement of a meaningful line versus continuum slope is also complicated by light travel-time effects, which imply that the Lyα flux measured at a particular time should be referenced to the continuum level at an earlier epoch. (If significant Lyα emission is generated over a wide range of radius, there is, in fact, no well-defined time delay appropriate for this comparison.) In the case of NGC 5548, Pogge & Peterson [2] found that correction for time delays in response significantly reduced the scatter of the line versus continuum correlation, but did not linearize it.

2. Balmer Continuum Opacity

Lyman-α emission may also respond nonlinearly to continuum variations if the conditions in the radiating plasma deviate substantially from our assumptions for a simple nebula. Destruction of Lyα photons via Balmer continuum absorption represents one mechanism that can perturb our idealized nebula. Following Ferland & Netzer [3], we can note that the density n_2 of hydrogen in the $n = 2$ state will scale with the recombination rate, while the total depth of ionized gas in an optically thick cloud (the Strömgren depth R_S) is related inversely to recombination rate and directly to the ionizing photon flux $\phi(H)$. Balmer continuum opacity τ_{BC} is proportional to $n_2 R_S$, implying that $\tau_{BC} \propto \phi(H)$, to good approximation. We thus expect destruction of

Lyα photons in hydrogen ionizations to grow as the ionizing photon density increases.

The preceding arguments are supported by detailed calculations employing the photoionization code CLOUDY. Figure 1 shows predicted $EW(Ly\alpha)$ as a function of ionization parameter U (the ratio of ionizing photon to nucleon densities) for optically thick clouds with the indicated hydrogen densities n_H, irradiated by a representative AGN continuum. The continuum here is described by an optical-ultraviolet power law with $\alpha = -0.5$ ($f_\nu \propto \nu^\alpha$), exponentially cutoff at $E_C = 20$ ev to an X-ray power law with $\alpha = -0.7$, with normalization such that α_{OX} (the 2500 Å − 2 keV spectral index) is -1.4. The resulting $EW(Ly\alpha)$ tends to decrease as n_H is increased at fixed U, or U increases at fixed n_H. By definition, $n_H U = \phi(H)$, indicating that the Lyα line is increasingly destroyed at increasing flux levels, consistent with our expectation for $\tau(BC)$.

3. Implications

The results shown in Fig. 1 illustrate a nonlinear behavior of Lyα that is consistent with that seen in variable Seyfert nuclei. If this effect is the predominant cause underlying the observed nonlinearity, the line versus continuum slope can be used to estimate $\phi(H)$ and limit the permitted combinations of n_H and U describing the BLR clouds. The observed values of $1 - \beta$ (-0.5 to -0.3) appear consistent, for example, with values of $U = 0.1$ and $n_H = 10^{11}$ cm^{-3} now commonly adopted for the Lyα–CIV zone.

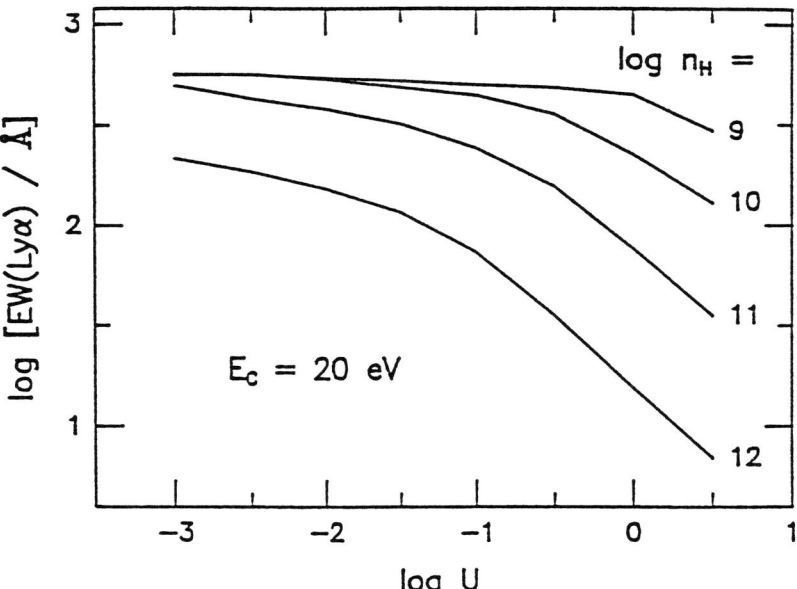

Fig. 1. Equivalent width of Lyα as a function of U. Separate curves correspond to the indicated values of log n_H.

Additional calculations demonstrate that the shapes of the curves shown in Fig. 1 are insensitive to choice of EUV cutoff energy E_C, although their normalization increases with increasing E_C. The dependence of normalization on E_C is easily understood by analogy with Zanstra methods for determining stellar temperatures, i.e., $EW(\text{Ly}\alpha)$ is a measure of the hardness of the ionizing continuum. For comparison with Seyfert galaxy observations, predictions such as those shown in Fig. 1 must also be reduced in normalization by the covering factor $f_c \leq 1$ of emission-line clouds surrounding the central continuum source. Given the results in Fig. 1, an observed $EW(\text{Ly}\alpha)$, and a measured $1 - \beta$, knowledge of E_C thus specifies f_c, and vice-versa. For NGC 5548, for example, $E_C = 20 - 60$ eV corresponds to $f_c = 0.8 - 0.5$. A specific constraint that follows is that E_C can be no less than the value at which $f_c = 1$. An interesting application of this bound is offered by Fairall 9, which exhibits a luminosity-dependent CIV/Lyα ratio that has been invoked as evidence for a very soft E_C (≈ 10 eV) [4,5]. The observed $EW(\text{Ly}\alpha)$ behavior for this object implies that $f_c = 1$ for $E_C \approx 20$ eV; a softer E_C would be difficult to understand.

The inverse relationship between $EW(\text{Ly}\alpha)$ and luminosity observed in variable Seyfert nuclei bears a qualitative resemblance to the Baldwin effect, a similar pattern seen in ensembles of AGNs (originally noted for CIV, but also seen in Lyα [6]). While the similarity between the individual and ensemble behavior is intriguing, it does not necessarily imply a common physical origin. For example, in variable nuclei we can be confident that U is changing as the source brightens, while the same statement is not obviously true for the ensemble. X-ray studies furthermore provide some indications that the covering factor of the central source is inversely related to luminosity within ensembles [7], which provides one possibility for explaining the Baldwin effect. In contrast, individual sources very likely have f_c independent of luminosity, at least to first order. Categorizing the individual and ensemble dependence of $EW(\text{Ly}\alpha)$ on luminosity as a single phenomenon should be avoided unless a firmer connection is drawn between the physical processes underlying the two cases.

References

[1] Edelson, R. A., Krolik, J. H., & Pike, G. F. 1990, ApJ, 359, 86

[2] Pogge, R. W., Peterson, B. M. 1992, AJ, 103, 1084

[3] Ferland, G. J., & Netzer, H. 1979, ApJ, 229, 274

[4] Binette, L., Prieto, A., Szuszkiewicz, E., & Zheng, W. 1989, ApJ, 343, 135

[5] Clavel, J., & Santos-Lleó, M. 1990, A&A, 230, 3

[6] Kinney, A. L., Rivolo, A. R., & Koratkar, A. P. 1990, ApJ, 357, 338

[7] Lawrence, A., & Elvis, M. 1982, ApJ, 253, 410

Implications of Broad Line Profile Diversity among AGN

A. Robinson *

Abstract

The broad emission lines of active galactic nuclei exhibit a wide variety of profile shapes and widths. A simple model is used to illustrate the possible significance of this diversity and to gain some insight into its physical origin.

1. Introduction: from igloos to Eiffel Towers

The broad emission lines are often the most prominent features in the optical-UV spectra of active galactic nuclei. Understanding how they are formed is therefore a key problem, not least because the emitting gas is thought to be closely associated with the fundamental energy source and so must be strongly influenced by its radiation field and the dynamical forces it produces. It is generally believed that the great widths of the line profiles arise mainly from large bulk velocities of the emitting gas. However, the structure and dynamics of the broad line region (BLR) are not understood in detail, although many possible models have been proposed (Section 2).

The shape of the emission line profile is one of the principle observational constraints on theories of the BLR. Initially, it was thought that the observations were adequately described by logarithmic profiles ($L_\lambda \propto -\ln[\Delta\lambda]$), as expected for radiation pressure acceleration (Blumenthal & Matthews 1975). However, as the quality and quantity of data have increased, it is becoming clear that other forms (e.g. power laws—van Groningen 1983; Penston et al. 1990) are often more appropriate. Indeed, casual inspection of various collections of AGN spectra published in recent years reveals that the Hα and Hβ lines exhibit a wide variety of profile shapes and a large range in width (e.g. Osterbrock & Shuder 1982; de Robertis 1985; Crenshaw 1986; Stirpe 1990; Jackson & Browne 1991; Miller et al. 1992; Boroson & Green 1992). Many profiles have classical, concave wings ("Eiffel Towers") but others appear triangular or bell-shaped and some are broad domes ("igloos"). Profile widths range over at least an order of magnitude; in the sample studied by Boroson & Green, for example, Hβ FWHM varies from $\sim 10^3$ to 10^4 km s^{-1}. A similar diversity in shape and width seems to be present among the ultra-violet lines (Lα, CIV, CIII]; e.g. Steidel & Sargent 1992; Francis et al. 1992).

---

* Institute of Astronomy, Madingley Rd, Cambridge, CB3 0HA

It is well known that the broad emission lines in the spectra of individual AGN often vary in shape, width and velocity shift (e.g. Espey, this volume and references therein). Broad line profiles are also commonly asymmetric (de Robertis 1985; Stirpe 1991; Boroson & Green 1992). However, I will concentrate here on the diversity in the basic profile shape (and width) among the AGN population. I will argue that this diversity is an important observed property which may have important implications for other aspects of AGN and whose physical origin we should therefore seek to understand.

2. The importance of profile diversity

If we accept that profile broadening is due to the macroscopic velocity dispersion of the emitting gas, then it is likely that the observed diversity of the line profiles among AGN reflects a corresponding range in the structure and dynamics of the BLR or in the orientation of some physically important axis (c.f. Wills & Browne 1986). This raises some important questions. Do profile differences indicate a continuous variation in one or two dominant parameters (as in unified schemes) or are families of AGN distinguished by fundamental structural differences? Similarly, does the diversity arise from fixed intrinsic differences in the properties of the BLR or is it associated with changes in an evolving or stochastically fluctuating system? Clearly, understanding the origin of the observed diversity and in particular, identifying the underlying physical variable(s), would represent a major advance in our understanding of AGN.

In principle, the geometry and kinematics of the BLR can be studied directly by analysing emission line variability using the reverberation mapping techniques reviewed by Peterson elsewhere in this volume. However, reverberation mapping demands lengthy and intensive monitoring campaigns and so, in practice, it can only be applied to relatively few objects. Moreover, to maximize the chances of success, the targets are likely to be selected because they are already known to be strongly variable. If profile diversity indicates a variation in BLR properties, objects selected mainly on the basis of their variability may not be representative of the AGN population as a whole. Evidently, it is necessary to understand the origin of profile diversity in order to appreciate the general significance of reverberation studies

Another way of tackling this problem is to carry out statistical studies of large samples of objects. Two recent studies have applied principal components analysis (PCA) to large sets of quasar spectra. Francis et al. (1992) analysed the UV spectra of over 200 quasars by treating the emitted flux in each wavelength interval as a separate variable, thus avoiding the need to make possibly subjective measurements from complex spectra. They propose a 2-dimensional classification scheme for quasar spectra based on the major sources of variance in their sample: the strength of the core component of the emission line profile and the slope of the UV continuum. Boroson & Green (1992), on the other hand, studied a sample of 87 low redshift quasars by applying PCA to a range of properties measured from Hβ spectra or compiled from the literature. They identify the principle source of variance as an anti-correlation between the strengths of the (broad) FeII and the (narrow) [OIII] lines, objects with weak FeII but strong [OIII] tending to have broader Hβ profiles. Radio-loud quasars appear to represent extreme cases rather than a distinct sub-population. These studies reveal some intriguing relationships between observed properties. Nevertheless, the problem remains of how these empirical correlations are related to the underlying physical variables.

Unfortunately, there is as yet no generally accepted theoretical framework to aid the interpretation of such results. Models of the BLR are almost as diverse as the line profiles. Accretion disks (e.g. Chen et al., 1989; Dumont & Collin-Souffrin 1990), bi-conical jets (Zheng, Binette & Sulentic 1991), winds (Perry & Dyson 1985; Smith & Raine 1988), combinations of disks and winds (Collin et al.; Cassidy & Raine 1992), Keplerian ensembles of clouds (Kwan & Carroll 1982; Bradley & Puetter 1986) or irradiated stars (e.g. Kanzanas 1989; Penston et al. 1990) and supernovæ (Terlevich et al. 1992) have all been discussed in recent years. Although some models indicate possible candidates for the underlying parameters and others have stochastic features which might be expected to result in a variety of profile shapes, there is no well-established 'standard model' for the structure and dynamics of the BLR which can be confidently employed to interpret observational data.

3. A parametric model

Rather than attempt to assess the merits of the various existing models I will try instead to illustrate the potential importance of profile diversity and (hopefully) gain some insight into its physical origin using a simplified model based on a parameterization of some basic properties of the BLR, namely, its volume emissivity, radial velocity distribution and radial depth.

Suppose that the BLR is a system of discrete cloudlets surrounding the continuum source and confined within well-defined boundaries r_0 and r_s, so that its radial depth is measured by the ratio $y = r_s/r_0$. Profile broadening results from the velocity dispersion of the cloud ensemble. I assume that the system is characterized by a radial velocity law, $v = v_0 (r/r_0)^p$, and that the volume emissivity in a given line may be similarly specified as $\varepsilon_V = \varepsilon_0 (r/r_0)^\beta$. In many dynamical models these properties are related (e.g. if mass is conserved within the cloud system, the cloud volume density, and hence the line emissivity, are functions of velocity) but the precise nature of the relationship is model-dependent and for current purposes it is more appropriate to treat y, p and β as independent parameters. Note also that the precise configuration of the emitting gas (i.e. whether it is distributed in clouds or more coherent structures) is not too important; the fundamental assumption of the model is that its properties are smooth, single-valued functions of radius.

In the case of a spherically symmetric cloud ensemble with a radial or chaotic (i.e., velocity vectors randomly oriented) velocity field the line profile is given by

$$L = constant \times \left(\frac{v_{max}}{v_0}\right)^\eta \left[1 - (\max(x_{min}, x))^\eta\right]$$

where $x = c|\lambda - \lambda_0|/v_{max}\lambda_0$, v_{max} and v_{min} are, respectively, the maximum and minimum radial velocities, $x_{min} = v_{min}/v_{max} = y^{-|p|}$ and $\eta = (\beta + 3)/p - 1$. The profile has a flat-topped core of $1/2$-width x_{min} ($x \leq x_{min}$) and wings ($x_{min} < x \leq 1$) whose shape is determined by the parameter η; concave wings resulting for $\eta < 1$ but convex wings when > 1 (Fig. 1a). In the special case $\eta = 0$, the profile has the logarithmic wings, $L_\lambda \propto \ln(1/x)$ predicted by several dynamical models (e.g. Capriotti et al. 1980). Families of profiles for rotating disks and bi-conical outflows can be similarly obtained (Figs. 1b,c). Unlike the spherical shell, both of the latter configurations produce double-peaked profiles (in fact, the bi-conical profile is split into two halves). In the bi-conical model the profile shape also depends on the inclination of the axis to the sight-line and, to a lesser extent, the cone opening angle. In the simple disk model, the only effect of inclination is to scale the line-

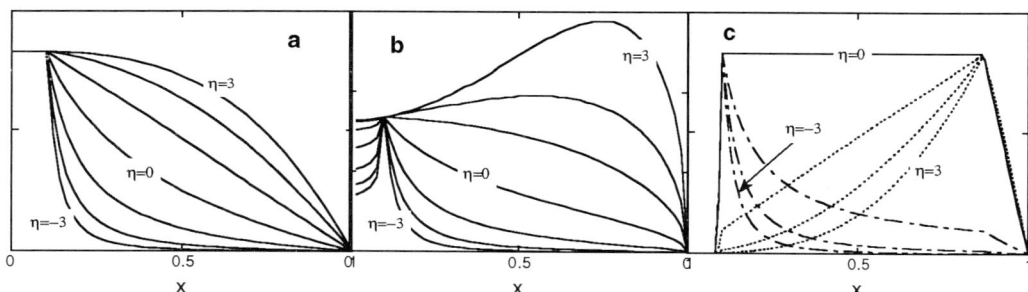

Figure 1. Line profiles for (a) shell (b) disk and (c) bi-conical models.

of-sight velocity and hence it does not influence the profile shape.

There are a variety of physical effects not considered here (e.g. obscuration and scattering; relativistic effects; variability) which are capable of modifying the profile shape or introducing asymmetries. Nevertheless, for each geometrical configuration, both the basic structure and the perceived width of the profile (i.e., FWHM or similar) are mainly determined by the parameters x_{min} and η. Thus, the profile is broad and "stumpy" when $\eta > 1$ and $x_{min} \to 1$, but relatively narrow, with prominent concave wings, when $\eta \approx -1$ and $x_{min} \to 0$. It seems reasonable, therefore, to assert that the observed diversity is the result of a variation in core width (x_{min}) and/or wing curvature (η) among the AGN population. A casual examination of published spectra suggests that this idea is at least plausible, while the few existing studies of profile wing shapes show that curvature does indeed appear to vary from object to object (van Groningen 1983; Penston et al. 1990).

A proper investigation of this proposal must await a systematic and sophisticated analysis of a large sample of high quality spectra, which would have to overcome several serious practical difficulties, notably the problem of line blending. Nevertheless, we can attempt to set constraints on x_{min} and η using existing empirical profile parameterization schemes such as that based interpercentile velocity widths (IPV) proposed by Whittle (1985). The advantage of the IPV scheme is that it is based on integrated intensity and is thus rather more robust than one based on simple fractions of peak intensity. The only existing data set for which IPV's are measured for the broad lines is that of Stirpe (1990, 1991). This comprises high quality spectra of Hα and Hβ profiles for 30 low redshift AGN and, using ratios of IPV widths to crudely characterize profile shape, these data can be compared with theoretical grids calculated for the spherical shell model outlined above (Fig. 2). The diversity in IPV shape and width is essentially independent of continuum luminosity (this also appears to be true of Boroson & Green's much larger sample) and nor is it significantly reduced if the most asymmetric or lumpy profiles are removed, suggesting that this comparison is not greatly influenced by the 'modifying effects' which have been omitted from the model. Assuming, therefore, that the observed ranges in IPV shape and width are entirely due to variations in core width and wing curvature, the Hα and Hβ profiles in the sample correspond to the ranges $0.03 \leq x_{min} \leq 0.3$ and $-2 \leq \eta \leq 1$, with a maximum velocity $v_{max} \approx 1.2 \times 10^4$ km s^{-1}. Note that the correlation between shape and

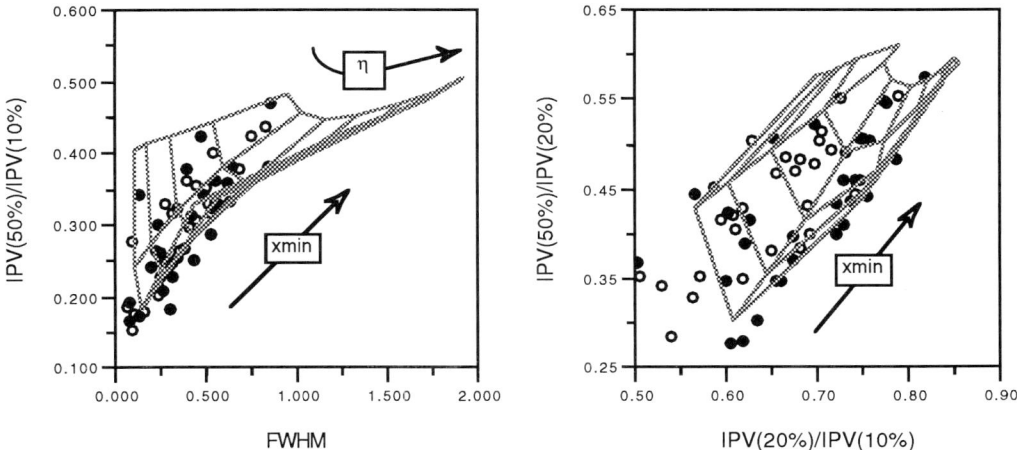

Figure 2. Comparison of interpercentile widths and ratios for Stirpe's (1990) sample of Hα (open circles) and Hβ (filled circles) profiles with model grids ($\eta - x_{min}$) for a spherical shell.

FWHM suggests that the dispersion in FWHM is mainly determined by η and x_{min}, i.e. profile shape, rather than by a variation of v_{max} within the sample.

Similar conclusions hold for the other geometrical configurations, albeit with somewhat tighter restrictions on the ranges in x_{min} and η for the bi-conical case. In general, however, the IPV parameterization does not allow a clear distinction between the three geometries; the respective model grids are rather similar over much of the relevant ranges in x_{min} and η. While the apparent rarity of double peaked profiles (Jackson et al., 1991) would seem to argue against disk and bi-conical configurations for most AGN, these features merge at small x_{min} and in disks they are suppressed for $\eta \geq 0$. Double-peaked profiles can also be disguised by, for example, emission from the narrow line region, or the presence of random turbulence. However, it seems unlikely that axial orientation is the principle cause of profile diversity in this (mostly radio-quiet) sample. Even for the bi-conical model where both the shape and width of the profile depend on the angle of inclination, the effect is relatively small for the values of x_{min} and η which are of interest.

4. Implications for BLR structure and dynamics

If this interpretation of the observed profile diversity is even partially correct, the inferred ranges in x_{min} and η imply corresponding ranges in one or more of β, y, p and thus a dispersion in the basic properties of the BLR among the AGN population (even assuming a universal geometry). The ranges in profile core width and wing shape can be explained by a variation in the velocity law only, since both of these properties are functions of p. In this case, a relationship between the two is predicted, moderately concave wings ($\eta \rightarrow -1$) being associated with narrow cores ($x_{min} \rightarrow 0$). Conversely, if p is fixed, then x_{min} and η are independent and variations in both the emissivity distribution (β) and the radial extent (y) of the BLR are implied. Note that, unless the velocity law is rather steep ($|p| \geq 2$), the inferred range in x_{min} implies that the BLR has a significant radial depth. For example, if $p = -0.5$, as for Keplerian dynamics, $y \geq 10$ (but more typically ~100). Dispersions in β and y will then generate a large range in the luminosity-weighted radius, R, of the BLR (i.e. the effective size — Robinson & Pérez 1990), even if the inner radius, r_0, is fixed; i.e. irrespective of the absolute scale. This, in turn, has notable consequences for a number of observable properties of the BLR.

The transfer function which, according to the conventional reverberation picture, governs the response of the broad emission lines to ionizing continuum fluctuations (Blandford & McKee 1982), itself depends on the volume emissivity distribution and the radial extent of the BLR (e.g. Robinson & Pérez 1990). It follows that the emission line light curve will be related to the profile shape[1]. The centroid of the transfer function (which is simply proportional to the luminosity-weighted radius; $\tau_{cen} = R/2r_s$) is inversely related to the profile FWHM (Fig. 3). Thus broader profiles ($x_{min} \rightarrow 1$; $\eta > 0$ if $p < 0$ or $\eta < 0$ if $p > 0$) are

---

[1] The profile shape may also change in response to ionizing continuum fluctuations — e.g., Pérez, Robinson & de la Fuente (1991).

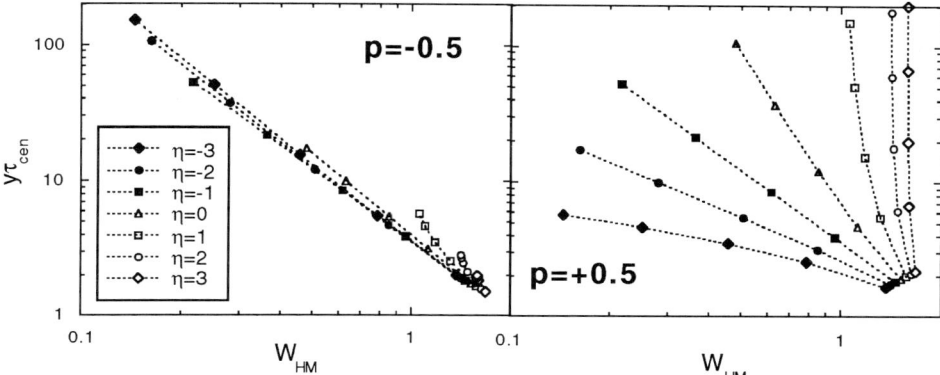

Figure 3. Relationship between the transfer function centroid (normalized to r_0) and profile FWHM for the spherical shell model.

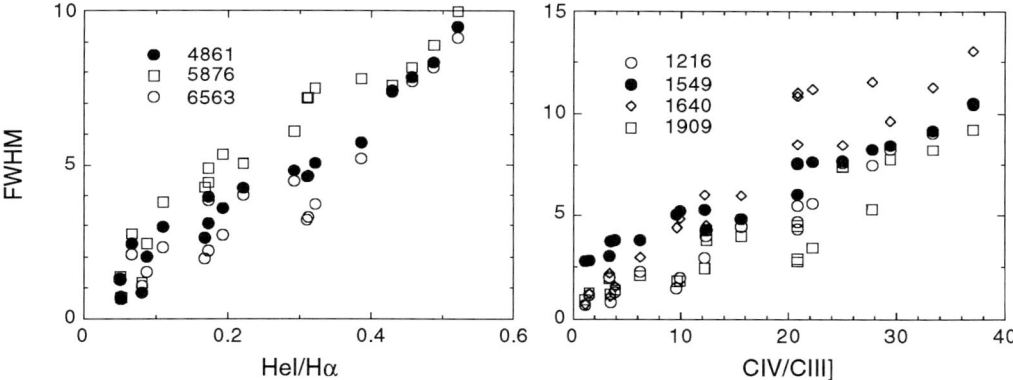

Figure 4. Relationship between line intensity ratios and profile FWHM for cloud ensemble photoionization models with parameters corresponding to the inferred ranges in η and x_{min}.

associated with small τ_{cen} and R and hence a rapid, large amplitude response to continuum fluctuations. Conversely, narrow profiles correspond to a relatively slow, weak response. For a fixed absolute scale (i.e. constant r_0) the inferred range in x_{min} and η corresponds to a factor 10 range in τ_{cen}. For the standard photoionization model, in which $R \propto r_0 \propto L_{ion}^{1/2}$, this range in effective size would correspond to a factor 100 in continuum luminosity.

Profile diversity may also have implications for the observed level of excitation as indicated by the relative intensities of the broad emission lines. The relative intensities reflect the emissivity-weighted ionization parameter which will be a function of β and y, unless the gas density $n \propto r^{-2}$. If these are the underlying variables, therefore, the observed diversity in the profile shape implies a range in excitation level among AGN. In general, the radial variation of the line reprocessing efficiency within the BLR will differ from line to line (e.g. Rees et al. 1989), modifying. the volume emissivity distribution (i.e. the effective value of β) with the result that the profile shapes of various emission lines (e.g. Lα, CIV, Hα) differ significantly in individual objects. Nevertheless, correlations between ionization-sensitive line intensity ratios and, say, the FWHM, of any given line can be expected (Fig. 4). Furthermore, since the transfer function also depends on β, there will be a relationship, in individual AGN, between profile shape (and hence width) and τ_{cen} (and hence variability amplitude and timescale). Since high ionization lines (e.g. CIV) will generally have steeper emissivity distributions than low ionization lines (e.g. MgII, Hα), the former will be associated with smaller R, have broader profiles (if $p < 0$) and will be more variable than the latter. It is worth emphasizing that such trends, which have already been observed in some objects (Clavel et al. 1987; Krolik et al. 1991), arise naturally in radially extended regions with smoothly varying properties, as in the model considered here, and do not necessarily imply that various lines arise in physically distinct components of the emitting gas.

5. Conclusions

Understanding the physical origin of the observed diversity in the broad emission line profile shapes and widths among the AGN population is an important goal. The simple model discussed here suggests that this diversity can reasonably be interpreted in terms of variations in profile core width and wing curvature. This would imply significant variations in either the radial velocity law or the line emissivity distribution and the radial depth of the broad line region. If the latter two properties are the underlying variables, then the effective size of the BLR has a wide range, and both the variability behaviour and the relative intensities of the broad lines will be related to the profile shape.

References

Blandford, R. & McKee, C.F., 1982. *Astrophys. J.*, **255**, 419.
Blumenthal, G.R. & Matthews, W.G., 1975. *Astrophys. J.*, **198**, 517.
Bradley, S.E. & Puetter, R.C., 1986. *Astr. Astrophys.*, **165**, 31.
Boroson, T.A. & Green, R.F., 1992. *Astrophys. J. Suppl. Ser.*, **80**, 109.
Capriotti, E., Foltz, C. & Byard, P., 1980. *Astrophys. J.*, **241**, 903.
Cassidy, I. & Raine, D.J., 1992. *Mon. Not. R. astr. Soc.*, in press.
Chen, K., Halpern, J.P. & Filippenko, A.V., 1989. *Astrophys. J.*, **339**, 742.
Clavel, J., et al., 1987. *Astrophys. J.*, **321**, 251.
Collin-Souffrin, S., Dyson, J.E., McDowell, J.C. & Perry, J.J., 1988. *Mon. Not. R. astr. Soc.*, **232**, 539.
Crenshaw, D.M., 1986. *Astrophys. J. Suppl. Ser.*, **62**, 821.
de Robertis, M., 1985. *Astrophys. J.*, **289**, 67.
Dumont, A.M. & Collin-Souffrin, S., 1990. *Astr. Astrophys.*, **229**, 313.

Francis, P.J., Hewett, P.C., Foltz, B.C. & Chaffee, F.H., 1992. *Astrophys. J.*, **398**, 476.
Jackson, N.J. & Browne, I.W.A., 1991. *Mon. Not. R. astr. Soc.*, **250**, 414.
Jackson, N.J., Penston, M.V. & Pérez, E., 1991. *Mon. Not. R. astr. Soc.*, **249**, 577.
Kanzanas, D., 1989. *Astrophys. J.*, **347**, 74.
Krolik, J.H., Horne, K., Kallman, T.R, Malkan, M.A., Edelson, R.A. & Kriss, G.A., 1991. *Astrophys. J.*, **371**, 541.
Kwan, J. & Carroll, T.J., 1982. *Astrophys. J.*, **261**, 25.
Miller, P., Rawlings, S., Saunders, R. & Eales, S., 1992. *Mon. Not. R. astr. Soc.*, **254**, 93.
Osterbrock, D.E. & Shuder, J.M., 1982. *Astrophys. J. Suppl. Ser.*, **49**, 149.
Penston, M.V., Croft, S., Basu, D. & Fuller, N., 1990. *Mon. Not. R. astr. Soc.*, **244**, 357.
Pérez, E., Robinson A. & de la Fuente, L., 1992a. *Mon. Not. R. astr. Soc.*, **256**, 103.
Perry, J.J. & Dyson, J.E., 1985. *Mon. Not. R. astr. Soc.*, **213**, 665.
Robinson, A. & Pérez, E., 1990. *Mon. Not. R. astr. Soc.*, **244**, 138.
Rees, M.J., Ferland, G.J. & Netzer, H., 1989. *Astrophys. J.*, **347**, 640.
Smith, M.D. & Raine, D.J., 1988. *Mon. Not. R. astr. Soc.*, **212**, 425.
Steidel, C. & Sargent, W.L.W., 1992. *Astrophys. J.*, **382**, 433.
Stirpe, G.M., 1990. *Astr. Astrophys. Suppl. Ser.*, **85**, 1049.
Stirpe, G.M., 1991. *Astr. Astrophys.*, **247**, 3.
Terlevich, R., Tenorio-Tagle, G., Franco, J. & Melnick, J., 1992. *Mon. Not. R. astr. Soc.*, **255**, 713.
van Groningen, E., 1983. *Astr. Astrophys.*, **126**, 363.
Whittle, M., 1985. *Mon. Not. R. astr. Soc.*, **213**, 1.
Wills, B.J. & Browne, I.W.A., 1986. *Astrophys. J.*, **302**, 56.
Zheng, W., Binette, L. & Sulentic, J.W., 1991. *Astrophys. J.*, **365**, 115.

Emission Line Studies of AGN

Brian R. Espey [*]

Abstract

The Hubble Space Telescope (HST) has been producing data from regularly scheduled observations for some time now. At this juncture, it is an appropriate time to review some of the areas where instruments such as the Faint Object Spectrograph (FOS), when coupled with ground-based data, may improve studies of Active Galactic Nuclei (AGN) and, hopefully, stir the imaginations of the theorists! We show examples of objects where the data indicate that the emission line properties of high- and low-luminosity objects are similar, lending support to relatively simple models in which the emission line regions of AGN scale homologously. We note, however, that the definition of what the 'normal' properties of AGN are rather than individual peculiarities will have to await complete samples of objects with data covering as large a wavelength range as possible.

1. Line Ratios.

It has become feasible in recent years to study the Balmer lines of high redshift QSOs in some detail and to determine velocity-resolved ratios such as those of Lyα/Hα – see, for example, [1],[2], [3]. Even more recently, it has become possible to regularly obtain UV data with the HST and to perform more detailed analysis than was possible using IUE spectra. As an example, in Figure 1 we show a comparison between the Lyα/Hα line ratios in two radio-loud objects of similar linewidth and velocity shifts.

The 3C 273 data are from [4] and the PKS 1448 − 232 are taken from [2]. It is readily apparent that the line ratios in these two objects are in remarkable agreement in shape (and probably also in intensity given the slight uncertainties in the Galactic extinction correction) despite the 2.5 magnitude difference in luminosity. This result supports arguments for the similarity of the Broad Emission Line Regions (BELRs) in these objects both in terms of ionization and velocity structure and lends support to simple models of AGN in which objects of differing luminosity represent merely scaled versions of each other. This suggestion is certainly not new [5], but confirmation is certainly important in an age when the number of creatures in the AGN 'zoo' has increased markedly.

---

[*]Department of Physics and Astronomy, University of Pittsburgh, Pittsburgh, PA 15260. USA.

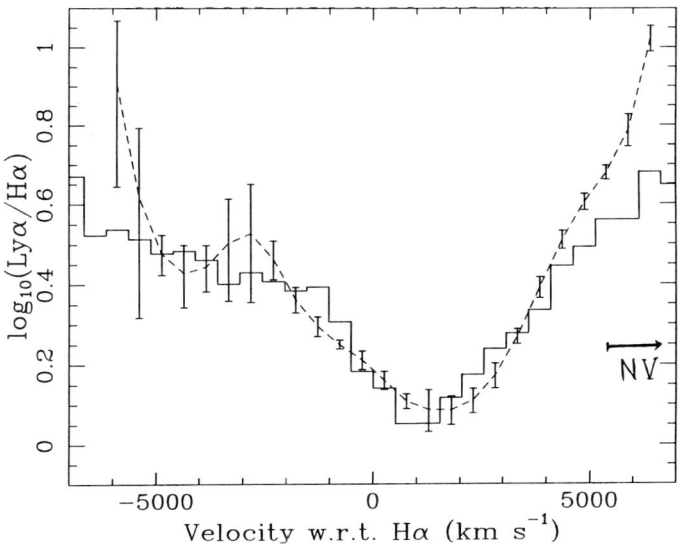

Figure 1. Data for 3C 273 (histogram) and PKS 1448−232 (solid curve with error bars). These objects have similar linewidths, profile shapes and velocity shifts between low- and high-ionization lines and hence are good candidates for comparison. Note the similarity of the velocity-resolved Lyα/Hα ratios over the region covered.

2. Velocity Field.

We expand on the theme of similar velocity fields in Figure 2: Figure 2a shows a plot of linewidth *vs.* critical density for a sample of objects selected from the literature together with more recent data. Straight lines have been fitted and an offset for each object calculated so that the lines would intersect at the left-hand edge of the figure. The dispersion in the data is larger than the typical error, due to the contribution of more than one emitting component to each line. It is apparent from Figure 2b that the best-fit slope remains roughly constant over $\approx$ 2000 range in luminosity. The generally accepted view of this relationship is that it is due to infalling optically thick clouds in parabolic orbits [6],[7] with the smaller than expected slope due to the contamination of the linewidths by material at very different densities and distances. Some observations of variable AGN support the infall model [8], but models involving radial outflow have also been postulated [9].

Although the basic result of a relationship between density and linewidth is not new, it does show two things: 1) that a relationship discovered to hold in low luminosity objects also holds in more luminous systems, and: 2) that this relationship holds over the range of radii from the Narrow Emission Line Region (NELR) right into the more ionized parts of the BELR. The lack of any pronounced influence of the ionizing spectrum is interesting as it can be readily shown that optically thin clouds would be dominated by the radiation pressure of the central object. If clouds are to fall inwards then they must be thick and not subject to too much disruption. Detailed study of the FWHM *vs.* n_{crit} relationship, as well as the relation of the derived velocity field to the line asymmetries and line shifts seen quite pronouncedly in some objects, but not in others, should well repay analysis.

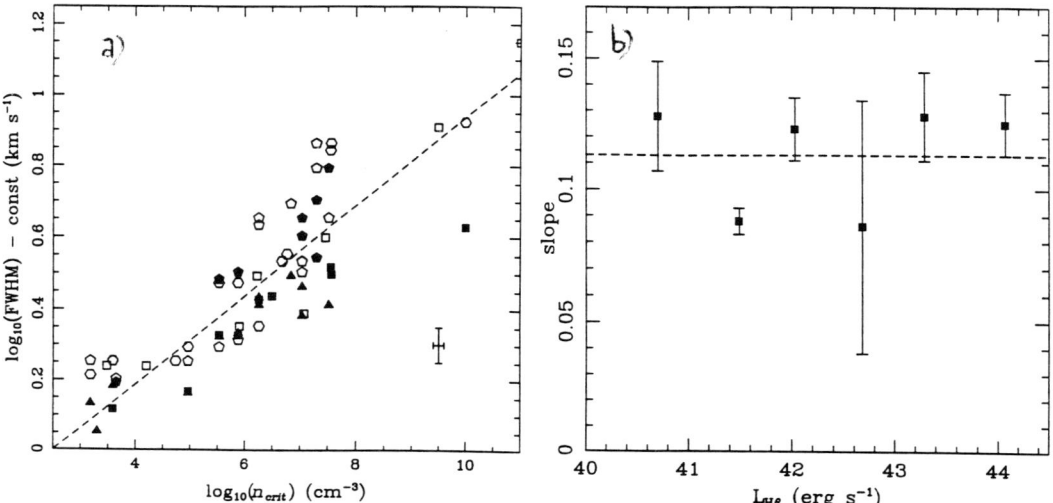

Figure 2. **a)** Collected data for a number of objects ranging from LINERs to QSOs. Note that the data scatter about a roughly similar line. A line with slope 0.125 is shown for comparison. **b)** This figure shows the distribution of the best-fit slopes to the data plotted in Figure 2a. Note that there is no trend for the slope to alter as the luminosity increases.

3. A Very Broad Line Region(?)

A natural extension of the relationship between linewidth and density is the implication that there might exist a region interior to the conventional BELR which has been called the 'Very Broad Line Region' by Ferland and his co-workers [10]. Penston *et al.* [11] and Zheng [12] amongst others have shown that the far wings of the Hβ lines in their respective samples may be well fitted with power-law profiles (but see [13]), which suggests that the material is of low angular momentum and probably infalling. From variability arguments, Zheng argues in favour of a luminosity-dependent velocity field with non-variable optically thin material occurring preferentially in more luminous objects.

4. Line Ratios and Profile Shapes.

One of the problems with current models for AGN is that they have difficulties in predicting the structure of the most luminous objects. One case in point is the question of accretion disk models in which the luminosity of the central source has a major influence on the structure of the accretion disk itself. As an example, we note that the BELR Lyα/Hα line ratio in high luminosity objects is similar to that seen in objects of lower luminosity, although in the accretion disk model, the 'disk' is very probably broken up under the action of self-gravity. Nevertheless, the approach of simultaneously attempting to explain the emission line strengths and profiles self-consistently is a direction in which future work should go. As examples, see the work of Rokaki et al. [14] and Kallman et al. [15].

References

[1] Espey, B.R. PhD Thesis, University of Cambridge.
[2] Espey, B.R., Carswell, R.F., Bailey, J.A., Smith, M.G. and Ward, M.J. 1989. *A.J.* **342**, 666.
[3] Carswell, R.F. et al. *Ap.J.Letters* **381**, L5.
[4] Burbidge, E.M. et al. 1991. *B.A.A.S.* **23**, 1425.
[5] Davidson, K. and Netzer, H. 1979. *Rev. Mod. Phys.*
[6] Fillipenko, A.V. 1985. *Ap.J.* **289**, 475.
[7] Carroll, T.J. and Kwan, J. 1983. *Ap.J.* **274**, 113.
[8] Koratkar, A. and Gaskell, C.M. 1991. *Ap.J.Letters* **370**, L61.
[9] Rosenblatt, E.I. and Malkan, M.A. 1990. *Ap.J.*, **350**, 132.
[10] Ferland, G.J., Korista, K.T. and Peterson, B.M. 1990. *Ap.J.Letters* **363**, L21.
[11] Penston, M.V., Croft, S., Basu, D. and Fuller, N. 1990. *M.N.R.A.S.* **244**, 357.
[12] Zheng, W. 1992. *Ap.J.* **385**, 127.
[13] van Groningen, E. 1983. *Astron. & Astrophys.* **126**, 363.
[14] Rokaki, E., Boisson, C. and Collin-Souffrin, S. 1991. *Astron. & Astrophys.* **253**, 57.
[15] Kallman, T.R., Wilkes, B.J., Krolik, J.H. and Green, R. 1992. *Ap.J. accepted*.

A Search for Velocity Shifts in QSO Broad Lines

Todd A. Small & Wallace L.W. Sargent [*]

Abstract

We have embarked on a search for temporal velocity shifts, on timescales of a year or so, in QSO broad lines. Eighteen quasars in a total sample of thirty have been analysed so far, and no evidence for velocity shifts has been found. If no shifts are found in the remaining twelve QSOs, then our results will call into question the identification of certain QSO pairs as lensed systems and the limits on the sizes and spatial distribution of Lyman α clouds derived from such systems.

1. Introduction

Steidel and Sargent (1991) recorded high signal–to–noise ratio spectra of the QSO pairs Q1634+267A,B and Q2345+007A,B in order to determine whether the systems were gravitationally lensed. They concluded that the systems were both lensed on the strength of detailed comparisons of the line profile and continuum shapes. However, they also discovered that the Ly α, N V $\lambda1240$, C IV $\lambda1549$, and Si IV $\lambda1400$ emission lines of the two images of Q1634+267A,B exhibit a relative velocity shift of as much as ~ 1000 km s^{-1}. The authors favored the explanation that the redshifts of individual lines in QSO spectra vary with time and that, due to the roughly one year time delay between the light paths of the two images, one is seeing the individual QSO at two different times. Here, we describe our efforts to verify this prediction. We have reobserved 30 objects from the Mg II $\lambda2800$ survey of Steidel and Sargent (1992), being careful to use an identical instrument configuration and to obtain similar quality spectra. We have performed a preliminary analysis of 18 objects and have not found evidence for dramatic variations. In fact, of ten the agreement between the spectra taken at two different epochs—and reduced by a different person using a different reduction package—is striking.

---

[*]Palomar Observatory, 105–24, California Institute of Technology, Pasadena, CA, 91125. TAS was partially supported by an NSF Graduate Fellowship, and WLWS was partially supported by NSF Grant AST89-19791.

2. Observations

Our sub-sample consists of QSOs preferentially chosen to be bright and to have $z \sim 1$. All observations were made with the Double Spectrograph mounted at the Cassegrain focus of the Palomar Observatory's Hale 5.08 m telescope. Our setup covered the wavelength range from $\sim 3100 - 4700$ Å with the blue camera and from $\sim 4600 - 7000$ Å with red camera, sufficient to observe redshifted C IV $\lambda 1549$, C III] $\lambda 1909$, and Mg II $\lambda 2800$. The resolution was ~ 4 Å on the blue side and ~ 6 Å on the red side.

3. Preliminary Results

We have made preliminary comparisons of the spectra of 18 QSOs. We have found not one convincing case for a velocity shift so far. On the other hand, we have found startling agreement between the two epochs for several objects. An example is shown below.

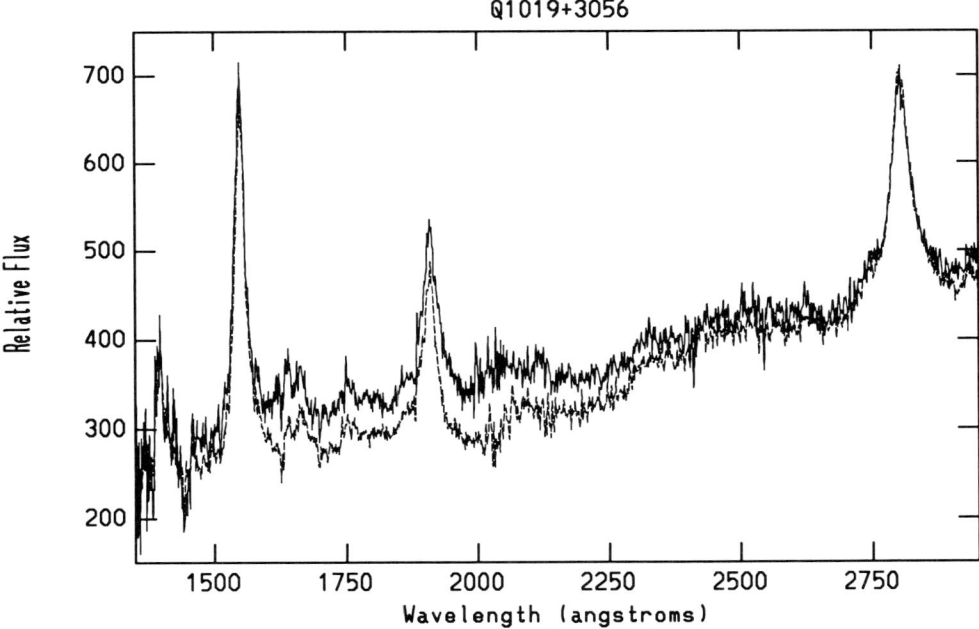

Figure 1. Spectra of Q1019+3056 recorded on 1990 January 22 (dotted line) and on 1992 January 31 (solid line). The wavelength scale has been transformed to the QSO rest frame $z=1.316$

If broad line velocity shifts continue to be absent in the twelve remaining objects, it will, at the least, raise significant questions about the nature of Q1634+267A,B and

Q2345+007A,B as lensed systems. Moreover, limits on the sizes and distributions of Lyman α clouds derived on the assumption that these systems are lensed will have to be revised.

References

Steidel, C. S. and Sargent, W. L. W., 1991. *Astron. J.*, **102**, 1610.
Steidel, C. S. and Sargent, W. L. W., 1992. *Astrophys. J. Suppl. Ser.*, **80**, 1.

The Broad Line Region Structure from Profile Shapes

Keith L. Thompson [*]

1. Model

The broad line profiles of AGN in the standard model represent our view of the v_z components of a system of emission clouds: a collapse of the cloud emissivity $E(\vec{v}, \vec{r})$ in phase space onto v_z. If we assume that all radii emit with a velocity distribution shape S depending only on r, we can write the profile $P(v_z)$ as the integral of S multiplied by a cloud emission function $F_r(r)$,

$$P(v_z) = \int F_r(r) S(v_z; r) dr.$$

Assuming S and measuring P, we can invert the equation to determine F, the net line emission from the clouds as a function of r. This is similar to Puetter and Hubbard's (1987) approach, although we assume a particular relationship between radius and velocity (gravitational dynamics), and allow only trivial forms of cloud emission asymmetry. Note also that F is closely related to the transfer function derived from reverberation mapping.

The models chosen for this exercise are clouds in circular orbits confined to a plane; clouds in parabolic orbits distributed with spherical symmetry; clouds in orbits approximating star clusters with a central massive black hole. Using these models and a few assumptions we can derive F_r. The three models give similar F_r functions. All models include relativistic beaming and gravitational redshifting, which involves a small degree of asymmetry (see Chen, Halpern, and Filippenko 1989). The cluster model has $S(v_z, r)$ derived from the model of Bahcall and Wolf (1976).

Figures 1 and 2 show the cluster model emission functions F_r for two QSOs. Note the shapes can be approximated by broken power laws. For most objects, the first index is between about -0.7 and -1.0, followed by a turndown generally steeper than -2. Since the two sides of the profile are both used to determine F_r, the residuals represent one check on the model (shown with the profiles below the F_r plots). The two Hα profiles displayed are among the best examples; several in the sample are not sufficiently symmetric for a good fit. The most extreme example is 1803+676, representing a complete failure of the model, though improper deblending of Fe II might explain other poor fits.

---

[*]University of Texas at Austin and California Institute of Technology. Partial support for this work came from NSF grant AST-9117100. Current address: Naval Research Laboratory, Code 7210, 4555 Overlook Ave. SW, Washington, D.C. 20375-5351, USA

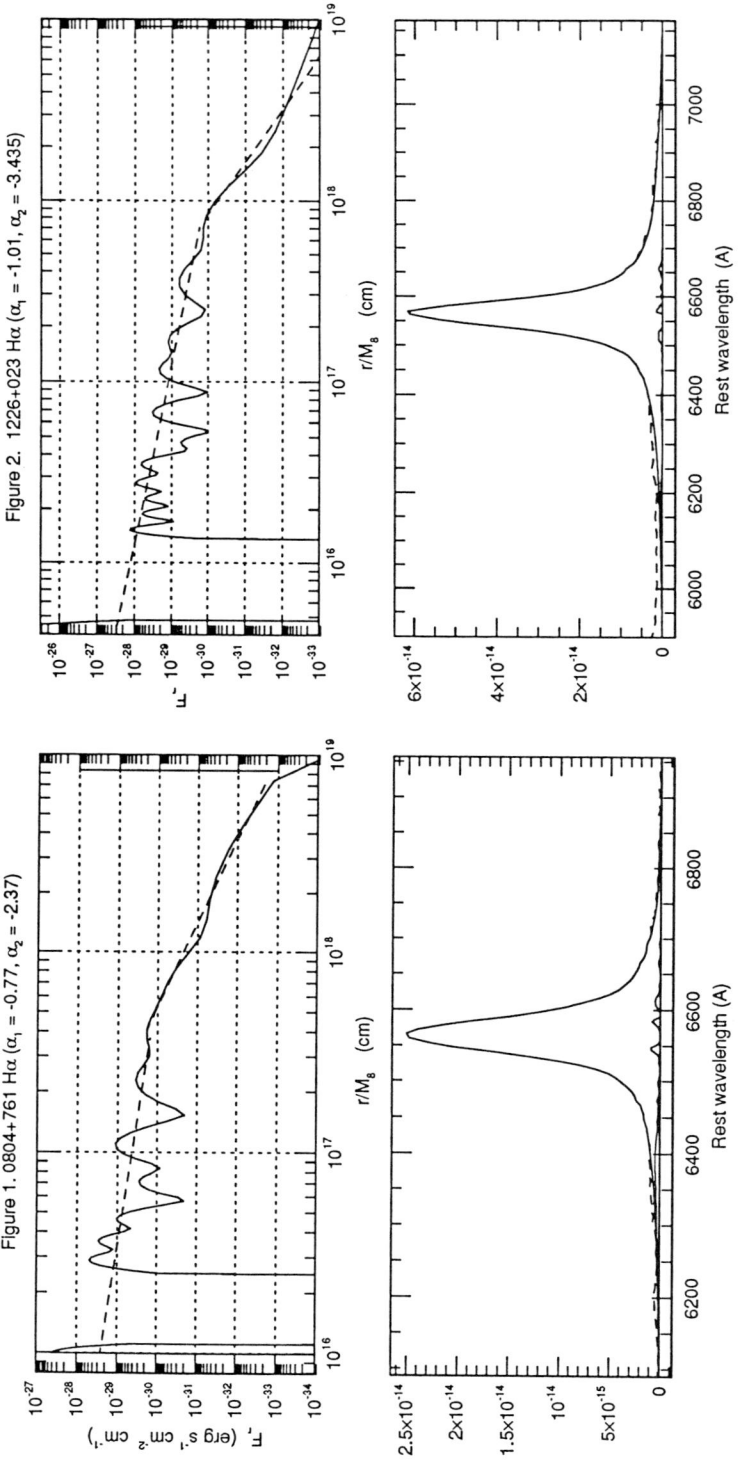

Figure 1. 0804+761 Hα ($\alpha_1 = -0.77$, $\alpha_2 = -2.37$)

Figure 2. 1226+023 Hα ($\alpha_1 = -1.01$, $\alpha_2 = -3.435$)

2. Discussion

If we assume the cloud number density we can compute the effective cross section. Bahcall and Wolf (1976) derived the star number density in their cluster to be proportional to $r^{-7/4}$. Since the typical $F_r \sim r^{-0.85}$ or so, the individual cloud cross sections should go as $\sim r^{+0.9}$, i.e. effective cloud area roughly proportional to radial distance. This closely matches the cloud size prediction of the rudimentary stellar wind model considered by Kazanas (1989).

The emission functions may turn over past some r because the cloud number density falls off past that radius, or the cloud emissivity falls off, perhaps from a decrease in cross sections. The dynamical model requires some continuity, so we prefer the latter explanation. Furthermore, we might associate this break in F_r with the transition between photon and gas pressure dominance in the clouds (Ferland and Persson 1989).

A severe disadvantage of the model as presented is the inability to explain the strong asymmetry seen in many profiles. Steady state asymmetry can be generated by an asymmetric distribution of clouds in phase space or by asymmetry in individual cloud emission. Ferland et al. (1992) argue that the emission from clouds should be nearly zero in the direction away from the central source. To generate an asymmetric profile, the radial velocity distribution must be biased. Alternately, the clouds might be affected by the intercloud medium, so that the emission properties depend on whether we see the part of the cloud facing into or away from its direction of motion.

References

Bahcall J.N. and Wolf R.A. 1974, ApJ, 209, 214.
Chen K., Halpern J.P., and Filippenko A.V., 1989, ApJ, 339, 742.
Ferland G.J. and Persson S.E. 1989, ApJ, 347, 656.
Ferland G.J., et al. 1992, ApJ, 387, 95.
Kazanas D. 1989, ApJ, 347, 74.
Puetter R.C. and Hubbard E.N. 1987, ApJ, 320, 85.

X-rays and Accretion Disks

X-ray Variability in AGN

T. Jane Turner *

1. Introduction

X-ray emission provides $\sim 10\%$ of the bolometric luminosity of a typical Seyfert galaxy and as X-ray photons of energy $> 2\,\mathrm{keV}$ can penetrate column densities of $> 1 \times 10^{22}$ atoms cm^{-2}, such observations provide the best observational probe available (with current instrumentation) of the active nucleus, and its immediate environment.

Variations cannot be observed faster than the light travel time, thus $\delta t \sim r/c$ gives an upper limit to the size of the emitting region. A variability timescale may be associated with the dynamical (orbital) timescale of the inner accretion disk. If the X-rays arise from $r \sim 5r_s$ and $r_s = 2GM/c^2$ then $t \sim 50 M_6 s$ (M_6 are units of 10^6 solar mass). This is reasonable if a persistent period were found, but could be misleading otherwise.

Two other parameters of interest are efficiency η (of mass to energy conversion) and compactness l. Efficiency, $\eta \sim 5 \times 10^{-43} dL/dt$, if $\eta > 0.1$ an exotic mechanism is required (such as relativistic beaming). Compactness, $l = L\sigma_T/Rm_e c^3$ (Svensson 1986; Guilbert, Fabian & Rees 1983). When $l > 10$ the source becomes optically thick to γ-rays and pair production becomes important (assuming the spectrum extends to $\sim > 1 MeV$), affecting the X-ray spectrum and the temporal behaviour (e.g., Mosalik & Sikora 1986; Fabian et al. 1986).

2. Historical

Ariel V and HEAO-1 established long term (days to years) variability as a property of AGN (e.g., Marshall, Warwick & Pounds 1981), with variations in amplitude of factors of ~ 7 or so being common. Only two sources showed evidence for rapid variability in the HEAO-1 or *Einstein* data; NGC6814 with a 100 second event (Tennant & Mushotzky 1983) and NGC4051 with a factor 2 change in ~ 2000 seconds (in IPC data; Marshall et al. 1983). The detection of significant short term variability is a function of signal-to-noise and observation duration, hence the lack of detection of rapid variability prior to *EXOSAT* observations. For example when the highly variable source NGC5506 was observed by HEAO-1 A2 no variability was observed, whilst the *EXOSAT* ME instrument revealed its true nature.

---

*Laboratory for High Energy Astrophysics, NASA/Goddard Space Flight Center, Greenbelt, MD 20771, U.S.A. This author was supported by the Universities Space Research Association

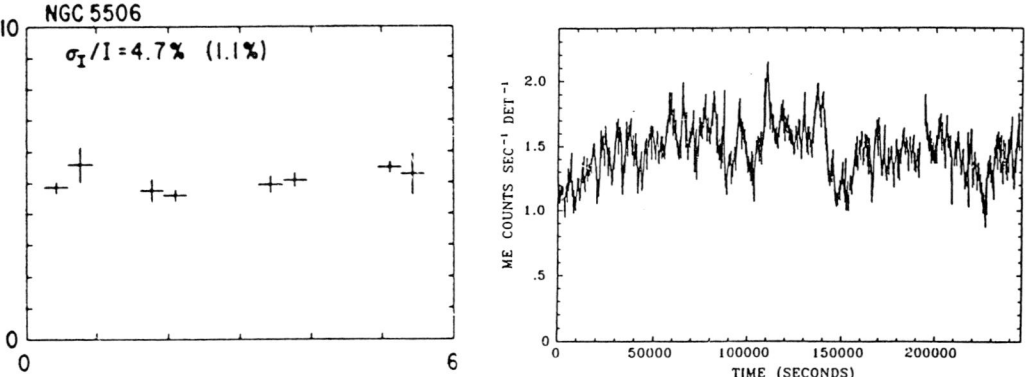

Figure 1. a) The HEAO-1 A2 light curve of NGC5506 in 12,000 second bins b) The *EXOSAT* ME light curve of NGC5506 in 400 second bins. Both taken from Urrey (1991); originals appeared in Tennant & Mushotzky (1983) & McHardy & Czerny (1987).

Long duration, uninterrupted X-ray observations of AGN were first afforded by *EXOSAT*. Its highly elliptical orbit eliminated the earth occultation and SAA passage constraints. Examination of the light curves from short ($\sim$ 20,000 sec) observations showed that rapid variability was common, contrary to the conclusions from low earth orbit satellites. Lawrence et al. (1985) reported rapid variability for NGC4051, shortly followed by MCG-6-30-15 (Pounds, Turner, & Warwick 1986a) and Mkn335 (Pounds et al. 1986b). In a study of a sample of 48 Seyfert galaxies, 30% showed marked rapid X-ray variability (Turner 1988; as some observations were signal- to-noise limited, this is a lower limit). These results prompted the *EXOSAT* "long look" (1–3 days) observations of several Seyfert type AGN, examples light curves are shown in Figure 2.

Barr & Mushotzky (1986) suggested that the doubling timescale of a source was inversely proportional to its luminosity, small objects taking less time to vary than larger ones (Bassini, Dean & Sembay 1983). Whilst the effect is probably real, the doubling timescale can be misleading when derived by extrapolation of lower amplitude events.

3. Power Spectra

Power Spectra are the most intuitively obvious of the Fourier techniques, constructed by breaking the light curve into cosine components of different frequency. The square of amplitude of these components as a function of frequency (Hz) is the power density spectrum (PDS). When the amplitude at any time is independent of the amplitude at any other time, the intensities in the light curve are a random distribution about

a mean and the PDS is a constant, i.e. has equal power at all frequencies, known as

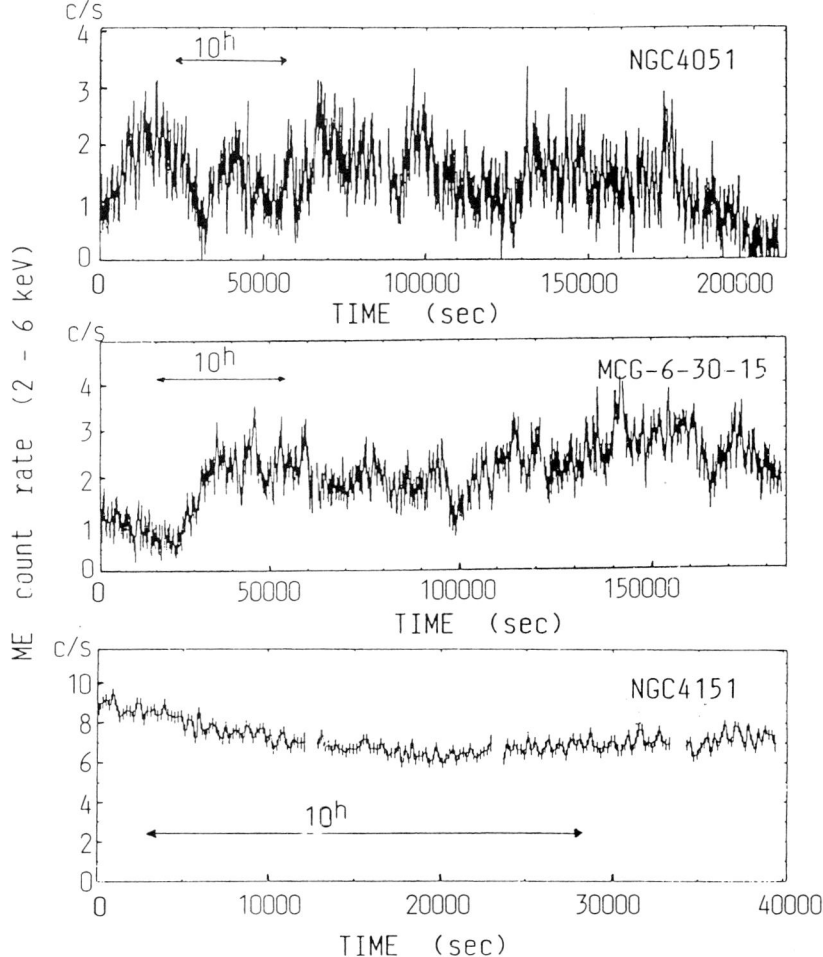

Figure 2. Light curves from *EXOSAT* long look observations a) NGC4051; b)MCG-6-30-15 and c) NGC4151

a "white noise" spectrum. In the case where the amplitude of each point is a random displacement from the adjacent point (rather than the mean) then the intensities are on a random walk, the PDS has a f^{-2} shape, known as a "red noise" spectrum. The observed $\sim f^{-1}$ noise is between these two and is characteristic of many systems in nature, unfortunately the origin of $\sim f^{-1}$ is not well understood.

The orbits of many satellite missions have meant frequent interruptions to the data train of observations and in many cases uneven sampling. Standard FFT computations of power spectra and correlation functions require interpolation over gaps in the data, assuming source behaviour when not actually observed. Some algorithms can use only observed data (Deeming 1975; Scargle 1982 and Lomb 1976), but these require even sampling. To properly analyse unevenly sampled data one needs to perform many simulations of known forms through the observation window, to fully understand the distortions in derived products (PDS, ACF,CCF) from the convolution of the window function and the source variability.

The *EXOSAT* orbit allowed uninterrupted observations and avoided windowing problems. The "long look" observations allowed determination of PDS in the range $\sim 10^{-5}$ to $10^{-3} Hz$. Those PDS showed a power-law fit to high frequencies of slope $f^{-\alpha}$ with $\alpha \sim 1 - 2$ (Mchardy 1989). PDS is expected to break as $\sim f^{-1}$ would be divergent. At the low frequency end the PDS must turn over, since the overall amplitude of AGN variations does not keep growing with new observations. At the high frequency end the fastest variability expected is related to the light crossing time of the region, so the PDS must go to zero beyond some finite limit. Fitting in the range 10^{-4} to 10^{-3} Hz shows $\alpha > 1$, therefore it seems that the intrinsic power to high frequencies is converging (Mchardy 1989). *Ginga* observations of NGC4051, however, extend the power spectrum down to ~ 100 seconds, and there is still no evidence for high frequency steepening indicative of a "characteristic" timescale (Pounds & McHardy 1988).

Signal-to-noise constraints meant that *EXOSAT* LE PDS were only obtained for a very few sources. These were steeper than the ME PDS, suggesting that there is less high frequency variability in the soft X-ray regime, consistent with the idea that some significant fraction of soft X-ray emission is originating in a larger region than the nucleus (Mchardy 1989).

One commonly discussed simplistic model is the shot-noise model of randomly spaced pulses of equal decay time. This predicts a PDS which has f^{-2} at high frequencies and is flat at low frequencies, with the spectral change occurring sharply (i.e., within a decade of frequency). This model is, however, inconsistent with observations (e.g.,. MCG-6-30-15 and NGC4051). A modification is to allow a range of pulse decay times, justified by inferring a medium with a range of densities. The shapes of the summed power spectra depend on whether low or high frequency processes dominate (Lehto 1989).

The *EXOSAT* "long looks" also provided information on the phases of the frequency components. Krolik, Done & Madejski (1992) find no strong correlation between phases of different frequency components in any source, implying a large number of independent events are responsible for the observed variability.

In an alternative characterisation of light curves by fractal dimensions (McHardy & Czerny 1987), the "length" of a structure (degree of variability) depends on the size of the "ruler". A large "ruler" (or data sampling size) averages out small scale structure, a finer measure gives a longer measured length. If a light curve contains fine structure, the "degree of variability" determined will depend on the sampling time. This is illustrated in Figure 1.

3.1 Comparison with Galactic Black Hole Candidates

The phenomenon of f^{-1} noise in X-ray light curves is already well known for Galactic sources. Cygnus X-1 has a f^{-1} PDS with a knee at 5×10^{-2} Hz and a few seconds (Meekins et al. 1984; Nolan et al. 1981). This knee frequency is speculated to be related to a dynamical timescale in an accretion disk surrounding a 10 solar mass black hole. Inclusion of archival data shows a "knee" in the power spectrum of NGC5506 at $\sim 10^{-7}$ Hz, the power spectrum remains flat down to $\sim 10^{-9}$ Hz, beyond which there are no further data available. By comparison with Cyg X-1 the central black hole mass of NGC5506 is implied to be $< 10^6 M_o$. Examination of the X-ray light curve of Cyg X-1 shows remarkable similarity to that of AGN, as well as having a strong similarity in observed X-ray spectrum. Does this imply a similarity in basic physics?

3.2 NGC6814

NGC6814 exhibits the most rapid X-ray variability of any Seyfert observed to date, with a factor of 2 decrease in 50 seconds (Kunieda et al. 1990). In addition to this NGC6814 shows a strong correlation between iron line and continuum flux variations, with a limit on the lag of the line of < 256 seconds, placing the reprocessing region very close to the continuum emitting region. Detailed timing analysis (Done et al. 1992) shows an f^{-1} power spectrum descending into the noise at ~ 300 seconds, plus a confirmation of the 12,000 second period first seen in the *EXOSAT* data (Mittaz & Branduardi-Raymont 1989). The periodic component in the *Ginga* data accounts for at least 75% of the observed variability in this source, has an amplitude of $\sim 36\%$ and includes most of the extreme events. Sections from one of the *Ginga* light curves are shown in Figure 3.

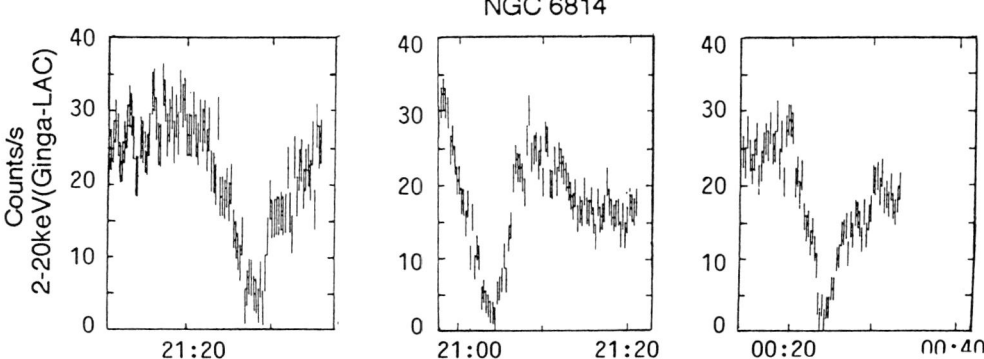

Figure 3. Sections of a *Ginga* light curve from NGC6814 (Kunieda et al 1991)

If the period is related to an orbital timescale, such as that of the inner accretion disk, then periodic occultation of the continuum/reprocessing region by vertical structure in the inner disk is a possibility (Done et al. 1992; as suggested for the Galactic black hole candidate LMC X-1; Ebisawa, Mitsuda, & Inoue 1989). Recent work also shows evidence for quasi-periodicity in some other AGN (see Papadokis et al -these proceedings), as well as for several Galactic black hole candidates.

4. Spectral Variability

Ginga data have shown the hard X-ray spectra of Seyfert galaxies to comprise a steep power-law (photon index $\sim 1.9 - 2.0$), an iron K emission line observed at ~ 6.4 keV (with tentative evidence for an associated edge) and a flattening to high energies (> 10 keV) interpreted as a "Compton reflected" component (Lightman & White 1988; George & Fabian 1991; Nandra 1991). In the soft X-ray regime (< 2 keV) sources exhibit a variety of spectral forms, partly due to the effects of photoelectric absorption by material either intrinsic to the source or along the line-of-sight. This material may be partially ionized, or only partially covering the source, giving rise to absorption features (e.g., Nandra & Pounds 1992), or emission lines (Turner et al. 1991; Turner, George, & Mushotzky 1992). In addition to these effects, some sources have a distinct steep soft continuum component, consistent with an origin in the inner regions of an accretion disk. Observations of Mkn335 have clearly shown rapid and uncorrelated variability in the soft and hard X-ray components (Turner & Pounds 1988), in support of such a model. Extended soft X-ray emission regions has also been observed (Elvis, Briel & Henry 1986; Wilson et al. 1992; Halpern 1991). Many of the claims of spectral variability which have appeared in the literature may be explained by variations in the relative normalizations of the aforementioned spectral components.

For example, simple power-law fits to the *EXOSAT* and *Ginga* observations of NGC-5548 show the photon index to increase with hard X-ray flux. When the data are modelled with a power-law plus reflection component, however, the spectral variability can be entirely accounted for by differential variations of a constant slope power-law and a Compton reflection hump (Nandra et al. 1989). Similar behaviour has been observed for MCG-6-30-15 (Pounds, Nandra, & Stewart 1991).

Few multiwaveband campaigns have been carried out with good coverage in the X-ray band. UV and X-ray observations of NGC5548 suggest a correlation, with UV lag $0 < t < 6$ days. This correlation breaks down during strong UV flares (Clavel et al. 1992). In another campaign, NGC4051 revealed factor of three variations in X-ray flux whilst the optical flux varied by less than 0.7% (Done et al. 1990), showing that these two cannot be explained by a single emission region, nor by a single electron population.

5. AGN models

Any model for the AGN phenomenon must explain the amplitudes and timescales of X-ray variability (as extreme as $dL/dt \sim 10^{41} ergs^{-2}$) and the character of the X-ray light curves. Current debate is whether Seyfert galaxies are powered in the same way as starburst galaxies or quasars. Observations of quasars suggest that for some, their radiation is relativistically beamed towards us. The issue is complex, as some authors infer different emission mechanisms for radio loud and radio quiet sources. The high magnetic fields inferred for some quasars, however, are consistent with a black hole scenario, and inconsistent with a starburst powered source. The gradation of properties between quasars and Seyfert galaxies suggests that some degree of unification exists between AGN, and so accretion onto a black hole may also power Seyfert galaxies. The appeal of the starburst model, that one does not need to invoke poorly understood phenomena, is reduced when one must invoke a black hole model for somewhat similar sources further along the luminosity scale. Whilst the starburst model (Terlevich et al. 1991) can explain some of the observed properties of AGN, it does not appear able to explanation of the large amplitude and short duration dips observed in the X-ray light curves of some Seyferts (e.g., Figure 3), nor the periodicity. Furthermore, X-ray observations of starburst galaxies (to date) have not shown the rapid variability observed in Seyfert galaxies.

6. Future Missions

The X-ray Timing Explorer (*XTE*), scheduled for launch in 1995, will have $1\mu s$ resolution with the proportional counter (2-60 keV), and $10\mu s$ with the crystal scintillator (2-200 keV), allowing determination of AGN power spectra down to $\sim 60s$.

Astro-D, scheduled for launch in Feb 1993 will have a similar sampling pattern to *Ginga*. Its imaging capability and high spectral resolution will help in time resolved spectral studies.

Spectrum-X, scheduled for launch in 1995, will have an orbit allowing continuous observations of most sources for up to ~ 3 days. The CCD detector will have $E/\Delta E > \sim 50\%$ around 7 keV and cover a bandpass of 0.3–10 keV. XMM and AXAF will also allow uninterrupted observations of the order of days duration, ideal for time resolved spectroscopy.

References

Barr, P. & Mushotzky, R.F. 1986. Nature, 320, 421.
Bassini, L., Dean, A.J., & Sembay, S. 1983. Astron. Astroph., 125, 52.
Clavel, J. et al. 1992. Ap.J. In press.
Deeming, T.J., 1975. Astrophys. Space Sci. Rev., 36, 137.
Done, C. et al. 1990. MNRAS., 243, 713.

Done, C. et al. 1992, Ap.J.in press.
Ebisawa, K., Mitsuda, K. & Inoue, H., 1989. PASJ., 41, 519.
Elvis, M., Briel, U.G. & Henry, J.P. 1986. Ap.J., 268, 105.
Fabian, A.C. et al. 1986. MNRAS., 221, 931.
Guilbert,P.W., Fabian, A.C. & Rees, M.J. 1983. MNRAS, 205, 593.
George, I.M. & Fabian, A.C. 1991. MNRAS, 249, 352.
Halpern, J. 1991. In "Testing the AGN Paradigm", ed. S.S.Holt, S.G.Neff & C.M.Urry, AIP.
Krolik, J., Done, C. & Madejski, G. 1992. Ap.J.In press.
Kunieda, H. et al. 1990. Nature, 345, 786.
Kunieda, H. et al 1991. In "Iron Line diagnostics in X-ray sources", ed. A. Treves, G.C. perola, & L. Stella, Springer-Verlag.
Lawrence, A. et al. 1985. MNRAS, 217, 685.
Lehto, H. 1989. In 23rd ESLAB Symp., ESA SP-296, 1, 499.
Lightman, A.P. & White, T.R. 1988. Ap.J., 335, 57.
Lomb, N.R. 1976. Astrophys. Space Sci., 39, 447.
Marshall, F.E. et al. 1983. Ap.J., 269, L31.
Marshall, N., Warwick, R.S. & Pounds, K.A. 1981. MNRAS, 194, 987.
Meekins, J.F. et al. 1984. Ap.J., 278, 288.
Mchardy, I.M. 1989. In 23rd ESLAB Symp., ESA SP-296, 2, 1111.
McHardy, I.M. & Czerny, B. 1987. Nature, 325, 696.
Mittaz, J.P.D. & Branduardi-Raymont, G. 1989. MNRAS, 238, 1029.
Mosalik, P. & Sikora, M. 1986. Nature, 319, 649.
Nandra, K. et al. 1989. MNRAS, 236, 39p.
Nandra, K & Pounds, K.A. 1992. Nature, in press.
Nandra, K. 1991. PhD Thesis, University of Leicester, UK.
Nolan, P.L. et al. 1981. Ap.J., 246, 494.
Pounds, K.A., Nandra, K & Stewart, G.C. 1991. Active Galactic Nuclei, eds.H.R. Miller and P.J. Wiita, Heidelberg: Springer-Verlag,p.257.
Pounds, K.A. et al. 1986b.MNRAS, 224, 443.
Pounds, K.A., Turner, T.J. & Warwick, R.S.1986a, MNRAS, 221, 7p.
Scargle, J.D. 1982. Ap.J. 263, 835.
Svensson, R. 1986. In "Radiation-Hydrodynamics in stars and compact objects". IAU Coll. 89, 325. eds. D. Mihalas & K. Winkler, Springer-Verlag, Berlin.
Tennant, A.F. & Mushotzky, R.F. 1983. Ap.J., 264, 92.
Terlevich, R. et al. 1991 MNRAS, 255, 713.
Turner, T.J. 1988, PhD Thesis, Univeristy of Leicester, UK.
Turner, T.J. & Pounds, K.A. 1988, MNRAS, 232, 463..
Turner, T.J. et al. 1991. Ap.J., 381, 85.
Turner, T.J., George, I.M. & Mushotzky, R.F. 1992. MNRAS, submitted.
Urry, C.M. 1991. Active Galactic Nuclei, eds. H.R. Miller and P.J. Wiita, (Heidelberg: SPringer-Verlag), p. 257.
Wilson, A.S. et al. 1992. Ap.J. Lett, 391, L75.

Thermal Reprocessing of X-Rays in NGC 5548

J. Clavel [*]   K. Nandra [†]   K. Pounds [†]   W. Wamsteker [‡]

Abstract

A set of 11 contemporaneous IUE and GINGA observations of NGC 5548 reveal the existence of correlated variations of its hard X-ray (2–10 keV) and Ultraviolet (1200–3300 Å) flux over time scales of 2 days to one year. This is best explained in the framework of a model where the X-rays irradiate a cold (T $\sim 10^5$ K) accretion disk. Only a tiny fraction of the irradiating flux is compton reflected back to the observer in the form of a hard X-ray tail while the bulk of the X-rays are absorbed in the disk and eventually re-emitted as thermal radiation in the Ultraviolet. The absence of a detectable phase delay between the two bands together with the absence of rapid ($\sim$ hours) fluctuations of the UV flux further constrain the X-ray source to lie between 200 and 1400 Schwarzschild radii above the disk. The thermal reprocessing model provides a natural explanation for the simultaneity of the optical and UV variations in NGC 5548 and may solve most of the problems facing the accretion disk model.

1. Introduction

The presence of a strong Iron K_α line near 6.4 keV and a "hard tail" above 10 keV is a common property of the X-ray spectrum of Seyfert 1 galaxies [1]. This has been interpreted as evidence for reprocessing of the X-rays by a "cold" (T $\sim 10^5$ K) accretion disk [2]. In this model, less than 10 % of the irradiating photons are compton scattered back to the observer forming a spectral hump centred near 15 keV while the bulk of the X-rays are absorbed in the disk. The absorbed X-rays generate heat inside the disk and are eventually re-emitted as thermal radiation. Since the effective temperature of the disk is about 10^5 K, this re-emission will take place in the UV and optical range. At the high densities inside the disk, the absorption and re-emission are essentially instantaneous. Hence, the reprocessing model implicitly predicts that the variations of the X-ray and UV flux should be correlated. In this contribution we report simultaneous X-ray and UV observations of NGC 5548 which strongly support this scenario.

---

[*]ISO Science Operations, ESTEC/SAI, Postbus 299, 2200-AG Noordwijk, The Netherlands.
[†]Department of Physics & Astronomy, University of Leicester, Leicester LE1 7RH, U.K.
[‡]ESA IUE Observatory, P.O. Box 50727, 28780 Madrid, Spain

2. The Observations

The nucleus of NGC5548 was observed at 11 different epochs during May-July 1990 with the *IUE* satellite. The acquisition procedure and data reduction techniques used were standard. The average continuum in 40 Å bins was measured at 1350 Å. The 11 X-ray observations were performed with the Large Area proportional Counter (LAC) aboard the *Ginga* satellite. Nine of these observations were carried out over from May 24 to July 5 1990. The remaining two observations were obtained in 1989. All are simultaneous to within a day with the UV observations. The 2–10 keV fluxes were determined assuming the best fitting power-law model for each observation. A full account of the UV and X-ray observations is given in [3].

3. The pattern of variability

Figure 1 shows the 2–10 keV X-ray flux as a function of the 1350 Å continuum for the 11 epochs of simultaneous X-ray and UV observations. As can be seen, F_{2-10} and F_{1350} are well correlated. The linear correlation coefficient $r = 0.89$, ($P(r \geq 0.89) = 3.62\,10^{-4}$). A non-parametric Spearman Rank correlation test yields a $4.7\,10^{-2}$ probability for the correlation to arise from chance. It is worth noting that the intercept of the best-fit regression line is zero within the uncertainties, implying that the X-ray flux is strictly proportional to F_{1350}.

Figure 2 shows the cross-correlation (CC) and Discrete Correlation of F_{2-10} with F_{1350}. Both peak at $\Delta t = +0 \pm 3$ day with a maximum correlation coefficient: $r_{max} = 0.85$. The formal uncertainty $\Delta t = \pm 3$ days probably underestimates the true error on the delay. We have adopted a conservative upper limit of ± 6 days.

4. Implications

There are two categories of models which can potentially account for the correlation of the X-ray and UV flux. In the first scenario, the UV photons are Compton scattered on a hot ($T > 10^7$ K) plasma up to X-ray energies. In the second one, the X-rays are absorbed in a cold ($T \sim$ a few 10^5 K) optically thick slab and re-emitted as thermal radiation in the ultraviolet. The first scenario leaves the strong Iron line and the hard X-ray "tail" unexplained unless a significant fraction of the X-ray flux is back scattered onto the slab. However, to be consistent with existing observations such a model requires that virtually all the power be dissipated directly within the hot layer [5]. In such a case, the "seed" UV photons become completely negligible in the energy balance of the hot phase and cannot drive the variations of its flux. By contrast, the second scenario comes as a natural consequence of the X-ray reprocessing model which was developed for independent reasons. For these reasons, we favour it over the hot phase reprocessing scenario.

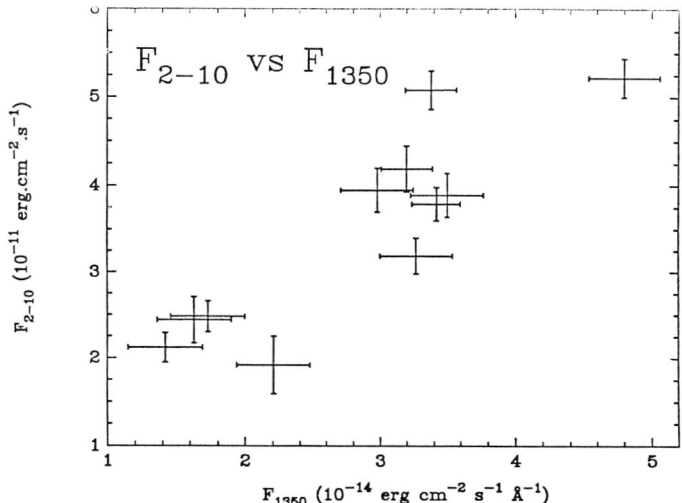

Figure 1. The observed 2–10 keV flux plotted as a function of the 1350 Å continuum intensity for all epochs of simultaneous measurement. The errors quoted correspond to a 72% confidence level for F_{1350} and a 90% confidence level for F_{2-10}. The two parameters are highly correlated.

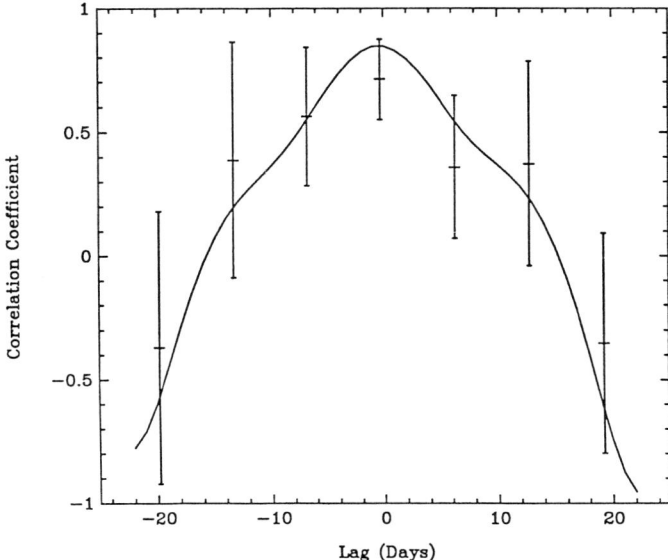

Figure 2. The cross-correlation of the observed 2–10 keV flux with the 1350Å continuum intensity. The discrete correlations are also plotted with their associated errors. The correlations peak at or near zero lag, implying that the variations of the X-ray and UV flux are simultaneous within the accuracy of the cross-correlation techniques, ±6 days.

In the thermal reprocessing model, the bulk of the variations of the optical and UV flux are driven by fluctuations of the external X-ray source. One does not expect a perfect correlation between the two bands. As a matter of fact, those X-ray variations whose time scale is not commensurate to the travel time of the X-ray photons to the reprocessing medium are filtered-out from the UV light-curve. This argument can be used to set a lower limit on the distance between the X-ray and UV sources. The fastest variations observed to date correspond to 30 % fluctuations of the X-ray flux in ~ 5 hours [6]. Such rapid variations are not observed in the ultraviolet, the PDS of the UV light-curve being steep with no discernible power on time scales shorter than a day [3]. This implies that the travel time of the X-ray photons is at least of the order of a few times 5 hours, say one day. If we combine this lower limit with the 6 days upper limit imposed by the absence of a detectable lag of the UV flux, we obtain $2.5\,10^{15} \leq H \leq 1.5\,10^{16}$ cm, where H is the height of the X-ray source above the disk. Under the assumption that the gas is gravitationally bound, the emission line delays measured in [3] together with the average line width yields $M = 3.7\,10^7\,M_\odot$ for the mass of the black-hole in NGC 5548. Hence, the reprocessing region lies at distance comprised between 225 and 1350 R_s, where $R_s = 1.1\,10^{13}$ cm is one Schwarzschild radius. This is one to two orders of magnitude larger than the theoretical size of the 1350 Å emitting region in a standard α-disk without reprocessing around a Schwarzschild Black hole $\sim 20R_s$.

The fact that the 1350 Å flux is directly proportional to F_{2-10} suggests that, at least in 1990 when NGC 5548 was relatively faint, the *entire* UV flux originated from thermal reprocessing. This is energetically possible. Assuming an isotropically emitting X-ray source with a power-law spectrum of energy index, $\alpha = 0.8$ [6] sited above an optically thick disc, the available X-ray flux integrated from 0.0136 to 20 keV is $\sim 10^{-10}$ erg cm^{-2}s^{-1}, and roughly 3 times more if one extends the integration up to 1 MeV. This is sufficient to account for the dereddened UV–optical continuum flux in NGC 5548, from 5000 Å to 1 Ryd ($\sim 8\,10^{-11}$ erg cm^{-2} s^{-1}).

It is customary to explain the optical-UV spectral energy distribution (SED) of AGN's as thermal emission from a geometrically thin accretion disk around a massive black-hole. However, the geometrically thin disk model is plagued with serious problems. In the LTE approximation at least, the emergent spectrum should show a strong absorption at the Lyman limit of Hydrogen. Such a discontinuity is not observed [7]. Another problem is the SED which, in at least one object, the very luminous QSO HS 1700+6416 extends up to 3 Rydbergs making the disk largely super-Eddington. For the same reason, the strong soft X-ray excesses, a common feature of most if not all AGN's including NGC 5548 [6] are also difficult to account for in the framework of standard disk models. Last and most important in the present context, the simultaneity of the variations at 1350 Å and ~ 5000 Å observed in NGC 5548 [4], NGC 4151 [8], Fairall 9 [9] essentially rules-out the thin disk model in its standard form. In the thin disk model, the effective radius of the optically emitting region is about seven times larger than that of the UV emitting layer. Since the two regions cannot communicate on a time scale which is shorter than the sound crossing time, the optical flux should lag the UV by ≥ 10 years in NGC 5548.

The problems listed above can be solved if the bulk of the UV-optical flux originates from thermal reprocessing of the externally supplied X-rays. X-ray heating from above will make the temperature gradient in the upper layers of the disk atmosphere much shallower than in a disk model without reprocessing. This may suppress the Lyman absorption edge. Adding externally supplied energy means that the UV-optical energy distribution of AGN's can be accounted for with smaller Black-Hole masses and therefore a hotter disk than in models without reprocessing. Hence, the disk spectrum may extend up to $\sim 100\,\mathrm{eV}$ and therefore account for soft X-ray excesses in Seyfert 1 galaxies. Moreover, if the X-rays are generated outside the accretion flow – in a jet for instance – their luminosity is not constrained by the Eddington limit. Hence, even in a very luminous QSO, such as HS 1700+6416, thermal reprocessing can potentially explain the high frequency extension of the UV bump. It therefore appears that reprocessing has the potential to solve most of the problems which plague the accretion disk model.

References

[1] Nandra, K., Pounds, K. and Stewart, G.C. 1991 *Iron line diagnostics in X-ray source Springer-Verlag, p.177*
[2] George, I.M. and Fabian, A.C. 1991 *MNRAS 249, 352.*
[3] Clavel, J. *et al.* 1992 *Ap.J. 393, 113*
[4] Clavel, J. *et al.* 1991 *Ap.J. 366, 68.*
[5] Haardt, F. and Maraschi, L. 1991, *Ap.J. 380, L51.*
[6] Nandra, K. *et al.* 1991 *MNRAS 248, 760.*
[7] Antonucci, R.R.F., Kinney, A.L. and Ford, H.C. 1989, *Ap.J. 342, 64.*
[8] Clavel, J. *et al.* 1990 *MNRAS 246, 668.*
[9] Clavel, J., Wamsteker, W. and Glass, I. *et al.* 1989 *Ap.J. 337, 236.*

New Ginga Observations and Model of NGC 6814 Periodicity

Sachiko Tsuruta *   *Karen Leighly* [†]   *Ran Sivron* *

Abstract

Recent detailed data analysis of the last two Ginga observations of the Seyfert nucleus NGC6814 (Leighly et al. 1992, hereafter LKT92) has not only reconfirmed the periodicity of the fastest X-ray variability reported earlier for this source (Done et al. 1992, hereafter DMM92), but also has shown several very unique, definite characteristics which severely constrain any acceptable models. Consequently, various existing models have been ruled out (LKT92). Here we present, as a natural and self-consistent physical model which satisfies these detailed observational constraints, the occultation of the central X-ray source by matter overflowing the Roche lobe of a low mass star orbiting around a supermassive black hole. The importance of careful, detailed comparison of this type of model with further observations is emphasized, because the result may lead to strong circumstantial evidence for the presence of a supermassive black hole in the central engine of active galactic nuclei (AGN).

NGC6814 is among the most interesting Seyfert galaxies in the sense that its nuclear X-ray emission was found to be most rapidly variable, with the timescale of ~ 300 seconds, and moreover the fastest variability exhibits periodicity of ~ 12000 seconds (DMM92). The recent detailed data analyses of the last two Ginga observations, in April and October 1990, respectively, have shown several new detailed characteristics, such as the spectral variability and lags in flux between different energy bands (LKT92). Lags are shorter for the April data than for the October data. During flux decrease (ingress) the hard flux lags while during flux increase (egress) the soft flux lags. In both cases the lags decrease with increasing energy. Associated significant hardening of the spectrum during the ingress was observed in both April and October data. The April flux decreases nearly to zero at the bottom of the dips, while the October dips do not. These detailed, new observational constraints have ruled out several suggested models, but it was found that a variable absorption model could be fit explicitly to the detailed data (LKT92). In this model the dips are caused by

---

*Department of Physics, Montana State University, Bozeman, Montana 59717, USA. This author was partially supported by NASA grants NAGW-2208 and NAG-783.

[†]Laboratory for High Energy Astrophysics, NASA, Goddard Space Flight Center, Greenbelt, Maryland 20771, USA. This author was partially supported by NASA grants NAGW-2208 and NAG-783.

the occultation of the central X-ray source by absorbing matter with variable column density. To be consistent with the variable column density it was shown that the ionization parameter of the matter has to be less than ~ 100 (LKT92). Combined with the constraint imposed by the observed column density, of $\sim 10^{23}$ cm^{-2}, the implication is that the absorbing matter has to be relatively dense and sheet-like in form (LKT92). Although various models have been proposed where some orbiting matter is responsible for the periodicity, many of these have been shown to be incompatible with observation (LKT92).

Here we present a physical model which naturally satisfies all the existing observational constraints as outlined above. Several independent observational arguments lead to a relatively low mass for the central black hole in NGC6814 of $M_h \sim 10^6 M_\odot$ (LKT92, Kunieda et al. 1990). In our model the Roche lobe overflow from a low mass star orbiting around a black hole of $10^6 M_\odot$ acts as the variable absorber which periodically occults the X-ray source as it comes between the source and the observer. The corresponding gravitational radius of the black hole is $R_G \sim 3 \times 10^{11}$ cm. The corresponding Keplerian orbital radius R_k for the observed ~ 12000 sec period is $27 R_G$. It has been shown (Syer, Clarke and Rees 1990, hereafter SCR90) that a star can be captured into such a very tightly bound orbit around a black hole through the repeated interactions with an existing accretion disc around the hole. For a low mass black hole of order $10^6 M_\odot$ the original stellar capture most likely takes place through a binary disruption, but once captured its orbit will be circularized and the star will be grounded through repeated interactions with the disc (SCR90). We assume that the orbit of a star captured in this manner has already been circularized but the orbital plane is still inclined to the disc plane, a situation expected with an appropriate choice of the original relative inclination of these planes when the star was captured (SCR90). The mass of the star is determined from the requirement that the star has already filled its Roche lobe so that a steady Roche lobe overflow has been taking place. It is $m_* \sim 0.37 M_\odot$ and $0.15 M_\odot$ for a Schwarzschild and an extreme Kerr black hole, respectively. In our calculations as a representative example the stellar mass of $m_* = 0.3\ M_\odot$ is adopted.

In our model the matter overflowing from the star trails behind the star after passing through the Lagrangian point L_1 due to the orbital motion. The matter is compressed to a sheet-like structure due to the combined effects of radiation pressure by the X-rays from the central source, gravity and magnetic fields (Celotti, Fabian and Rees 1992). This sheet-like matter acts as the variable absorber which periodically occults the X-ray source. As the star moves across the line of sight in its orbital motion the trailing sheet starts to cover the source. The sheet is narrow near the L_1 point where it starts, but spreads fan-like away from the star, and hence becomes broader with distance d away from the Lagrangian point. The X-ray source may be circular in cross section, but for simplicity we have calculated the occultation of a square shaped source of size R_x by the fan-shaped sheet. Near the L_1 point where the absorbing sheet starts, its dimension is smaller than the X-ray source and the flux decrease due to the occultation is small, even though the gas is opaque. However, with increasing d, the width of the occulting sheet R_s increases. From the cusp-like shape of the hard X-ray dip to nearly zero in the April data the opaque portion of the absorbing matter and the hard X-ray source should have approximately the same size when the total occultation takes place. Then the size of the hard X-ray source can be found, from simple geometrical considerations, to be 10^{12} cm (Tsuruta, Sivron and Leighly 1992, hereafter TSL92). Note, on the other hand, that the Roche lobe radius r_* of a $0.3 M_\odot$ star is only 2.7×10^{10} cm. Since the flux drops to zero the sheet must be opaque (the optical depth $\tau > 1$) during the total eclipse. Subsequent increase of flux after the zero point is reached is explained by the decreasing density of the sheet with distance d. If we further take into account the density decrease with the angle α, wider light curves are obtained for lower energy X-rays. Including the density decrease along both d and α, the light curves due to the occultations of the source have been calculated. Considering the simplicity of the model, the fits are found to be excellent (TSL92).

In our model the soft flux leads the hard flux during the ingress while the hard leads the soft during egress, because the densities are less at larger angles α. Near the outer edge of the fan-shaped sheet, therefore, lower energy X-rays are still partially absorbed while this part of the sheet is transparent to harder X-rays. In the April data the dip drops to zero because when the source is completely covered the sheet is opaque ($\tau \sim 1$), causing a total eclipse. The dip is cusp-like at the highest energy window because the density decrease along d causes $\tau < 1$ after this critical point has passed. The dip does not go to zero in the October data. It could be that the overflowing matter itself changed the shape and density (e.g. spreading further and becoming less dense). However, even when the configuration of the sheet has not significantly changed, the change of the shape of the observed light curves can be caused by the change of the orbital inclination, due e.g. to the precession effect (Sikora and Begelman 1992). If the inclination is changed, the line of sight will deviate from the centre of the sheet, and the total eclipse will no longer take place. The spectral hardening during the ingress is due to the greater absorption of the lower energy flux by the variable absorber sheet expected in our model.

By comparing the periodic X-ray variability observed by the EXOSAT and Ginga it has been suggested that the period varied by ~ 1 percent during the five year interval between these two satellite missions (DMM92). In our model the orbiting star, during the earlier stages after the original capture, will go through the gradual migration toward the centre and consequent period decrease due to the repeated interactions of the star with the geometrically thin, but optically thick disc (SCR90). However, theory predicts that due to thermal instability such a thin disc would terminate at a certain critical radius r_{crit}, and the accretion flow inside r_{crit} would be an optically thin, geometrically thick torus (Shapiro, Lightman and Eardley 1976). The mechanism which is responsible for the inward migration of the star and consequent period decrease is then absent within r_{crit}, since the disc density is too low. For a black hole of 10^6 M$_\odot$, $r_{crit} \sim R_k \sim 27R_G$. Therefore, it is logical to assume that the inward migration of the star and the period decrease due to the star-disc interactions stopped at $r_{crit} \sim R_k$, and the long-range stability of the orbital period has been established. For our relatively low mass star-black hole system timescales for other potential causes of the period decrease, such as the gravitational radiation, are much longer than the constraint imposed by observations (TSL92).

We have shown that all the existing observational constraints are naturally satisfied by our current orbiting star model. Comparison of this type of model with further observations through ROSAT, ASTRO-D, etc. would be invaluable. For instance, the fundamental change in the average shape of the light curves from April to October, if found to repeat periodically with the timescale of several months to a few years, may be due to the effect of precession around a black hole (Sikora and Begelman 1992). Discovery of such a precession may be interpreted as a strong circumstantial evidence for the presence of a supermassive black hole in the central engine of AGN.

References

Celotti, A., Fabian, A.C., and Rees, M.J., *Mon. Not. R. Astr. Soc.*, 255 (1992), 419.
Done, C., Madejski, G.M., Mushotzky, R.F., Turner, T.J., Koyama, K., and Kunieda, H., *Astrophys. J.*, in the press (1992).
Kunieda, H., Turner, T.J., Awaki, H., Koyama, K., Mushotzky, R.F., and Tsusaka, Y., *Nature*, 786 (1990), 345.
Leighly, K., Kunieda, H., Tsusaka, Y., Awaki, H., and Tsuruta, S., *Astrophys. J.*, submitted (1992).
Shapiro, S.L., Lightman, A.P., and Eardley, D.M., *Astrophys. J.*, 204 (1976), 187.
Sikora, M., and Begelman, M.C., *Nature*, 356 (1992)
Syer, D., Clarke, C.J., and Rees, M.J., *Mon. Not. R. Astr. Soc.*, 250 (1989), 505.
Tsuruta, S., Sivron, R., and Leighly, K., in preparation (1992).

Power Spectrum Fits to EXOSAT Long Looks

I. E. Papadakis and A. Lawrence *

Abstract

We present results from a power spectrum analysis of EXOSAT long looks at AGN. In most cases a power law is a good fit to the power spectra. The slope of the power law appears to be the same for all objects, $\alpha = 1.55$. The amplitude of variations decreases with luminosity but more slowly than expected from simple scaling relations. We also report the detection of quasi-periodic oscillations (QPOs) in the ME power spectrum of NGC 5548.

1. Introduction

We have performed a power spectrum analysis of 11 EXOSAT long looks (the duration of all the observations is > 70000 sec), using a new technique for estimating red noise power spectra (Papadakis and Lawrence 1992). This technique provides us with estimates of the logarithm of the power spectrum which are unbiased, independent, have known variance and are (approximately) normally distributed. Consequently these estimates can be used in a χ^2 model fitting procedure. Before the estimation of the power spectra we normalized the light curves to their mean values. In this way we can compare power spectra of objects with different luminosities. The model that we fitted to the power spectra is the following,

$$\log[P(\nu)] = \log[A(10^4\nu)^{-\alpha} + C],$$

where $P(\nu)$ is the power spectral density at frequency ν, A is the amplitude of variability at 10^4 Hz, α is the spectral slope and C represents the constant Poisson noise level. In this report we present results from the power spectrum analysis of the ME light curves only.

2. Results of model fits

A power law model is a good fit to the power spectra of almost all the objects. In figure 1 we plot the log of the best fit slopes of the ME power spectra against the log of the (2-10 kev) luminosity of the objects (assuming $H_0 = 50$ km s^{-1} Mpc^{-1}).

---

*Department of Physics, Queen Mary and Westfield College, University of London, Mile End Road, London E1 4NS.

The best-fit line to this plot has a slope consistent with zero. Therefore, there is no correlation between α and L_X. Furthermore, the hypothesis that all the power spectra have the same slope (equal to the mean value $\overline{\alpha} = 1.55$) is consistent with the data. In figure 2 we plot the log of the normalized amplitude of variations at 10^{-4} Hz against $\log L_X$. A strong anticorrelation is evident. This confirm the claims of Barr and Mushotzky (1986) and McHardy (1989) but puts the effect on a firm quantitative basis. A straight line does not fit this plot well but a fitted straight line has a slope of -0.53 ± 0.02. A linear relationship, $A \propto L_X^{-1}$, is strongly excluded.

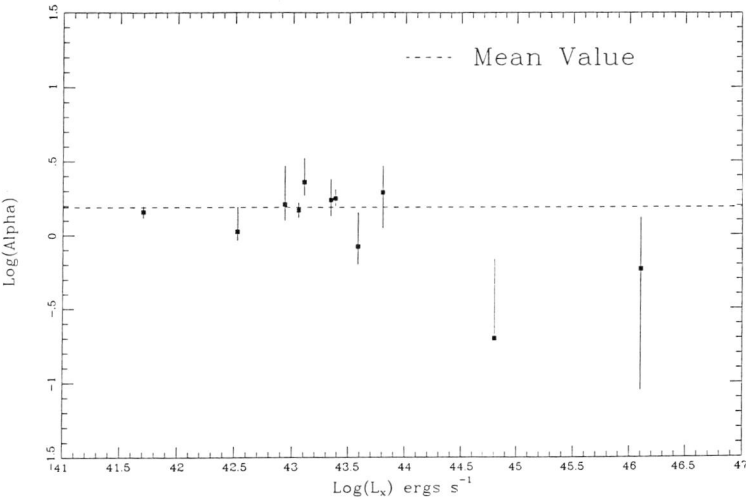

Figure 1. Plot of the log of the power spectrum slopes against the log of luminosity

It has been suggested many times in the past that the X-ray variability in AGN is due to "flare" like events (shots) that occur randomly. Traditional models with exponential shots are rejected, as these will give $\alpha = 2$. Furthermore, any model where the shots have the same amplitude in all the objects but more shots occur in the more luminous of them (for example the starburst model) predicts that $A \propto L^{-1}$ which is not consistent with our results. A shot noise model where the rate of occurrence and the amplitude of the shots scale as $L^{0.5}$ is consistent with the fact that $A \propto L^{\sim -0.5}$. If the shape of the shots is $\propto t^{-0.25}$ (where t denotes time) then the slope of the power spectrum is -1.5.

It is now tempting to ask whether variability timescales are consistent in general with the idea that AGN are all the same except of having black holes of differing masses. However to address this question, we must be much more precise than we have been in the past about exactly what we mean by "variability timescale". If we assume that AGN power spectra in general vary as $L^{-\beta}\nu^{-\alpha}$ then the frequency at which we see a fixed spectral power varies as $L^{-\beta/\alpha} = L^{-0.34\pm0.02}$. Suppose however

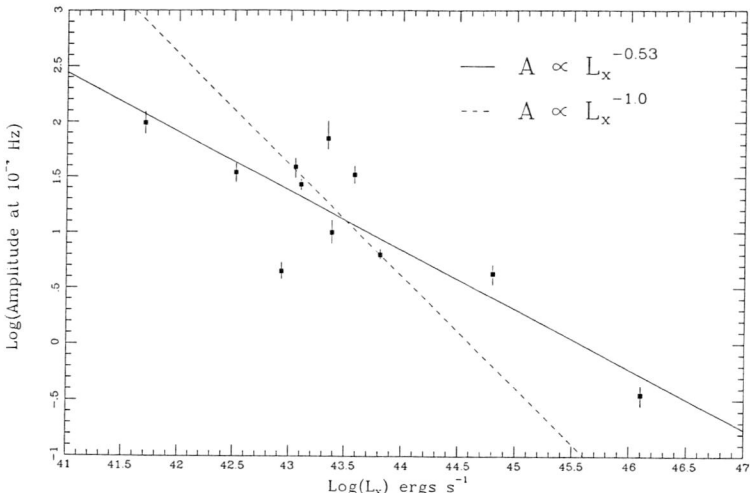

Figure 2. Plot of the log of the normalized variability amplitude against the log of luminosity

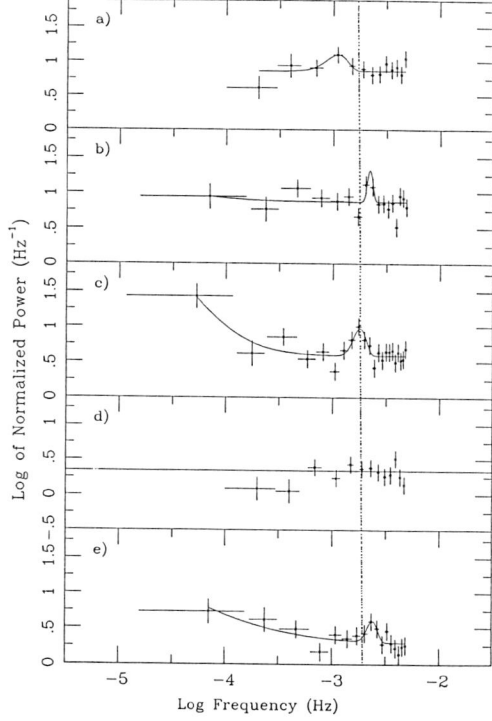

Figure 3. Power spectra of different ME observations of NGC 5548. The power spectra are plotted in order of increasing source intensity from top to bottom. The dotted line indicates the fitted centroid frequency of the middle observation.

we ask a different question: if we observe a given AGN for a length of time T, what will be the spread $< \Delta L/L >$ in our measurements of its luminosity? This involves contributions from all Fourier components from $\nu = 1/T$ upwards, and so we would find $< \Delta L/L >$ varies as T^d where $d = (\alpha - 1)/2) \approx 0.28 \pm 0.05$. Across AGN of differing luminosities, we can likewise ask what observation length T would give us a fixed $< \Delta L/L >$. We then find that T varies as $L^{\beta/(\alpha-1)} \approx L^{0.96\pm 0.16}$. These scaling relations present a serious challenge to any model of the X-ray emission from AGN; the onus is now on the modellers to make explicit predictions.

3. Detection of QPOs in NGC 5548

In the longest EXOSAT observation of NGC 5548 we see a significant broad peak, similar to the QPO peaks that appear in the power spectra of the Low Mass X-ray binaries (LMXB). Power spectrum analysis of other EXOSAT observations of NGC 5548 revealed similar QPO peaks in some of them (figure 3). In order to quantify the characteristics of the QPO we fitted the different spectra with a power law plus a Gaussian model (or a constant and a Gaussian when the red noise component was not apparent). A grouped χ^2 test shows that a single power law model cannot describe the three power spectra where red noise is apparent. Inclusion of the QPO feature significantly improves the fit.

As is obvious from figure 3 QPOs do not appear all the time. Where they appear, as the source brightens the centroid frequency increases and the strength of the QPO decreases. This behaviour is very similar to the behaviour of the so called horizontal branch QPOs in some of the brightest LMXB. Those galactic binary systems are powered from accretion onto a neutron star. It seems likely that QPOs are a generic feature of accretion onto compact objects. However the QPO here is worryingly fast; if we identify the time scale with the orbital time scale at $3R_{Sch}$, the mass of the black hole in NGC 5548 must be $< 10^6 M_\odot$ where BLR studies suggest $4 \times 10^7 M_\odot$ (Clavel et al 1992).

References

Barr, P. and Mushotzky, R. F. *Nature* **320**, 421, 1986
Clavel, J. et al *Ap. J.*, in press
McHardy, I. M. in *X-Ray Astronomy (Proc.. of the 23rd ESLAB Symposium)*, ed. N. E. White **2**, 1111, 1989
Papadakis, I.E. and Lawrence, A. *MNRAS*, in press

The Dramatic X-ray Spectral Variability of Mkn 841

Ian M. George [*]   Paul Nandra [†]   Andy C. Fabian [†]
T. Jane Turner [*]   Chris Done [‡]   Chas S.R. Day [*]

We present the results from a detailed analysis of the X-ray spectral variability of Mkn 841 based on *EXOSAT*, *Ginga* and *ROSAT* observations over the period 1984 June to 1990 July.

Variability is apparent in both the soft (0.1–1.0 keV) and medium (1–20 keV) energy bands (Fig. 1). Above 1 keV, the spectra are adequately modelled by a power-law with a strong emission line (of equivalent width ~ 450 eV). The energy of the line (~ 6.4 keV) is indicative of K-shell fluorescence from neutral iron, leading to the interpretation that the line arises via X–ray illumination of cold material within the accretion flow. In addition to the flux variability, the continuum shape also changes in a dramatic fashion, with variations in the apparent photon index $\Delta\Gamma \sim 0.6$.

The large equivalent width of the emission line suggests an enhanced reflection component in this source, compared to other Seyferts observed with *Ginga*. The spectral changes are interpreted in terms of variable power-law continuum superimposed on the flatter refection component. For one *Ginga* observation, the reflected flux appears to dominate the medium energy X–ray emission, resulting in an unusually flat slope ($\Gamma \sim 1.0$).

The soft X-ray excess reported by Arnaud et al. (1987), is found to be highly variable by a factor ~ 10. No evidence is found for an intrinsic column density $n_H \lesssim$ few 10^{20} cm^{-2}.

The implications of these results for the physical models of the emission region(s) in this and other X-ray bright Seyferts are briefly discussed in George et al. (1992).

References

Arnaud, K.A. *et al.*, 1987. MNRAS 217, 105.
George, I.M., Nandra, K., Fabian, A.C., Turner, T.J., Done, C. & Day, C.S.R., 1992. MNRAS, in press.

---

[*]Laboratory for High Energy Astrophysics, Code 668, NASA/Goddard Space Flight Center, Greenbelt, MD 20771, U.S.A., and Universities Space Research Association.
[†]Institute of Astronomy, Magingley Road, Cambridge, CB3 0HA, U.K.
[‡]X-ray Astronomy Group, Department of Physics & Astronomy, University of Leicester, Leicester, LE1 7RH, U.K.

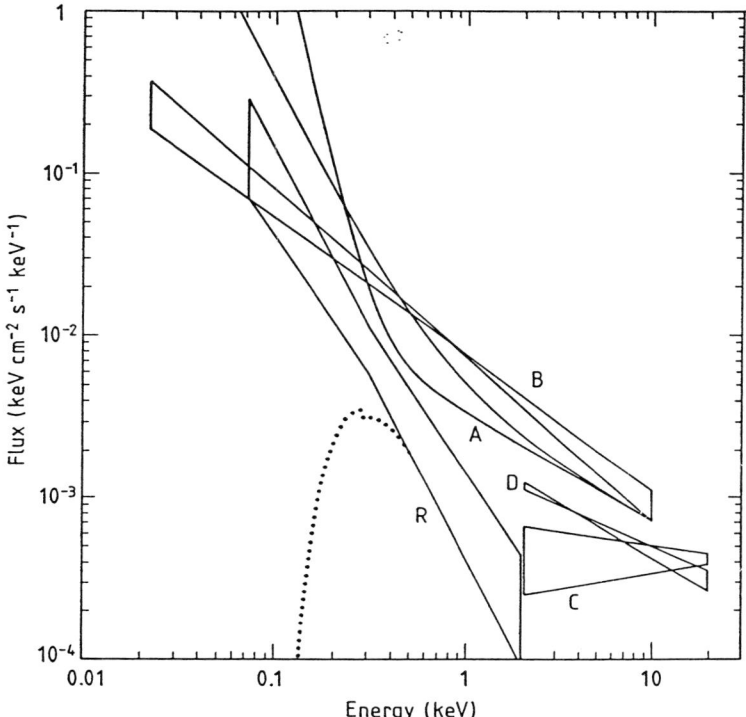

Figure 1. Schematic diagram showing the best-fitting spectra (corrected for the effects of low-energy absorption) from the 5 observations of Mkn 841 reported in George et al. (1992). The regions represent the 90% confidence regions for the best-fitting spectral models to the continua (the iron $K\alpha$ line has been excluded for clarity), and extend over the energy range of the respective datasets. The dramatic spectral variability exhibited by Mkn 841 in both the soft and medium X-ray bands is clearly apparent, as is the general steepening of the X-ray spectrum to lower energies. The soft X-ray excess in Observation A represents the 90% lower limit on the steep power-law ($\Gamma_S = 3.4$) and $\Gamma_S = 5$. The effect of uncertainties in the diffuse XRB on Observation C is evident, as is the usually flat spectrum observed during this epoch. The effect of an equivalent hydrogen column density equal to the Galactic HI value from 21cm measurements is shown dotted in the case of the $ROSAT$ PSPC observation (R). Observation R also includes an assumed ±25% systematic error in the normalization to take into account the uncertainty in the degree of shadowing by the PSPC window support structure.

Thermal and Non-Thermal Emission from Accretion Discs

A.C. Fabian *   R.R. Ross †

Abstract

Luminous accretion discs around black holes are expected to be optically thick and radiate much of their emission in the EUV and soft X-ray bands. Quasi-blackbody emission consistent with such discs is observed in many Seyfert 1 galaxies and from Galactic black hole candidates such as Cygnus X-1. The harder, rapidly variable, X-rays from such objects must originate above the disc, probably from non-thermal processes involving magnetic fields. The disc is therefore irradiated by a hard X-ray continuum. Backscattering and fluorescence from the disc produce a reflection spectrum, which is now observed in X-rays. Features in the reflection spectrum act as a diagnostic of the geometry and conditions of the inner disc, offering the strong possibility that it can be mapped in the near future.

1. Introduction

We begin by reviewing the case for the presence of accretion discs in many Active Galactic Nuclei (AGN), such as the Seyfert 1 galaxies. Here we are concentrating on the inner disc within radii $R \lesssim 100 R_S$, where R_S is the Schwarzschild radius of the central object (assumed here to be a black hole). Such discs were first detected from the UV excess and in particular by the variable soft X-ray emission that they produce. Further rapid progress has been hindered by the unfortunate coincidence that most of the direct thermal radiation produced by accretion discs around massive objects is emitted in the EUV, where photoelectric absorption by the interstellar medium of our Galaxy is strong.

A disc is however detectable in a passive way by scattering, absorbing and re-emitting any incident X-ray emission to make what is known as the 'reflection' spectrum. In the following Sections we discuss the properties of the reflection spectrum and of 'plain' discs, in which the irradiation from the X-rays does not significantly change the ionization structure of the disc surface. We then outline the spectra of irradiated discs, for which the incident X-ray flux is strong enough to change the ionization

---

*Institute of Astronomy, Madingley Road, Cambridge CB3 0HA

†Institute of Astronomy, Madingley Road, Cambridge CB3 0HA and Physics Department, College of the Holy Cross, Worcester, MA, USA

state of the matter near the surface. The emission lines produced from an irradiated accretion disc, in particular the iron line, are a very powerful diagnostic and test of the properties of the disc. X-ray observations, particularly with ASTRO-D, should soon enable us to confirm and refine our models of these regions.

We conclude with some brief remarks on the X-ray similarities between many AGN and Cygnus X-1, the Galactic Black Hole candidate. The similarities encourage us to believe that the structure of the inner disc and X-ray emission mechanism in both classes of objects are similar, apart from the obvious changes of scale. This review of the inner regions of discs is complementary to that of Collin-Souffrin [6], which concentrates on the outer parts.

2. The case for discs

Reprocessing is rife in the spectra of AGN. Little of the power generated directly escapes to the observer. Most of it is absorbed by matter in and around the engine which degrades it, re-emitting it at longer wavelengths. The clearest view of the central engine should be in the (spectrally) hardest band which carries a significant fraction of the power and shows the most rapid variability, which means the X-ray band. The observation of strong rapid variations seen in the X-ray observations of many Seyfert 1 galaxies argues very strongly that we are dealing with a process at least as efficient as accretion onto a massive black hole. Stellar processes and supernova remnants are just not efficient enough in a mass $\rightarrow$ energy conversion sense. Too much obscuring or scattering matter is left around if the matter $\rightarrow$ energy conversion efficiency is low, as in stellar processes or supernova shocks (see arguments in [12]).

Even in the case of the accreting black hole, which is the most efficient 'steady' source of power expected, the accreting gas tends to obscure the central engine when the power is high, *i.e.* when the luminosity $L \rightarrow L_{Edd}$. In spherical geometry,

$$\tau_T \approx 10\epsilon_{0.1}^{-1}\left(\frac{\dot{M}}{\dot{M}_{Edd}}\right)\left(\frac{R_{in}}{R_S}\right)^{1/2}\left(\frac{v}{v_{ff}}\right)^{-1}.$$

where R_{in} is the inner radius of the region considered and v_{ff} is the free-fall velocity of the infalling matter. If, as observed, there is a significant fraction of the power observed in variable X-rays, then the accretion rate $\dot{M} \ll \dot{M}_{Edd}$ or the flow is aspherical. It is very easy to envisage $\tau_T \gg 10$ if the infall velocity $v < v_{ff}$ which would cause any rapid variations (such as observed) to be smeared out.

The expected angular momentum in most accretion flows should easily lead to a disc geometry. Then if the hard X-ray emission originates above the disc, a low scattering and X-ray absorbing column along the line-of-sight can allow rapid X-ray variability to be observed. The strong similarities (scaled) between the variability and spectrum of Cyg X-1 and many AGN (Fig. 1a,b) suggest that such a model is sensible. Cyg X-1 is certainly not powered by starbursts or supernovae!

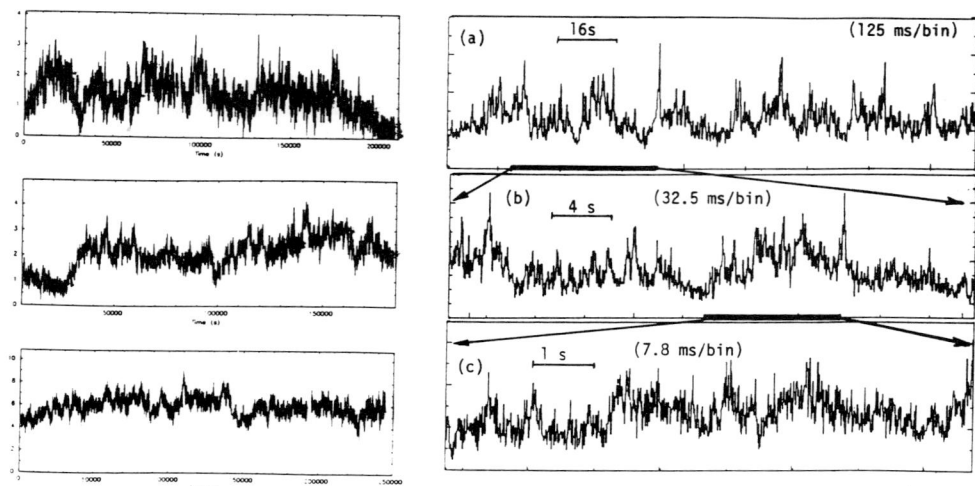

Fig. 1a. X-ray variability of 3 AGN, NGC 4051 (top left), MCG-6-30-15 (middle left) and NGC 5506 (bottom left) from [30] (the time units are 50000s) to be compared with Cygnus X-1 on different timescales (right hand panels) from [22].

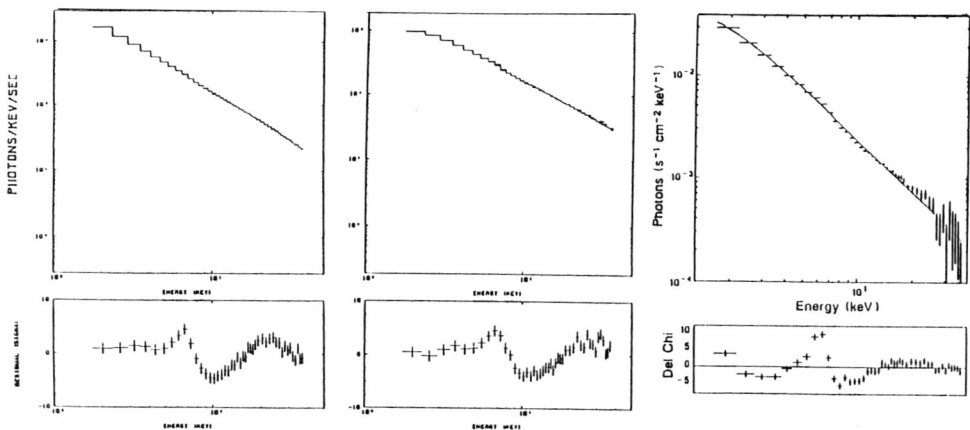

Fig. 1b. X-ray spectrum of Cygnus X-1 (left panel) over the range 1 – 50 keV to be compared with the X-ray spectrum of the sum of 12 AGN (right panel) from [41]. Note that the lengths of the y-axes are different. The lower panels show the residuals after subtracting the best-fitting power-law continuum. Note the similar 'wiggles' in all the spectra. Is the middle spectrum an AGN or a Galactic object?

For reasons that should become apparent, the multitude of disc models that have been introduced over the past 20 years in which the disc is very hot, 2-phase, or thick etc. will be ignored here. We adopt the standard Shakura-Sunyaev (1973, hereafter SS73) α-disc model here (see also the review by [32]).

3. X-ray reflection and emission from the standard disc

We adopt the disc description given in SS73 and concentrate on the radiation-pressure-dominated inner region which, for massive black holes, has approximately constant thickness and for which the Thomson depth is several 100 (increasing with R). If the radiation is thermalized in the disc (as appears to be roughly correct) then the peak temperature

$$T \approx 4 \times 10^5 M_6^{-1/4} \, \text{K}.$$

Most of the emission should therefore emerge in the EUV. Observations of the 'big blue bump' in the UV were the first strong evidence for a disc [36],[21]. A spectral excess observed in soft X-rays agrees with this picture [3],[47],[40], showing that there is a high luminosity (much of the total power) in the EUV band. Variability of the soft X-ray excess demonstrated that it was a key part of the central engine.

The hard X-rays observed from many Seyfert 1 galaxies presumably originate from above and below the disc. They therefore irradiate it and in so doing produce a characteristic X-ray spectrum of reflected and fluoresced X-rays [15],[20]. This 'reflection spectrum' was discovered in EXOSAT [27] and GINGA spectra of Seyfert 1 galaxies [31],[24]. What is observed then consists of the direct emission from the X-ray source(s) above the disc, plus the scattered and emitted spectrum from the disc itself. In the 2–20 keV band the reflection spectrum increases with energy as photoelectric absorption weakens relative to electron scattering, and shows fluorescent iron (and nickel) lines around 6.4 keV The spectrum drops at higher energy as Compton recoil becomes important (Fig. 2).

The reflection spectrum is detected in the spectra of many AGN, mostly Seyfert 1 galaxies (which are the ones with data strong enough to show such features). Nandra [28] has recently analysed 29 GINGA observations of 19 AGN which give strong confirmation of the reflection spectrum. Iron lines are detected in 17 out of 19 objects with a mean energy $\bar{E} = 6.43 \pm 0.07$ keV, indicating that the iron in the reflecting matter (the disc) is cold (*i.e.* less ionized than FeXVII). The average equivalent width of the line is 140 eV which implies that the reflector covers about 2π sr of sky at the source. This is consistent with a flat disc [14],[25]. 6 of the observations require that the line is broad ($\sigma \sim 0.7$ keV) which suggest the relativistic broadening (Doppler + gravitational) expected in a disc about a black hole [11]. None of the observations exclude such a linewidth.

Having given the good news it should be noted that a minority of objects show weak lines (50 eV; 3C273), strong lines (NGC6814, [16]; MCG 5-23-16, [29]; Mrk

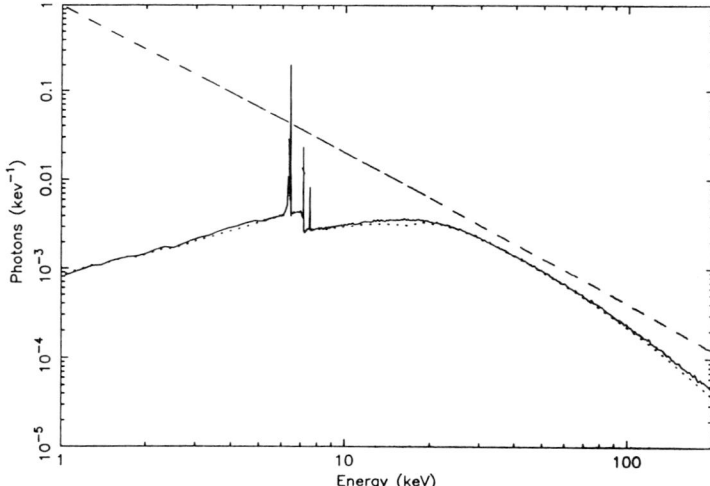

Fig. 2. Monte Carlo simulation of a reflection spectrum from [14]. The incident spectrum is a power-law continuum (dashed). Note the iron $K\alpha$ and β lines produced by fluorescence. The drop in albedo at low energies is due to photoelectric absorption, the drop at high energies is due to Compton recoil.

841 [8]), or narrow lines (NGC4151, [23], but see [46]). Most of the sources show a spectral hardening above 10 keV (the 'reflection bump', Fig. 2) and are well fitted by a reflection spectrum. A few however are not and 'partial covering' appears better ([24],[28]). The strength of the iron emission line in most objects is not consistent with the accreting blob picture proposed in the original work of Guilbert & Rees [15], unless most of the emission is generated from the outer surface of the mess of blobs. The equivalent width of the iron line in transmission spectra, if there is an added component of direct emission sufficient to compare with observed spectra, is less than the observed equivalent width. The overall fit with a flat irradiated disc is remarkably good.

Broad emission lines are expected from the doppler and gravitational effects in the disc model [11],[37],[17]. Most of the observed lines are consistent with an irradiated disc model (see [26] for diagrams of equivalent width, energy and broadening as functions of disc inclination). Confirmation of the irradiated black-hole disc model is anticipated after the launch of ASTRO-D in 1993 which has detectors with 100 eV spectral resolution. Observations of the response of the emission lines to changes in the irradiating spectrum should reveal the absolute size of the region and thus allow the mass of the central object to be determined.

The emission spectrum of a 'plain' accretion disc around a massive black hole has been calculated by many workers (*e.g.* [35],[7],[45] [18],[38]). It should be a quasi-blackbody at optical-UV wavelengths, but more complicated in the soft X-ray band where opacities are low and photons diffuse out from large Thomson depths. Comptonization should play a role.

We have recently carried out full, non-LTE calculations of the radiative transfer in the inner parts of such a disc [33]. The only assumption is that the density is constant with depth and that the energy is deposited (via viscosity) uniformly with depth, The temperature, ionization and excitation structure are determined by the radiation fields. We confirm that H and He are highly ionized and that the soft X-ray spectrum is mainly shaped by the view to the photosphere at different frequencies (Fig. 3). For $\alpha = 0.1$, abundant soft X-rays (at energies above 0.15 keV) are obtained when $L > 0.1 L_{Edd}$, $M < 10^8 \, M_\odot$. A model in which $L = 0.28 L_{Edd}$, $M = 2 \times 10^7 \, M_\odot$ appears to fit both the soft X-ray and IUE band excess emission in Mrk 841.

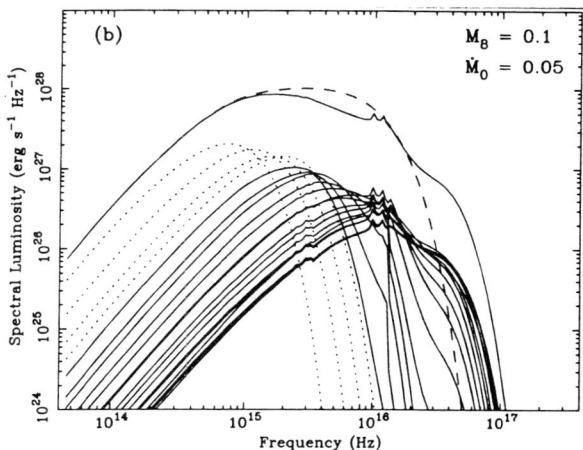

Fig. 3. Spectra of an accretion disc (solid line) with central mass of $10^7 \, M_\odot$ and accretion rate corresponding to $L/L_{Edd} = 1/6$. The dashed curve shows the total luminosity from the blackbody approximation, the lower curves show the contribution from individual annuli and the dotted curves show the blackbody approximations used for outermost annuli. From [33].

There are however some well known observational problems with plain discs; a lack of any H Lyman edge [1], a lack of the polarization expected from such a flat scattering atmosphere and optical-UV time lags which are too short. These may be overcome by X-irradiation of the disc (or a hot corona); by free-free opacity and relativistic effects [19] or Faraday depolarization (Phinney – this meeting), and by making the optical emission originate from the inner parts of the disc [6]. Further work is need before a conclusive statement can be made about these problems.

4. X-ray reflection and emission from an X-ray irradiated disc

In the previous section we saw that the X-ray data are generally well fit by reflection spectra, although there are some instances when the iron line is too strong or too weak for the standard (flat) model. Our purpose now is to investigate irradiated discs to see whether this explains the odd sources. Also, since we know that the disc must be irradiated to obtain the reflection spectrum, we wish to see the general effects of the X-ray heating on the ionization structure of the disc.

We have therefore calculated the X-ray emission from a disc, as described in the previous section but with its surface irradiated by an X-ray continuum source [34]. Since it is the ionization structure of the outermost layers of the disc that determines the reflected spectrum, we concentrate on these layers and approximate the disc by a slab of thickness $\tau_T \sim 5$. An isotropic flux, F_h, of power-law X-rays is incident onto its top face and a blackbody flux, F_s, onto its lower face; the density, n and blackbody temperature are appropriate for the required radius in an SS disc. The ionization state of the disc (for elements beyond He) is essentially determined by the ionization parameter

$$\xi = \frac{4\pi F_h}{n}.$$

If we assume that $F_h \propto \dot{M}$ and $F_h = F_s$ then

$$\xi \propto \frac{\dot{M}^3}{R^{9/2}},$$

except near the innermost radii. $\dot{M}$ is here the ratio of the mass flow rate in the disc to the Eddington-limiting value. If $L = fL_{Edd}$ then $\xi = 7 \times 10^4 f^3$ at $7R_s$, so the gas should become highly ionized as $f \to 1$.

Generally, we find that the disc is highly reflective up to the carbon edge for all ξ and to higher energies as ξ increases and oxygen and other elements become more highly ionized. The emergent spectrum is shown in Fig. 4 where strong recombination lines of oxygen and iron are apparent. It is essentially all reflected emission above 1 keV since the temperature of the layer ($T < 10^7$ K) is to low for any significant X-ray production above those energies. The equivalent width of the iron Kα line initially drops relative to the cold gas value as ξ rises to few 100, then increases steadily to a peak of about 600 eV when ξ reaches about 2000. It drops again at yet higher values of ξ.

The initial drop in equivalent width occurs when iron in the surface layers is predominantly Fe XVIII – Fe XXIII. Since there is then a vacancy in the L-shell of the ion, the iron line then has a high optical depth to resonance absorption, followed by a probably of only about 30% of re-emission, since the Auger process dominates. This means that the line is rapidly destroyed by resonance scattering. Lower ionization stages have a filled L-shell so resonance scattering does not occur; higher stages do not have a second L-shell electron which can be ejected in the Auger process.

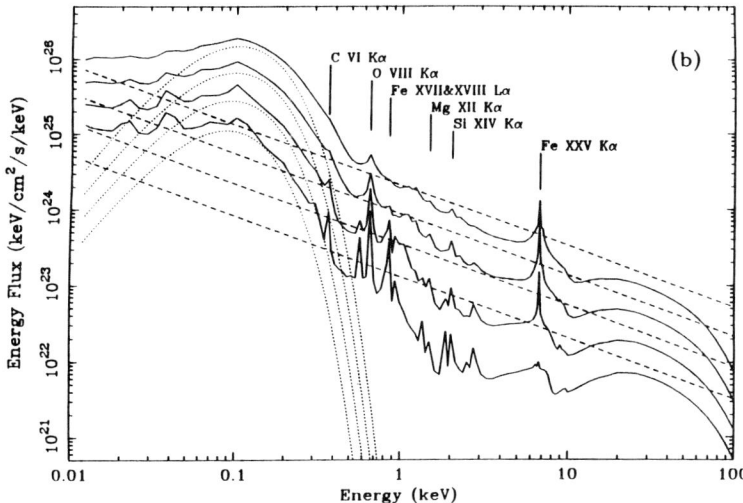

Fig. 4. Model reflection spectra [34] for the conditions appropriate to a hard power-law flux of photon index 1.8 incident on a section of an accretion disc at $7R_S$ around a $10^7 \, M_\odot$ black hole. The illuminating hard flux equals the soft flux produced locally by thermal emission in the disc. Accretion rates correspond to 0.15, 0.2, 0.25 and 0.30 of $\dot{M}_{Edd}$. The spectra have been separated vertically for ease of presentation.

The ranges of equivalent width that we have found encompass those found in AGN so it seems now possible that irradiated discs are capable of explaining all observations (c.f., Fig. 1b). An interesting bonus is that the weak iron line, and large iron absorption edge, seen in Cygnus X-1 (and other Galactic black hole candidates; [41],[10],[9]) is also readily explained since the hotter disc around a stellar mass black hole will always maintain iron in the state where resonance absorption followed by Auger destruction dominates (Fig. 5).

The detailed spectrum from an irradiated disc depends upon the range of ξ across the disc, which depends upon the distribution and power of the incident X-ray continuum. Most reasonable discs will have ξ decreasing outward, which opens up the possibility of a strong iron line from the central regions, a weak iron line zone and then an outer region in which the 'cold' value dominates. The mean energy and shape of the line are then determined by relativistic effects (see paper by G. Matt) which are greatest where the line is strongest. The outer layers of the disc are now much hotter than for a 'plain' disc and some of the hydrogen Lyman edge problem may be reduced.

The oxygen (and iron-L) emission lines predicted by X-irradiation may already have been seen in Einstein Observatory SSS spectra [43]. Preliminary fits of some combined

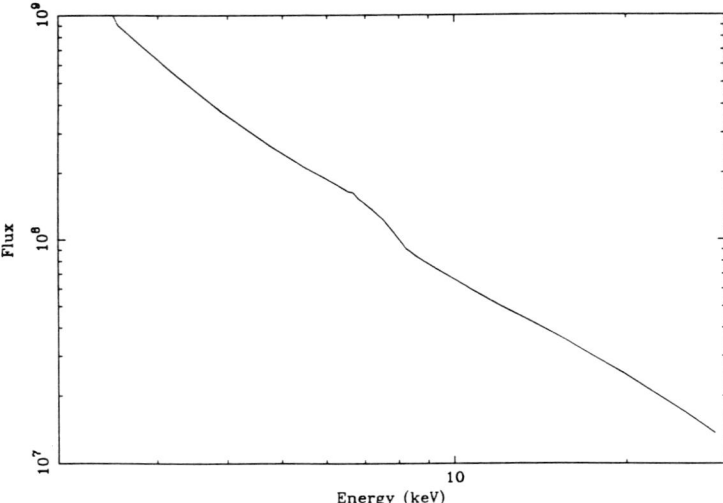

Fig. 5. Model reflection spectra for the conditions appropriate to Cyg X-1, with a hard power-law flux of photon index 1.8 incident on a section of an accretion disc at $7R_S$ around a $10\,M_\odot$ black hole. The illuminating hard flux equals the soft flux produced locally by thermal emission in the disc. The accretion rate corresponds to 0.05 of $\dot{M}_{Edd}$. Note the weak iron line and large iron edge above 7 keV.

ROSAT-Ginga spectra of AGN such as Fairall 9 also seem to be well fit by spectra such as those in Fig. 4 (Nandra et al., in preparation). There is much yet to be done in this area – the number and range of parameters are large – but there is a good chance that we can use the detailed X-ray spectra and their time variability to map out the innermost parts of the accretion flows around massive objects.

5. The Hard X-ray Flux

So far we have shown that a standard SS73 disc, modified by the irradiation of a power-law X-ray continuum, can explain may of the spectral features of observed AGN. What we have not shown, nor do the present models account for, is the hard X-ray continuum.

The observation that the reflected spectra do fit the data strongly suggests that there is no optically-thick corona above the disc, which would further reprocess that spectrum. The emission region must be optically thin and the emission must therefore be due to very hot, probably relativistic, electrons. The likely acceleration process for the electrons are electric fields produced by Faraday's law from wound-up, moving, magnetic fields in the disc, which pop out of the surface (Parker's instability) when they reach equipartition with the disc pressure. They are the likely source of viscosity (see paper by J. Hawley) and the surface of a disc may be covered in flares of which solar ones could be a very pale analogy. (Flare models have been proposed by [13] and others.)

If this is the case, the magnetic fields must be strong and synchrotron radiation should dominate the observed IR/opt/UV emission. Such strong fields can however trap cool gas, which would then be very dense and free-free absorb much of the emission below $\sim 10^{15}$ Hz [4]. The emission region in the central engine of AGN may indeed be very complicated, with only the bare essentials apparent so far. Further progress requires all of the X-ray emission components to be identified and understood, together with their implications for other wavebands.

6. Summary

The X-ray reflection spectrum together with the EUV bump strongly support the simple, thin, accretion disc picture. There are some exceptions which may be understood in the context of irradiated thin discs. So far the simple SS73 α–disc model is proving adequate. An important result is that the picture explains both the spectra of AGN and of Cygnus X-1, underlining the similarity between these objects.

Observations of the extremely broad iron line region (the iron EBLR) with good sensitivity and spectral resolution should begin next year with ASTRO-D. We may then begin the next stage of understanding the mass, geometry, flows and properties of the central engine of the nearest and brightest AGN.

References

[1] Antonucci, R., 1988. In: *Supermassive Black Holes*, p. 26, ed. Kafatos, M., Cambridge University Press, Cambridge
[2] Antonucci, R. R. J., Kinney, A. L. & Ford, H. C., 1989. apj, **342**, 64
[3] Arnaud, K. A., Branduardi-Raymont, G., Culhane, J. L., Fabian, A. C., Hazard, C., McGlynn, T. A., Shafer, R. A., Tennant, A. F. & Ward, M. J., 1985. MNRAS, **217**, 105
[4] Celotti, A., Fabian, A.C., Rees, M.J., 1992, MNRAS, 255, 419
[5] Collin-Souffrin, S. & Dumont, A. M., 1990. AaA, **229**, 292
[6] Collin-Souffrin, S., 1992. These Proceedings
[7] Czerny, B. & Elvis, M., 1987. ApJ, **321**, 305
[8] Day, C.S.R., Fabian, A.C., George, I.M., Kunieda, H., 1990, MNRAS, 247, 15P
[9] Done, C., Mulchaey, J.S., Mushotzky, R.F., Arnaud, K.A., 1992, ApJ, 375, 295
[10] Ebisawa, K., 1991, Ph.D. Thesis, Univ. of Tokyo
[11] Fabian, A.C., Rees, M.J., Stella, L., White, N.E., 1989, MNRAS, 238, 729
[12] Fabian, A.C., 1992, in Holt, S.S., Neff, S.E., Urry, C.M., Testing the AGN Paradigm, AIP Conf. Proc. 254, p. 657
[13] Galeev, A.A., Rosner, R., Vaiana, G.S., 1979, ApJ, 229, 318
[14] George, I.M., Fabian, A.C., 1991, MNRAS, 249, 352
[15] Guilbert, P.W., Rees, M.J., 1988, MNRAS, 233, 475

[16] Kunieda, H., Turner, T.J., Hisamitsu, A., Koyama, K., Mushotzky, R., Tsusaka, Y., 1990 Nature, 345, 786
[17] Laor, A., 1990. ApJ, **376**, 90
[18] Laor, A. & Netzer, H., 1989. MNRAS, **238**, 897
[19] Laor, A., Netzer, H. & Piran, T., 1990. MNRAS, **242**, 560
[20] Lightman, A.P., White, T.R., 1988, ApJ, 335, 52
[21] Malkan, M. A. & Sargent, W. L. W., 1982. ApJ, **254**, 22
[22] Makishima, K, 1988, in Tanaka, Y., ed. Physics of Neutron Stars and Black Holes, Universal Acad press, Tokyo, p 175
[23] Marshall, F.E., et al, 1992. in Holt, S.S., Neff, S.G., Urry, C.M., Testing the AGN Paradigm, AIP, New York, p 183
[24] Matsuoka, M., Piro, L., Yamauchi, M., Murakami, T., 1990, ApJ, 361, 440
[25] Matt, G., Perola, G.C., Piro, L., 1991, AaA, 247, 25
[26] Matt, G. *et al.*, 1992, in preparation
[27] Nandra, K., Pounds, K.A., Stewart, G.C., Fabian, A.C., Rees, M.J., 1989, MNRAS, 236, 39P
[28] Nandra, K., 1991, Ph.D. Thesis, Univ. of Leicester
[29] Piro, L., Matsuoka, M., Yamauchi, M., 1991, in Duschl W.J., Wagner S.J., Springer-Verlag, Physics of Active Galactic Nuclei, in press
[30] Pounds, K.A., McHardy, I.M., 1988, in Tanaka, Y., ed., Physics of Neutron Stars and Black Holes, Universal Acad press, Tokyo, p 285
[31] Pounds, K. A., Nandra, K., Stewart, G. C., George, I. M. & Fabian, A. C. 1990. *Nature*, **344**, 132
[32] Pringle, J.E., 1981, ARA&A, 19, 135
[33] Ross, R.R., Fabian, A.C., Mineshige, S., 1992, MNRAS, 258, 189
[34] Ross, R.R., Fabian, A.C., 1992, MNRAS, in press
[35] Shakura, N. I. & Sunyaev, R. A., 1973. AaA, **24**, 337
[36] Shields, G., A., 1978, Nat, 272, 706
[37] Stella, L., 1990, Nat, 344, 747
[38] Sun, W. H. & Malkan, M. A., 1987. In: *Astrophysical Jets and Their Engines*, p. 125, ed. Kundt, W., Reidel, Dordrecht
[39] Sun, W. H. & Malkan, M. A., 1989. ApJ, **346**, 68
[40] Turner, T. J. & Pounds, K. A., 1989. MNRAS, **240**, 833
[41] Tanaka, Y., 1991, in Treves A., Perola G.C., Stella L., Iron Line Diagnostics in X-ray Sources, Springer-Verlag, Berlin, 98
[42] Turner, M.J.L., 1990, MNRAS, 244, 310
[43] Turner, T.J., Weaver, K.A., Mushotzky, R.F., Holt, S.S., Madejski, G.M., 1991, ApJ, 381, 85
[44] Turner, T.J., Done, C., Mushotzky, R., Madejski, G.M., Kunieda, H., 1992, ApJ, 391, 102
[45] Wandel, A. & Petrosian, V., 1988. ApJ, **329**, L11
[46] Weaver, K.A., 1992, in Holt, S.S., Neff, S.G., Urry, C.M., Testing the AGN Paradigm, AIP, New York, p 192
[47] Wilkes, B. J. & Elvis, M., 1987. ApJ, **323**, 243

Ultra-soft X-ray Emission in AGN

E. M. PUCHNAREWICZ AND K. O. MASON

Mullard Space Science Laboratory, University College London, Holmbury St. Mary, Dorking, Surrey RH5 6NT, UK.

1 INTRODUCTION

A soft X-ray excess below 2 keV is a common feature in the X-ray spectra of AGN (Turner & Pounds 1989; Masnou et al. 1992). A popular interpretation is that it represents the high energy tail of the big blue bump for which the models include accretion disks and reprocessing in cold matter. *ROSAT* provides us with the opportunity to study the parameters of the soft X-ray excess for the first time.

In this paper, we discuss new *ROSAT* PSPC spectra (covering 0.1-2.4 keV) of four AGN taken from the USS survey which selected the softest sources in the *Einstein* IPC database (Córdova et al. 1992; Puchnarewicz et al. 1992a, hereafter C92 and P92a): these AGN are E1346+266, E0845+378, E0844+377 and E2034−228.

2 E1346+266 - A HIGH REDSHIFT ULTRA-SOFT X-RAY AGN

Observationally, soft X-ray AGN are generally found at low redshifts; all of the soft excess AGN in the Turner & Pounds and Masnou et al. samples have redshifts below 0.2, leading to the suggestion that a redshift of 0.5 is sufficient to make any soft X-ray component undetectable (Masnou et al. 1992; P92a). This implies an upper limit to the effective temperature of the soft component in the rest-frame, providing an observational constraint on models for the big blue bump.

E1346+266 has a redshift of 0.92, much higher than the suggested z=0.5 cut-off, yet the *ROSAT* PSPC spectrum confirms earlier indications from the *Einstein* data that this object has a strong soft X-ray excess. The presence of this ultra-soft X-ray excess implies both a high intrinsic temperature and a high luminosity for the soft X-ray component and poses a severe test for models of its production. Full details are given in Puchnarewicz et al. (1992b; hereafter P92b).

The soft X-ray spectrum is very steep; for a single power-law fit to the data, the best-fit energy index $\alpha=2.44^{+0.43}_{-0.30}$ with no evidence for any absorption instrinsic to the AGN. (We define the energy index α such that $F_\nu \propto \nu^{-\alpha}$.) In terms of a simple blackbody, the soft component has a high effective temperature with a kT_{eff} of 110^{+40}_{-60} eV in the AGN rest-frame. This is significantly higher than for the Masnou et al. sample and the average for the low redshift (z<0.5) USS AGN (~10 eV; C92).

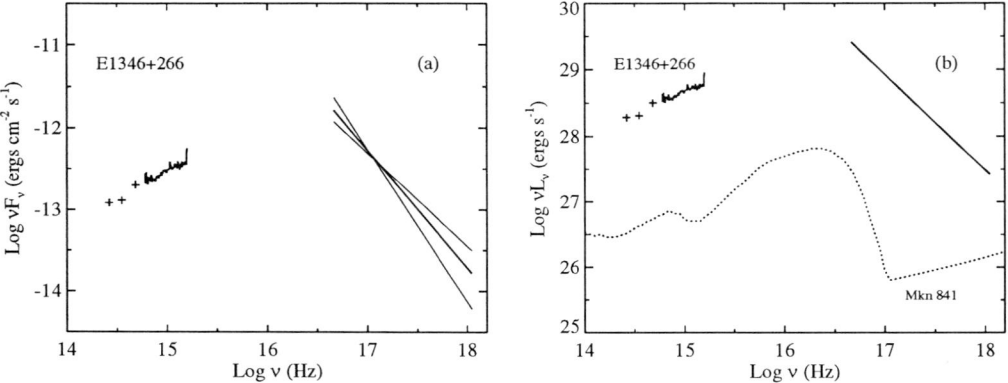

Figure 1 : The multiwavelength spectrum of E1346+266 from IR to X-rays. (a) The IR-to-optical spectrum with the single power-law model fitted to the PSPC data, plotted in the rest-frame. The thick line represents the best-fit model and the thin lines represent the 90% confidence limits on α. (b) The IR-to-optical spectrum plotted with the best-fit single power-law model (plotted as a solid line); also plotted for comparison as a dotted line, is the multiwavelength spectrum of Mkn 841 (Arnaud et al. 1985). These spectra are plotted in terms of rest-frame luminosities (νL_ν).

The best-fit single power-law model is combined in Figure 1 with J, H and K IR magnitudes and optical spectroscopy to produce a multiwavelength spectrum. In Figure 1b, the spectrum of E1346+266 is plotted in terms of its intrinsic luminosity and compared with the model fitted to the spectrum of Mkn 841 by Arnaud et al. (1985). Mkn 841 was chosen because its soft X-ray excess is thought to be typical of those found in nearby AGN; when represented by a blackbody, the soft component in Mkn 841 (z=0.037) was best fit with a kT_{eff} of 17 eV (Arnaud et al. 1985). The diagram illustrates that the soft X-ray component of E1346+266 lies at a much higher temperature (blackbody kT_{eff}=110 eV in the rest-frame) relative to Mkn 841.

The soft component of E1346+266 is perhaps the strongest identified so far, in terms of its luminosity and its ratio with the hard component at 0.2 keV (P92b). Thus it presents an important observational test for models of the soft X-ray excess and the big blue bump. If the optical and X-ray data of E1346+266 mark the low and high energy tails of the big blue bump, we estimate a luminosity of $\sim 10^{47}$ ergs s^{-1}, which implies a black hole mass, $M \geq 10^9 \, M_\odot$ for a system radiating at or below its Eddington limit.

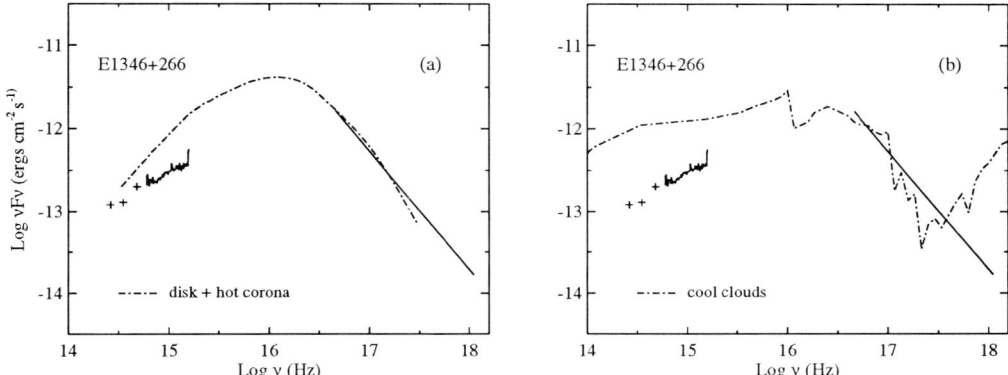

Figure 2 : The multiwavelength spectrum of E1346+266 compared with two different models (shown as dot-dash lines): (a) the Czerny and Elvis thick disk ($M = 10^8\ M_\odot$, $\dot{M} = 10\ M_\odot$ yr^{-1}) surrounded by a hot corona (corona temperature = 120 keV, $\tau = 0.1$): (b) the Ferland and Rees (1988) cool clouds for a density $N = 10^{14}$ cm^{-3} and volume filling factor 10^{-5}.

Thin accretion disk models (eg. Sun & Malkan 1989) for the big blue bump in E1346+266 may be ruled out since the implied accretion rate is too high (a few orders of magnitude times the Eddington rate). Thick disks with a surrounding hot, optically thin corona (eg. Czerny & Elvis 1987; see Figure 2a) or reprocessing in cool clouds (eg. Guilbert & Rees 1988; Figure 2b) can reproduce the X-ray spectrum, although published spectra of both models produce an excess of IR/optical flux relative to the X-rays.

3 E0845+378

The USS AGN E0845+378 is a 'secure' identification (ie. the soft component has a relatively high statistical significance - see C92 for details) and lies at a redshift of 0.31. It was observed by the *ROSAT* PSPC for 12.4 ksecs and the image reveals a faint companion approximately 1' to the west. The PSPC spectrum of E0845+378 is well-represented by a single power-law model ($\chi^2_\nu = 1.3$ for $\nu = 16$, see Figure 3a). The best-fit energy index $\alpha=1.9^{+0.5}_{-0.3}$ with little or no intrinsic absorption (Figure 3a). Thus the spectrum is softer than the average for other low-redshift AGN ($\alpha=1.4$) detected with *ROSAT* (Fink 1992). The spectrum of the companion is much harder, $\alpha=0.6^{+1.3}_{-0.3}$ (assuming Galactic N_H).

4 E2034−228 AND E0844+377

Neither of these sources show evidence for a soft excess in the ROSAT data (Figure 3b and c). E2034−228 has a faint companion approximately 30" to the east (further from the AGN) whose spectrum *is* very soft. Due to the poorer resolution of the *Einstein* IPC, these two sources may not have been resolved, *ie.* the soft companion may have been mistaken for the apparent 'soft component' in the IPC X-ray source. The evidence for a soft excess in the Einstein data on E0844+377 was always weak due to the small number of counts detected (it was a non-secure USS source in C92). Interestingly, the optical continuum of E0844+377 rises steeply to the blue (P92a) suggesting the presence of a strong big blue bump.

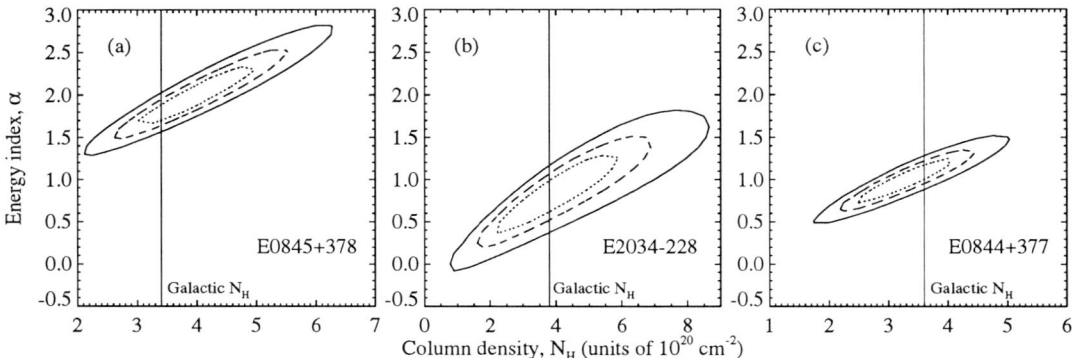

Figure 3 : Confidence contour plots for the three USS AGN, (a) E0845+378, (b) E2034−288 and (c) E0844+377. Contours are drawn at 68% (dotted), 90% (dashed) and 99% (solid) confidence for two interesting parameters. The Galactic column density in each case is indicated by the solid, vertical line.

References

Arnaud, K. A. *et al.* 1985, M.N.R.A.S., **217**, 105.

Córdova, F. A. *et al.* 1992, Ap. J. Suppl., **81**, 661 (C92).

Czerny, B. and Elvis, M. 1987, Ap. J., **321**, 305.

Ferland, G. J. and Rees, M. J. 1988, Ap. J., **332**, 141.

Fink, H. H. 1992, *in* Space Science with particular emphasis on High Energy Astrophysics, ed. J. Trumper, in press.

Guilbert, P. W. and Rees, M. J. 1988, M.N.R.A.S., **233**, 475.

Masnou, J. L. *et al.* 1992, A.&A., **253**, 35.

Puchnarewicz, E. M. *et al.* 1992a, M.N.R.A.S., **256**, 589 (P92a).

Puchnarewicz, E. M., Mason, K. O. and Córdova, F. A. 1992b, M.N.R.A.S., (submitted) (P92b).

Sun, W.-H. and Malkan, M. A. 1989, Ap. J., **346**, 68.

Turner, T. J. and Pounds, K. A. 1989, M.N.R.A.S., **240**, 833.

Highly Ionized Gas in Seyfert Galaxies

K. Nandra *

Abstract

Recent observations with *ROSAT* have revealed an absorption feature in the spectra of some active galaxies, which is associated with the K–edge of highly ionized oxygen. This confirms that a large column of partially ionized material, the so-called warm absorber, which had been inferred from previous X–ray observations at higher energies, is present in the line–of–sight.

1. Introduction

X–ray evidence for highly ionized material in the line–of–sight to AGN has been accumulating over the past few years. Halpern (1984) originally suggested that highly ionized gas might be responsible for the soft X–ray absorption observed in some AGN, and cause spectral variations, as the opacity of the material changes with the ionizing flux. Further evidence was found using *EXOSAT* with many workers presenting evidence for flux–correlated changes in the soft X–ray absorption (e.g., Pan et al. 1990). However, given the widely recognized complexity of Seyfert spectra and the many candidate processes for producing the spectral changes, such observations were inconclusive. Subsequently, analysis of the *Ginga* spectra showed evidence for absorption at the iron–K edge, at an energy apparently higher than that expected from cold iron at 7.1 keV, in $\sim 50\%$ of a sample of Seyfert galaxies. Nandra (1991) found a mean energy $\sim 8 - 8.5$ keV, depending on the assumed continuum model. Again, however, it was difficult to establish the ionization state, or even the significance of the edge feature, in many individual cases, given the modest resolution of the LAC detector.

The launch of *ROSAT*, with the high sensitivity of the PSPC to soft X–rays (0.1 – 2.0 keV) has provided an additional opportunity to search for the signatures of the warm absorber in the soft X–ray band. In particular, the degree of ionization and optical depth suggested by the *Ginga* data imply a substantial edge arising from oxygen–K, which should occur at 0.53 – 0.87 keV, depending on the ionization state.

2. The ROSAT data

A spectral analysis of the PSPC data for two Seyfert nuclei, NGC 5548 and MCG-6-30-15 is presented. For further details see Nandra et al. (1992) and Nandra & Pounds (1992).

---
*Institute of Astronomy, Madingley Road, Cambridge CB3 0HA.

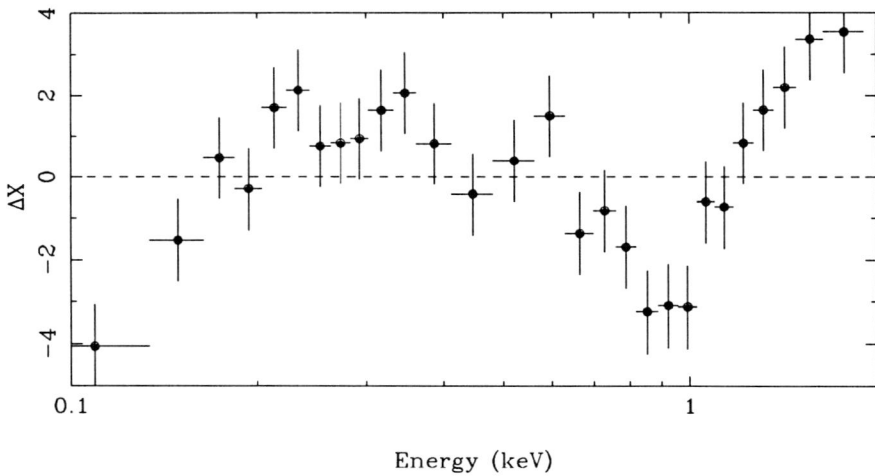

Figure 1. Data–minus model residuals derived from a power law fit to the PSPC spectrum of NGC 5548. A deficit of counts at $\sim 0.8\,\mathrm{keV}$ is indicative of K–shell absorption by highly ionized oxygen.

2.1 NGC 5548

NGC 5548 was observed between 1990 July 16 and 1990 July 20 (Principal Investigator: A.C. Fabian) with the *ROSAT* PSPC. In the total exposure time of 17917s, a mean observed flux 1.48×10^{-11} erg cm^{-2} s^{-1} was found, implying a relatively weak state. After extraction the spectrum was rebinned so that at least 1000 photons were present in each bin and compared to various trial models. Two per cent systematic errors were added, to account for the remaining calibration errors. A simple power law gives a very poor fit ($\chi^2_\nu = 4.3$ for 25 d.o.f.), with the main residual being a deficit of counts around $\sim 0.8\,\mathrm{keV}$ (see Fig. 1). Fitting this with an absorption edge, as expected from highly ionized oxygen, improves the fit dramatically, with the F-statistic (F=26.7) showing a significant feature at the $> 99\%$ level. The confidence contours for the best fit of Nandra et al. (1992) (which also includes a blackbody soft excess component) are shown in Fig. 2. This clearly demonstrates that the oxygen responsible for the absorption feature is highly ionized.

2.2 MCG-6-30-15

The *ROSAT* PSPC observation of MCG-6-30-15 took place on 1992 Jan 29, with an exposure time of 8500s. Again the source was found to be in a low state, with observed $0.1 - 2.0\,\mathrm{keV}$ flux 2.4×10^{-11} erg cm^{-2} s^{-1}. The fits to this data-set are shown in Fig. 3. In this case the data have been re-binned with 500 photons per channel, as the exposure time is shorter. A clear edge is present (see Fig. 3) with best–fit parameters $E_{th} = 0.79 \pm 0.03$ and $\tau = 1.2 \pm 0.2$, again indicating oxygen which

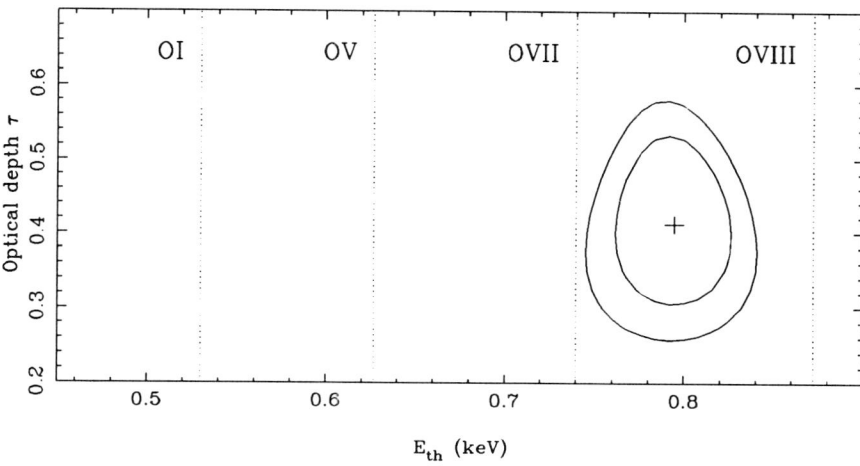

Figure 2. Confidence contours derived from fitting an absorption edge to the *ROSAT* PSPC spectrum of NGC 5548. The contours represent (outwards) 68 and 90% confidence for the edge optical depth and energy. The cross marks the position of minimum χ^2. The dotted vertical lines show the expected energies for various ionization states of oxygen. Highly ionized oxygen is strongly preferred by these data, with a mixture of OVII and OVIII being most likely.

is very highly ionized, consistent with the *Ginga* warm absorber measurements.

3. Discussion

Warm material, responsible for absorption signatures at the oxygen and iron K–edges, has been found in the line of sight to two Seyfert galaxies. Given that Fe K–edge detections by *Ginga* are common in these objects (Nandra 1991), it seems likely that the warm absorber is present in most active galaxies. The location of the highly ionized material is unclear, but the variability observed at soft X–ray energies suggests that it responds quickly to changes in the ionizing flux, implying a high density and a location close to the central source. There it could be responsible for substantial re-processing (e.g., Ferland, Korista & Peterson 1990) and may be intimately involved with the accretion process.

Acknowledgments The author thanks Andy Fabian and many other colleagues, especially Ian George, Ken Pounds, Richard Saxton, Steve Snowden, Gordon Stewart, Jane Turner and Tahir Yaqoob, for many discussions.

References

Halpern, J.P., 1984, ApJ, 281, 90.

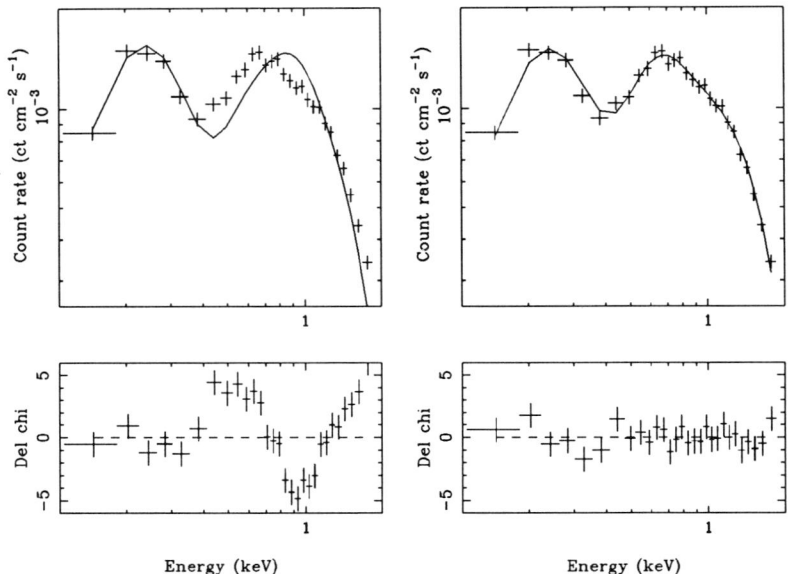

Figure 3. Spectral fits to the *ROSAT* PSPC spectrum of MCG-6-30-15. The top panels show the data and folded best–fit model (solid line) for (a) a power law and (b) power law plus edge. The lower panels show the residuals. The power law fit is clearly inadequate ($\chi^2_\nu=9.2$ for 26 d.o.f), with a very large deviation at the oxygen edge. Adding an edge (b), reduces χ^2 by 220, leaving the fit entirely acceptable, with $\chi^2_\nu=0.8$ for 24 degrees of freedom.

Ferland, G., Korista, & Peterson, B.M., 1990, ApJ, 363, L21.
Nandra, K., 1991. Ph.D. Thesis, Leicester University.
Nandra, K., Fabian, A.C., George, I.M., Branduardi–Raymont, G., Lawrence, A., Mason, K.O., McHardy, I.M., Pounds, K.A., Stewart, G.C., Ward, M.J. & Warwick, R.S., 1992, MNRAS, in press.
Nandra, K. & Pounds, K.A., 1992, Nature, 359, 215.
Pan, H.-C., Stewart, G.C. & Pounds, K.A., 1990, MNRAS, 242, 177.

EUV Observations of Seyfert 1 Galaxies and Quasars

P. M. Gondhalekar and B. J. Kellett *

1. Introduction

It is widely believed (but not proven) that accretion of matter onto a supermassive black hole is the primary source of energy for active galactic nuclei. Features commonly observed in AGN spectra such as the width of the broad emission lines and the ultraviolet to soft X-ray excess have been attributed to the accretion process. Many attempts have been made to fit accretion disk model spectra to the continuum energy distribution (CED) of AGNs to refine and/or verify these models. Thin disk models have had some success in explaining the CED of Seyfert 1 galaxies and low redshift quasars (Malkan & Sargent 1982; Sun & Malkan 1989; Laor 1990) but these fits have been made only to the optical/ultraviolet part of the CED. For the accretion rates and the black hole masses expected in AGNs the disk emission is expected to peak in the extreme ultraviolet/soft X-ray region and observations close to this peak would put tighter constraints on the thin disk models. EUV observations of four AGNs are described in this paper.

2. The EUV Observations

ROSAT/WFC survey images have been analysed to identify EUV detections of Seyfert 1 galaxies and quasars. Four AGNs were detected (Table 1). The image of Q1821+64 may be a blend of the quasar and a nearby white dwarf K1-16 (Grauer & Bond 1984) which is 100" from the quasar and would not be resolved by *WFC*. The AGNs were only detected in the S1a filter of *WFC* (FWHM 74.3–121.8 Å) and all detections are better than 3σ.

An accretion disk spectrum (Czerny & Elvis 1987) was used to convert the count rates to fluxes — only a face on disk and a viscosity parameter $\alpha = 0.1$ were considered. The disk spectrum was convolved with the absorption coefficient for the galactic hydrogen column (Stark *et al.* 1984) and the instrumental response function to obtain the count rates. The count rates and the gas column are given in Table 1. The central mass and the accretion rate were altered to obtain agreement with both the *WFC* count rate and the the optical/UV spectrum. The optical, UV and the EUV data were not obtained simultaneously and only a visual agreement between a model spectrum and the observations was considered sufficient. The fits are shown in Fig. 1 and the

---

*Astrophysics Group, Rutherford Appleton Laboratory, England

ID	α_{cat} α_{wfc}	δ_{cat} δ_{wfc}	m_v	z	M_{abs}	N(h) $\times 10^{20}$ cm^{-2}	t sec	S1a ct/s	σ
3C273	12:29:06 12:29:07	+02:03:08 +02:03:08	12.8	0.158	-27.0	1.91	955	0.014	0.003
NGC5548	14:17:59 14:17:59	+25:08:12 +25:08:11	13.73	0.017	-21.3	2.46	1630	0.015	0.002
Mkn478	14:42:08 14:42:06	+35:26:22 +35:26:04	14.58	0.079	-23.8	1.23	2560	0.043	0.006
Q1821+64	18:21:59 18:21:59	+64:20:36 +64:21:08	14.2	0.297	-27.1	5.62	5060	0.021	0.002

Table 1. Seyfert 1 galaxies and quasars observed with *ROSAT/WFC*

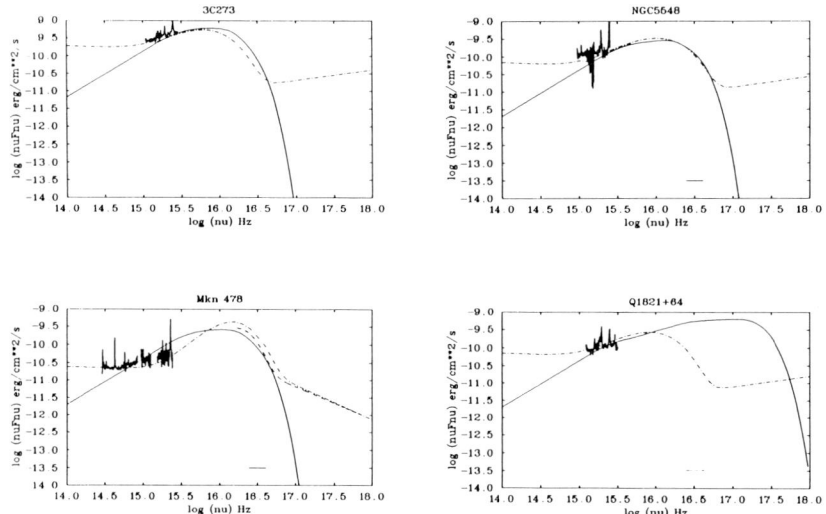

Figure 1. The accretion disk model (*full line*) required to reproduce the observed *WFC* count rate and the *IUE* UV spectrum (*histogram*). The CED required to reproduce the observed $CIV/Ly\alpha$ ratio is shown by the *dot-dash* curve. The *PSPC* spectrum of Mkn478 is shown by the *dashed* curve. The horizontal bar is the *FWHM* of the *WFC/S1a* detector.

central mass and the accretion rate required for the fit are given in Table 2. But note that an underlying power-law has not been added to the disk spectrum, inclusion of such a power-law would alter the disk parameters. Also shown in Fig. 1 are the CEDs required to reproduce simultaneously the $Ly\alpha$ luminosity and the $CIV/Ly\alpha$ line ratio observed in these AGNs (Gondhalekar 1992). The photoionization model calculations were made with the code *ION*. The close agreement between the observations and the CED required to reproduce the $CIV/Ly\alpha$ ratio would suggest that we see the

ID	$\log(Ly\alpha)$ erg s^{-1}	$CIV/Ly\alpha$	M $\times 10^7 M_\odot$	$\dot{M}$ $M_\odot/yr$
3C273	45.43	0.46	3.45	1.75
NGC5548	42.75	0.95	1.20	0.825
Mkn478	44.16	0.17	1.35	0.70
Q1821+64	45.55	0.52	0.397	2.50

Table 2. Central mass and accretion rate

same ionizing radiation as the broad line region clouds.

3. Conclusion

The UV and EUV observations of four AGNs have been fitted with the model emission spectrum of a thin accretion disk. Acceptable fits have been obtained for black hole masses in the range $M \sim 0.4 - 3.5 \times 10^7\ M_\odot$ and an accretion rate in the range $\dot{M} \sim 0.7 - 2.5\ M_\odot/yr$. However, these numbers should be treated with caution as the ratio of the UV/soft X-ray luminosity to the Eddington luminosity for 3C273 and Mkn478 are of the order of eight and three respectively and the assumption of a optically thin disk in the model of Czerny and Elvis (1987) is not valid for these objects. For NGC5548 the model may be valid as in this case the ratio of luminosities is ~ 0.12.

Acknowledgements PMG would like to thank H. Netzer for the photoionization code *ION*.

References

Czerny, B. & Elvis, M. 1987, *Astrophys. J.* **321**,305.
Gondhalekar, P.M. 1992, *Mon.Not.R.astr.Soc.* **255**,663.
Grauer, A.D., & Bond, H.E. 1984, *Astrophys. J.* **277**,211.
Laor, A. 1990, *Mon.Not.R.astr.Soc.* **246**,369.
Malkan, M.A., & Sargent, W.L.W. 1982, *Astrophys. J.* **254**,22.
Stark, A.A., Heiles, C., Bally, J., & Linke, R. 1984, Bell Labs, private communication.
Sun, W-H., & Malkan, M.A. 1989, *Astrophys. J.* **346**,68.

0.1–20 keV Spectra of 3C273 and E1821+643

Richard D. Saxton [*]   *Martin J. L. Turner*   *Anthony Lawson*

The high sensitivity and improved spectral resolution ($\Delta E/E = 0.4$ at $1\,\mathrm{keV}$) of *ROSAT* has enabled high quality spectra to be obtained in the soft X-ray (0.1–$2\,\mathrm{keV}$) band. When added to simultaneous *Ginga* data, spectra extending up to $20\,\mathrm{keV}$ are achieved.

3C273 was observed by *ROSAT* in a pointed observation in June 1990 and again in December 1990 during the survey, when it was also observed by *Ginga*. The December observation yielded a 3σ detection in the S1 filter of the Wide Field Camera. E1821+643, benefits from being at a high galactic latitude and was observed for a total of 40 days during the *ROSAT* survey, with an effective exposure time of 9ks. *Ginga* observed the source during this period. It was also observed serendipitously during the pointed phase in February 1991.

Both sources need a two component model to fit their combined 0.1–$20\,\mathrm{keV}$ spectrum (see Tables 1 & 2). The spectrum of 3C273 has a significant soft excess which exists below $1.2\,\mathrm{keV}$ (source rest frame). This may be adequately parameterised by a $200\,\mathrm{eV}$ Bremsstrahlung or a power law of $\alpha = 3.4$, however, a single temperature Blackbody or Raymond-Smith model is not a good fit to the data, neither are elementary disc models.

E1821+643 has a strong, steep soft excess which rises below $0.65\,\mathrm{keV}$ (source rest frame) in both *ROSAT* observations. The source possesses a strong Iron line of equivalent width $260 \pm 120\,\mathrm{eV}$, which has not varied since first being observed in 1987. It is not possible to constrain the spectral shape of the soft component, perhaps because of its limited coverage by *ROSAT* (channels: 8-50). A Bremsstrahlung parameterisation has a best fit energy of $\sim 100\,\mathrm{eV}$.

The soft excess component in 3C273 was seen to vary independently of the medium energy power law and increased by a factor of three during the six months between the two *ROSAT* observations. Previous *EXOSAT* observations showed that it was variable on time scales of ~ 10 days, however, its mean level has not changed significantly since it was observed by *Einstein* in 1979. No variability was observed in the soft component of the E1821+643 spectrum during the 40 days of the survey. The measured flux is broadly consistent with that seen by *EXOSAT*.

---

[*]Department of Physics and Astronomy, University of Leicester, Leicester, LE1 7RH, England.

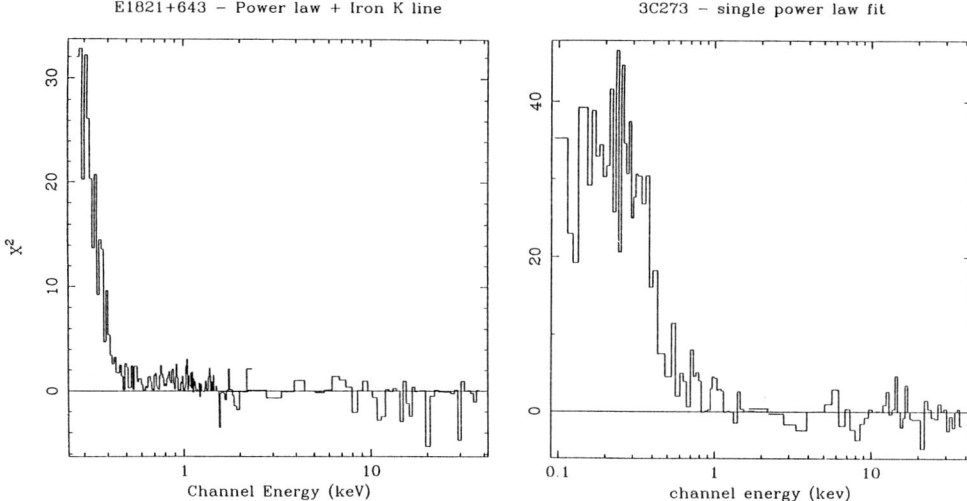

Figure 1. Residuals from extrapolating the fit to the *Ginga* observations into the *ROSAT* band. Note the very soft excess of E1821+643.

Table 1: Spectral fits to 3C273 survey data

Model[a]	PLAW index	Temp(eV) / Index	Norm.	S.E. flux[b]	χ^2/ν
Power Law	1.60			–	909/100
PLAW + BBODY	1.58	73	5.8×10^{-4}	4.6	122/98
PLAW + BREM	1.58	208^{+36}_{-38}	0.18	5.8	111/98
PLAW + PLAW	1.56	$3.38^{+0.32}_{-0.24}$	2.7×10^{-3}	9.1	118/98
PLAW + SSS LINE	1.58	770[c]			872/99
PLAW + RAY-SMITH	1.58	120	0.022	4.1	128/98

Table 2: Spectral fits to E1821+643 survey data

Model[d]	PLAW index	Temp(eV) / Index	Norm.	S.E. flux[b]	χ^2/ν
Power Law	1.98			–	224/185
PLAW + BBODY	1.91	40	1.4×10^{-4}	1.0	95/183
PLAW + BREM	1.90	105	0.1	1.3	95/183
PLAW + PLAW	1.89	4.52	1.4×10^{-4}	4.0	93/183
PLAW + RAY-SMITH	1.91	66	0.014	2.0	95/183

[a] All fits have N_H fixed at the galactic value of $1.4 \times 10^{20} \text{cm}^{-2}$

[b] 0.1–2 keV flux (source rest frame), units of $10^{-11} \text{ergssec}^{-1}$

[c] Fit to the soft excess using the best fit model from the *Einstein* SSS data, a fixed energy 0.77 keV line, with a max. width of 150 eV.

[d] Fits include a variable strength, narrow Iron K line of energy 4.82 keV. In addition they have N_H fixed at $3.4 \times 10^{20} \text{cm}^{-2}$, measured in May 1991 (Pedlar, p.comm.)

Iron Lines from Ionized Discs

G. Matt * A.C. Fabian * R.R. Ross *†

Abstract

The properties of the iron Kα line emitted by a photoionized accretion disc have been calculated for different source geometries.

The properties of the iron Kα fluorescence line emitted by an α-viscosity accretion disc illuminated by an external X-ray source have been calculated for different values of the disc accretion rate $\dot{m}$. The vertical ionization structure of the matter has been computed by using the numerical code described in [1]. Two different source geometries have been studied: a point source located at $20\,r_g$ ($=GM/c^2$) above the disc on its symmetry axis, and an extended source above the innermost part ($r = 6 - 50\,r_g$) of the disc. Assuming $\alpha=0.1$, a Schwarzschild black hole and a hard luminosity equal to the disc luminosity, we find that for large values of $\dot{m}$ ($\gtrsim 0.2$, in units of the critical value) the matter can be significantly ionized, and the iron line equivalent width can reach values as high as 250 eV for the point source, and up to about 400 eV for the extended source (while for neutral matter it is $\sim$150 eV for a face-on disc [2,3]). The line centroid energy, in the emitting rest frame, is significantly higher than 6.4 keV, the value for neutral iron. A further increase of $\dot{m}$ ($\gtrsim 0.5$) leads to a strong decrease of the line intensity, because the iron becomes fully stripped in the inner region of the disc. In the figure we show the equivalent width W_α and the centroid energy $E_{c\alpha}$ as a function of μ_i, the cosine of the disc inclination angle, for the point source (circles) and the extended source (squares), assuming inner and outer radii equal to 6 and $10^4\,r_g$ respectively, a black hole mass 10^6 times the solar mass, and $\dot{m}$=0.4; the kinematic and relativistic corrections have been taken into account. The results obtained with the extended source and a disc seen almost face-on are consistent with the observed values for the Seyfert galaxies MCG-5-23-16 [4] and NGC 6814 [5].

More details on the computational method and on the results can be found in [6].

References

[1] Ross, R.R., Fabian, A.C., 1992, MNRAS, in press

---

*Institute of Astronomy, University of Cambridge, Madingley Road, Cambridge CB3 0HA, England.
†Physics Department, College of the Holy Cross, Worcester, MA 01610, USA.

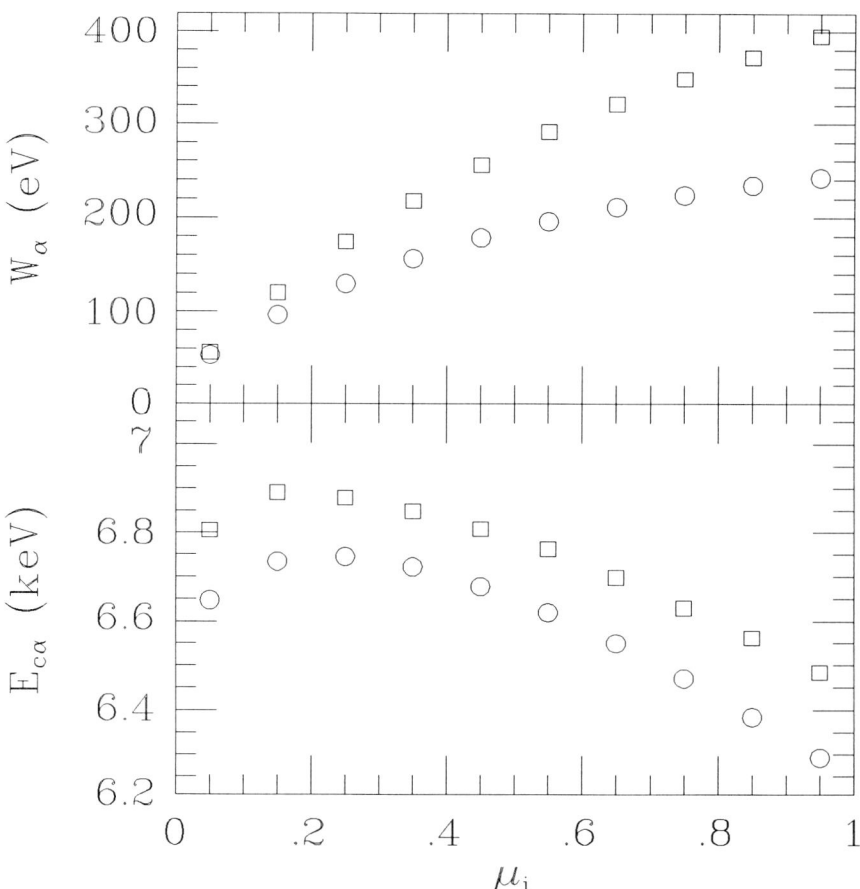

[2] George, I.M., Fabian, A.C., 1991, MNRAS, 249, 352
[3] Matt, G., Perola, G.C., Piro, L., 1991, A&A, 247, 25
[4] Piro, L., Matsuoka, M., Yamauchi, M., 1991, in Proc. of the conference on Physics of Active Galactic Nuclei. Heidelberg, June 1991, in press
[5] Turner, T.J., Done, C., Mushotzky, R., Madejski, G. 1992, ApJ, 391, 102
[6] Matt, G., Fabian, A.C., Ross, R.R., 1992, MNRAS, submitted

Reflection Effects in Realistic Accretion Discs

Carlo Burigana *

The Compton reflection of X-rays by low temperature electrons has received much attention in the recent past. It is a useful framework for the explanation of some features of the spectra of AGNs and a basic ingredient for understanding the X-ray background spectrum, together with the cosmological evolution of AGN. Since the basic works of Illarionov et al. (1979) and of Lightman et al. (1981), principally based on a semi-analytical approach to the problem, many authors have developed careful Monte Carlo methods to include the effects of the Klein-Nishina cross-section and to extend the study up to the hard X- and γ-ray regions. However, the semi-analytical approach has some useful advantages. Firstly it allows us to obtain directly the reflected spectrum in the energy region and at the inclination angle of interest; in addition it may be versatile enough to link with other methods for the computation of the spectrum in the UV and IR regions, where other physical processes are important, but where the X-rays reprocessed by thermal matter may be an important part of the overall luminosity.

The basic assumption of this approach is the separability of the spatial and energy transport problems and of the absorption due to bound-free transitions. In the most general case the Green functions may be written as a sum on n of the terms $A_n(\mu, \mu_0) F_n(y, y_0, T_e) W_n(y, y_0)$; here F_n is the distribution function of photons after n scatterings, y_0 and y being the incident and emergent dimensionless wavelengths $(m_e c^2 / h\nu)$ and T_e the electron temperature; A_n is the probability of escape after n scatterings and depends on the cosines μ_0 and μ of the incident and emergent angles with respect to the outward normal; W_n is a correction factor to include the absorption and to consider the case $y < y_0$, as is necessary if T_e is not negligible, so generalizing the method of Lightman & White (1988).

The reflected spectrum is the convolution of the incident spectrum with the Green function: an incident power law spectrum with an energy spectral index of 0.7 has been considered here . In this preliminary study some results based on a detailed numerical code are described and a first comparison with the studies based on Monte Carlo simulations (George and Fabian 1991; Matt et al. 1991) is presented.

The distribution functions F_n can be calculated for any value of the energy of the incident photon under different approximations. If the matter is neutral, the non-relativistic approximation works quite well up to $\approx 10 keV$, but the first order Klein-Nishina correction can be easily included to calculate more accurately the F_n up to a few $\times 100 keV$. If the matter is ionized, the electrons will have a maxwellian distribution of velocity; in this case the effect of the electron temperature in the

---

*Dipartimento di Astronomia, Vicolo dell'Osservatorio 5, I-35122 Padova, Italy

scattering may be very important and can be included to the first order, together with the Klein-Nishina correction, numerically calculating the scattering kernel (F_1) and the distribution function F_2. In any approximation, for $n > 2$, a gaussian distribution for the functions F_n is quite good and the mean wavelength shift and the wavelength dispersion can be computed using suitable recursion relations. For many realistic values of T_e (typically if $T_e \gtrsim 10^4 K$) the temperature effect on the F_n is always stronger than the relativistic corrections up to $\approx 50 keV$; of course the importance of the former grows as the wavelength increases.

An important step in the calculation of the functions A_n is the relaxation of the two-stream approximation that produces an error of $\approx 5 - 15\%$. For $n \gtrsim 100$ a very accurate analytical approximation is available, but for $n \lesssim 100$ a detailed numerical calculation is necessary.

The probability of escape as a function of n depends on the specific geometry; the usual choices presented in the literature are: a) a semi-isotropic illuminated slab and b) a centrally illuminated disc. In case a) the average reflected spectrum obtained when the temperature effect is negligible has been studied in many papers and has been employed extensively in the comparison with the observations of X-ray sources and of the X-ray background. At $E \lesssim 15 keV$ the usual formula is valid only in the two-stream approximation. In this approximation the probability P_1 of escape after one scattering is found to be overestimated by $\simeq 12\%$; in this energy range the opacity is quite high, so that P_1 is very important and the approximation overestimates the reflected spectrum by $\approx 5 - 12\%$. In case b) the reflected spectra at various angles of sight are found to be in good agreement with the results of Matt et al. (1991); small differences are present at $E \gtrsim 35 keV$, but they are partly due to some little different assumptions and are within the limits of accuracy of the two computational methods. In case b) the dependence of the reflected spectrum on the angle of sight is found to be less strong than in case a).

Finally, the overall effect of the temperature on the continuum reflected spectrum is unimportant up to $T_e \approx 10^7 K$. For greater values of T_e, the reflected spectrum becomes a little lower at $\approx 20 keV$ and consequently a little higher at $E \gtrsim 35 keV$; the line edges become less sharp. Of course the opacity at $E \lesssim 10 keV$ can be strongly modified, particularly in the soft X-ray region, and a fully satisfactory treatment may be difficult; fortunately, for $E \lesssim 10 keV$, even a small fraction of the direct component is always greater than the reflected one.

References

George & Fabian, 1991. MNRAS, **249**,352
Illarionov et al, 1979. ApJ, **228**, 279
Lightman et al., 1981. ApJ, **248**, 738
Lightman & White, 1988. ApJ, **335**, 57
Matt et al., 1991. A&A, **247**, 25

X-ray Polarization Properties in the Two-Phase Model for AGN

Giorgio Matt [*] *Francesco Haardt* [†]

Abstract

The polarization properties of a two-phase model, recently proposed to explain the X-ray emission of Active Galactic Nuclei, have been calculated for different values of the model parameters. An important signature of the model is the orthogonality between the UV/soft X-ray and hard X-ray polarization.

Recently, a two-phase model in which hot, thermal electrons in an optically thin layer comptonize the soft photons coming from an underlying cold, optically thick accretion disc, has been proposed to explain the X-ray emission of Active Galactic Nuclei [1,2].

Assuming a plane-parallel geometry, and isotropic and unpolarized disc thermal radiation, we have calculated the polarization properties as a function of the energy and of the inclination angle, for different values of τ_0, the optical depth of the hot phase (which, in the adopted model, is related to the electron temperature). This was done by solving the well-known equation of radiative transfer [3] by separating the different scattering orders [4]. The polarization of the X-rays reflected from the disc [5] has also been taken into account. In the figure we show the degree of polarization as a function of the energy for different values of the inclination angle (at the two extremes of the energy range $|P|$ increases with it). The assumed energy shape of the thermal radiation is a black-body with $T=50$ eV. Note that the hard X-rays have a negative polarization (i.e. the polarization vector lies in the meridian plane), while the polarization of the UV/soft X-rays is positive (i.e. the polarization vector is perpendicular to the meridian plane). The Stellar X-ray Polarimeter aboard the Spectrum-X-Gamma mission [6] should be able to test the model.

More details on the computational method and results can be found in [7].

---

[*]Institute of Astronomy, University of Cambridge, Madingley Road, Cambridge CB3 0HA, England, and Istituto Astronomico dell'Università di Roma "La Sapienza", Via Lancisi 29, I-00161, Roma, Italy.

[†]International School for Advanced Studies, S.I.S.S.A., via Beirut 2-4, I-34014 Trieste, Italy.

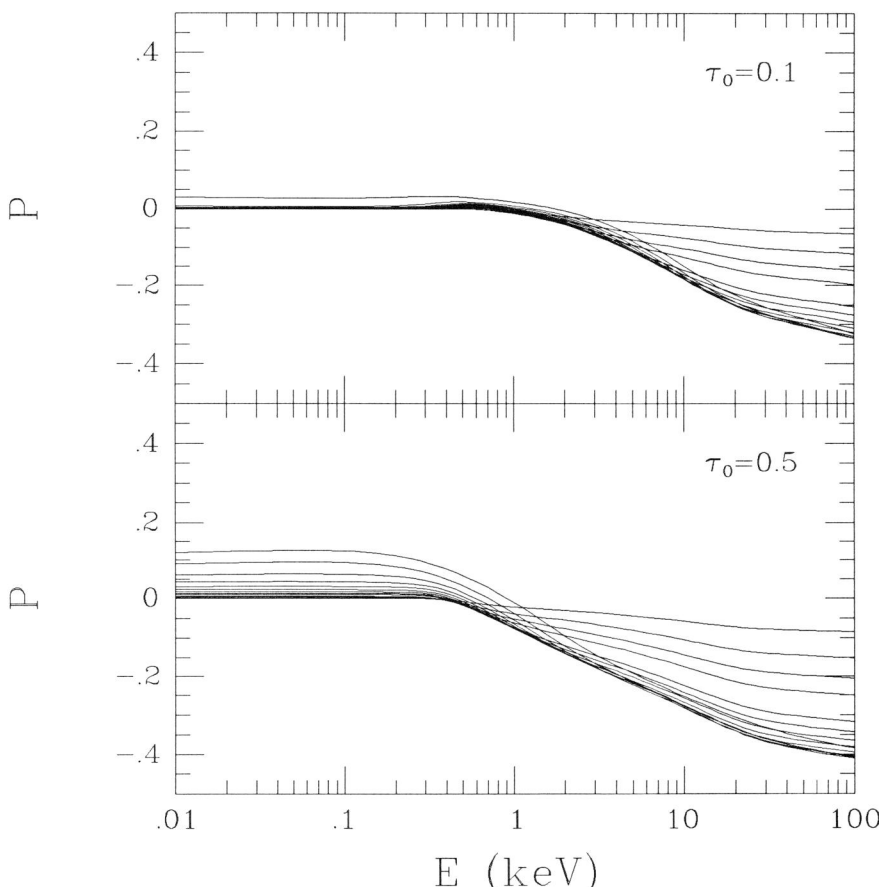

References

[1] Haardt F., Maraschi L., 1991, ApJ, 380, L51
[2] Haardt F., Maraschi L., 1992, ApJ, submitted
[3] Chandrasekhar S., 1960, Radiative Transfer. Dover, New York
[4] Sunyaev R.A., Titarchuk L.G., 1985, A&A, 143, 374
[5] Matt G., 1992, MNRAS, in press
[6] Kaaret P., *et al.*, 1991, in Production and Analysis of Polarized X-Rays. SPIE vol.1548, p.106
[7] Haardt F., Matt G., 1992, MNRAS, in press

X-Ray Reprocessing and UV Continuum in NGC 4151

G.C. Perola [*]   L. Piro [†]

Abstract

The correlation observed in NGC 4151 between the O, UV and X-ray fluxes is explained in terms of reprocessing of hard X-rays by a thick disk that reradiates the incoming energy into O and UV photons. The flatness of the UV spectrum and the upper limits on X-ray reprocessed components (high energy bump, variable part of iron line) demand tight limits on the mass of the central object, the luminosity (absolute and relative to the Eddington one) and the extension of the spectrum in the γ-ray region.

1. Introduction

Reprocessing of X-rays by a thick medium has been called for to explain two X-ray features observed in several Seyfert galaxies: the iron line and the high energy bump [1]. The fact that reprocessing of hard X-rays may play an important role also in the optical and UV, an idea firstly advanced in [2,3], has been recently proposed to account for the short time-scale correlation of optical and UV light curves, too short to be explained by processes directly connected to accretion by a disk [4,5,6,7].

Similar considerations apply to the case of NGC 4151, where the optical, UV and X-ray fluxes are correlated down to a time scale of 1 l.d. [4,8] - although the correlation between UV and X-ray breaks at higher UV luminosities (we will comment on this behaviour in the following). In this object, however, the absence of an high energy bump [9] as well as of a broad and variable iron line [8,10] apparently argues against the presence of a thick reprocessor near the central source. We will show as this apparent contradiction with the reprocessing scenario can be recovered if the reprocessor is a disk in a nearly edge-on configuration.

2. Reprocessing by disk

We modelled the system by a flat, optically thick disk illuminated by a hard X-ray source of luminosity L_0, located above the disk at a height H equal to the inner

---
[*]Istituto Astronomico, Universita' di Roma
[†]Istituto Astrofisica Spaziale, C.N.R., Frascati

radius of the disk. X-rays impinging on the disk are reradiated with a black body spectrum with a temperature given by $\sigma T^4 = F_{abs}$ where F_{abs} is the flux absorbed by the disk. The spectrum is obtained by integrating over the disk, and depends on L_0, the cosine of inclination angle μ and H (Perola and Piro, in preparation). We obtain the following results and implications for the system parameters.

- Optical and UV variations are well explained by a variable L_0.

- The flatness of the UV spectrum in the band 1455–1715Å over the full range of UV luminosities implies an upper limit on the source size of $10^{13}/\sqrt{\mu}$ cm and a lower limit on the high state luminosity of $6\ 10^{43}/\mu$ $erg\ s^{-1}$.

- The requirement that the O–UV emission is dominated by reprocessing instead of accretion processes, fulfilled when $R/R_S \geq 40$ in the case of accretion onto a Schwarzschild black hole, implies an upper limit on the mass of $\simeq 10^6/\sqrt{\mu}$ solar masses and thus $L_0(high)/L_{EDD} \geq 0.5/\sqrt{\mu}$.

- The upper limit on the X-ray reprocessed components are satisfied if the disk inclination is $\mu \leq 0.3$. Therefore in the high state $L_0 \geq L_{EDD}$.

- Since $L_0 >> L(2-10\ keV)$, the bulk of energy comes from the high energy part of the spectrum, that should extend up to 0.5–1 MeV.

- A change in the energy cut-off or spectral shape at high energies [11] can thus explain a different f(UV)-f(2-10 keV) correlation such as that observed in 1979 when the UV flux was about two times greater than the 1983–85 data but the X-ray flux was in an intermediate level [8]. Furthermore, since in the high state of the '83–85 campaign we derive that $L_0(high) \simeq L_{EDD}$, we speculate that this change may be triggered by a transition from a sub-Eddington to a super-Eddington regime.

References

[1] Matt G. Perola, G.C. and Piro L. 1991 Astron. Astrophys. 247, 25
[2] Lightman A.P., White T.R. 1988, Ap.J., 335, 57
[3] Guilbert P.W., Rees M.J. 1988, M.N.R.A.S., 233, 475
[4] Ulrich M. et al. 1991, Ap. J, 382, 483
[5] Molendi S., Maraschi, L. and Stella L. 1992, Astron. Astrophys. 255, 27
[6] Collin-Souffrin S. 1991, Astron. Astrophys. 249, 344
[7] Clavel J. et al. 1992, Ap.J., 393, 113
[8] Perola G.C. et al. 1986, 306,508
[9] Maisack M. and Yaqoob T. 1991, Astron. Astrophys, 249, 25
[10] Marshall et al. 1991, Nagoya conf. on *Frontiers of X-ray Astronomy*
[11] Jourdain E. et al. 1992, ICRC OG3.3.12

Dense Clouds near the Centres of Active Galactic Nuclei

Ran Sivron [*]   *Sachiko Tsuruta* [*]

Recently some interesting spectral features, such as an Fe line-edge system and a hard X-ray bump, have been observed in many Seyfert nuclei (e.g. Nandra *et al.*, 1991, hereafter NPS91). These observed characteristics are naturally explained by the 'two-component' model which consists of a non-thermal hot plasma and a 'cold' plasma which reprocesses and reflects the non-thermal radiation. The presence of such cold matter in the vicinity of the central region is a natural consequence of the accretion model of active galactic nuclei (AGN) (e.g. Guilbert and Rees 1988, hereafter GR88). It is also supported by X-ray observations of many Seyfert nuclei by the EXOSAT and Ginga missions (e.g. NPS91, and references therein), for instance, the simultaneous rapid variations of the X-ray continuum and Fe line fluxes (Kunieda *et al.*, 1990). The cold component is envisioned to be in a form of either a thin disc (or slab) or an assembly of clouds distributed over a three-dimensional configuration (GR88; Celotti, Fabian and Rees 1992, hereafter CFR92). For convenience, we shall call the former kind 'the disc model' and the latter 'the cloud model'.

Here a model is presented which assumes the existence of 'cold' dense clouds near the central engine of AGN. In order for the cold clouds to exist near the central engine they must be sufficiently dense, and some physical pressure is required to confine them. This pressure may be supplied by equipartition magnetic fields (Rees 1987, CFR92). The spectral and temporal behaviour of the outcoming radiation depends on the geometry of the radiating and reflecting plasmas. So far, numerical calculations for the reflection and reprocessing of the non-thermal radiation by the cold matter have been carried out only for the disc (or slab) geometry (e.g. George and Fabian 1990, hereafter GF90). Here we extend such calculations to an assembly of cold, dense clouds. The luminosity, fastest variability timescale and absorption column density of Seyfert nuclei are used to derive self-consistent equations for such a cloud system, in which the temperature, density and covering factor depend on the distance from the central source. For simplicity we assume spherical symmetry and the presence of a central power-law source (Sivron and Tsuruta 1992, 1993, hereafter ST92 and ST93). The effects of such clouds on the observed spectra and light curves are calculated, using the Monte Carlo integration methods (ST93).

Generally the spectra from our cloud model are qualitatively similar to those in GF90, but there are some important differences, as outlined below. In our model the equiv-

---

[*]Department of Physics, Montana State University, Bozeman, Montana 59717, USA. This author was partially supported by NASA grants NAGW-2208 and NAG-783.

alent width of the Fe line can be large while the ionization edge is small, and the hard X-ray bump is present only when the covering factor f is large. Also our model is more flexible because f can take any value between 0 and 1, and hence can produce a negligible to strong line-edge feature. On the other hand in the conventional thin disc model f is fixed at 0.5. In our model the extreme UV luminosity is significantly higher than in a disc model. These effects are due to the obscuration of the reprocessed components by the clouds between the reflecting clouds and the observer. The variability in the cloud model and the disc model is significantly different. In the standard disc model the variations at different wavelengths are closely correlated with each other. On the other hand in the cloud model the optical and IR variations are found to be many orders of magnitude longer and less intense than in the x-ray window, a result consistent with the multi-frequency observations of NGC4051 (Done et al., 1990). The observation of only weak polarization and a lack of Lyman edges also favour the cloud model (ST93). In conclusion our cloud model has been shown to be consistent with the complicated observed spectral and variability behaviour of the most extensively studied Seyfert nuclei (ST92).

References

Celotti, A., Fabian, A.C., and Rees, M.J., *Mon. Not. R. Astr. Soc.*, 255 (1992), 419.
Done, C., Ward, M.J., Fabian, A.C., Kunieda, H., Tsuruta, S., Lawrence, A., Smith, M.G., and Wamseker, W., *Mon. Not. R. Astr. Soc.*, 243 (1990), 713.
George, I.M., and Fabian, A.C. *Mon. Not. R. Astr. Soc.*, 249 (1988), 352.
Guilbert, P.W., and Rees, M.J., *Mon. Not. R. Astr. Soc.*, 233 (1990), 475.
Nandra, K., Pounds, K.A., Stewart, G.C., George, I.M., Hayashida, K., Makino, F., and Ohashi, T., *Mon. Not. R. Astr. Soc.*, 248 (1991), 760.
Kunieda, H., Turner, T.J., Awaki, H., Koyama, K., Mushotzky, R.F., and Tsusaka, Y., *Nature*, 786 (1990), 345.
Rees, M.J., *Mon. Not. R. Astr. Soc.*, 228 (1987), 47p.
Sivron R., and Tsuruta S, *Astrophys. J.*, 402 (1993).
Sivron R., and Tsuruta S, *in preparation* (1992).

Accretion Discs in the AGN Context: Hints Toward Non -Standard Discs?

Suzy Collin-Souffrin [*]

Abstract

The structure and emission properties of accretion discs are briefly reviewed, with special attention given to geometrically thin "standard" accretion discs. Different approximations used to compute their emission spectrum are summarized, and it is shown that the bulk of the luminosity is emitted in the EUV range (the "UV bump"). However, according to recent optical, UV and X–ray monitoring campaigns of Seyfert galaxies, it appears that a large fraction of the observed optical–UV and even X–ray continuum is reprocessed by the disc from a primary source, probably of hard X–ray continuum, sitting near the black hole and heating the farther regions of the disc. A first consequence is that the disc must also contribute some fraction of the line emission. A second consequence is that standard accretion discs drawing their energy from viscosity and radiating it locally are ruled out in AGN, and another type of model must be built.

1. Generalities on accretion discs

For the reader who wishes to know more than just the basic properties that are recalled below, several reviews on accretion discs can be recommended: Pringle (1981), Frank, King & Raine (1992; revised version), Begelman (1985), Shields (1990), Treves, Maraschi & Abramovicz (1988).

If the matter accreted onto the black hole possesses angular momentum, it will settle in an accretion disc, in which matter is transported inward and angular momentum outward. This may be achieved through viscous friction, or through other stresses such as produced by a magnetic field. In the viscous case, gravitational energy is converted locally into heat, which is radiated away at the surface of the disc. The efficiency depends on the nature of the central object: for rotating black holes it can reach $\eta = 30\%$, while for non rotating ones $\eta \sim 0.5\%$.

The accretion rate $\dot{M}$ is scaled by the *"critical accretion rate"*, which we define here as $\dot{M}_{\rm crit} = L_{\rm Edd}/\eta c^2$. Depending on the value of the ratio $r = \dot{M}/\dot{M}_{\rm crit}$, the disc will

---

[*]DAEC, Observatoire de Paris, Section de Meudon, 92195, Meudon, France, and Institut d'Astrophysique, 98bis Bld. Arago, 75014, Paris, France

adopt different thickness to radius ratios, H/R. In the range $0.01 < r < 0.3$, the inner regions of the disc are *"geometrically thin"*, i.e. $H/R \ll 1$. According to their central masses (estimated by dynamical methods) and to their observed luminosities, Seyfert nuclei lie within this range of r, so they probably contain thin discs, while luminous AGN (quasars) could have "thick discs" or "slim discs". In this review, we shall consider only thin discs.

2. Structure of thin accretion discs and standard "alfa" discs

We consider here *stationary* discs, in which the mass transfer rate is constant within time scales equal to a fraction of the viscous time (c.f., below). Several other assumptions are implicit in these computations: no self-gravity (this is not justified for large accretion rates and/or at large radii), no magnetic field, vertical heat transport only with energy released by viscous dissipation balancing the radiative losses, negligible radial pressure gradient, and therefore keplerian orbital velocity. Hydrostatic equilibrium leads to: $c_s/V_K \sim H/R \ll 1$ where c_s is the sound velocity and V_K the keplerian velocity. It shows that the keplerian velocity (i.e. the disc velocity) is strongly supersonic.

The problem of viscous dissipation is generally circumvented using the Shakura and Sunayev prescription (1973), which assumes that the viscosity is $\nu = \alpha c_s H$, with α being an *ad hoc* constant, of the order of or less than unity. Then the radial velocity is largely subsonic, as shown by $V_R = \alpha c_s H/R \ll c_s$.

The temporal variations of a thin disk can be characterised by three different time scales: 1. the dynamical time t_{dyn}, or the time for a dynamical perturbation to propagate vertically through the disc: $t_{\text{dyn}} = H/c_s = R/V_K$; 2. the viscous time t_{visc}, or the time for a perturbation of the mass rate to propagate radially through the disc: $t_{\text{visc}} = R/V_R = t_{\text{dyn}}/\alpha \, (R/H)^2$; 3. the thermal time t_{therm}, or the time for the disc to evacuate its thermal energy: $t_{\text{therm}} = t_{\text{dyn}}/\alpha$. Note that these times obey $t_{\text{dyn}} \leq t_{\text{therm}} \ll t_{\text{visc}}$. Any variation affecting globally the disc structure requires a time equal to a fraction of t_{visc} to establish itself, while a perturbation at a given radius is settled in a dynamical time. Take for instance an AGN with the following parameters: $L_{\text{bol}} = 10^{44} \, \text{erg s}^{-1}$, $L_{\text{bol}}/L_{\text{Edd}} = 0.1$ (consequently $R_G = 2 \, 10^{12} \, \text{cm}$), which should adequately represent an object like NGC 4151. After solving the disc structure, one finds, with $\eta = 0.1$ and $\alpha = 1$, that for $R = 100 R_G$, $t_{\text{dyn}} = 0.5$ days and $t_{\text{visc}} = 1$ year, and that for $R = 1000 R_G$, $t_{\text{dyn}} = 15$ days and $t_{\text{visc}} = 100$ years. We shall use these values in § 3.

The "standard disc"

In spite of all simplifications above, the problem is still difficult to handle, and requires in particular the solution of the vertical transfer problem, and therefore knowledge of the vertical heat deposition. This is why it has been simplified a step further, with the additional assumption that the disc can be described by vertically averaged quantities: it is then called the "standard α-disc".

When this is done for the transport of energy, assuming that it is purely radiative and that the disc is optically thick (as it is in the regions we consider later on), equality between the radiative flux emitted at each face and the release of gravitational energy gives:

$$F_{\rm rad} = f \frac{3GM\dot{M}}{8\pi R^3} \qquad (1)$$

with $F_{\rm rad} \sim \sigma T^4/\tau$, where τ is a frequency averaged optical thickness, T is a vertically averaged temperature, close to the equatorial temperature, and f takes into account the boundary conditions, and therefore depends on the angular momentum of the black hole (cf. Novikov & Thorne 1973, and Page & Thorne 1974).

It is then possible to solve the set of equations by dividing the disc into concentric rings, and assuming that each ring behaves like an independent plane-parallel radiating slab. Equation 1 shows that it is the opacity law which determines the temperature, and therefore the thickness of the disc. It is necessary to check *a posteriori* that the assumed opacity is dominant in the physical conditions prevailing at each point. For instance free-bound opacity, which is not taken into account in the Shakura-Sunayev solution, exceeds free-free opacity over a large fraction of the disc.

Several regions can be distinguished in a disc (actually the radial structure depends strongly on the assumptions made to compute the emission spectrum and the opacity, so we shall come back to this point in the next section; however the following results remain valid whatever the assumptions). Radiation pressure and Thomson opacity dominate in the inner regions. The scale height of the disc is constant in this region and is $\sim (3R_{\rm in}/4\eta) \, f \, (\dot{M}/\dot{M}_{\rm crit})$. In the outer regions, gas pressure and atomic or molecular opacity dominate, and the disc is "flaring", i.e. its H/R ratio increases with R (c.f., Collin-Souffrin & Dumont 1990). Owing to mass conservation in the accretion flow, the surface density decreases with radius and the disc becomes optically thin at large radii. Furthermore, it is no longer geometrically thin, because H/R becomes larger than unity (this region can be identified with the "molecular–dust torus" of the Unified Scheme). The temperature can then increase again, because of external irradiation. In between, there can be an intermediate region where gas pressure and Thomson opacity dominate.

3. The emission spectrum of thin discs

The emitted spectrum is computed using various approximations, ordered here by increasing complexity:

1 - a superposition of blackbodies with no relativistic corrections

The disc is assumed to emit at each radius a blackbody spectrum of effective temperature given by equating σT_{eff}^4 to the radiative flux (eq. 1). T_{eff} decreases as $R^{-3/4}$, and the blackbody radiation peaks at the wavelength:

$$\lambda \sim 200 \left(\frac{\eta}{0.1}\right)^{1/4} \left(\frac{L}{L_{\text{Edd}}}\right)^{-1/2} \left(\frac{R}{10 R_G}\right)^{3/4} L_{44}^{1/4} \text{ Å}, \qquad (2)$$

where L_{44} is expressed in 10^{44} erg s^{-1}. For the typical AGN considered previously, far UV radiation is emitted near $100 R_G$ and optical radiation near $1000 R_G$. The flux at Earth integrated over the disc is:

$$F_\nu = \frac{2\pi \cos i}{D^2} \int_{R_{\text{in}}}^{R_{\text{out}}} B_\nu(T_{\text{eff}}) \, R \, dR \qquad (3)$$

where D is the luminosity-distance, i is the inclination of the disc axis on the line of sight, R_{in} and R_{out} are respectively the inner and outer radii. It leads to the characteristic shape of the disc spectrum, proportional to ν^2 at low energy, to $\nu^{1/3}$ at intermediate energy, and with an exponential cut-off at high energy. The low energy break and the high energy cut-off correspond to the maximum of flux at $T_{\text{eff}}(R_{\text{out}})$ and $T_{\text{eff}}(R_{\text{in}})$ respectively. Note that the inclination has no influence on the spectrum, but only on the absolute value of the flux through the $\cos i$ term. With this approximation, the spectrum depends only on two parameters, the central mass and the accretion rate.

2 - a superposition of blackbodies with relativistic corrections

The introduction of relativistic effects is important, especially for rotating black holes seen at large inclinations. Cunningham (1975) has computed the observed spectrum for a non rotating and for a maximally rotating ($a/M = 0.998$) black hole and has provided useful formulae for scaling it to a given mass and accretion rate. One finds that νf_ν peaks at the wavelength (for a face-on disc):

$$\lambda \sim [600 - 200] \; M_8^{1/4} \left(\frac{\dot{M}}{\dot{M}_{\text{crit}}}\right)^{-1/4} \text{ Å} \qquad (4)$$

respectively for a non rotating and for a maximally rotating black hole. The spectrum for a rotating black hole is harder, and moreover the hardness increases considerably at large inclinations, owing to light bending which is more important at small radii. For a maximally rotating black hole, between a disc seen edge-on and a disc seen pole-on, the frequencies of the peak of the spectrum differ by one order of magnitude, while there is only a factor two between the peak fluxes (one order of magnitude for a non-rotating black hole). Note that in this approximation, the observed spectrum already depends on four parameters, the black hole angular momentum, the disc inclination, the central mass and the accretion rate. In these two approximations *the spectrum does not depend on the value of the viscosity*, but only on the assumption that the gravitational energy is deposited and radiated locally.

3 - a vertically averaged solution, with suitable opacity laws

Departures from a local blackbody spectrum are expected when atomic opacity or Thomson (or Compton) scattering dominates the free-free opacity. Czerny & Elvis (1987), Wandel & Petrosian (1988), have taken into account electron scattering and shown that it is important in the inner regions of the disc for $\dot{M}/\dot{M}_{\mathrm{crit}} \geq 0.1$, and that it increases the X-ray flux and decreases the UV flux. This approximation requires knowledge of the average density, which is determined by the viscosity parameter. In the case of an α disc, one must therefore specify the value of α, in particular if it takes into account the radiation pressure stresses. It is generally assumed to be of the order of unity. A smaller value of α would lead to a larger density, with the opacity less easily dominated by Thomson scattering.

4 - complete solution, with vertical structure

This treatment takes into account the radiation transfer and solves the vertical structure in thermal and hydrostatic equilibrium. Laor & Netzer (1989) have performed such computations for a disc atmosphere, assuming LTE, and Ross et al. (1992) have made non-LTE computations valid for the inner radiation dominated regions. Note that they require *additional knowledge of the vertical heat deposition.* Laor & Netzer get a Lyman edge in emission or in absorption, depending on the assumptions concerning this (completely unknown) point. This illustrates that the computation of the emission spectrum, for given central mass and accretion rate, involves strong uncertainties due to the (often arbitrary) assumptions made.

Several people have preferred to stay within the first or the second approximation, when fitting an observed continuum to the emission of the disc. It yields the simplest and least model dependent solutions, and this may not be a bad choice. In any case, a characteristic of the spectrum, whatever the approximation (and this is also valid for thick discs) is that *the bulk of the emission takes place in the UV and/or soft X-ray range, and no model based on the local heat dissipation assumption is able to account for the emission of a hard X-ray continuum,* even when it does not dominate the bolometric luminosity.

There are however interesting variants to the standard disc model, where an X-ray continuum can also be emitted by the disc, and can constitute an important fraction of the bolometric luminosity. One is the "two phase" accretion disc, in which the disc contains both a hot optically thin and a cold optically thick phase. In a model proposed by Haardt & Maraschi (1991), a substantial fraction of the gravitational power is dissipated via buoyancy and reconnection of magnetic fields in a corona above the disc; the hot and cold media interacting through inverse and direct Compton processes. In an alternative model (Wandel & Liang, 1991), the hot medium is located near the centre at $R < 30R_G$.

4. A new ingredient one should take into account: heating of the disc by external radiation

Several observations have recently led us to modify the "standard disc" picture. They stress the importance of a different phenomenon, namely radiation reprocessing.

X–ray observations: the "reflection model"

Seyfert galaxies reveal the existence of several features. First an absorption edge near 8 keV is attributed to photoelectric absorption in a "hot" (10^7K) medium with a column density of order 10^{22-24}cm^{-2} surrounding the X–ray source. A 6–7 keV iron line and a high energy "hump" are associated with partial reprocessing of a hard X-ray continuum (Pounds et al. 1990, and subsequent works). The emission line is due to fluorescence, and the hump to Compton reflection, both produced in a "cold" ($< 10^6$K), optically thick medium, presenting a large covering factor to the X-ray source. The evidence is growing that the reflecting-reprocessing medium is a disc-like configuration located very close to the nucleus, which can be identified with the inner regions of the accretion disc. About 10% of the primary continuum is thus "Compton reflected" in the observer's direction, and a question which follows immediately is: what is the fate of the corresponding fraction — $\sim$ 20%, according to Compton diffusion computations — of the incident flux hitting the disc which is absorbed inside the disc? A part of the answer is given by the optical and UV continuum flux variations.

Optical and UV observations

Several monitoring campaigns recently carried out on Seyfert galaxies in UV and optical ranges have shown a striking similarity between the light curves of lines and of continuum, and between the light curves of the continuum at different wavelengths, the most astonishing result being the absence of a measurable time delay between the UV and optical continuum light curves. This phenomenon is in contradiction with the standard disc model. We have seen that optical and UV continua are emitted by regions of the disc located typically at $10^2 R_G$ and $10^3 R_G$, and the time delay between their variations should be much larger than observed, if energy is transferred by viscous processes ($\sim$ 1 year). The observed limits on the time delay imply actually that local temperatures in the disc are causally linked with the transmission of information at the speed of light. This would be the case if the disc were radiatively heated by a central source, as proposed by Collin-Souffrin (1991) and by Molendi et al. (1992). This external supply of energy is reprocessed as thermal radiation at frequencies corresponding to the effective temperature of the disc, i.e. as UV and optical continuum emission. A confirmation of this is given by the similarity between the Hβ light curve and that of its underlying continuum (with a delay equal to the light travel time from the source to the BLR). Since Hβ is produced in a region heated mainly by hard X–ray photons (Collin-Souffrin et al. 1986), it means that *all continuum bands from optical to hard X–ray are tightly correlated within a very small time lag.*

In summary both X-ray and optical-UV continuum observations can be explained by a model where a central source of X–ray continuum involving a large fraction of the bolometric luminosity, is reprocessed in the disc and subsequently either radiated away or absorbed.

Line emission produced by the disc

A first consequence is that line emission should be produced by the disc. Indeed the fate of the radiation absorbed by the disc depends on the radius. At small radii, ($R \leq 10^2 R_\mathrm{G}$), the density and temperature in the disc are large, so the visible/X–ray opacity ratio is ≥ 1, and X–rays are absorbed under the photosphere. As a consequence, the incident radiation is reprocessed into UV continuum. At large radii ($R \geq 10^3 R_\mathrm{G}$), the visible/X–ray opacity is ≤ 1, so a "photoionized chromosphere" is created above the disc, which re-emits the incident X–ray continuum mainly as a low ionization line spectrum (Dumont & Collin-Souffrin 1990). At intermediate radii, there is a mixing of line and continuum emission. According to our model, the line emission peaks at $10^3 R_\mathrm{G}$, owing to saturation and thermalisation, and we have seen from the study of the continuum emission that the disk extends at least that far. One expects therefore the typical line widths to be close to the keplerian velocity at this radius, i.e. 10^4 km s^{-1}, and this is indeed the observed value.

What is the intensity of this line spectrum? It depends on the portion of the central luminosity that the outer regions of the disc are able to absorb. If it amounts to 10% or more, almost all the line spectrum is emitted by the disc. There are several reasons why the disc could absorb such a large amount of primary radiation. The flaring of the disc, the "returning" radiation due to relativistic bending of light rays, and the finite size of the central source, already provide a small illumination of the disc surface. But above all, the hot medium, whose existence is deduced from X–ray measurements and whose Thomson opacity is of the order of a few percent or more, can backscatter the required amount of radiation towards the disc. The disc may also be warped, as suggested by Sanders et al. (1989), and finally the central source may lie at a finite height above (and below!) the disc. These last configurations allow the central source to be "seen" directly by the disc through a solid angle of almost 2π, so that a large fraction of the central radiation hits the disc. These ideas have been applied to NGC 5548 (Rokaki et al. 1992, and Rokaki's contribution in this volume). It is found that about 70% of the broad Hβ line can be accounted for by disc emission in this object. The central mass, the radius of the central source and that of the diffusing medium, and the disc inclination, are deduced from the fitting of the line intensity and line profile, and from their correlation with the UV continuum flux. Note that a fundamental assumption of this model — the X-ray light curve is in phase with the UV light curve and has the same amplitude of variations — has been observationally confirmed by a campaign of simultaneous hard X-ray and UV observations (Clavel et al. 1992).

Consequences for the viscous release

A second consequence of the reprocessing is that accretion rates deduced from fitting the optical–UV continuum with a standard thin accretion disc drawing its energy only from viscous dissipation (Malkan 1982, and subsequent papers) is strongly overestimated. For instance in the case of NGC 5548, at least 80% of the optical–UV continuum emission is provided by external heating, and the integrated "viscosity luminosity" is at least one order of magnitude smaller than the luminosity of the central source. A first interpretation is that the accretion rate through the disc is very low, and that the rest of the accretion occurs spherically. In this case the accretion disc plays mainly the role of a "fluorescent screen" able to absorb the radiation of the spherical accretion flow! A second interpretation is that the accretion rate through the disc is one order of magnitude larger than deduced from the viscosity luminosity, which implies that gravitational energy is not converted locally into heat, but into another form of energy which can be released in the central region. Several mechanisms may be invoked such as magnetic storage and reconnection or hydrodynamic winds. In this case, however, the accretion disc physics would be far from standard and would need severe revision.

5. Conclusion

The recent optical and UV monitoring campaigns of NGC 4151 and NGC 5548 have shown that a large fraction of the optical–UV and X-ray continuum is not produced in the disc through gravitational viscous release, but is reprocessed from a primary source. According to Ginga observations, this source consists of a hard X-ray continuum involving a large fraction of the bolometric luminosity, sitting near the black hole and heating the surface of the disc. It means that, either accretion through the disc represents only a marginal portion of the total accretion rate, or the disc accretion rate supply is mainly converted into X-rays, and not into EUV radiation as it would be for the case of a standard accretion disc. It implies, moreover, that the disc must also contribute to a fraction of the line emission. This fraction can be dominant for some disc configurations (warped or flaring discs), or if the "warm" absorber, whose existence is inferred from X-ray absorption, backscatters efficiently the central continuum onto the outer regions of the disc.

References

Abramovicz, M., Calvani, M., Nobili, L., 1980, ApJ 242, 772
Begelman, M.C., 1985, in the proceedings of the Santa Cruz workshop on "Active Galactic Nuclei and QSO", Ed Miller
Clavel J. et al., 1992, ApJ, in press
Collin-Souffrin, S., Joly M., Pequignot D., Dumont S., 1986, A & A 166, 27
Collin-Souffrin S., Dumont A.M., 1990, A & A 229, 292
Collin-Souffrin S., 1991, A&A 249, 344
Cunningham C., 1975, ApJ 202, 788
Czerny, B., Elvis, M., 1987, ApJ 321, 243
Dumont A.M., Collin-Souffrin S., 1990, A & A 229, 313
Frank, J., King, A.R., Raine, D.J., 1992, "Accretion power in astrophysics", Cambridge University Press
Haardt F., Maraschi L., 1991, ApJ 380, L51
Laor A., Netzer H., 1989, MNRAS 238, 897
Malkan M.A., 1983, ApJ 268, 582
Molendi S., Maraschi L., Stella L., 1992, MNRAS 255, 27
Novikov I.D., Thorne K.S., 1973, "Black Holes", ed. C. DeWitt & D. DeWitt, Eds Gordon & Breach, p.342
Page D.N., Thorne K.S., 1974, ApJ 191, 499
Pounds K.A., Nandra K., Stewart G.C., Leighly K., 1989, MNRAS, 240, 769
Pringle, J.E., 1981, Ann. Rev. Astr. Ap. 19, 137
Rees, M.J., Begelman, M.C., Blandford, R.D., Phinney, E.S., 1982, Nature 295, 17
Rokaki E., Collin-Souffrin S., Magnan C., 1992, A & A, in press in A & A
Ross R.R., Fabian A.C., Mineshige S., 1992, preprint
Sanders D.B., Phinney E.S., Neugebauer G., Soifer B.T., Matthews K., 1989, ApJ 347, 29
Shakura, N.I., Sunayev R.A., 1973, A & A 24, 337
Shields G.A., 1990, Annals of the NY Academy of Sciences 571, 110
Treves A., Maraschi L., Abramowicz M., 1988, PASP 100, 427
Wandel, A., Petrosian V., 1988, ApJ 319, L11
Wandel, A., Liang, 1991, preprint

Accretion Disk Instabilities

John F. Hawley & Steven A. Balbus [*]

Abstract

Accretion disks are the preferred central engine for AGN, but theoretical progress has long been hampered by the unknown nature of the angular momentum transport mechanism. An obstacle preventing the general acceptance of turbulent disk theories has been the intransigent stability of thin Keplerian disks to hydrodynamic perturbations. This difficulty is overcome by the result discussed here, that a weak magnetic field renders a disk *locally unstable* whenever the angular velocity decreases outward. The salient and remarkable feature that distinguishes this instability is that the maximal growth rate (which is of order the angular velocity) is independent of the strength of the field, even as the latter vanishes. The instability can be derived straightforwardly from the equations of orbital mechanics and a spring-like magnetic interaction. We list several key questions regarding the instability's implications for realistic disks, and review the results from two-dimensional numerical simulations.

1. Introduction

Many of the scenarios created to explain the spectra from Active Galactic Nuclei (AGN) involve, either implicitly or explicitly, the release of gravitational energy through accretion. The classical paper of Lynden-Bell (1969) was the first to suggest that the heart of an active nucleus consisted of an accretion disk surrounding a supermassive black hole. The appeal of this model (and its direct descendents) has always been one of energetics: no other proposed mechanism is as efficient and as compact. Gravitational potential energy release is hard to beat as an energy production mechanism. To the extent that there is any consensus within the AGN community about the black hole – accretion disk model, it ends with those two elements. The reason that no specific model has gained ascendency is not mysterious: black hole accretion physics is enormously complex. It involves multidimensional, time-dependent gas flows in which such processes as radiation transport, pair production, and magnetohydrodynamics play important roles. These topics are difficult enough when examined in isolation, so it is not surprising that no one has assembled them into a rigorously self-consistent whole. This complexity leaves the black hole disk model so

---

[*]Department of Astronomy, Virginia Institute for Theoretical Astronomy, University of Virginia, Charlottesville, VA 22903, USA. This work was partially supported by NSF grants PHY-9018251, AST-901348, and NASA grants NAGW-1510 and NAGW-2376.

thoroughly unconstrained as to give the broadest possible scope to the ingenuity and imagination of the theorist. Here, we wish to point to an effect that may have as its consequence the tightening of a notoriously loose constraint.

Why do accretion disks accrete? Motion in a central force law conserves angular momentum, yet this must be shed locally before a disk fluid element can find its way to the gravitational center. Since ordinary molecular and radiative viscosities are far too small to provide the necessary couple, the standard approach is to assume turbulence within the disk, as in the "alpha" (α) model of Shakura and Sunyaev (1973). Here α represents the ratio of the momentum stress to the disk pressure, and its functional form is generally taken to be either a power law in radius, or else simply constant. The long-recognized difficulty with this approach is that since the angular momentum increases outward in a disk, the classical Rayleigh criterion guarantees local stability. The point has often been made that the Reynolds number in the disk is likely to be large, and that Taylor vortices in laboratory Couette flows are seen under such circumstances, but ultimately there has never been a compelling reason to believe that supersonic accretion disks with very different boundary conditions from laboratory flows will exhibit this sort of behavior. It is this absence of a generic linear hydrodynamic instability that has always cast some doubt on the turbulence picture, and led to the sometimes contentious debate surrounding the origin of the anomalous viscosity. Even if one chooses to assume that *some* mechanism will inevitably lead to turbulence within the disk, the lack of knowledge about the basic physics involved condemns the standard accretion disk model to the mercy of an unfettered free parameter, α. The implications of this shortcoming are obvious for attempts to model, say, optical or X-ray variability in terms of viscous instabilities, or coherent nonaxisymmetric structure.

Recent developments in the study of the magnetohydrodynamic (MHD) stability of accretion disks lead us to suggest that the answers lie with magnetic fields. Magnetized accretion disk models have a long history (reviewed, for example, by Blandford 1989), and while the magnetic fields play a essential role in many of these models (e.g., Eardley & Lightman 1975; Coroniti 1981), most relegate magnetic fields to second-class status. For example, the kinematic dynamo mechanism depends on *pre-existing* turbulence to amplify the fields to dynamically important levels, usually defined as equipartition strength or greater. In that sense magnetic fields are an addendum to a hydrodynamic disk model, brought in as required to explain nonthermal emission in some classes of objects, or to launch and focus bipolar jets. In contrast to this point of view, we maintain that magnetic fields are fundamental to the evolution and structure of accretion disks. We have found that, the Rayleigh criterion notwithstanding, subthermal magnetic fields in accretion disks lead to violent instability on wavelengths $\lambda \sim v_A/\Omega$, with growth rates on order the rotation frequency Ω (Balbus & Hawley 1991).

Although this result is rather surprising, a form of this instability has been in the literature for some time. Velikhov (1959) considered the local stability of a magnetized Couette flow and found that for sufficiently weak magnetic fields, such flows were unstable when $d\Omega^2/dR < 0$. Chandrasekhar (1960) found that Velikhov's result was not sensitive to the velocity rotation law. Despite some interest within the fluid mechanics community in the early '60s, the instability was almost universally ignored by the astronomical community because the results appeared to be dependent upon Couette boundary conditions on infinite cylinders. However, a local dispersion relation for restricted magnetic geometries was derived for fields within rotating stars by Fricke (1969), who used it to rule out certain internal field configurations. But in general, throughout the 1970s and 1980s, as accretion disk theory flourished, there seems to have been no mention of MHD instabilities. In fact, we have been unable to find a single reference in the accretion disk literature to either Velikhov (1959) or Chandrasekhar (1960) prior to 1991. It was therefore a pleasant and unexpected result to find that accretion disks are a natural venue for the appropriate generalization of this instability. All that is required is some weak seed field, good magnetic coupling to the fluid, and $d\Omega^2/dR < 0$ in a rotationally supported system. These conditions are easily and generally met; indeed, in disk theory they have nearly universally been taken for granted.

2. The MHD Instability: A Simple Derivation

The physics of the instability is simple and straightforward. The most unstable disturbances are displacements perpendicular to the disk normal, which are unaffected by pressure forces in a thin accretion disk. Hence, orbital mechanical arguments alone are sufficient to understand how displaced fluid elements are destabilized. To begin then, let us consider the accelerations, a_r and a_ϕ at the disk midplane in cylindrical coordinates, (r, ϕ),

$$a_r = \ddot{r} - r\dot{\phi}^2 \tag{2.1a}$$

$$a_\phi = r\ddot{\phi} + 2\dot{r}\dot{\phi}. \tag{2.1b}$$

For a central force $a_\phi = 0$, and equilibrium corresponds to a circular orbit at $r = R$ with angular velocity Ω, $a_r = -R\Omega^2$. Now, consider small excursions from the circular orbit, let $x = r - R$ and $y = R(\phi - \Omega t)$, and substitute those definitions into the above equations. By retaining the linear terms, we obtain the local equations of motion (Hill 1878; Toomre 1990)

$$\ddot{x} - 2\Omega\dot{y} = -x\frac{d\Omega^2}{d\ln R} \tag{2.2a}$$

$$\ddot{y} + 2\Omega\dot{x} = 0. \tag{2.2b}$$

If one assumes solutions of the form $x, y \propto e^{\pm i\kappa t}$ then one obtains epicyclic motion with the epicyclic frequency defined as

$$\kappa^2 = 4\Omega^2 + \frac{d\Omega^2}{d\ln R} = \frac{2\Omega}{R}\frac{dl}{dR}, \tag{2.3}$$

where l is the mass specific angular momentum, $R^2\Omega$. For Keplerian orbits, the epicyclic frequency is simply Ω^2. Stability requires

$$\kappa^2 \geq 0. \tag{2.4}$$

In other words, the angular momentum $l = R^2\Omega$ must increase outwards. This is the usual Rayleigh stability criterion, and it is easily satisfied within an accretion disk.

To prepare the way for the inclusion of a magnetic field, consider next the addition of an attractive linear force for small displacements from the equilibrium position, i.e.,

$$f_x = -Kx, \qquad f_y = -Ky, \tag{2.5a}$$

where K is a "spring constant"; thus

$$\ddot{x} - 2\Omega\dot{y} = -\left(\frac{d\Omega^2}{d\ln R} + K\right)x, \tag{2.5b}$$

$$\ddot{y} + 2\Omega\dot{x} = -Ky. \tag{2.5c}$$

The system (2.5) may be thought of describing the motions of a pair of orbiting masses connected by a spring. Assuming solutions of the form $e^{i\omega t}$ leads to the relation

$$\omega^4 - \omega^2(\kappa^2 + 2K) + K(K + \frac{d\Omega^2}{d\ln R}) = 0. \tag{2.6}$$

Stability requires

$$(\kappa^2 + 2K) - \left[(\kappa^2 + 2K)^2 - 4K(K + \frac{d\Omega^2}{d\ln R})\right]^{1/2} \geq 0 \tag{2.7}$$

which implies

$$K \geq -\frac{d\Omega^2}{d\ln R}. \tag{2.8}$$

If $d\Omega^2/d\ln R < 0$, a sufficiently small spring constant K will lead to instability.

We conclude that a system consisting of two orbiting masses connected by a spring is unstable, provided the spring is *weak*. If this seems counterintuitive, recall that in orbital mechanics a body that is accelerated in the direction of its orbit will gain angular momentum and move to a higher orbit, resulting in a *decrease* in angular velocity. Thus, if we displace one of our masses to a slightly higher orbit, it will begin to drop behind the lower mass, stretching the spring. If the tension is sufficiently strong, the spring will keep the masses together. If the spring is weak, however, the tension merely acts to decelerate the lower mass and accelerate the higher mass, in effect transferring angular momentum outward. The lower mass drops to a still lower orbit, the higher mass moves higher, and the displacement grows.

We are now in a position to apply these results to the magnetized accretion disk. Consider a rotationally supported, incompressible, perfectly conducting fluid threaded by an equilibrium magnetic field. If the equilibrium flow experiences a small Lagrangian displacement ξ in the plane of the orbit of a given fluid element, i.e., $\xi \propto e^{i(kz-\omega t)}$, then there are no pressure or buoyancy forces, and the magnetic field acts only through tension forces $(\mathbf{B} \cdot \nabla)\mathbf{B}/4\pi\rho$. This translates to a local return tension force $-(\mathbf{k} \cdot \mathbf{v_A})^2\xi$, where $\mathbf{v_A}$ is the Alfvén speed ($= \mathbf{B}/\sqrt{4\pi\rho}$), and $\mathbf{k}$ is the wavevector. The resulting equations for Lagrangian fluid displacements in a rotating system are then identical to equations (2.5) with an effective spring constant of $(\mathbf{k} \cdot \mathbf{v_A})^2$. The perturbation analysis proceeds exactly as before, and we find that the displacements are unstable when

$$(\mathbf{k} \cdot \mathbf{v_A})^2 < -\frac{d\Omega^2}{d\ln R}. \tag{2.9}$$

The necessary criterion for instability becomes an outwardly decreasing *angular velocity*, not *angular momentum*. The Rayleigh criterion is irrelevant for astrophysical accretion disks!

For Keplerian orbits, the disk will be unstable for all wavenumbers such that

$$\frac{(\mathbf{k} \cdot \mathbf{v_A})^2}{\Omega^2} < 3. \tag{2.10}$$

As long as $\mathbf{v_A}$ is less than the sound speed, wavelengths satisfying equation (2.10) will fit comfortably within a disk scale height. Note also that the radial and azimuthal field components are immaterial (as long as they are weak), since the magnetic field strength enters solely through the dot product of the Alfvén speed with the (z-oriented) wavevector $\mathbf{k}$. The strength of the magnetic field serves only to set a characteristic wavenumber scale, which explains why magnetic effects do not weaken even as the magnetic field vanishes. (Eventually the characteristic wavenumber will become so large that dissipative physics, e.g., resistivity, will limit growth. But this is important only for extraordinarily weak fields that are unlikely to be present in fully ionized disks.) It is a straightforward matter to establish that by tuning the spring constant $K = (\mathbf{k} \cdot \mathbf{v_A})^2$, equation (2.6) yields a maximum growth rate that is independent of field properties, and is simply the local Oort A value of the disk,

$$|\omega_{max}| = -\frac{R}{2}\frac{d\ln\Omega}{dR}, \tag{2.11}$$

or $\frac{3}{4}|\Omega|$ for a Keplerian disk (Balbus & Hawley 1992a). This is in excess of a hundredfold amplification per orbit! The basic properties of the instability carry over into the nonaxisymmetric analysis (Balbus & Hawley 1992b); even purely toroidal magnetic fields prove to be unstable, although with longer growth times.

3. Consequences and Questions

The existence of a powerful local magnetic shear instability is a compelling argument for the presence of anomalous viscosity in accretion disks. In this context, it is of interest to note that laboratory Couette experiments (Donnelly and Ozima 1960; discussed in Chandrasekhar 1961) with magnetized mercury used the sudden measured increase in viscosity as the defining hallmark for the triggering of the MHD instability. However, the full consequences of the MHD instability for the structure and evolution of an accretion disk have yet to be fully explored. One cannot hope to understand MHD disk turbulence in detail, but one can aim for an understanding at the level of, say, convective turbulence, where scaling laws and gross energetics are thought to be known. A few of the more important questions include: (1) What is the physical mechanism that produces nonlinear saturation of the instability, and at what amplitude does saturation occur? (2) Does the instability constitute a coherent field generation mechanism? (3) What is the net angular momentum transport rate, i.e., what is α? (4) Where does dissipation occur, in the disk or outside, perhaps in a corona, wind, or a jet? (5) What types of disk models can be ruled out?

The most important issue is what the expected magnetic field strength will be in an unstable disk. The stability limit from the linear analysis indicates that all wavelengths shorter than $2\pi v_A/\sqrt{3}\Omega$ are stable. This means that when the Alfvén speed is greater than the sound speed, this critical wavelength exceeds the disk scale height. Taken at face value this implies that accretion disks are likely to be fully magnetic, with $\alpha \sim 1$. But the most compelling route to understanding nonlinear evolution is clearly via numerical solutions of the full MHD equations.

We have carried out a series of such simulations in a two-dimensional axisymmetric shearing sheet system using the magnetized version of Hill's equations (Toomre 1990, Hawley & Balbus 1992). Loss mechanisms that have the potential to limit the instability growth are buoyancy and reconnection. Density stratification is not included in the current simulations, although it is clearly an important complication to be considered later. A proper treatment of buoyancy and its related instabilities (e.g., Parker instabilities) requires three-dimensional calculations; thus these simulations can consider only the role of reconnection as a field limitation mechanism. Reconnection is due to numerical resolution effects, and the reconnection length scale is set by the size of an individual grid zone. The effects of numerical reconnection are monitored by running a given model at multiple resolutions. Regardless, the numerical simulations almost certainly overemphasize the role of magnetic reconnection. Despite this list of limitations, these first computations have revealed several intriguing and unexpected effects.

The outcome of a given simulation depends dramatically upon whether or not the average value of the magnetic field vanishes when calculated over the computational domain. Consider the initial condition of a vertical field passing through the disk.

When such a field is present, regardless of its initial strength (so long as the unstable wavelengths are resolved on the numerical grid), perturbed flow evolves into a *nonlinear* exponentially growing solution consisting of two coherent channels on the largest scales available within the computational domain. One channel has fluid with low angular momentum flowing inward, the other fluid with excess angular momentum flowing outward. Magnetic field amplitudes quickly increase above equipartition values. Clearly the resulting magnetically-driven flow is quite different from a traditional turbulent cascade! The channel solution suggests, for example, that an accretion disk embedded in a neutron star magnetosphere may be a much more active system than implied by a passive-field scenario. Similarly, reconnection events outside a disk may be important for determining the nature of the transport *within* the disk. Global simulations of the evolution of vertical field geometry have been carried out recently by Stone & Norman (1992). They find that an initially Keplerian disk threaded by a uniform z-field collapses on an orbital timescale, and generates a significant wind along the magnetic field lines due to the increase in magnetic pressure. It remains to be seen from three-dimensional simulations whether these channel solutions are nonaxisymmetrically stable.

If the initial magnetic field has no net dipole moment over the computational domain, the evolution is quite different. Although the linear phase proceeds as before with exponential field amplification, the nonlinear flow is much closer to turbulence, and power is transferred to all available wavenumbers. The average angular momentum flux is proportional to the poloidal magnetic pressure. In terms of the alpha parameter we have $\alpha = P_{mag}/P_{gas}$. This suggests that the α disk should be replaced with the β disk, where β is the ratio of the gas pressure to the magnetic pressure. Since it is not yet possible to calculate an average field strength throughout the disk, moving from α to β may not seem like much of an improvement! But at least a physical theory for the origin of the turbulence has been presented, pointing to a very promising direction for future progress.

It is possible, at least, to understand the qualitative behavior exhibited by the two-dimensional simulations. In axisymmetry, the fate of the field is governed by the "anti-dynamo theorem" (see, e.g., Moffat 1978). The poloidal magnetic field is determined solely by the ϕ-component A of the vector potential. The induction equation takes on a particularly simple form (Hawley & Balbus 1992):

$$\frac{dA}{dt} = 0 \qquad (3.1)$$

where the time derivative follows a fluid element. In other words, A labels fluid elements. It is then a simple matter to show that for incompressible flow, if the flux of A^2 through the domain boundaries is zero, the volume integral of A^2 is conserved in ideal MHD, or decays with time if resistivity is present. Magnetic energy B^2 can grow only by increasing the A-gradients on ever smaller scales, leading naturally to a situation conducive to dissipative reconnection. What we see in the simulations is that after turbulent motion begins, the total magnetic energy decays at a rate determined by the grid resolution. (The sensitivity of the flow to the value of the small dissipative length scale indicates that the flow is not truly turbulent. Turbulent flow is insensitive to the precise value of the dissipational scale.) The instability and reconnective dissipation compete, and in two dimensions, dissipation must be the inevitable victor. But the field evolution is likely to be very different in three dimensions, where dynamo activity is possible (e.g., Tout & Pringle 1992).

The origin and maintenance of cosmic magnetic fields is a vast, classical problem, and a matter of some importance is to determine the role of the instability in establishing or assisting dynamo activity. Several key questions await investigation. Does the instability lead, in and of itself, to a turbulent, mean-field dynamo? It is generally thought (e.g., Cowling 1981) that a nonvanishing mean helicity in the turbulent flow is an important ingredient for field amplification. Under what conditions does the rotational magnetic shear instability lead to a turbulent velocity field with this property? If the answer is "never", does the instability lead to field amplification by some other route that does not involve mean helicity turbulence, or is the ultimate role of the instability merely to inject short wavelength power to be seized upon by an essentially independent process? Clearly global field topology is important; in two dimensions a weak vertical field is amplified without dissipation, closed field lines are not. In three dimensions one can imagine a simple scenario in which poloidal field generates toroidal field (in addition to more poloidal field) via the instability, and this, in turn, causes the growth of more poloidal field via a nonaxisymmetric version of the instability. Does this feedback loop comprise a successful dynamo?

Three dimensional MHD simulations hold the promise of some very exciting and interesting answers.

References

Balbus, S.A. & Hawley, J.F. 1991, *Astrophys. J.*, **376**, 214.
Balbus, S.A. & Hawley J.F. 1992a, *Astrophys. J.*, **392**, 662.
Balbus, S.A. & Hawley, J.F. 1992b, *Astrophys. J.*, **400**, 610.
Blandford, R.D. 1989, in *Theory of Accretion Disks*, ed. Meyer, F., Duschl, J., Frank, J., and Meyer-Hofmeister, E. (Dordrect: Kluwer), 35.
Chandrasekhar, S. 1960, *Proc. Nat. Acad. Sci.*, **46**, 53.
Chandrasekhar, S. 1961, *Hydrodynamic and Hydromagnetic Stability*. (New York: Oxford)
Coroniti, F. V. 1981, *Astrophys. J.*, **344**, 587.
Cowling, T.G. 1981, *Ann. Rev. Astron. Astrophys.*, **19**, 115.
Donnelly, R.J. & Ozima, M. 1960, *Phys. Rev. Letters*, **4**, 497.
Eardley, D. M. & Lightman, A. P. 1975, *Astrophys. J.*, **200**, 187.
Fricke, K. 1969, *Astron. & Astrophys.*, **1**, 388.
Hawley, J.F. & Balbus, S.A. 1992, *Astrophys. J.*, **400**, 595.
Hill, G.W. 1878, *Am. J. Math*, **1**, 5.
Lynden-Bell, D 1969, *Nature*, **223**, 690.

Moffat, H.K. 1978, *Magnetic Field Generation in Electrically Conducting Fluids* (Cambridge: Cambridge Univ. Press)
Shakura, N.I. & Sunyaev, R.A. 1973, *Astron. & Astrophys.*, **24**, 337.
Stone, J.M. & Norman, M.L. 1992, preprint.
Toomre, A. 1981. *The Structure and Evolution of Normal Galaxies*, ed. S. M. Fall & D. Lynden-Bell (Cambridge Univ. Press), 111.
Tout, C. & Pringle, J. 1992, *Mon. Not. Roy. Astr. Soc*, in press.
Velikhov, E.P. 1959, *Soviet JETP*, **36**, 995.

Compton–Heated Winds from Accretion Disks

Christopher F. McKee *†   *D.T. Woods* †‡   *J.I. Castor* ‡
R.I. Klein †‡   *J.B. Bell* ‡

1. Introduction

The intense X–ray emission of AGN (active galactic nuclei) can heat the gas in these objects to high temperatures, driving a wind from regions in which the thermal velocity is comparable to or greater than the escape velocity (Begelman et al. 1983). Other mechanisms, such as heating due to dissipation of magnetic fields, or acceleration by rotating magnetic fields or radiation pressure, can also produce winds in AGN; thus, X–ray heated winds may be considered to be the minimum required by observation. These winds are important both because they can alter the accretion rate onto the central object by extracting mass, and because they provide important diagnostics of the distribution and dynamics of gas in AGN (Begelman and McKee 1983).

The nature of the wind is determined by the geometry of the gas relative to the source of the X–rays. The variability of the X–ray emission in AGN indicates that the source of the emission is compact (e.g., Turner and Pounds 1988). The gas may be distributed around this compact source in several possible ways: First, it could be in an accretion disk, although direct observational evidence for this assumption is lacking at present; by contrast, there is good evidence for accretion disks in many binary X–ray sources in the Galaxy. A wind will be driven from an accretion disk either if the disk flares (as it does in the standard α disk—Shakura and Sunyaev 1973) or if the source of the X–rays is above the disk (as in Compton reflection models—Fabian, this volume). Second, Antonucci and Miller (1985) have provided convincing observational evidence for the existence of obscuring gas tori in Seyfert galaxies. X–ray heating of these tori can be expected to produce strong winds; for Seyferts with obscured nuclei, such as NGC 1068, such winds can reflect the radiation from the central object into our line of sight by Compton scattering (Krolik and Begelman 1986). Finally, on a somewhat more speculative note, Sanders et al. (1989) have suggested that the IR spectra of AGN can be understood if the IR emission is due to reprocessing of the radiation from the central source by a warped disk. This model will almost necessarily be accompanied by a wind, since the warping occurs in a region where the binding energy of the gas is relatively low.

---

*Department of Physics, University of California, Berkeley CA 94720
†Department of Astronomy, University of California, Berkeley CA 94720
‡Lawrence Livermore National Laboratory, Livermore CA 94550

2. Theory of Compton–heated Winds

The effect of X–rays on a gas is determined by the ionization parameter $\Xi \equiv P_{\rm rad}/P_{\rm gas}$ (Krolik, McKee and Tarter 1981). For small values of Ξ, the gas can cool effectively and the temperature is usually $\sim 10^4$ K, typical of a photoionized nebula. At the other extreme (large Ξ) the ions in the gas are fully stripped, and the heating is due to Compton scattering. The temperature in this case is proportional to the frequency–weighted intensity, $T = T_{IC} \equiv (4kJ)^{-1} \int h\nu J_\nu d\nu$, where J_ν is the mean intensity of the radiation. For typical AGN spectra, $T_{IC} \sim 10^7$ K (Fabian et al. 1986).

The cool, nebular phase of the X–ray heated gas can exist only for ionization parameters less than some maximum, $\Xi_{c,\max}$. In a disk in hydrostatic equilibrium, the gas pressure decreases upward and the ionization parameter rises. When the ionization parameter rises above $\Xi_{c,\max}$, thermal balance is no longer possible at nebular temperatures, and the gas heats up, forming a corona or a wind. A rough estimate for the distance at which a corona develops into a wind can be obtained by equating the thermal velocity to the escape velocity, which yields the critical radius $R_{IC} \equiv GM\mu/(kT_{IC}) = 1.0 \times 10^{18} M_7 T_{IC,7}^{-1}$ cm, where M is the mass of the central object, M_7 is in units of $10^7 \, M_\odot$, and μ is the mean mass per particle.

Analytic calculations show that the wind actually commences at a distance of about $0.1 R_{IC}$ (Begelman et al. 1983). Comparison of the heating and dynamical times for the wind reveals three cases: (1) Isothermal wind. If the luminosity is sufficiently high and the distance to the central source not too great, the wind heats up to a temperature T_{IC} in less than a dynamical time, so that the wind is approximately isothermal. A necessary condition for this to occur is that the luminosity L exceed $0.1 T_{IC,7}^{-1/2} L_{\rm Edd}$, where $L_{\rm Edd}$ is the Eddington luminosity. (2) Steadily heated, free wind. At greater distances from the central source, the heating time remains less than the dynamical time, but is not short enough for T to reach T_{IC}. (3) Gravity–inhibited wind. Finally, if the luminosity is low enough that the heating time exceeds the dynamical time, then the gravitational field reduces the mass loss rate in the wind.

3. Numerical Computations

In order to obtain a greater understanding of the dynamics of Compton–heated winds, and in order to compare theoretically predicted spectra with observation, we have undertaken an extensive set of numerical hydrodynamic calculations (Woods et al. 1992, 1993). We have modified a 2–D code with adaptive mesh refinement (AMR; Berger and Colella 1989; Klein et al. 1990) in cylindrical coordinates employing a 2nd–order, operator split, Godunov scheme to include the effects of central gravity, rotation, and energy sources and sinks. AMR is particularly useful in this problem because it allows resolution of the small scale structure in the inner disk in a calculation that follows the flow on large scales, all with a moderate number of cells. In a typical run, we utilize three levels of refinement with a 100×100 grid at the first level. Regions with rapid spatial variation are refined at the second level by dividing first level cells into 16 equal parts. Finally, those second level cells which still show substantial variation are refined by a further factor of 16, so that altogether the calculation is equivalent to a fixed grid calculation with 2.56×10^6 zones.

The heating and cooling have been calculated as a function of both T and Ξ by using the CLOUDY code, kindly supplied by Gary Ferland. We have also adopted his AGN spectrum, which has $T_{IC} = 1.3 \times 10^7$ K. The major approximation we have made is that the corona is optically thin, so that the mean intensity at any point is simply $J = L/(4\pi)^2 R^2$. Corrections for electron scattering opacity, which can be substantial, have been calculated by Ostriker et al. (1991). We have focused on the case in which $L \ll L_{\rm Edd}$, so that radiation pressure is negligible.

4. Results and Conclusions

We have studied the dynamics of Compton–heated winds for luminosities in the range $(0.01 - 0.3)L_{\rm Edd}$ and for disks which extend out to as much as $12 R_{IC}$. In this brief communication, there is no room to present detailed results, so they will be briefly summarized: (1) The analytic theory of Begelman et al. (1983) for the wind mass loss per unit area is confirmed to within a factor 3 for $R \lesssim 8 R_{IC}$. Agreement is slightly worse at larger radii for low luminosities. The reasons for the discrepancies have been identified, enabling significant improvements in the accuracy of the analytic results. (2) The mass loss in the wind can be comparable to the accretion rate, as noted by Begelman et al. (1983). (3) Expansion cooling lowers the temperature in the wind by a factor of a few from the simple analytic estimate. (4) Finally, X-ray spectroscopy provides a promising diagnostic for winds from accretion disks. In Figure 1 we show the column density of iron ions as a function of ionization stage for different inclination angles of the disk for the case of $L/L_{\rm Edd} = 0.3$. These results were obtained from our hydrodynamical calculation using CLOUDY. Evaluation of the column density of different iron ions shows that the Kα line should be observable in absorption over a wide range of ionization states for lines of sight close to the surface of the disk. For accretion disks which are nearly face on, the absorption is less and is confined to highly stripped iron. For the case we considered ($T_{IC} = 1.3 \times 10^7$ K), the maximum velocity in the wind is about $700\,{\rm km\,s^{-1}}$. Since the iron lines may also be in emission, observation of the absorption lines requires a spectral resolution high enough to resolve lines of this width.

Acknowlegements. This research is supported by NASA grant NAGW-3027. The research of CFM is also supported by NSF grant AST89-18573.

References

Antonucci, R. R. J. and Miller, J.S. 1985, ApJ, 297, 621
Begelman, M. C., and McKee, C. F. 1983, ApJ, 271, 89

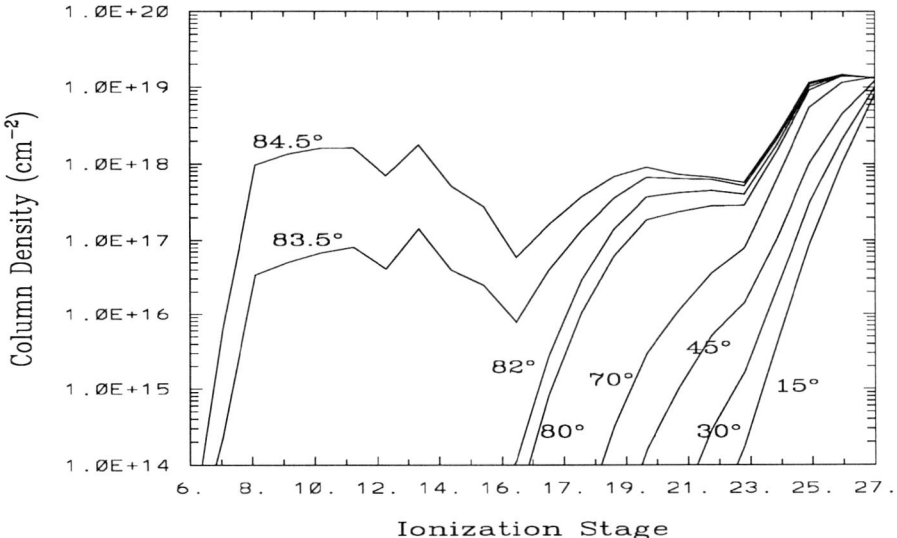

Figure 1. Column density of iron ions as a function of ionization stage for various inclination angles of the disk.

Begelman, M. C., McKee, C. F., and Shields, G. A. 1983, ApJ, 271, 70
Berger, M., and Colella, P. 1989, J. Comp. Phys., 82, 64
Fabian, A. C., Guilbert, P. W., Arnaud, K., Shafer, R., Tennant, A., and Ward, M. 1986, MNRAS, 218, 457 Klein, R., McKee, C.F., and Colella, P. 1990, in *The Evolution of the Interstellar Medium*, ed. L. Blitz (San Francisco: Astronomical Society of the Pacific), p. 117
Krolik, J. H., and Begelman, M. 1986, ApJ, 308, L55
Krolik, J. H., McKee, C. F., and Tarter, C. B. 1981, ApJ, 249, 422
Ostriker, E., McKee, C.F., and Klein, R.I. 1991, ApJ, 377, 593
Sanders, D. B, Phinney, E. S., Neugebauer, G., Soifer, B. T. and Matthews, K. 1989, ApJ, 347, 29
Shakura, N. I., and Sunyaev, R. A. 1973, A&A, 24, 337
Turner, T.J., and Pounds, K.A. 1988, MNRAS, 232, 463
Woods, D.T., Castor, J.I., Klein, R.I., McKee, C.F., and Bell, J.B. 1992, in *Testing the AGN Paradigm*, ed. S. Holt, S. Neff, and C.M. Urry (New York: AIP), p. 276
Woods, D.T., Castor, J.I., Klein, R.I., McKee, C.F., and Bell, J.B. 1993, ApJ, 000, 000

Determination of a Transonic Solution in a Stationary Accretion Disc

John Papaloizou [*] *Ewa Szuszkiewicz* [*]

Abstract

We discuss here conditions which determine a global solution of a one dimentional transonic accretion flow with non-zero angular momentum.

1. Introduction

There has been a considerable body of literature on transonic solutions in the theory of accretion discs (see Anderson 1989 for history and references). Studies of such stationary one dimensional discs have been restricted mainly to the case in which viscosity is modelled by the simplest standard form (viscous stress tensor proportional to the pressure). The flow is then described by a single first order ordinary differential equation with a critical point. Here we assume the viscous stress tensor to be proportional to the angular velocity gradient. The governing second order ordinary differential equation allows use of a powerful method for constructing the solution which overcomes difficulties met in previous studies (see Abramowicz *et al.* 1988).

2. A model

Stationary axially symmetric accretion flows are often described by a one dimensional model in which all the physical quantities are vertically integrated and the laws of mass, angular momentum and momentum conservation are expressed by a set of ordinary differential equations. The mass conservation equation can be integrated directly and it gives a relation between the surface density, Σ, the radial velocity, v_r, and accretion rate, c_1:

$$2\pi \Sigma v_r r = -c_1 \qquad (1)$$

The distribution of specific angular momentum, $l(r) = r^2 \Omega$, is determined by the requirement that the rate of flow of angular momentum through a circle of radius r be constant

$$c_1 l(r) + 2\pi r^2 \tau_{r\varphi} = c_1 c_2, \qquad (2)$$

---

[*]Astronomy Unit, School of Mathematical Sciences, Queen Mary and Westfield College, University of London, Mile End Road, London, E1 4NS, England. The authors acknowledge support from the SERC through grants GR/H09454 and GR/E/50247

where c_2 is the constant specific angular momentum associated with this flow, and $-\tau_{r\varphi}$ is the $r\varphi$ component of the viscous stress tensor. The equation of motion reads:

$$v_r \frac{dv_r}{dr} = -\frac{1}{\Sigma}\frac{dP}{dr} - \frac{d\Phi}{dr} + \frac{l^2}{r^3}, \qquad (3)$$

where P is the pressure. The gravitational potential due to the central object with a mass, M, is given by $\Phi = -GM/(r - r_G)$, the gravitational radius, r_G, being defined in the usual notation by $r_G = 2GM/c^2$. A polytropic equation of state $P = K\Sigma^\gamma$ is assumed with K and γ being constant. The sound speed is then given by $c_s^2 = dP/d\Sigma = \gamma P/\Sigma$. Combining the equation of motion with the continuity equation and the equation of state gives

$$\left(v_r - \frac{c_s^2}{v_r}\right)\frac{dv_r}{dr} = -\frac{GM}{(r - r_G)^2} + \frac{c_s^2}{r} + \Omega^2 r. \qquad (4)$$

This exhibits a critical point when the radial velocity reaches the sound speed.

2.1 Nature of the critical point

We assume a particular form of the usual "α" viscosity: $\tau_{r\varphi} = \nu\Sigma r(d\Omega/dr)$, with $\nu = \alpha c_s^2/\Omega_K$, Ω_K being the circular orbit frequency. Combining this with equations 2 and 4 one can obtain a set of three nonsingular equations:

$$\frac{du}{d\tau} = -\frac{m}{(x - x_G)^2} + \frac{\tilde{c}_s^2}{x} + \tilde{\Omega}^2 x, \quad \frac{dx}{d\tau} = \left(u - \frac{\tilde{c}_s^2}{u}\right), \quad \frac{d\tilde{\Omega}}{d\tau} = \frac{u(\tilde{c}_2 - \tilde{\Omega}x^2)\tilde{\Omega}_K}{\alpha\tilde{c}_s^2 x^2}\left(u - \frac{\tilde{c}_s^2}{u}\right),$$

where we have adopted as units of radius and velocity r_* and v_*, being respectively, the radius and velocity at the critical point. Thus $u = -v_r/v_*$ (making $u > 0$ for inflow) $x = r/r_*$, $x_G = r_G/r_*$, $\tilde{c}_s^2 = (ux)^{1-\gamma}$ is the square of dimensionless sound speed, $\tilde{c}_2 = c_2/(v_* r_*)$, $\tilde{\Omega} = \Omega r_*/v_*$, and $m = GM/(r_* v_*^2)$ is the Bondi parameter. Subscript $*$ denotes values appropriate to the critical point.

Regarding τ as a pseudo-time variable, the critical point is an equilibrium point with $x = u = 1$. Linearizing in the neighbourhood of this point, we find the type of critical point. For perturbations varying as $\exp(\lambda\tau)$, the non-zero eigenvalues, λ, satisfy the characteristic equation:

$$\lambda^2 = (\gamma - 1)^2 - (\gamma + 1)\left(-\frac{2m}{(1 - x_G)^3} + \gamma - \tilde{\Omega}_*^2 - 2\tilde{\Omega}_*\frac{\tilde{\Omega}_{K*}}{\alpha}(\tilde{c}_2 - \tilde{\Omega}_*)\right). \qquad (5)$$

Being either real with opposite signs or purely immaginary, the only physically acceptable type of critical point is a saddle. This is unlike the case with $\tau_{r\varphi} \propto P$, where both saddle and nodal types are possible. This result was also obtained by Abramowicz & Kato (1989) for the case of isothermal flow.

3. Construction of disc models

It is possible to derive a single nonlinear second order differential equation for $\tilde{\Omega}$ from the above equations which may be written

$$-\tilde{\Omega}^2 x - \frac{1}{u^{\gamma-1} x^\gamma} + \frac{m}{(x-x_G)^2} - \frac{1}{\gamma}\left[\frac{1}{u^{2\gamma-1} x^{\gamma-1}} - \frac{1}{u^{\gamma-2}}\right]\frac{du^\gamma}{dx} = 0, \qquad (6)$$

where $u^\gamma = \dfrac{\alpha \frac{d\tilde{\Omega}}{dx}}{\tilde{\Omega}_K x^{\gamma-3}(\tilde{c}_2 - \tilde{\Omega} x^2)}$ and $\left(\dfrac{d\tilde{\Omega}}{dx}\right)_{x=1} = \dfrac{\tilde{\Omega}_{K*}(\tilde{c}_2 - \tilde{\Omega}_*)}{\alpha}, \qquad (7)$

with the angular velocity at the critical point being given by $\tilde{\Omega}_* = \sqrt{m(1-x_G)^{-2} - 1}$. To solve eq. 6 exterior to the critical point, in addition to regularity it is only necessary to specify the derivative of $\tilde{\Omega}$ there. However, this must be chosen so that the flow velocity matches on to a standard Keplerian disc at large radii. Technically this is only possible if $\gamma > 3$. This is because it is necessary for the disc to be sufficiently cool at large radii for the matching to be possible. However, if the size of the disc is less than a Bondi radius the constraint on γ should not be needed in order to match to a small flow speed.

From the above discussion we expect that a steady disc model is specified once γ the viscosity parameter α, x_G, and Bondi parameter m are given. We solve eq. 6 by a relaxation method. We solve (explicitly) the time dependent diffusion equation

$$\frac{\partial \tilde{\Omega}}{\partial t} = D(x)\left[-\tilde{\Omega}^2 x - \frac{1}{u^{\gamma-1} x^\gamma} + \frac{m}{(x-x_G)^2} - \frac{1}{\gamma}\left[\frac{1}{u^{2\gamma-1} x^{\gamma-1}} - \frac{1}{u^{\gamma-2}}\right]\frac{du^\gamma}{dx}\right].$$

The steady state solution gives the required model. This problem is readily solved on a computational domain extending from the critical point to outer boundary where the angular velocity may be specified. Note that $\tilde{\Omega}$ is specified at the critical point. The function $D(x)$ can be chosen for numerical convenience to equalize the diffusion coefficient throughout the computational grid.

Note that in principle there is no restriction in the choice of x_G. However, for a thin (large m) low viscosity disc eq. 7 shows that the specific angular momentum carried by the accretion flow is roughly the Keplerian value at the critical point. If the interior supersonic flow is to carry material into a black hole we would expect this value to correspond roughly to the minimum one, so $x_G \sim 1/3$.

An example of the solution of this equation, with $m = 10$, $x_G = 0.1$, $\alpha = 0.1$ and $\gamma = 4$, obtained numerically by means of the above relaxation method is presented in Figure 1.

4. Summary

We have presented a powerful method for the construction of the part of a transonic solution exterior to the critical point in a one dimensional accretion disc which satisfies

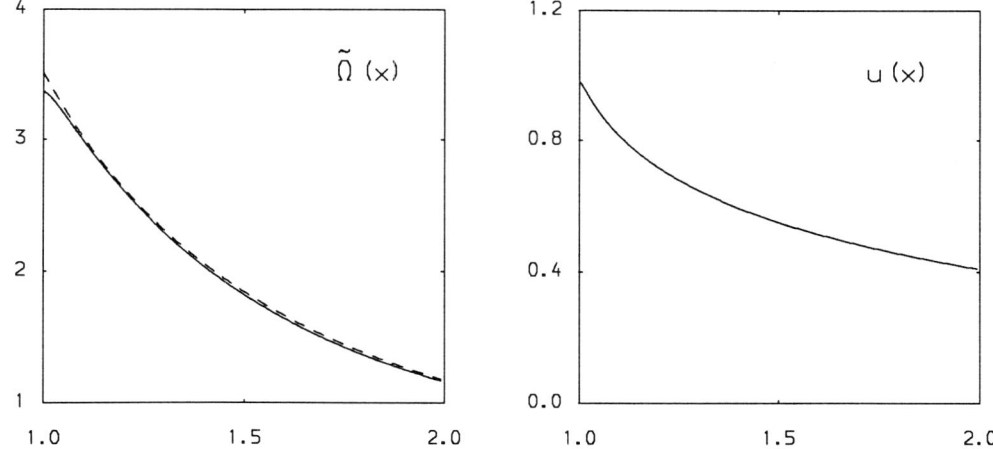

Figure 1. The angular, $\tilde{\Omega}(x)$, and radial, $u(x)$, velocities in the accretion disc with $m = 10$, $x_G = 0.1$, $\alpha = 0.1$ and $\gamma = 4$. The circular orbit frequency, denoted by the dashed line, is given for comparision.

the regularity conditions there and tends to a standard Keplerian disc at large radii. The relaxation method we used allowed us to find the solution with the required type of critical point without troubling about forbidden regions in the parameter space. The condition $\lambda^2 \geq 0$, (see eq. 5) is implicitly built into the procedure of constructing a solution, being here specified by γ, α, x_G, and Bondi parameter m. It should be possible to generalize this approach to more complicated non-adiabatic flows.

References

Abramowicz M.A., Czerny B., Lasota J-P., Szuszkiewicz E., *Astrophys. J.* 332 (1988) 646.
Abramowicz M.A., Kato S., *Astrophys. J.* 336 (1989) 304.
Anderson, M., *Mon. Not. R. astr. Soc* 239 (1989) 19.

Black Holes and Accretion Disks

Fredrik H. Wallinder [*]

Abstract

The possible relation between observed variability behaviour and slim disk stability properties is examined. It is argued that processes which give rise to QPOs in galactic sources are operative in AGN as well. Thus, it may be that unstable acoustic modes in the inner part of the accretion disk give rise to both the quasi-periodic short-term X-ray variability in NGC 6814 and the horizontal branch oscillations (HBOs) in X-ray binaries. The estimated central mass in NGC 6814 is $\sim 10^6\, M_\odot$.

1. Introduction

The majority of compact galactic and extragalactic sources seem to accrete at a rate $\dot{m} \sim 1$, where $\dot{m} = L/L_{\rm E}$, $L_{\rm E} = 10^{38}\, m$ erg s^{-1} being the Eddington accretion rate and where $m = M/M_\odot$ is the central mass in solar units. Some examples corresponding to AGN are shown in Figure 1. It follows that the standard Shakura-Sunyaev model (Shakura & Sunyaev 1973, 1976) is simply inadequate when it comes to a relevant description of, especially, the inner accretion disk in these sources, where the bulk of the luminosity is generated. One may also note that Shakura-Sunyaev disks contain an artificial singularity at the inner edge, due to an improper neglect of some inertial terms in the radial structure equations. Thus, any model which attempts to combine the effects of magnetic fields, electron/positron pairs, winds or whatever, with a Shakura-Sunyaev Keplerian disk, is bound to yield questionable results. The more appropriate slim disk models (Abramowicz et al. 1988, 1989) are able to include relevant basic physics, such as transonic radial motion, non-Keplerian rotation and pressure, velocity and entropy gradients. These effects have a significant influence on the slim disk stability. For instance, advection cooling is found to stabilize the disk against classical thermal and viscous instabilities when $(\alpha, \dot{m}) = (\ll 1, \sim 1)$, where α is the standard viscosity parameter (Wallinder 1991a). Moreover, the slim disks contain acoustic modes, not present at all in the Shakura-Sunyaev theory. If such mode instabilities manifest themselves in the emitted radiation from the disk, interpretation of observed variability patterns may provide information about basic accretion parameters, such as the central mass. Since the physics of the inner region is properly described even for small accretion rates, the same arguments should also apply to e.g., low-luminosity Seyferts. The only AGN which so far has shown clear

---
[*]NORDITA, Blegdamsvej 17, DK-2100 Copenhagen Ø, Denmark

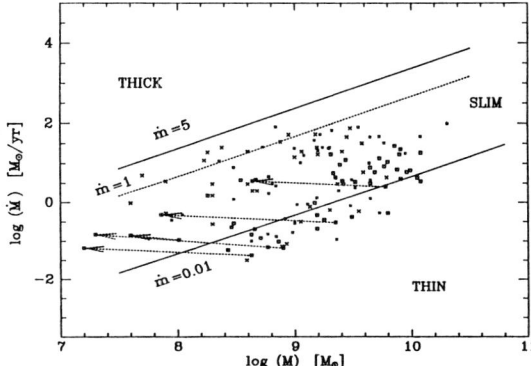

Figure 1. The observed relation between central mass and accretion rate in AGN. A majority of cases apparently accretes in the slim disk range. From Abramowicz et al. (1992a).

evidence of quasi-periodic X-ray variability is NGC 6814. The peaks in the power spectrum lie in the range $10^{-4\pm1}$ Hz. The "fundamental" peak at $\nu_0 = 8.2 \times 10^{-5}$ Hz has been shown to be rather stable over a period of several years (implying $|\dot{P}| \lesssim 10^{-6}$, Done et al. (1992)), during which time the X-ray luminosity and hence the accretion rate changed by about one order of magnitude. Another peak with significant power is the "fourth harmonic" at $\nu = 4 \times 10^{-4}$ Hz (Fiore et al. 1992). No indications of long-term coherence exist, even though the data are consistent with coherence over a few years. The low level of continuum emission implies that the source is photon-starved, which may be due to the accretion disk being edge-on. However, if the plane of the latter coincides with that of the galaxy, then the accretion disk should be approximately face-on (McHardy 1989). The inner accretion disk may then be accessible to observation, in contrast to other sources (Mittaz & Branduardi-Raymont 1989), whereas the low continuum level may at least partially be due to a low accretion rate. One important fact is that the topology of the folded light-curve changed from a single peak to a multi-peaked one between the *EXOSAT* and *Ginga* observations (Done et al. 1992). There are also some indications of a slight period increase, i.e. $\dot{P} > 0$ (Done & King 1993).

2. Method and basic results

The basic assumption of the method is that variability arises due to unstable modes in the inner region of a slim disk. The stability properties are inferred from a linear stability analysis, using the slim disk equilibria. The resulting fourth-order dispersion relation comprises four modes: two acoustic, one thermal and one viscous. The acoustic instability frequency is found to increase with $\dot{m}$ in the general case, as well as when the radius diminishes. The horizontal branch oscillations (HBOs) observed in X-ray binaries may be connected with this growing behaviour (Wallinder 1991a).

The same should also apply to AGN, since instability frequencies in general scale as m^{-1}. The slim disk luminosity typically has its maximum about $x \equiv r/r_g \simeq 5$, a value which seems rather independent of $\dot{m}$. It follows that the instability frequencies present in this region are likely to show up in the emitted radiation and hence in the observed variability.

3. Application to NGC 6814

Wallinder (1991b) suggested that the quasi-periodic variability in NGC 6814 should arise due to the same mechanism as the HBOs. The seemingly constant period would then be explained if the "typical" accretion rate is rather low, say $\dot{m} \sim 10^{-2}$. Figure 2 shows the influence of $\dot{m}$ on the acoustic frequency at various radii in the inner disk, when $(\alpha, m, \lambda/H) = (0.5, 4 \times 10^6, 5)$, where λ/H is the scaled perturbation wavelength. Obviously, the frequency is *independent* of $\dot{m}$ in certain ranges, and the accretion rate may vary at least one order of magnitude from a typical value without changing this result. The frequency increase referred to above apparently does not set in until $\dot{m} \gtrsim 0.1$. It is interesting to note that the frequency corresponding to e.g., the fourth harmonic does not occur too far away from the radius of maximal luminosity at $x \sim 5$.

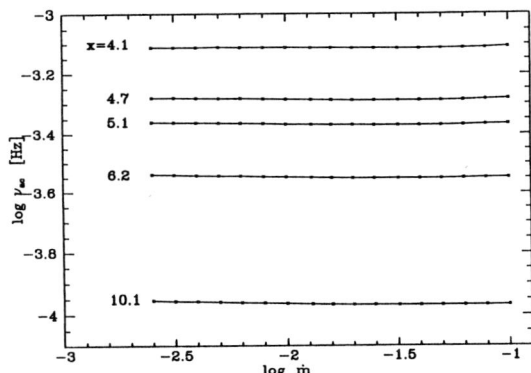

Figure 2. The relation between acoustic instability frequency and accretion rate. The frequency is evidently independent of the latter, for a particular radius and small accretion rates. This may partly explain the constant period seen in NGC 6814.

The relation between the acoustic slim disk frequency and the central mass is also of some related interest. For typical observed frequencies, consistency is obtained for $m \sim 10^6$, and so much mass concentrated within a volume only a few Schwarzschild radii across strongly supports the notion of a supermassive black hole in NGC 6814 (Wallinder 1991b). This interpretation is supported by the global, time-dependent slim disk simulations carried out by Honma et al. (1992). In addition, Nowak & Wagoner (1991) found global, trapped acoustic modes with about the same frequency in the inner disk, despite adopting quite different assumptions. The trapping occurs because the epicyclic frequency has a maximum in the inner disk, which is a general relativistic effect (Okazaki et al. 1987).

4. Discussion and conclusions

The suggested model predicts that the short-term variability in NGC 6814 arises through acoustic mode behaviour in the innermost part of an accretion disk around a supermassive object, presumably a black hole. The observations are consistent with typical parameter values $(\alpha, \dot{m}, m) \sim (0.5, 10^{-2}, 10^6)$, but it should be stressed that these are only estimates, which in addition do not arise from any severe fine tuning. The limited range of observed frequencies may be due to either the radial dependence of the luminosity or to trapping in the innermost disk, or to a combination of both.

Alternative models, based on rotation of a single spot on the accretion disk surface (Abramowicz et al. 1992b,c), collisions between a star and the accretion disk (Rees 1991; Syer et al. 1991; Zurek et al. 1992), and an orbiting screen in the outer region of the disk (Done et al. 1991; Tsuruta, this volume) may suffer from being fine tuned. Thus, the accretion disk may not be as highly inclined as required by both the spot and the screen models, which anyhow may be ruled out due to the change of topology of folded light-curve structure. In addition, the reason why a screen in the outer part would have an approximately stable structure seems obscure. The simplest version of the collisional model seems also ruled out, since the star probably moves ballistically. One modified version which combines star collisions and occultation by the ejecta from the encounters could be more promising. For instance, the change of folded light-curve structure may be connected with precession of the orbit of the star (Rees 1992). However, the observation that $\dot{P} > 0$ may impose major difficulties for this model, since the opposite effect is expected (Done & King 1993).

Thus, the only model which at present seems able to meet all observational constraints could be the pulsational one. This model also predicts that other sources should show the same behaviour, provided that the accretion rate is low enough and the inner accretion disk can be observed. A search for such sources should consequently concentrate on nearby, face-on and low-luminosity AGN, or their stellar galactic counterparts.

Acknowledgements. Financial support from the Royal Swedish Academy of Sciences, the Royal Physiographical Society of Lund and the Swedish Natural Science Research Council is gratefully acknowledged.

References

Abramowicz M.A., Czerny B., Lasota J.P., Szuszkiewicz E., 1988, ApJ **332**, 646
Abramowicz M.A., Szuszkiewicz E., Wallinder F., 1989, in: *Theory of Accretion Disks*, eds. F. Meyer, W.J. Duschl, J. Frank, E. Meyer-Hofmeister, NATO ASI Series Vol. 290,

Kluwer, p. 141

Abramowicz M.A., Malkan M.A., Szuszkiewicz E., 1992a, in preparation

Abramowicz M.A., Lanza A., Spiegel E.A., Szuszkiewicz E., 1992b, *Nature* **356**, 41

Abramowicz M.A., Bao G., Fiore F., Massaro E., Perola G.C., Spiegel E.A., Szuszkiewicz E., 1992c, in: *Physics of Active Galactic Nuclei*, eds. W.J. Duschl & S.J. Wagner, Springer, in preparation

Done C., Madejski G.M., Mushotzky R.F., Turner T.J., Koyama K., Kunieda H., 1991, in: *Frontiers of X-ray Astronomy*, ed. Y. Tanaka, Universal Academy Press, Tokyo, in press

Done C., Madejski G.M., Mushotzky R.F., Turner T.J., Koyama K., Kunieda H., 1992, ApJ, in press

Done C., King A.R., 1993, in preparation

Fiore F., Massaro E., Barone P., 1992, A&A, in press

Honma F., Matsumoto R., Kato S., 1992, PASJ, submitted

McHardy I.M., 1989, in: *ESLAB Symp. on Two Topics in X-Ray Astronomy*, ESA SP-296, p. 1111

Mittaz J.P.D., Branduardi-Raymont G., 1989, MNRAS **238**, 1029

Nowak M.A., Wagoner R.V., 1991, ApJ **378**, 656

Okazaki A.T., Kato S., Fukue J., 1987, PASJ **39**, 457

Rees M.J., 1991, personal communication

Rees M.J., 1992, personal communication

Shakura N.I., Sunyaev R.A., 1973, A&A **24**, 337

Shakura N.I., Sunyaev R.A., 1976, MNRAS **175**, 613

Syer D., Clarke C.J., Rees M.J., 1991, MNRAS **205**, 505

Wallinder F.H., 1991a, A&A **249**, 107

Wallinder F.H., 1991b, MNRAS **253**, 184

Zurek W.H., Siemiginowska A., Colgate S.A., 1992, in: *Testing the AGN Paradigm*, eds. S.S. Holt, S.G. Neff and C.M. Urry, AIP Conf. No. 254, p. 564

Testing the "disc X-ray reprocessing" in UV–optical continuum and line emission in NGC 5548

E. Rokaki *

The importance of the X-ray reprocessing in NGC 5548 is indicated both by the simultaneously, and highly correlated variable optical, UV and X-ray continuum emission and by individual features (the 6-7 keV Fe line and the "hard tail") of its hard X-ray spectrum. We model here the optical–UV continuum emission and its temporal variations in terms of the accretion disc surrounding the central black hole and X-ray source. The model also includes the "warm absorber" indicated by the absorption feature in the spectrum at 8-10 keV. Assuming electron scattering of the X-ray photons in the hot medium, the photoionized outer regions of the disc ($\sim 10^3 R_{\rm gr}$, where

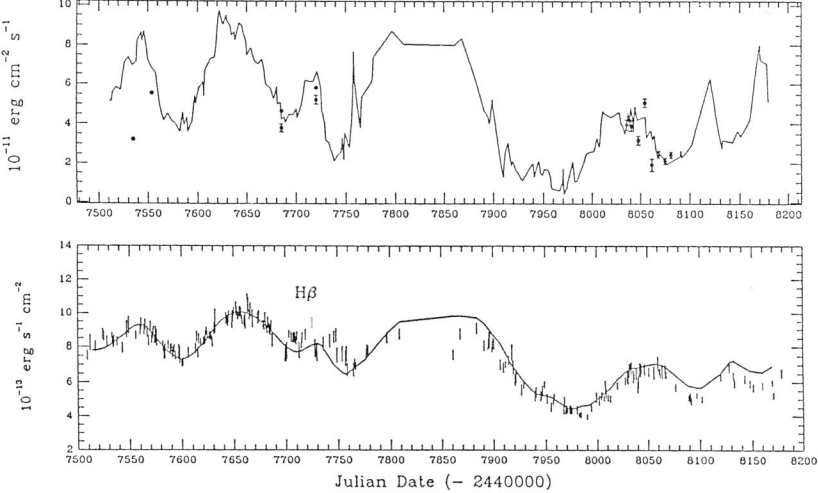

Figure 1. (a) The predicted 2–10 keV light curve for the two years monitoring. It is in phase and with the same amplitude of variations as the observed UV [5] during the first year, and in phase but amplified relative to the optical light curve [6] for the second year of the observations. The points are the observed 2-10 keV fluxes [7,8]. The total X-ray flux used in the following computations is 12 times the 2-10 keV flux. (b) Computed (solid lines) and observed (bars) Hβ light curves for Years 1 and 2.

---

*IAP, Paris and DAEC, Meudon, France.

$R_{gr}=2GM_{BH}/c^2$, M_{BH} is the mass of the black hole) emit significant fluxes of low ionisation lines as proposed by Collin-Souffrin and Dumont [1-4]. We take into account light travel delays to compute the continuum and Hβ light curves, and relativistic effects in the computations of line profiles. The fitting of the fluxes, profiles and light curve of Hβ indicate a "mass loaded flow" for the structure of the hot diffusing medium: a radial density dependence $\sim r^{-a}$, $a = 1 - 0$, located relatively close to the black hole ($\sim$10–15 light days).

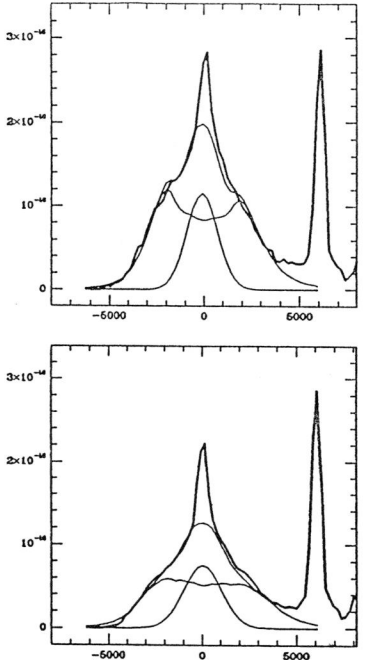

Figure 2. Decomposition of the broad Hβ profiles observed in the high (a) and low (b) state into the disc component and a gaussian semi-broad component. The semi-broad component has a σ of 1300–1100 km s^{-1}, and an integrated flux of $2 - 2.6\,10^{-13}$ erg cm^{-2} s^{-1} (the variations are of the order of the error bars).

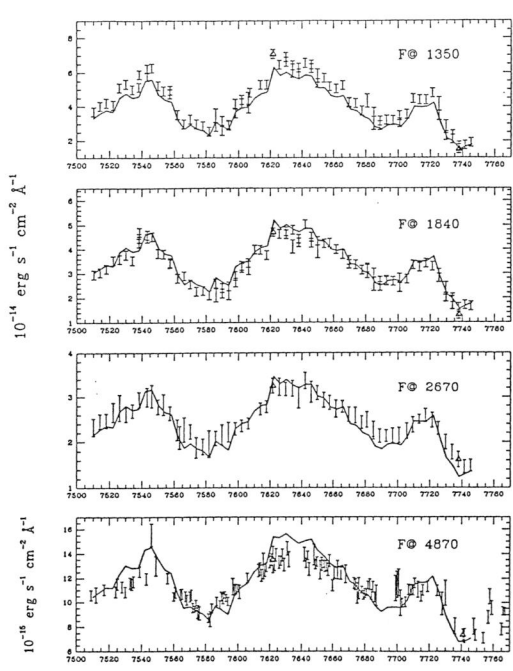

Figure 3. The computed (solid lines) and observed (bars) continuum light curves for Year 1, at four wavelengths.

References

[1] Collin-Souffrin, S., Dumont, A.M., 1990, A&A 229, 292
[2] Dumont, A.M., Collin-Souffrin, S., 1990, A&A 229, 302
[3] Dumont, A.M., Collin-Souffrin, S., 1990, A&A 229, 313
[4] Dumont, A.M., Collin-Souffrin, S., 1990, A&AS 83, 71
[5] Clavel J. et al., 1991, ApJ, 366, 64

[6] Peterson B.M. et al., 1992, in press
[7] Nandra K., et al., 1991, MNRAS 248, 760
[8] Clavel J. et al., 1992, ApJ, 393, 113

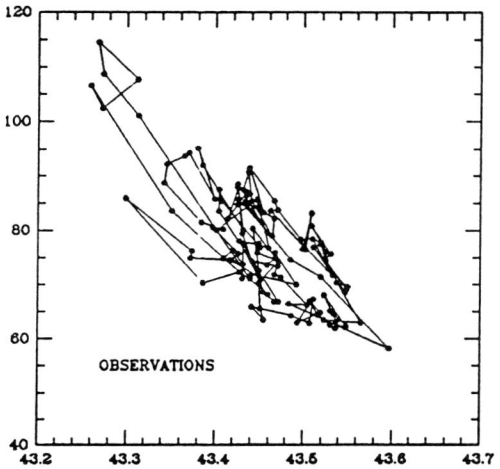

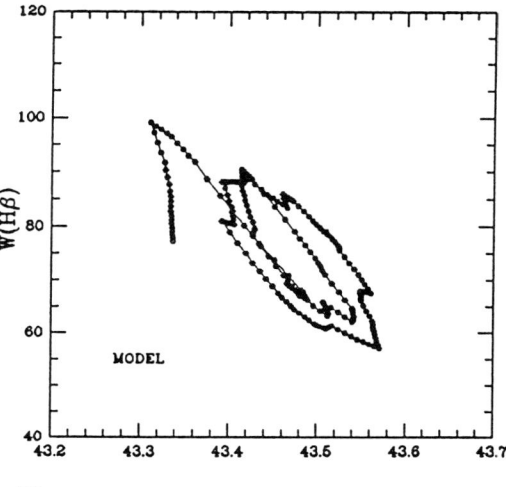

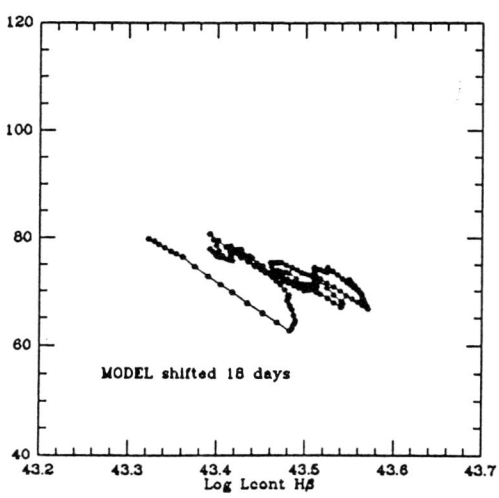

Figure 4. The Hβ equivalent width, versus the value of νL_ν for the continuum underlying Hβ. The two first panels give the values measured and computed at the same dates during Year 1. The "Baldwin effect" which is due mainly to the light travel time in the line emitting region is clearly visible on these two panels but it is absent on the last one.

AGN Accretion Discs in Realistic Potentials

R.J.R. Williams [*]   Judith J. Perry [†]

Abstract

A realistic stellar cluster potential may have an observable effect on the thermal structure of an AGN accretion disc. Optimization of model parameters has found an extremely good fit to the broad-band spectrum of 3C 273 for reasonable assumptions about the central black hole and surrounding cluster.

1. Introduction

Various authors have shown that compact stellar clusters around AGN Black Holes are important in determining the properties and evolution of Active Galactic nuclei, *e.g.* [1], [2]. We describe a way of directly determining the mass and size of such clusters in certain cases.

Radiation from an AGN accretion disc derives, at least in part, from the local release of binding energy through viscosity. Here we assume that accretion is the sole energy source: the spectrum of the disc is determined by the potential into which it is falling. We derive an estimate of the disc spectrum by assuming black-body emission.

In a forthcoming paper, [3], we find the spectrum expected from this model AGN, in which a dense, young stellar cluster surrounds a massive Black Hole. Here, we show a fit of this model to the continuum of 3C 273 (in its low state) [4].

2. A Model of 3C 273

In Perry & Williams, we derive cluster parameters from the parameters of the fit given by [4]. We have optimized the fit of our model to the data, (Fig. 1). In this figure, the different symbols identify different observations; the points near 10^{21} Hz are upper limits. The dot-dashed line is the luminosity of the disc alone; we add two log-normal components at high and low energies. The log-normal is the natural spectrum expected for self-Compton emission in regions of moderate optical depth[?]. The subsidiary peak at 10^{14} Hz is emitted at the cluster core radius; it is caused by the sudden change in orbital shear as the mass of the cluster takes full effect.

---

[*]Institute of Astronomy, Madingley Road, Cambridge CB3 0HA. This author was supported during this work by an SERC Research Studentship.
[†]Institute of Astronomy, Cambridge. This author was supported by the Leverhulme Trust.

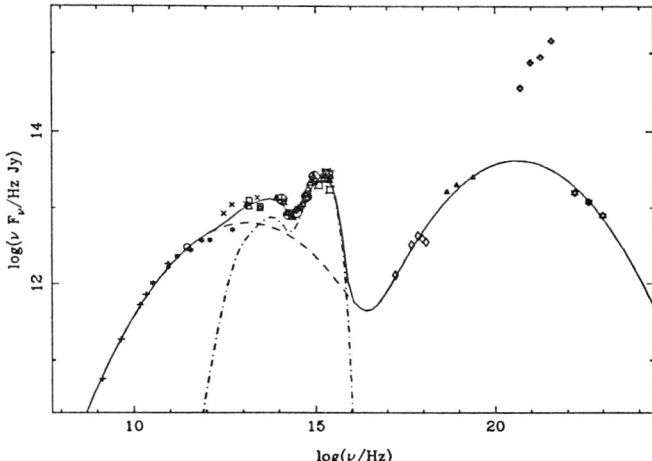

Figure 1. Disc + non-thermal continuum model of the spectrum of 3C 273 [4]. The dot-dashed line is the spectrum of the disc; the cluster produces the 10^{14} Hz peak.

The stellar cluster has core mass $2 \times 10^9 M_\odot$, core radius 0.21 pc, and is surrounding a $3 \times 10^8 M_\odot$ black hole. An accretion rate of $100 M_\odot$ yr^{-1} through the disc supplies the bolometric luminosity. Stellar evolution within a young cluster will provide a significant fraction of this fueling rate, but we expect that collisional mass loss will dominate. The effective collision time in the cluster is around 10^7 yr, suggesting that bright AGN may evolve on this timescale. For full details, see Perry & Williams [3].

References

[1] Perry, J.J. & Dyson, J.E., 1985. *Mon. Not. Roy. astr. Soc.* **213**, 665
[2] Scoville, N.Z. & Norman, C.A., 1988. *Ap. J.* **332**, 163
[3] Perry, J.J. & Williams, R.J.R., 1992. *Mon. Not. Roy. astr. Soc.*, in press
[4] Perry, J.J., Ward, M.J. & Jones, M., 1987. *Mon. Not. Roy. astr. Soc.* **228**, 623
[5] Cowsik, R. & Perry, J.J., 1987. Preprint

Test of the Accretion Disc Model and Orientation Indicator

Monique Joly *

Abstract

Here is presented a statistical test for the accretion disc model. The effect of different inclinations of the disc to the line of sight is shown. Then the possibility that the ratio R of the radio core luminosity to the radio lobe luminosity is a good orientation indicator is investigated. It is shown that R describes the intrinsic power of the central radio source.

1. A statistical test of the accretion disc model

The model adopted here is that, described by Collin and Dumont (1989), of an accretion disc illuminated by the central nonthermal source. The disc emits in its inner part the UV and optical continuum, and in its outer parts the low ionization lines (LIL; e.g., the Balmer lines) as proposed by Collin (1987). A second region emits the high ionization lines (HIL; e.g., Lα and CIV); it can have a spherical geometry. Both regions are illuminated by the central source. For the disc, illumination occurs through backscattering of the nonthermal radiation by a hot medium (halo). This model has been worked out by Dumont and Joly (1992).

Using the code of Dumont and Collin (1990) we have computed the line spectrum emitted by the illuminated disc as a function of the central source luminosity, the Eddington ratio (L_{bol}/L_{Edd}) and the characteristics of the scattering halo (density structure and optical thickness).

In figure 1 is plotted the computed line width (FWHM(Hβ)) versus line luminosity (L(Hβ)) from the model of Dumont and Joly (1992) which gives the best fit to observations (Eddington ratio of 0.1, optical depth of the halo of 0.5, power index of the halo density of 1.2). The model is computed with an inclination of the disc to the line of sight of 45°. On the right of the figure is shown the modification of the results induced by a variation of the inclination between 5 and 80°.

Observational data of the Hβ line reported for 165 AGN are superposed on the figure. A good correlation is observed with a significance of 10^{-7}. Note that the large scatter of the correlation can be easily explained by inclination effects.

---

*DAEC, Observatoire de Meudon, 92195 Meudon cedex, France.

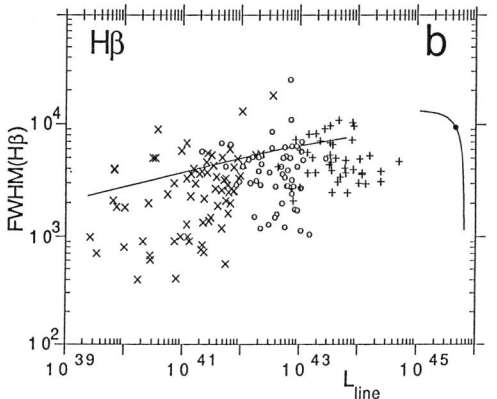 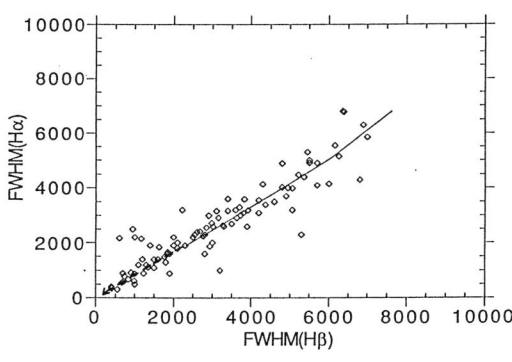

Figure 1. FWHM(Hβ) versus L(Hβ): results of an accretion disc model compared to observations; different symbols are for different optical luminosities (x: $L_o < 10^{44}$; o: $10^{44} \leq L_o < 10^{45}$; +: $L_o \geq 10^{45}$ ergs/s)

Figure 2. FWHM(Hα) versus FWHM(Hβ): results of an accretion disc model compared to observations; dashed line shows the effect of inclination.

On the contrary no correlation is observed between FWHM(CIV) and L(CIV), as is expected if CIV is emitted by a region different from Hβ and in particular, if the region is spherical.

Another prediction from the accretion disc model is that the width of Hα is smaller than the width of Hβ. Figure 2 shows the observed correlation between FWHM(Hα) and FWHM(Hβ) superposed on the theoretical results obtained for the same model. Very narrow lines can be explained by very small inclinations.

2. An orientation indicator

Considering the large scatter induced in the observed correlations by the inclination effects, it would help our understanding of AGN if we were able to measure the inclination in individual objects. An orientation indicator has been proposed: the ratio R of the radio core density flux to the lobe density flux (Orr and Browne 1982).

If R is a good measure of the inclination we expect some correlations with the line characteristics. We expect, for example, a correlation between R and FWHM(Hβ). Such a correlation has been claimed by Wills and Browne (1986), but figure 3 shows that, with a slightly different sample of AGN (it partly includes the Wills and Browne's sample), it is not convincing. Likewise, no correlation is observed between FW0I(Hβ) and R. On the other hand we do not expect any correlation between the equivalent width of Hβ and R since the line and the underlying continuum are supposed to be emitted by the disc (this is in agreement with observations), but

W(CIV) and W([OIII]) should decrease with R since these lines are emitted isotropically. For CIV such a trend is observed with low significance by Wills et al. (1991) but does not appear in my sample, while for [OIII] a variation is actually observed by Wills and Browne (1986) but Baker et al. (this volume) have shown that the correlation disappears if bright quasars are included. The situation, in conclusion, is quite confusing.

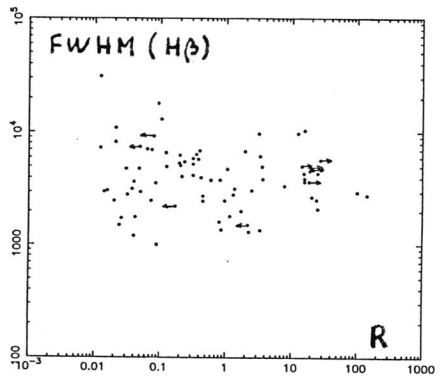
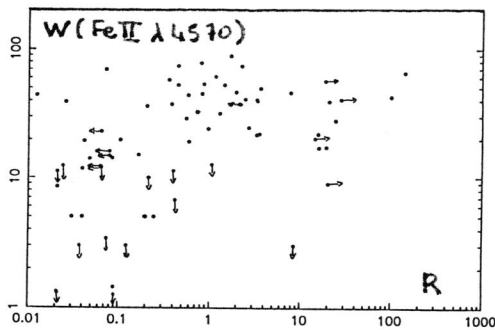

Figure 3. FWHM(Hβ) versus R observed in 80 AGN: no obvious correlation.

Figure 4. W(FeIIλ4570) versus R observed in 71 AGN with a significance $\sim 10^{-5}$.

A clue arises from the study of the FeII emission. Figure 4 shows the correlation observed between W(FeIIλ4570) and R; a similar relation stands for FeIIλ4570/Hβ and R. Although Jackson and Browne (1991) deny these correlations, they do observe an increase of FeII/[OIII] with R. Such correlations cannot be explained by inclination effects but they can be explained in the framework of the model of Norman and Miley (1984). These authors suggest that the emission region of FeII is closely associated with the jets responsible for the compact radio source. The higher the jet power, the denser the cocoon of the jet. This is in agreement with the results of collisional models which show that FeII/Hβ is an increasing function of the density and a decreasing function of the temperature (Joly 1987). Correlations between FeII and R are then explained as a function of an intrinsic variation of the radio core luminosity from object to object.

3. Conclusion

Statistical tests based on the lines Hα and Hβ support the accretion disc model. The large scatter in the correlations is due to inclination effects. Contrary to what is often assumed, R is probably not a good orientation indicator. In particular, it cannot be used to measure the inclination of the accretion disc. Indeed, if it partly reflects the orientation, the correlation with FeII emission demonstrates that it is also a function of the intrinsic radio power of the central compact source.

References

Collin S., 1987, Astron. Astrophys. 179, 60
Collin S., Dumont A.M., 1989, Astron. Astrophys. 213, 29
Dumont A.M., Collin S., 1990, Astron. Astrophys. 229, 313
Dumont A.M., Joly M., 1992, Astron. Astrophys. in press
Jackson N., Browne I.W.A., 1991, MNRAS 250, 422
Joly M., 1987, Astron. Astrophys. 184, 33
Norman C., Miley G., 1984, Astron. Astrophys. 141, 85
Orr M.J.L., Browne I.W.A., 1982, MNRAS 200, 1067
Wills B.J., Browne I.W.A., 1986, Astophys. J. 302, 56
Wills B.J., Fang D., Brotherton M.S., 1991, Heidelberg conf. on Physics of Active Galactic Nuclei, 3-7 june 1991

Orientation Effects in QSO Spectra

Paul J. Francis *

Abstract

Virtually all accretion disk models predict that QSOs observed from nearly edge-on should show extremely high equivalent-width emission lines. These are not seen. Either accretion disks must be significantly non-planar, or most edge-on QSOs must be concealed by an obscuring torus.

1. Model

If the UV-optical continuum emission of QSOs comes from an accretion disk, it will be emitted anisotropically. If in addition the line radiation is either isotropic, or anisotropic in a different way from the continuum radiation, then identical QSOs observed from different orientations will show different emission-line equivalent widths [2].

I assume that all QSOs have the same intrinsic line-to-continuum flux ratio, and that the line radiation is isotropic. Any magnitude-limited sample is strongly biased towards face-on QSOs, and this bias is taken into account using luminosity function information [1]. A wide variety of both thick and thin disk models have been used.

2. Results

A typical predicted equivalent-width distribution is compared with an observed distribution in the figure. Both are taken from [1]. Two discrepancies are evident. Firstly, the observed distribution has a broader, smoother peak than the prediction. This can easily be explained if there is an intrinsic dispersion in QSO equivalent-widths. Secondly, the model has a tail of very high equivalent-width QSOs not seen in the observations. This tail is significant at the 99% confidence level, for most accretion disk models, and for Lyman-α, C III] and Mg II as well as C IV.

3. Interpretation

The observed deficit of high equivalent-width lined QSOs cannot be a selection effect — such QSOs are the easiest to find using objective prism techniques. Introducing an

---

*Steward Observatory, University of Arizona, Tucson, AZ 85721, USA. The author was supported by a SERC/NATO Advanced Fellowship.

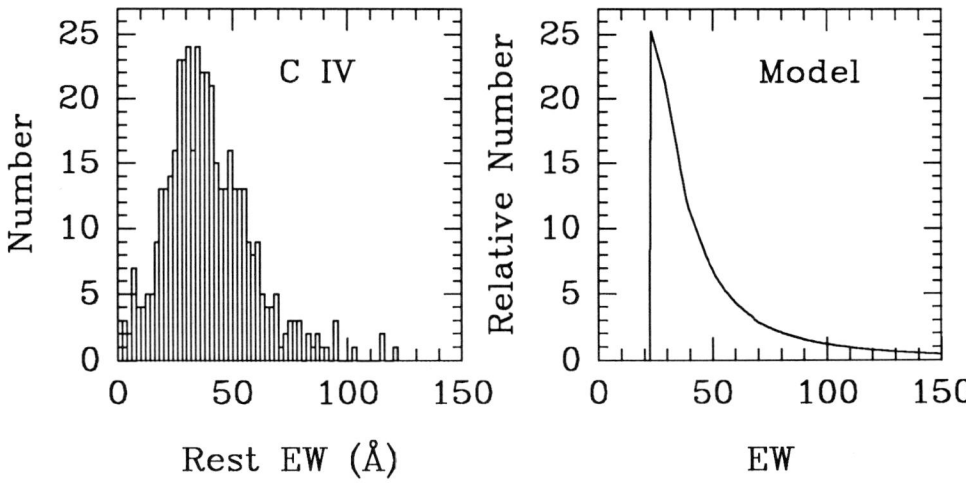

Figure 1. Observed (left) and predicted (right) rest-frame equivalent-width distributions for C IV (1549Å).

intrinsic dispersion in equivalent-widths only worsens the discrepancy. Line radiation anisotropy would only help if it were in the same sense as the continuum anisotropy, *i.e.*, if the line radiation appears strongest when the QSO is observed face-on. Most models with optically thick clouds predict anisotropy in the opposite sense, the exception being models in which the line emission comes from the accretion disk. Such models cannot however explain high ionisation lines.

If both the continuum source and broad-line region of most edge-on QSOs are concealed by an obscuring torus, the discrepancy would not occur. Without such a torus, the only accretion-disk models which do not overpredict the numbers of high equivalent-width QSOs are the least anisotropic, those in which the ratio of face-on to edge-on fluxes does not exceed ~ 5. Both thick and thin disk models are too anisotropic: what is required is a reprocessing atmosphere or an extremely non-standard disk.

References

[1] Francis, P. J., 1992, ApJ in press
[2] Netzer, H., Laor, A., & Gondhalekar, P. M., 1992, MNRAS, 254, 15

The Luminosity-Colour Distribution of Quasar Accretion Disks

David M. Caditz [*]

Abstract

Much effort has gone into analysing the now well-known luminosity evolution of the quasar population. Theoretical models, such as the accretion disk model, however, predict a spectral evolution which is more complicated than a simple overall luminosity shift [3], and more detailed spectral information is necessary to compare physical models with the data. We investigate the distribution and evolution of quasars on the color-flux plane in order to test for consistency with the thin accretion disk model.

1. Introduction

Much effort has gone into analysing the now well-known luminosity evolution of the quasar population. Theoretical models, such as the accretion disk model, however, predict a spectral evolution which is more complicated than a simple overall luminosity shift [3], and more detailed spectral information is necessary to compare physical models with the data. We investigate the distribution and evolution of quasars on the colour-flux plane in order to test for consistency with the thin accretion disk model.

2. Quasar Colours

It has been known since the first quasar surveys that active galactic nuclei (AGN) occupy a specific region of the two colour diagram, separate from most other galactic and extragalactic sources [1]. Consequently, UBV photometry has been used to select AGN in numerous surveys up to a redshift $z \sim 2.2$ where the L_α emission line and absorption edge are redshifted into the observed bands. Comparisons between colour-selected AGN and AGN selected by other criteria, e.g., radio flux, show that the region occupied by AGN on the colour diagram is intrinsic to the AGN population, and it is not artificially introduced by selection effects [4].

Characteristic quasar and Seyfert photometric colours are due to the large UV bump which many researchers have attributed to composite blackbody emission from an accretion disk surrounding a massive black hole [5],[7]. The strength and position

---

[*]Department of Physics, Montana State University, Bozeman, MT, 59715, USA.

of the UV bump produced by an accretion disk is a function of the central mass, M, and accretion rate, $\dot{M}$. It is useful to define a new parameter, the dimensionless accretion rate, $\dot{m} \equiv \dot{M}c^2/L_{Edd} = 4.4\dot{M}/M_8$ where, for the last equality, $\dot{M}$ is in units of $M_\odot/\text{yr}$ and M_8 is in units of 10^8 $M_\odot$. The flux and $U - B$ colour expected from a standard thin accretion disk are: $F(\nu) \propto \nu^{\frac{1}{3}} M_8^2 (\frac{\dot{m}}{M_8})^{\frac{2}{3}} I(\nu)$ and $U - B = -1.05 + 2.5 \text{Log}[\frac{I(\nu_B)}{I(\nu_U)}]$ where $I(\nu) \equiv \int_{x_{min}}^{x_{max}} \frac{x^{\frac{5}{3}} dx}{e^{(x/\phi^{\frac{1}{4}})} - 1}$. $x(r) = 0.081 \nu_{15} (\frac{M_8}{\dot{m}})^{\frac{1}{4}} r^{\frac{3}{4}}$. The dimensionless radius, $r = Rc^2/GM$, ranges from the innermost stable orbit around the black hole ($r_{min} = 6$ for a Schwarzschild black hole) to an outermost orbit where the disk is presumably unstable ($r_{max} \sim 1000$). The observed frequency ν_{15} is in units of 10^{15} Hz, $x_{min} = x(r_{min})$ and $x_{max} = x(r_{max})$. $\phi(x)$ is the relativistic correction factor for angular momentum transport at the inner disk edge. It is interesting to note that since x_{min} in the B band is less than x_{min} in the U band, there is a minimum colour such that $U - B \geq -1.05$.

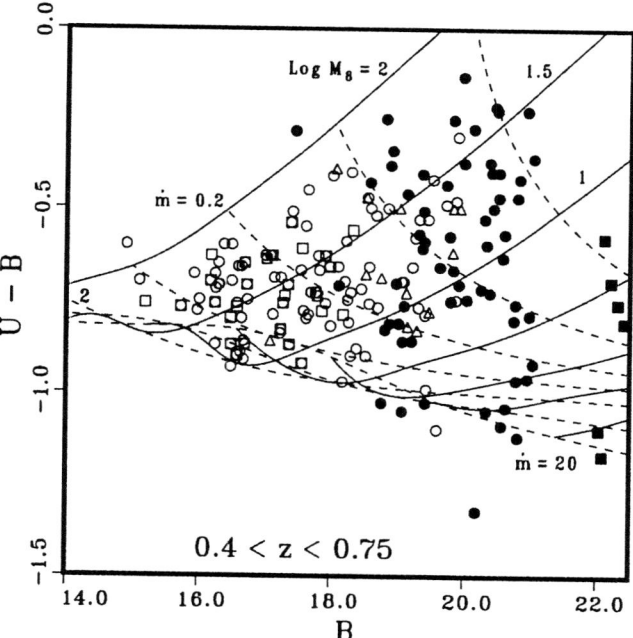

In Figure 1 we plot a grid of accretion disk $U-B$ colours and B fluxes for Schwarzschild thin disks of various central masses, M_8, and dimensionless accretion rates, $\dot{m}$. The solid lines are for constant mass from $M_8 = 10^{-1.5}$ to $M_8 = 10^{1.5}$ and the dashed lines are for constant dimensionless accretion rate from $\dot{m} = .002$ to $\dot{m} = 20$. A central mass, $M_8 = 10^{1.5}$ is a natural upper limit because larger central masses do not produce sufficient tidal forces to disrupt infalling stars, and at $\dot{m} = 20$ the disk is radiating at the Eddington luminosity at which the thin disk model is no longer self-consistent.

Also plotted on Fig. 1 are 151 observed colours and fluxes for quasars with redshifts $0.4 < z < 0.75$. These sources are from the AAT survey [2] (closed circles) and the Véron catalogue [6] (open circles). Out of the 151 sources, 143 fall on the accretion disk grid, and the distribution shows abrupt cut-offs at the natural mass and accretion rate limits. Sources with $B > 21$ are missing due to observational cut-offs; however, selection effects cannot explain the maximum mass and maximum dimensionless accretion rate cut-offs. We consider Fig. 1 to provide new evidence that the thin disk model applies to most quasars at these redshifts.

References

[1] Adam, G. 1985, *Astron. Astrophys.*, **61**, 225
[2] Boyle, B. J., Fong, R., Shanks, T., and Peterson, B. A. 1990, *MNRAS*, **243**, 1
[3] Caditz, D., Petrosian, V., and Wandel, A. 1991, *Ap.J.*, **372**, L63
[4] Moles, M., Garcia-Pelayo, J. M., Masegosa, A., Aparicio, A., and Quintana, J.M. 1985 *Astron. Astrophys.*, **152**, 271
[5] Sun, W. H., and Malkan, M. A. 1989, *Ap.J.*, **346**, 68
[6] Véron-Cetty, M.P., and Véron, P. 1989 *A Catalogue of Quasars and Active Galaxies, 4th Edition*, European Southern Observatory, Scientific Report
[7] Wandel, A., and Petrosian, V. 1988, *Ap. J. (letters)*, **329**, L11

Beams, Jets and Blazars

Magnetic Propulsion of Jets in AGNs

Mitchell C. Begelman [*]

Abstract

Magnetic forces seem likely to play the dominant role in both the acceleration and initial collimation of relativistic jets in AGNs. I describe recent developments in the theory of hydromagnetic jets and winds.

1. Introduction

Hydromagnetic propulsion as a mechanism for accelerating jets has become attractive largely through a process of elimination. Other mechanisms, such as acceleration by gas or radiation pressure, have been examined and found inadequate. The observational case for relativistic flow velocities, on both pc and kpc scales, continues to mount (see, e.g., [14], [19], and [3] for recent reviews). Models involving acceleration by radiation pressure would have to be stretched to extremes in order to account for the Lorentz factors $\sim 2-10$ which are needed to explain most instances of one-sidedness and superluminal motion [22]. Losses due to catastrophic cooling place even more severe constraints on acceleration by gas pressure. Recent observations of "intra-day" radio variability [23], [24] may require Lorentz factors as high as ~ 100 if catastrophic Compton losses are to be avoided [4].

Magnetic propulsion has other attractive features besides the ability to produce the high speeds indicated by observations. Chief among these is the tendency of magnetically driven flows to "self-collimate" due to the development of a predominantly toroidal magnetic field. Thus, it may not be necessary to invoke a "funnel" or external confining medium to explain the collimation of jets. The fact that magnetic propulsion is the favoured mechanism for protostellar jets adds the cachet of universality to the mechanism.

As a challenge to theorists, the problem of hydromagnetic propulsion has proven to be particularly tricky. Nevertheless, concerted attacks by several groups during the past few years have yielded important advances. Below I outline my personal view of the current state of affairs, gleaned through my collaboration with Zhi-yun Li (at JILA) and Tzihong Chiueh (now at National Central University, Taiwan).

---

[*]Joint Institute for Laboratory Astrophysics, University of Colorado and National Institute of Standards and Technology, Boulder, Colorado 80309-0440, USA. Also at Department of Astrophysical, Planetary, and Atmospheric Sciences, University of Colorado. The author's research was supported in part by NSF Grants AST88-16140 and AST91-20599, and NASA Grant NAGW-766.

2. Ingredients of Magnetic Propulsion

Qualitatively, the process of magnetic acceleration can be broken down into four stages — injection and crank, centrifugal sling, toroidal twist, and magnetic spring. At the base of the flow there must be a *crank*, a rotating anchor for the magnetic field. The crank might be an accretion disk threaded by open field lines [5] or the ergosphere of a rotating black hole [6]. By processes which are not well-understood (see § 4 below), matter is injected onto open field lines and accelerates away from the crank. A number of effects may contribute to this initial injection and acceleration phase, such as heating by some unsteady process like magnetic reconnection. But such dissipative energy injection is not essential for magnetic propulsion to occur. Instead, the initial acceleration may be dominated by the *centrifugal sling* effect of the rotating magnetic field. In this case, the magnetic field lines may be regarded as stiff wires emanating from the crank. Parcels of gas must corotate with the wires, but are free to slide along them. Provided that the centrifugal force resolved parallel to the field exceeds the gravitational force, the material will be accelerated outward, even in the absence of any gas pressure. This will occur for flow from a Keplerian disk if the field lines make an angle of more than 30° with the vertical [5].

As material is flung to increasing distances by centrifugal force, the tension required to keep it in corotation with the field also increases. Eventually the inertial forces must overcome the tension, and the field lines are "swept back" in the direction opposite to the sense of rotation, hence the term *toroidal twist*. Beyond this point, which essentially corresponds to the "Alfvén point" (the term has a precise technical meaning which we need not worry about here), centrifugal acceleration is inefficient. It is primarily the conditions at the Alfvén point which determine the energy flux in the wind or, equivalently, the "load" on the crank. Suppose the crank turns with angular speed Ω and the (conserved) magnetic flux associated with the wind is Φ. At the Alfvén radius, R_A, the characteristic flow speed is $v_A \sim \Omega R_A$. The total energy flux carried by the wind is then given roughly by $\dot{E} \sim \Phi^2 v_A / R_A^2$. To proceed further, we need to estimate R_A in terms of wind parameters such as density ρ or mass flux $\dot{M}$. In the nonrelativistic limit ($v_A \ll c$), the Alfvén point occurs where $(\rho v^2)_A \sim (B^2)_A$ is satisfied. Substituting $\dot{M} \sim \rho v R^2$, we obtain $\dot{E} \sim \dot{M}^{1/3} \Omega^{4/3} \Phi^{4/3}$. In the relativistic limit, the Alfvén point is located close to the light cylinder radius c/Ω, implying $\dot{E} \sim \Omega^2 \Phi^2 / c$. Thus, in the relativistic limit, the load on the crank is independent of $\dot{M}$!

Beyond the Alfvén point, the magnetic field is dominated by the toroidal component B_ϕ. As the field becomes ever more tightly wound, the magnetic tension force tends to collimate the flow about the rotation axis. At the same time, gradients in the magnetic pressure can lead to further acceleration of the flow. In this last phase of the propulsion process, the field behaves like a *magnetic spring*.

3. Energy Partition and Collimation

Magnetically propelled winds carry two forms of energy — Poynting flux (i.e., electromagnetic energy) and kinetic energy flux. Estimates for $\dot{E}$ given above correspond to the *total* energy of the flow; determining the relative amounts of the two forms of energy requires a much more subtle analysis. It turns out that relativistic flows have nearly all of their energy in electromagnetic form as they cross the Alfvén point. The subsequent conversion of Poynting flux to kinetic energy flux depends on the competition between magnetic pressure and magnetic tension forces, a competition which depends in turn on the geometry of the flow. The pressure force wants to separate the coils of the magnetic spring, thus accelerating the flow. However, because the flow is not perfectly collimated to a cylinder, there is a component of the tension force which tries to retard the acceleration. For highly relativistic flows with a perfectly radial (i.e., conical) geometry, the pressure and tension forces cancel almost exactly. This feature led early researchers (e.g., [20], [9], [12]) to conclude that magnetic propulsion alone could not produce relativistic flows dominated by kinetic energy. However, the case of radial flow geometry proves to be a singular one (e.g., [16] and references therein). Even a small degree of collimation towards the rotation axis (accompanied by local divergence of the flux surfaces at a rate which exceeds that of nested cones) leads to rough equipartition between electromagnetic and kinetic energy, at the very least. For certain types of flow geometry, nearly all of the energy ends up in kinetic form. This effect is illustrated in Fig. 1, devised by my collaborator Zhi-yun Li, where we compare the forces acting on a radial (Fig. 1a) and a collimated (Fig. 1b) flow. The radial case is only modestly relativistic, with Michel's "magnetization parameter" σ [20] set equal to 10. Yet, even here the pressure and tension forces are closely balanced, and only 17.6% of the energy goes into kinetic form. For the collimated case, we arbitrarily chose the shape of the magnetic flux surfaces to have the form $Z/R = (100 - R/R_L)^{-1}$, where R and Z are cylindrical coordinates about the rotation axis and R_L is the light cylinder radius. The dominance of pressure over tension in this case leads to significant acceleration outside the Alfvén point, with 99% of the total energy ultimately going into kinetic form.

Of course, the geometry of the flow is far from arbitrary. It is determined by the Grad-Shafranov (GS) equation, which describes force balance perpendicular to magnetic flux surfaces [17]. Although the GS equation is well-known in pulsar and Tokamak circles, its complexity is such that only a few exact solutions exist for boundary conditions appropriate to jet models (e.g., [5], [25], [7]). However, much has been learned through asymptotic analyses of the GS equation, which have been performed recently in both the nonrelativistic [10] and relativistic [8] limits. Asymptotes can be classified according to the asymptotic value of $B_p R^2$, where B_p is the poloidal magnetic field strength, and the asymptotic poloidal current density, j (Fig. 2). An important result of the asymptotic analysis is that "horizontal" and "conical" asymptotes (i.e., with $R/Z \to \infty$ and $R/Z \to const.$, respectively) are forbidden, effectively implying that

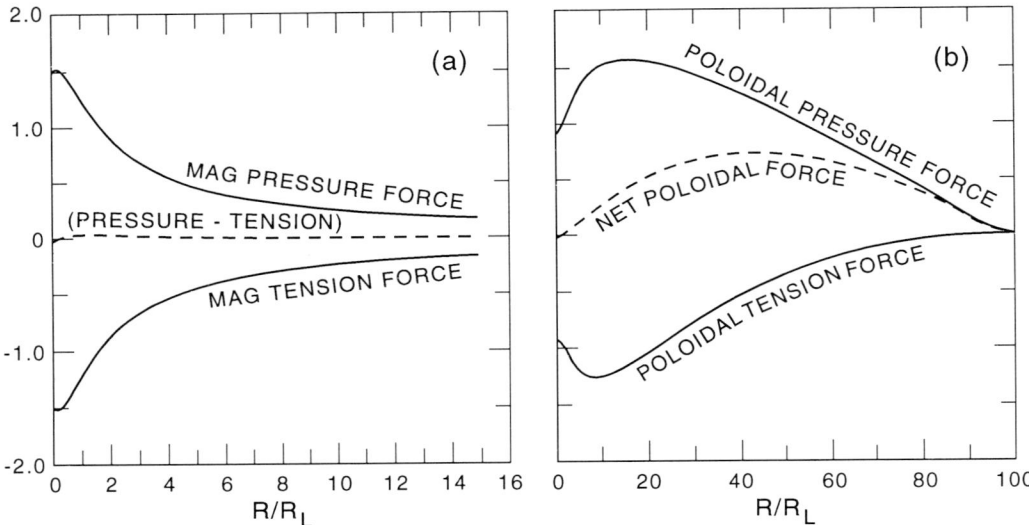

Figure 1. Comparison of forces acting on relativistic MHD winds with (a) radial and (b) collimated geometry. Details are given in text. Near balance between pressure and tension force in the radial case suppresses acceleration and leads to small ratio of kinetic energy to Poynting flux.

magnetically propelled flows must self-collimate. The only asymptotes which seem to be consistent with realistic boundary conditions are the ones labelled "cylindrical" and "current-free ($j = 0$) paraboloidal." In this case, "cylindrical" means that R is bounded, and does not exclude the possibility that the asymptotic flow is pinched [5] or is otherwise unsteady. "Paraboloidal" means that both R and Z tend to infinity, but that $R/Z \to 0$. Although the relationship of these asymptotes to the imposed boundary conditions is still unclear, my hunch is that the cylindrical asymptote applies when there is a sufficient external confining pressure (decreasing more slowly than Z^{-2} at large distances), whereas the wind will approach the paraboloidal asymptote when the external pressure is negligibly small.

A solution which approached the current-free asymptote at large distances would have very interesting properties. First, each flux surface would collimate towards the axis, but only very slowly — as a logarithm of the distance. Second, *all* of the energy in the flow would ultimately be converted into kinetic form, but again only at a logarithmic rate. We suspect that this is exactly what is happening in the wind from the Crab pulsar, which has been modelled as a magnetically propelled flow [11]. Finally, the acceleration in such a wind would be rather gradual, a feature which would obviously have important implications for jets in blazars. In the cylindrical case, the main uncertainties centre around stability, with violent kink or pinch modes a possibility .

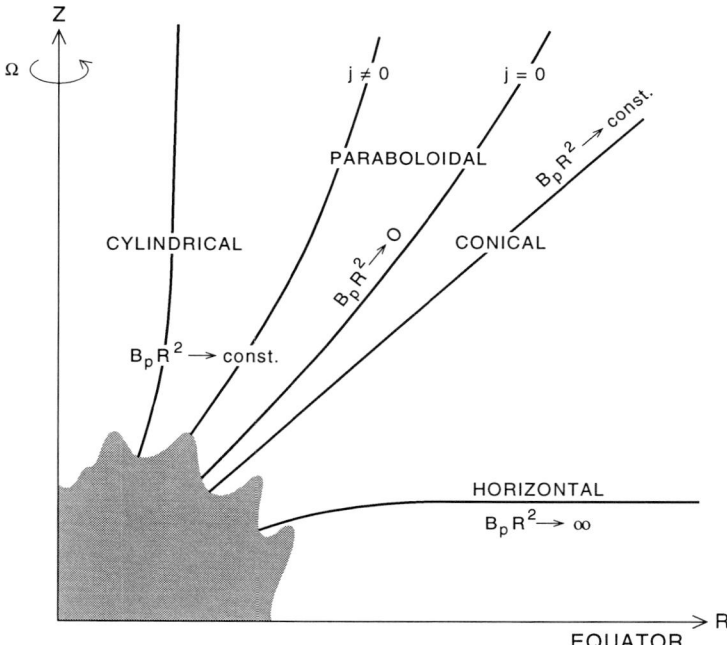

Figure 2. Schematic depiction of the possible asymptotic solutions of the Grad-Shafranov equation.

4. Production of Jets

Of the ingredients for jet production listed in §2, the "injection and crank" phase remains the most mysterious. Phinney [21], concentrating on flows propelled by a spinning black hole, suggested that the injected matter would consist of electron-positron pairs created by gamma-rays. Most other investigators have assumed that the flow is launched from rest near the surface of an accretion disk, using a self-similar wind solution as the starting point [5], [13], [18]. If the "centrifugal sling" is to overcome gravity, the field lines must come off the disk at an angle greater than 30° from the vertical. This creates a problem if the disk is treated as an infinitely thin current sheet [5], since a poloidal flux tube would then have to bend by a rather large angle while traversing the disk. A strongly curved flux tube is subject to a strong tension force, which will try to straighten it out. Königl and Lovelace et al. have ingeniously balanced the tension of the curved field by the drag due to field line slippage, either as a result of anomalous resistivity or ambipolar diffusion. They have shown that it is possible to construct self-similar models in which the interior structure of the disk is matched onto a wind solution. But I am not convinced that such a finely tuned balance will be established naturally. Time-dependent numerical simulations of these systems should be done as soon as possible.

Alternatively, the centrifugal sling might play only a supporting role in launching the flow. It is likely that the net magnetic flux advected inward by the disk contributes only a small fraction of the total magnetic energy present close to the disk. The rest would reside in a turbulent field with zero net flux, perhaps generated through instabilities inside the disk such as that described by Balbus and Hawley [1]. Emergence of a self-collimated jet from the messy dynamo field above the disk would require the reconnection and dissipation of most of the dynamo field; one might imagine that the energy liberated by reconnection could be used to launch the jet, by literally flinging blobs of gas onto open field lines. Since the gas would then start out with substantial kinetic energy, one need not worry about the angle at which the field lines join the disk.

An attractive conjecture is that the magnetized wind carries away most of the energy and angular momentum liberated during accretion [5]. To do this at each radius in a self-similar model requires that the poloidal magnetic field scale with radius as $B_p \propto R^{-5/4}$. But how does one obtain a field which increases so steeply with decreasing R? If the magnetic field is frozen into the disk and is advected inward, then one would expect $B_p \propto \sigma$, where σ is the disk surface density. In a steady-state disk with accretion rate $\dot{M}$, σ is related to the inflow speed v_R by $\dot{M} \sim \sigma v_R R$. To obtain the required scaling, the inflow speed would have to scale with R as $R^{1/4}$. Note that the standard α–model for disk viscosity predicts $\sigma \propto \alpha^{-1}(H/R)^{-1}R^{-1/2}$ (where H is the disk scale height), which falls far short of the required scaling for simple assumptions about H and α. The situation gets worse if one allows the field lines to slip.

The constraints are not quite so severe if one is simply interested in producing a powerful jet from the central regions of an accretion disk, or from the black hole ergosphere. Flux dragged inward by the accretion flow over a period of time might accumulate in the central regions. This flux cannot accumulate forever, since the pressure of the confined field would eventually stop the flow. But it is not clear that a steady-state solution in which outward slippage balances inward advection, as has been advocated by Königl and Lovelace et al., can be established in general. Instead, one might imagine a highly unsteady mode of field evolution, characterized by episodic reconnection of large regions of flux. It is tempting to identify the large-scale reconnection events with the production of superluminal knots in AGN jets [2].

5. Concluding Remarks

Twisted magnetic field lines provide an efficient way of extracting rotational energy from the central engines of active galactic nuclei. The resilience of the magnetic propulsion process is perhaps best illustrated by the response of a relativistic magnetized wind to radiation drag [15], which poses a particularly severe problem for other acceleration mechanisms. As radiation drag retards the acceleration of a flow near its source, it also increases the energy stored in the coiled magnetic field. Under certain conditions, the stored energy can be recovered once the flow reaches a region where the drag is small, giving back much of the acceleration that is initially robbed by the drag. Under other conditions, the drag may quench the flow entirely. It is an intriguing possibility that this kind of process might explain the production of well-collimated jets in a minority of AGNs, and their failure to form in the majority.

References

[1] Balbus, S. A., & Hawley, J. F. 1991, ApJ, 376, 214
[2] Begelman, M. C. 1992, in Astrophysical Jets, STScI Symp. Series Vol. 6, ed. D. Burgarella, M. Livio & C. O'Dea (Cambridge: Cambridge Univ. Press), in press
[3] Biretta, J. 1992, in Astrophysical Jets, STScI Symp. Series Vol. 6, ed. D. Burgarella, M. Livio & C. O'Dea (Cambridge: Cambridge Univ. Press), in press
[4] Blandford, R. D. 1992, in Astrophysical Jets, STScI Symp. Series Vol. 6, ed. D. Burgarella, M. Livio & C. O'Dea (Cambridge: Cambridge Univ. Press), in press
[5] Blandford, R. D., & Payne, D. G. 1982, MNRAS, 199, 883
[6] Blandford, R. D., & Znajek, R. L. 1977, MNRAS, 179, 433
[7] Camenzind, M. 1989, in Accretion Disks and Magnetic Fields in Astrophysics, ed. G. Belvedere (Dordrecht: Kluwer), 129
[8] Chiueh, T., Li, Z.-Y., & Begelman, M. C. 1991, ApJ, 377, 462
[9] Goldreich, P., & Julian, W. H. 1970, ApJ, 160, 971
[10] Heyvaerts, J., & Norman, C. A. 1989, ApJ, 347, 1055
[11] Kennel, C. F., & Coroniti, F. V. 1984, ApJ, 283, 694
[12] Kennel, C. F., Fujimura, F. S., & Okamoto, I. 1983, Geophys. Astrophys. Fluid Dyn., 26, 147
[13] Königl, A. 1989, ApJ, 342, 208
[14] Laing, R. A. 1992, in Astrophysical Jets, STScI Symp. Series Vol. 6, ed. D. Burgarella, M. Livio & C. O'Dea (Cambridge: Cambridge Univ. Press), in press
[15] Li, Z.-Y., Begelman, M. C., & Chiueh, T. 1992, ApJ, 384, 567
[16] Li, Z.-Y., Chiueh, T., & Begelman, M. C. 1992, ApJ, 394, 459
[17] Lovelace, R. V. E., Mehanian, C., Mobarry, C. M., & Sulkanen, M. E. 1986, ApJS, 62, 1
[18] Lovelace, R. V. E., Wang, J. C. L., & Sulkanen, M. E. 1987, ApJ, 315, 504
[19] Marscher, A. P. 1992, in Astrophysical Jets, STScI Symp. Series Vol. 6, ed. D. Burgarella, M. Livio & C. O'Dea (Cambridge: Cambridge Univ. Press), in press
[20] Michel, F. C. 1969, ApJ, 158, 727
[21] Phinney, E. S. 1983, Ph.D. thesis, University of Cambridge
[22] _____. 1987, in Superluminal Radio Sources, ed. J. A. Zensus & T. J. Pearson (Cambridge: Cambridge Univ. Press), 301
[23] Quirrenbach, A., Witzel, A., Krichbaum, T., Hummel, C. A., Alberdi, A., & Schalinski, C. 1989, Nature, 337, 442
[24] Quirrenbach, A., Witzel, A., Qian, S. J., Krichbaum, T., Hummel, C. A., & Alberdi, A. 1989, A&A, 226, L1
[25] Sakurai, T. 1985, A&A, 152, 121

MHD accretion-ejection model: X- and γ–rays and formation of relativistic pair beams

Guy Pelletier [*]   *Gilles Henri* [*]   *Jacques Roland* [†]

Abstract

Gamma ray emission from extragalactic sources is interpreted as the Doppler boosted annihilation and Inverse Compton radiation from a relativistic electron-positron beam in the frame of the two-flow model. In the case of 3C279, the high luminosity and the rapid variability of gamma ray emission suggest a relativistically moving source, but even so the compactness cannot be smaller than unity at light week scale with a reasonable Doppler factor. This supports the two-flow model of extragalactic radio sources, where the small scale emission comes from a relativistic electron-positron beam, heated by a MHD jet responsible for the large-scale (kpc) radio structures.

1. Introduction

The GRO satellite has detected intense gamma ray emission from several Active Galactic Nuclei and quasars. Remarkably, all of them are associated with a flat spectrum radio source, whose radio spectral index α_r is smaller than 0.5, and half of them exhibit known or probable superluminal motions (the others have not been observed at different epochs in VLBI). Just like the commonly invoked Doppler beaming amplification of radio emission, the high γ–ray luminosity suggests also that the emitting source is moving relativistically.

For 3C279 in particular, the spectrum reported by Hermsen & al. show a maximum emission per logarithmic energy interval around 10 MeV, with a photon spectral index of approximately 1.5 below the turn-over frequency and approximately 2 above it. A rapid flare has been observed with an increase of the luminosity by a factor 5 on a time scale of 2 days. We show that these observations imply a compactness of order unity even with a Lorentz factor of 10, and suggest the existence of a more compact region

---

[*]Laboratoire d'Astrophysique de l'Observatoire de Grenoble, BP 53X, F38041 Grenoble Cedex, France

[†]Institut d'Astrophysique, 98 bis bd Arago, F75014 Paris, France and Leiden Observatory, P.O. Box 9513, NL2300 RA Leiden, The Netherlands

where pair production can take place. This is interpreted in the frame of the two-flow model, where the high energy and the small scale radio emission are produced by a relativistic electron-positron beam, whereas the large scale (kpc) emission is due to a subrelativistic electron-proton jet emitted by a magnetohydrodynamic process from an accretion disk.

2. A model for the 3C279 γ-ray emission

The two-flow model basically explains the structure of extragalactic radio sources by the existence of two different fluids:

a) a relativistic $e^+ - e^-$ beam with a Lorentz factor γ_b between 3 and 20, giving rise to superluminal motion at VLBI (pc) scale.

b) a subrelativistic $p^+ - e^-$ jet carrying most of the kinetic luminosity at a speed $v_j \simeq 0.4c$, responsible for the large scale structure and the formation of hot spots.

The subrelativistic jet can be produced by a MHD ejection-accretion structure from an magnetized accretion disk around a central black hole. MHD instabilities can easily produce a strong Alfvèn turbulence capable of accelerating electrons at very high Lorentz factors, of the order 10^3. In the soft photon field of the accretion disk, these particles will cool by Inverse Compton (IC) process, producing enough X and γ photons to get a high compactness. Thus intense pair production can take place. We have shown recently [HP92] that if the turbulence is distributed along the jet far enough from the central source, the pair plasma can be efficiently heated and eventually escape with a relativistic velocity, overcoming the Compton drag problem.

Close to the central source, the pair plasma is opaque to γ photons and pair annihilation is inefficient, since most of the produced γ photons will create new pairs. However, the expansion will cause both the soft photon density to decrease and the transverse radius of the jet to increase, until the beam becomes transparent to γ-rays. At this place, pair annihilation takes place and one expects intense γ-rays emission both from annihilation and IC process. This ceases when most of the pairs have annihilated, defining the size of the emitting region.

Assuming $H_o \simeq 100$ km s^{-1} Mpc^{-1} and $q_o \simeq 0.5$, the apparent integrated luminosity of 3C 279 between 100 MeV and 10 GeV is [Ha92] $L_{\gamma,ap} \simeq 2 \times 10^{48}$ erg s^{-1}.

This value corresponds to the high activity state, for which the luminosity is at least five times as large as in the low activity state. A flare has been observed with a time scale δt of 2 days by the EGRET team [K92]. COMPTEL observations have measured a luminosity of about 10^{47} erg.s^{-1} in the range 1 MeV- 3 MeV. Taking an upper limit of $c\delta t$ for the size of the emitting region, these raw data would lead to a very high compactness for $\sim 1\,\mathrm{MeV}$ photons

$$l = \frac{L_\gamma^< \sigma_T}{4\pi m_e c^4 \delta t \Delta\epsilon} \simeq 50. \tag{1}$$

$L_\gamma^<$ being the plasma frame luminosity of few MeV photons in the interval $\Delta\epsilon\, m_e c^2$.

Such a high value can be avoided if the source is moving relativistically towards the observer with a Lorentz factor γ_b and a corresponding Doppler factor $\delta = \gamma_b(1 - (1 - \gamma_b^{-2})\cos\theta)^{-1}$. Then the observed frequency is blueshifted by a factor δ and the observed integrated luminosity is related to the intrinsic luminosity by $L_{\gamma,ap} = \delta^4 L_\gamma$. In the case of 3C279, a superluminal motion in the wide range $v_{ap} = 2 - 10c$ has been observed.

Assuming that the emitting region is a portion of cylinder of radius r_b and height h in the observer frame, it would have in the comoving plasma frame a height $\frac{h}{\gamma_b}$, a lateral area $S_b = 2\pi r_b h \gamma_b^{-1}$ and a volume $V_b = \pi r_b^2 h \gamma_b^{-1}$. For such a geometry, the optical depth for few MeV photons with respect to photon-photon pair production can be written as

$$l = \frac{L_\gamma^< \sigma_T \gamma_b}{2\pi h m_e c^3 \Delta\epsilon}. \tag{2}$$

Putting the constraint $l \simeq 1$ gives

$$h \simeq \frac{L_\gamma^< \sigma_T \gamma_b}{2\pi m_e c^3 \Delta\epsilon} \simeq 1.6 \left(\frac{L_{\gamma,ap}^<}{4\ 10^{47}}\right) \left(\frac{\delta}{5}\right)^{-3} \left(\frac{\gamma_b}{10}\right) \text{ light days.} \tag{3}$$

The variation time scale can in turn be interpreted as the time taken by the relativistic pairs to annihilate. Assuming that only low energy pairs do annihilate with a rate $R_{ann} = \frac{3}{8}c\sigma_T$, this constrains the low energy pair density:

$$n_\pm \simeq \frac{8\gamma_b}{3h\sigma_T} \simeq 10^9 \left(\frac{\gamma_b}{10}\right)\left(\frac{1}{h\ l.d.}\right) \text{ cm}^{-3} \tag{4}$$

The luminosity in the annihilation line is

$$L_{ann} \simeq \left(\frac{3}{8}\sigma_T n_\pm^2 c\right)(\pi r_b^2 h \gamma_b^{-1})(m_e c^2) \tag{5}$$

Combining equations (3),(4) and (5), and noting that $L_{ann} < L_\gamma^<$, one gets an upper limit of the radius of the beam, $r_b < \gamma_b^{-1} h$, that is, the emitting region must be a rather thin beam with an aspect ratio $\rho_a \equiv r_b/h < \gamma_b^{-1}$. Conversely, it could appear nearly spherical in the plasma frame. The annihilation radiation would appear as a broad structure around 0.5δ MeV, which is compatible with present observations of 3C279 and 3C273.

The high energy power law spectrum detected by EGRET can be produced by IC boosting of soft photons, either from the accretion disk [D92] or from the synchrotron emission [M92]. For an optically thin source, a γ-ray spectral index of 1 (in energy) would be produced by a power law distribution of electron Lorentz factors with an index $p = 3$. This can be obtained in the downstream flow of a strong shock by the radiation cooling of a γ^{-2} distribution. However, opacity effects can modify this value: indeed, if the region is marginally thin for photons with $\epsilon \sim 1$, one would expect it to be optically thick to higher energy photons, which interact predominantly with X photons of energy $\frac{1}{\epsilon}m_e c^2$ with an optical depth $\tau_{\gamma\gamma} \sim \frac{n(1/\epsilon)}{\epsilon}\sigma_T r_b$. In the latter case, the observed γ-ray spectral index is $\alpha_\gamma = 2\alpha_X$ [S84]. The observed value $\alpha_\gamma \simeq 1$ should correspond to $\alpha_X = 0.5$, close to that observed, and would be produced by a distribution of index $p = 2$. Such a power law could be due to the Compton cooling of a monoenergetic injected distribution of pairs.

References

[D92]   Dermer, C.D., Schlickeiser, R., Mastichiadis, A., 1992, *Astr. Ap.*, **256**, L27

[Ha92]   Hartman, R.C., Bertsch, D.L., Fichtel, C.E., Hunter, S.D., Kwok, P.W., Mattox, J.R., Sreekumar, P., Thompson, D.J., Kniffen, D.A., Liu, Y.C., Michelson, P.F., Nolan, P.L., Schneid, E., Kanbach, G., Mayer-Hasselwander, H.A., von Montigny, C., Pinkan, K., Rothermel, H. & Sommer, M., 1992a, *IAU Circ.* 5477; *ibid.* + Chiang, J., 1992b, *IAU Circ.* 5519; *ibid.*1992c, *Ap. J. (Letters)* **385**, L1

[HP92]   Henri, G. & Pelletier, G., 1991, *Ap. J. (Letters)* **383**, L7

[He92]   Hermsen, W., Aarts, H.J.M., Bennett, K., Bloemen, H., de Boer, H., Collmar, W., Connors, A., Diehl, R., van Dijk, R., den Herder, J.W., Lichti, G.G., Lockwood, J.A., Macri, J., Mc Connell, M., Morris, D., Ryan, J.M., Schönfelder, V., Simpson, G., Steinle, H., Strong, A.W., Swanenburg, B.N., de Vries, C., Webber, W.R., Williams, O.R. & Winkler, C., 1992, *Astr. Ap. Suppl.* in press

[K92]   Kanbach, G., Mayer-Hasselwander, H.A., von Montigny, C., Pinkan, K., Rothermel, H., Sommer, M., Bertsch, D.L., Fichtel, C.E., Hartman, R.C., Hunter, S.D., Kniffen, D.A., Kwok, P.W., Mattox, J.R., Sreekumar, P., Thompson, D.J., Liu, Y.C., Michelson, P.F., Nolan, P.L. & Schneid, E., 1992, *IAU Circ.* 5431

[M92]   Maraschi, L., Ghisellini, G., Celotti, A., 1992, *Ap.J (Letters)* **397**, L5

[S84]   Svensson, R., 1984, in *X-ray emission from Active Galactic Nuclei*, Edited by J.Tümper and W.Brinkmann, Max-Planck Institut, Garching

Relativistic Electron-Positron Beams in AGN: Construction of Transonic Solutions

Gilles Henri and Guy Pelletier *

The existence of relativistic motion in extragalactic radio sources is directly established by the observation of apparent superluminal velocities on VLBI scales. However, there is no observational evidence for relativistic motion on kpc scales. The construction of fully relativistic jets meets considerable difficulties, due essentially to the Compton drag problem in the very dense photon field surrounding the central object. This has led Sol et al. (1989) to propose the two-flow model, which basically explains the structure of extragalactic radio sources by the existence of two different fluids:

a) a relativistic $e^+ - e^-$ beam with a Lorentz factor γ_b between 3 and 20, giving rise to superluminal motion at VLBI (pc) scale. It carries away only a small fraction of the total kinetic luminosity.

b) a subrelativistic $p^+ - e^-$ jet carrying most of the kinetic luminosity at a speed $v_j \simeq 0.4c$, responsible for the large scale structure and the formation of hot spots.

We have proposed recently Henri & Pelletier (1991) that the pair beam can be produced and accelerated by 2nd order Fermi acceleration in the Alfvèn waves turbulence spectrum generated by the MHD jet, and can account for the hard X- and γ–rays emission of AGN's Pelletier, Henri & Roland, this volume). Here we further investigate the hydrodynamics of such beams. Assuming that acceleration and cooling are much faster than the typical escape time, the isotropized distribution function $\rho(\gamma)$ of the pairs is a solution of the Fokker-Planck equation (Henri & Pelletier 1991):

$$\frac{\partial}{\partial \gamma}\gamma^2 D_{\rm a}\frac{\partial}{\partial \gamma}\frac{\rho}{\gamma^2} + \frac{\partial}{\partial \gamma}A\gamma^2 \rho = 0 \qquad (1)$$

where $D_{\rm a}$ is the stochastic energy diffusion coefficient and $A = \frac{4}{3}\sigma_{\rm T}cW_{\rm ph}$ describes the cooling in the soft photon field of energy density $W_{\rm ph}$. If the Alfvèn waves have a power law energy distribution of index α, then $D_{\rm a} = D_0 \gamma^\alpha$ and the solution is a "pile-up" distribution peaking at a characteristic energy $\bar{\gamma} = ((3-\alpha)D_0/A)^{\frac{1}{(3-\alpha)}}$, for which the acceleration time is equal to the cooling time. In a soft photon field of typical energy $\epsilon_s m_e c^2$, particles of energy above $\gamma_{\rm thr} = (2\epsilon_s/3)^{-1/2}$ will produce by the IC process γ photons above the pair creation threshold. For the above pile-up

---

*Laboratoire d'Astrophysique de l'Observatoire de Grenoble, BP 53X, F38041 Grenoble Cedex

distribution, one can derive a pair creation rate:

$$\nu^+ \equiv \frac{1}{n_*}\frac{dn_*}{dt} = A\gamma_{\text{thr}}^2 \left(\frac{\gamma_{\text{thr}}}{\bar{\gamma}}\right)^\alpha \exp\left(-\left(\frac{\gamma_{\text{thr}}}{\bar{\gamma}}\right)^{(3-\alpha)}\right) \quad (2)$$

where n_* is the rest particle density. This creation rate increases very rapidly with $\bar{\gamma}$.

In the case of a stationary quasi-1D flow along the z-axis of a Schwarzschild metric, with a cross-section S(z), the general relativistic hydrodynamics equations reduce to:

$$\frac{1}{S(z)}\frac{\partial}{\partial z}S(z)\sqrt{B}n_*\beta_b\gamma_b = \nu^+ n_*$$

$$(p+e)\frac{\partial}{\partial z}\ln\sqrt{B}\gamma_b + \frac{\partial p}{\partial z} = \frac{\mathcal{F}_r^z}{\gamma_b\sqrt{B}}$$

where p and e are respectively the pressure and energy density of the plasma, γ_b and β_b its bulk Lorentz factor and velocity, $\mathcal{F}_r^z$ the spatial component of the 4-force density acting on the fluid due to radiation pressure and $B = g_{00} = 1 - \frac{2GM}{c^2 z}$. For a highly relativistic fluid, $e = 3p \simeq n_*\bar{\gamma}$. This system exhibit a solar wind-like critical point, at which the regularity condition fixes the value of ν^+ around the typical cooling time. By virtue of (2), this fixes $\bar{\gamma} \sim \gamma_{\text{thr}}/(\ln\gamma_{\text{thr}})^{\frac{1}{(3-\alpha)}} \sim 20$ for $\epsilon_s = 10^{-4}$. The adjustment to this value can be physically controlled by the fact that an increase of pair density lowers the Alfvèn velocity, which in turn reduces the acceleration coefficient D_0 and thus $\bar{\gamma}$.

Once the sonic point is passed, a subsequent adiabatic expansion will produce an asymptotic Lorentz factor $\gamma_{b\infty} \simeq \bar{\gamma}$. If the Compton drag is too high however, it could reduce the value of $\gamma_{b\infty}$ down to 2 or 3 Phinney (1987). This would explain the range of observed superluminal velocities.

References

Henri, G. & Pelletier, G., 1991, *Ap. J. (Letters)* **383**, L7
Phinney, E.S. 1987 in *Superluminal Radio Sources*, ed. J.A. Zensus and T.J. Pearson (Cambridge University Press), 301
Sol, H., Pelletier, G., and Asséo, E., 1989 *MNRAS* **237**, 411

Properties of Relativistic Jets

Stefan Appl and Max Camenzind [*]

The most favoured models for the collimation of jets are magnetic (Begelman, these proceedings). Jets consist of nested magnetic surfaces which are held together by the stresses of the toroidal field, which is equivalent to a poloidal current. The plasma streams along these magnetic surfaces, which are asymptotically collimated towards the axis. Collimated jets are essentially one-dimensional objects, whose structure will be investigated. Details are given elsewhere (Appl and Camenzind, A&A in press).

The asymptotic force-balance. The structure of relativistic magnetized jets follows from the force-balance across the magnetic surfaces. Neglecting inertial and pressure forces the equilibrium is determined by electromagnetic forces alone, $\rho_e \mathbf{E} + \frac{1}{c}\mathbf{j} \times \mathbf{B} = 0$. In contrast to Newtonian jets the electric field plays an important rôle in determining the equilibrium. It arises from the rotation of the poloidal magnetic field, $E_\perp = R\Omega^F B_p/c$. It is perpendicular to the magnetic surfaces Ψ. The poloidal field is anchored in a crank, e.g. the inner part of an accretion disc, and rotates with the angular velocity Ω^F of the footpoints. Differential rotation of the crank is neglected. The light cylinder (LC) is given by $R_L = c/\Omega^F$. This leads to the following equation for the equilibrium,

$$\left(1 - \frac{R^2}{R_L^2}\right) \nabla_\perp \frac{B_p^2}{8\pi} - \frac{B_p^2 R}{2\pi R_L^2} \nabla_\perp R + \frac{1}{2\pi c^2 R^2} \nabla_\perp (cRB_\phi/2)^2 = 0. \quad (1)$$

Introducing dimensionless quantities for the cylindrical radius, $x = R/R_L$, poloidal magnetic pressure, $y = (B_p^2/8\pi)R_L^4/\Psi_{max}^2$, and current, $I = \frac{c}{2}RB_\phi/I_{max}$, the force-balance equation (1) becomes for $\partial_z \ll \partial_R$

$$(1 - x^2)\frac{dy}{dx} - 4xy + \frac{g}{4\pi x^2}\frac{d}{dx}I^2 = 0. \quad (2)$$

The various normalizing constants have been collected into the coupling constant g for the current, $g = 2I_{max}^2 R_L^2/c^2\Psi_{max}^2$, which cannot exceed unity by a large amount. Though in reality determined by the conditions at the foot of the jet, the current is treated here essentially as a free function, reflecting our poor knowledge about the central engine (CE). Equation (2) is supplemented by the defining equation for the flux surfaces, $\nabla_\perp \Psi = RB_p$, or in dimensionless form,

$$\frac{d\Psi}{dx} = x\sqrt{8\pi y}. \quad (3)$$

---

[*]Landessternwarte Heidelberg, Germany

Strictly speaking, the current in Eq. (2) is a function of the flux surfaces, $I(\Psi)$. However, when as is commonly the case in fusion research, the current is taken as a function of radius, $I(x)$, Eqs. (2) and (3) decouple, and the solution can be given in closed form. For jets that are wider than the LC, as is the case in real sources, Eq. (2) possesses a critical point at the LC. The requirement of regularity at $x = 1$ determines the equilibrium entirely. For jets remaining within the LC, pressure equilibrium with the ambient medium determines the structure of the jet. The general solution for both cases is given by

$$y(x) = \frac{g/4\pi}{(1-x^2)^2} \int_x^{\min(x_{\rm jet},1)} \frac{1 - x'^2}{x'^2} \frac{\rm d}{{\rm d}x'} I^2(x') \, {\rm d}x'. \qquad (4)$$

Integrating Eq. (3) then gives the solution for the magnetic surfaces. Ψ being normalized to $\Psi_{\rm max}$ leads to $\Psi(x_{\rm jet}) = 1 = \int_0^{x_{\rm jet}} x\sqrt{8\pi y}\,{\rm d}x$. Together with Eq. (4) this provides a relation between the jet radius $x_{\rm jet}$ and the coupling constant g. For increasing g, the jet radius decreases, i.e. we are dealing with a true pinch.

Constraints for the current. The fact that y is a non-negative quantity poses some restrictions on possible current distributions $I(x)$ for relativistic screw pinches: The current can nowhere in the body of the jet drop to zero (it can however in the boundary layer, which is not force-free), with the consequence that only part of the current can flow back within the jet. A maximum of the current, if present, has to be outside the LC. Monotonically increasing $I(x)$ and $I(\Psi)$, in turn, are not constrained. The constraints arise as a consequence of taking the current as a free function.

The plasma velocity. Sufficiently far outside the LC the plasma rotates with constant angular momentum, so that the toroidal motion can be neglected. The MHD condition then leads to a relation between the poloidal velocity and the magnetic structure, $B_\phi \approx -xB_{\rm p}/\beta_{\rm p} = E_R/\beta_{\rm p}$. Since observed VLBI jets have diameters of the order of 100 $R_{\rm L}$, this approximation does not imply a serious restriction. The poloidal plasma velocity can, with the help of solution (4), be given in closed form. The above relation can be used to write the equilibrium in another way, eliminating the electric field,

$$\frac{1}{8\pi} \frac{\rm d}{{\rm d}R} \left(B_{\rm p}^2 + B_\phi^2/\gamma^2 \right) + \frac{B_\phi^2/\gamma^2}{4\pi R} = 0. \qquad (5)$$

Compared to the Newtonian case the toroidal field is reduced by the Lorentz factor γ, and therefore the confining tension, too. The electric field counteracts the collimation.

The jet boundary. The sum of the stresses on a fluid element at the boundary has to vanish, $\frac{1}{8\pi}(B_z^2 + B_\phi^2 - E_R^2)_{\rm jet} = (P + \frac{1}{8\pi}B_\phi^2)_{\rm ext}$. The total flux defines the outer boundary of the jet, therefore there is no poloidal field in the ambient medium. Pressure equilibrium for jets wider than the LC can only be fulfilled by appropriately adjusting the toroidal field in the external medium, since their internal structure is completely determined by the regularity condition. This is equivalent to a current sheet in the boundary layer, which flows in the direction opposite to the jet current.

If the ambient pressure is high enough the entire current is screened by a surface current, and for even higher pressures the jet radius is no longer determined by the strength of the current g, but by the ambient pressure.

A scenario. Near the CE, where the interstellar pressure is higher, the jet might be confined by the latter, the entire current being screened. Such a current sheet is ideally suited for particle acceleration, leading ultimately to limb brightening of the jet. The jet diameter increases acccording to the external pressure stratification, until the pressure is small enough that the jet becomes self-confined by its toroidal field. The shape of the jet is then cylindrical.

A Massive Binary Black Hole in 1928+738?

Nico Roos [*]   *Jelle S. Kaastra* [†]   *Christian A. Hummel* [‡]

Abstract

We apply the binary black hole model to explain the wiggles in the milliarcsec radio jet of the superluminal quasar 1928+738 (4C73.18) observed with VLBI at 1.3 cm wavelength by Hummel *et al.* (1992). The period and amplitude of the wiggles can be explained as due to the orbital motion of a binary black hole with mass of order 10^8 solar masses, mass ratio larger than 0.1 and orbital radius $\sim 10^{16}$ cm. The jet's inclination to the line of sight should be small, confirming the standard interpretation of superluminal motion and one-sidedness as due to relativistic motion in a direction close to the line of sight. The small orbital radius suggests that the binary has been losing a significant amount of orbital energy during the last 10^7 years, possibly by interaction with the matter which is flowing through the active galactic nucleus.

1. Introduction

Galaxy mergers must have been a common phenomenon especially during the collapse and virialisation of rich groups and clusters of galaxies. These mergers lead to the formation of massive binary black holes in galactic nuclei if black holes of $10^{7-9} M_\odot$ are formed in the nuclei of most bright galaxies at redshifts of about 2. A massive binary black hole (MBBH) may manifest itself by Lens-Thirring precession of a jet emitted along the spin axis of one of the holes (Begelman *et al.* 1980, hereafter BBR). When the binary separation is small a more rapid (smaller scale) wiggling of the jet caused by the orbital motion may become apparent (Kaastra and Roos, 1992). Such a wiggle may recently have been observed in the milliarcsec jet of the quasar 1928+738 by Hummel *et al.* (1992).

2. Modelling the wiggles in 1928+73

Hummel *et al.* (1992) have monitored the milliarcsec (mas) core-jet structure in the core-dominated quasar 1928+738 ($z = 0.3$) during 5 years using VLBI at 1.3 cm. They find that the jet exhibits ballistic superluminal motion along a sinusoidally curved jet ridge line. They obtained a good fit to their observations using a sine

---
[*]Sterrewacht Leiden, Niels Bohrweg 2, 2300 RA Leiden, The Netherlands.
[†]SRON, Niels Bohrweg 2, 2300 RA Leiden, The Netherlands.
[‡]U.S. Naval Observatory, 34th & Massachusetts Avenue, Washington, DC 20390, U.S.A

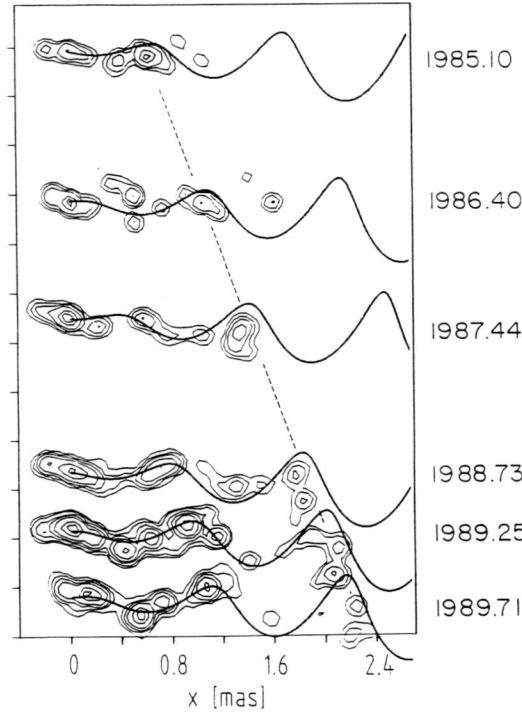

Figure 1. Model of a precessing relativistic jet superposed on the VLBI maps (at 1.3 cm, 0.1 mas Gaussian restoring beam) of the jet in the superluminal quasar 1928+738 (4C73.18), obtained at six different epochs by Hummel et al. (1992, see their figure 10). The precession is supposedly due to the orbital motion of the hole emitting the jet. The jet material moves along straight trajectories with velocity $v_{jet} = 0.95c$ and mean inclination to the line of sight $i = 15°$. The opening half angle of the precession cone is $\psi^{intr} = 2°$. The orbital period is 3.2 year, and the orbital phase at epoch 1989.71 was 75°.

wave model with wavelength $\lambda \sim 1.06$ mas, amplitude $A \sim 0.09$ mas and a phase shift $\mu \sim 0.28$ mas yr^{-1}. The observed phase shift implies an apparent jet velocity of $3.3c\ h^{-1}$ for a flat universe ($\Omega = 1$), where h is Hubble's constant in units of 100 km s^{-1}Mpc^{-1}. The observed period λ/μ is related to a proper period P in the rest frame of the quasar by $\lambda/\mu = P(1+z)$, yielding $P = 2.9$ year.

It seems highly unlikely that this rapid periodic motion would be due to geodetic precession of the spin axis of the hole emitting the jet. The precession period for a binary with mass ratio m/M and separation $r_{16} = r/10^{16}$ cm is given by $t_{prec} = 600\ r_{16}^{5/2} M_8^{-3/2}(M/m)$ yr, where $M_8 = M/10^8 M_\odot$. The timescale for loss of orbital energy via emission of gravitational waves is $t_{grav} = 2.9 \times 10^5 (M/m) M_8^{-3} r_{16}^4$ yr. A precession period of only 3 years would imply a gravitational lifetime for the binary which is extremely short. It is more realistic to assume that the observed period is associated with the orbital motion of a binary with separation $r_{16} = 1.4\,(M_8 + m_8)^{1/3}(P/3\,{\rm yr})^{2/3}$. We assume that the jet ejecta move on ballistic orbits with velocity $v_{jet} = \beta c$. The orbital motion of the hole from which the jet is emitted causes a wiggle which is due to both the motion of the origin of the jet and a periodic modulation of the jet velocity. The latter, which is most important when the inclination i of the jet to the observer's line of sight is small, causes an apparent precession of the jet over a cone-like structure with (intrinsic) opening half angle which can be approximated by $\sin\psi^{intr} = v_{orb}\cos\chi/v_{jet}$, where v_{orb} is the orbital velocity of the hole emitting the jet and χ is the angle between the orbital angular momentum and the jet velocity. For small $\psi^{intr} < i$, and a circular orbit, the observed opening angle $\psi^{obs} \approx \psi^{intr}/\sin i$, is given by

$$\psi^{obs} = 2.2°\ r_{16}^{-1/2} M_8^{1/2}(m/M)\left(\frac{M+m}{M}\right)^{1/2}\frac{\cos\chi}{\beta\sin i} \qquad (1),$$

where we have inserted for v_{orb} the orbital velocity of the most massive hole. Note that the factor m/M drops from the equation in the case where the jet is emitted by the secondary hole, or, more generally, in the case where the extra periodic component to the jet velocity is equal to the Kepler velocity at a distance r from the (primary) hole. In fig. 1 we give a model of a precessing relativistic ballistic jet. The parameter values of the model represent good guesses rather than a best fit to the observations.

We may invert eq. 1 to estimate i from the observed period and opening half angle ($\psi^{obs} \sim 8°$, see fig. 1), yielding $i \approx 15°\beta^{-1}\cos\chi(P/3\,{\rm yr})^{-1/3}(\psi^{obs}/8°)^{-1}M_8^{1/3}(m/M)$. The small inclination to the line of sight inferred for 1928+738 is consistent with the standard interpretation of the superluminal velocities observed in this quasar. This expression for i can also be used to argue that $m/M \gtrsim 0.1$ when the jet wiggle is due to the orbital motion of the primary hole, because otherwise i would have to be very small and we would be dealing with a very exceptional object.

3. Binary Evolution and Activity.

The formation of long-lived "wide" MMBHs as a result of mergers between galaxies containing MBHs has been discussed by (BBR, see also Roos, 1981). Such binaries have minimum separation of order $r_{lc} \sim 0.1 r_h$, where r_h is the radius of the cusp in the star distribution around the hole. Further evolution of the binary is impeded by the formation of a loss-cone in the velocity distribution of the stars around the binary.

The putative binary black hole in 1928+738 is a relatively short-lived phenomenon. Its orbital period implies a gravitational radiation time scale of only $10^6 (M/m) M_8^{-1/3}$ years. The fact that this short phase of rapid orbital evolution occurs simultaneously with an active period of the galactic nucleus strongly suggests a causal relation between central activity and binary evolution. It seems plausible that the binary has recently gone through a phase of rapid evolution driven by the flow of large amounts of matter into the nucleus of the host galaxy as suggested by Roos (1988). Adiabatic contraction of the binary orbit due to growth of the primary hole takes place on a timescale $t_{ev}^{ad} = M/\dot{M}$, which is still quite large compared to its present evolution time. However, the binary can evolve faster by slingshot interactions with stars passing through the binary orbit when the nucleus is perturbed, for instance during the late stages of a merger with another galaxy. Repopulation of loss-cone orbits may then cause a rapid evolution of the binary while a small fraction of the inflowing mass may be accreted by the hole causing the observed quasar activity. The evolution time of the binary is $t_{ev} \propto (M+m)/\dot{M}_{bin}$, where $\dot{M}_{bin}$ is the (stellar) mass that is flowing through the binary orbit. When the binary is surrounded by a stellar cusp of mass M^*_{cusp} with an isotropic velocity distribution we find, following Hills (1983)

$$t_{ev} \sim 2.5 \times 10^7 r_{16}^{-1} M_8^2 \sigma_{200}^{-5} \left(\frac{M}{M^*_{cusp}} \right) \text{ yr,} \qquad (2)$$

where σ_{200} is the velocity dispersion in units of 200 km s^{-1}. The binary evolution slows down as the binary orbit shrinks until t_{ev} reaches a maximum at a radius r_{grav} given by $t_{grav} = t_{ev}$, where loss of orbital energy by emission of gravitational waves becomes dominant. This radius is given by $r_{grav,16} = 2.4\ M_8 \sigma_{200}^{-1}(m/M^*_{cusp})^{1/5}$. A binary at this radius does not live very long. However, a new wide binary may settle in the nucleus if the evolution is driven by a merger event. These considerations suggest that many AGN may contain either wide ($r \sim r_{lc}$) or close ($r \sim r_{grav}$) binaries, which have maximal evolution times. It is intriguing that the characteristic radius r_{grav} for close binaries agrees very well with the radius inferred from the wiggle period in 1928+738.

References

Begelman, M. C., Blandford, R. D. and Rees, M. J. 1980, *Nature* **287**, 307.
Hills, J.G. 1983 *Astron. J.* **88**, 1269.
Hummel, C. A., Schalinski, C. J., Krichbaum *et al.* 1992, *Astr. Ap.*, in press.
Kaastra, J. S. and Roos, N. 1992, *Astr. Ap.* **254**, 96.
Roos, N. 1981, *Astr. Ap.* **104**, 218.
Roos, N. 1988, *Ap. J.* **334**, 95.
Roos, N., Kaastra, J.S. and Hummel, C. A. 1992, *Ap. J. Letters*, submitted.

Gamma-rays from Blazars: a comparison of 3C 279, PKS 0537–441 and Mrk 421

Laura Maraschi [*]   *Gabriele Ghisellini* [†]   *Alfredo Boccasile* [‡]

Abstract

The γ–ray emission from three of the sources detected by the COMPTON Observatory is discussed in the framework of relativistic jets emitting via the synchrotron self-Compton mechanism. The physical conditions in the three sources are derived and compared.

One of the most exciting discoveries of the COMPTON Observatory is the observation of γ–ray emission from 16 blazars. The broad band continuum (radio to X–ray) from these objects can be understood on the basis of the synchrotron self-Compton process in an inhomogeneous jet of plasma moving with relativistic speed (e.g. [1,2]). In this model the synchrotron emission usually extends up to the UV and possibly to the X-ray bands. It is therefore natural to expect γ–rays produced by first order self-Compton scattering. The strength of this component depends on the ratio between the radiation and magnetic energy densities, and on the optical depth to photon-photon absorption. The latter process can be avoided if the source is beamed, and independent evidence for beaming is indeed found from radio and X-ray observations [3].

Nevertheless the γ–ray intensity measured from 3C 279 was a great surprise, in that the γ–ray power exceeded that in all other bands. This requires the radiation energy density to be larger than the magnetic energy density, a situation usually called the "Compton catastrophe" since at first sight it gives rise to divergent Compton energy losses. Although the divergence is inhibited by the finite power available and by the Klein-Nishina limit, it is true that, in these conditions, continuous electron re-acceleration is needed and the available power is converted into γ–rays. In a previous paper [4] we applied a relativistic jet model to the case of 3C 279, explaining the IR to UV emission as synchrotron and the X-ray to γ–ray emission as Inverse Compton radiation.

Here we consider two other objects detected by GRO, PKS 0537–441 and Mrk 421, with the scope of discussing to what extent the inferred physical conditions differ in

---

[*]Department of Physics, University of Genova, Italy
[†]Osservatorio di Torino, Pino Torinese, Italy
[‡]Department of Physics, University of Milano, Italy

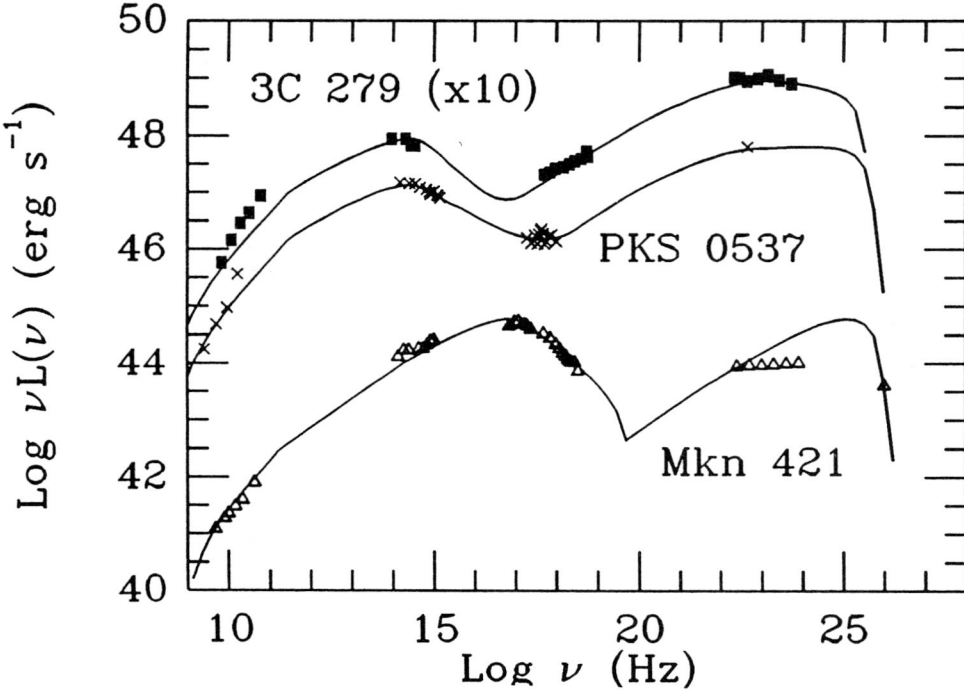

Figure 1. The overall energy distribution of the three sources is compared to the model. Model parameters are listed in Table 1, γ–ray data are from [10, 12, 13]

different objects. The overall energy distribution of PKS 0537–441 is similar to that of 3C 279, with the γ–ray power nearly dominating the bolometric luminosity. However the X–ray spectrum is variable and on average steeper than that of 3C 279 [5]. The observed emission lines (z=0.894) are broad but rather weak and at times disappear entirely [6]. Therefore this object is considered intermediate between HPQs and BL Lacs. Mrk 421 differs from the previous two sources, being a nearby, low luminosity object. As for the other sources the overall energy distribution of Mrk 421 shows two broad peaks (e.g. Fig. 1), but the first maximum falls in the UV–soft X–ray range, while for the other two sources it occurs in the far infrared. Moreover, at variance with the other two cases, for Mrk 421 the γ–ray power is less than the power at lower frequencies. Finally, Mrk 421 is of particular interest because it has been detected at TeV energies [7], thus imposing strong constraints on any model aiming at explaining the entire high energy spectrum.

We have applied the inhomogeneous relativistic jet model described in [4] and more fully in [8], to reproduce the entire energy distributions of the three sources. The model is also constrained by the observed variability and by the measured super-luminal velocity of the radio knots. The relativistic electron density, the magnetic field and the bulk Lorentz factor vary along the jet as power laws (see Tab. 1). The jet is parabolic in the inner, accelerating part ($r \propto R^{1/2}$) and conical in the outer, constant velocity part ($r \propto R$). The spectrum from the whole jet is obtained by integrating the local synchrotron and Inverse Compton emissivities over the volume of the jet. Our best results for the 3 sources are shown in Fig. 1, and the inferred parameters are reported in Table 1. For 3C 279 we have used X-ray and optical–UV data taken simultaneously with the γ-ray data [9]. For PKS 0537–441, the X-ray and optical data [4] are within few months of the γ-ray observations [10]. For Mrk 421 the X-ray data are from a simultaneous ROSAT GINGA observation [11], preceding the EGRET observation by about one year.

Table 1

	3C 279	PKS 0537	Mkn 421	Notes
R_0	2.1×10^{15}	2.8×10^{15}	2×10^{14}	[cm], initial scale length of parabol.
$R_{max,par}$	6.3×10^{18}	7.2×10^{18}	2.4×10^{17}	[cm], max. length of paraboloid
$R_{max,cone}$	4.2×10^{21}	5.6×10^{21}	5×10^{19}	[cm], max. length of cone
B_0	4.8	6.7	28	[G], magnetic field at R_0
K_0	3.1×10^6	3.9×10^5	10^5	[cm^{-3}], relat. electron density at R_0
ν_{max}	3.9×10^{16}	5×10^{17}	1.6×10^{19}	[Hz], max. synchro. frequency at R_0
α_0	0.54	0.52	0.5	local spectral index
m	0.96	1.0	1.0	$B = B_0(r/r_0)^{-m}$
n	1.5	1.3	0.9	$K = K_0(r/r_0)^{-n}$
e	0.36	0.67	0.37	$\gamma_{max} = \gamma_{max,0}(r/r_0)^{-e}$
Γ_0	2	2	2	initial bulk Lorentz factor
a	0.15	0.13	0.09	$\Gamma = \Gamma_0(R/R_0)^a$
θ	5°	3°	15°	viewing angle
Γ_{max}	6.6	5.6	3.8	final value of bulk Lorentz factor
U_r/U_B	166	114	9	ratio at R_0
β_{obs}	4–18	....	3.8	apparent superluminal velocity, $H_0 = 50$
β_{model}	5.6	2.9	3.7	predicted value
$\tau_{\gamma\gamma 5\ GeV}$	6×10^{-2}	2×10^{-2}	5×10^{-4}	γ–γ optical depth for 5 GeV photons

A comparison of the derived parameters shows that for the two high luminosity objects the magnetic field, the size, the Lorentz factor and the viewing angle are extremely close. The main difference is the maximum synchrotron frequency which must be higher in PKS 0537–441 in order to account for the steep component of the X-ray spectrum observed by ROSAT as synchrotron emission.

Mrk 421 is different in many respects: the size is smaller by an order of magnitude and the field is higher by a factor 5. The viewing angle is larger in agreement with other independent evidence (see e.g. [3]). In addition the maximum synchrotron frequency is larger by 1–2 orders of magnitude with respect to the other two objects. The TeV data force the model to predict a flat spectrum in the 0.1–5 GeV range, only marginally consistent with the recent spectral determination by EGRET. This may prove to be a difficulty for the model if the TeV flux is confirmed to be a persistent feature of Mrk 421.

In conclusion the entire broad band energy distributions of 3C 279, PKS 0537–441 and Mrk 421 can be interpreted as synchrotron (radio to X–rays) and self–Compton emission (X–rays to γ–rays) from a relativistic jet. In our model the physical parameters vary smoothly along the jet and are described with power laws. While this approximation may be sufficient for a description of the "average" spectrum, it is quite possible that a superposition of discrete components (analogous to the radio knots) or a less regular variation along the jet are required to fit the spectra in detail. We think, however, that the success of the model in reproducing the two broad humps in the energy distribution argues strongly in favour of the SSC mechanism as the basic emission process. Simultaneous multifrequency observations will be extremely valuable to further test the proposed model.

References

[1] Maraschi, L., 1992, in *Variability of Blazars*, eds. E. Valtaoja & M. Valtonen, Cambridge Univ. Press., p. 447.
[2] Marscher, A.P., 1992, in *Testing the AGN Paradigm*, eds. S. Holt, S. Neff & M. Urry, AIP conf. proceed. **254**, p. 377.
[3] Ghisellini, G., Padovani, P., Celotti, A, & Maraschi, L. *Ap. J.*, in press.
[4] Maraschi, L., Ghisellini, G. & Celotti, A., 1992 *Ap.J.*, **397**, L5.
[5] Treves, A., et al., 1992. *Ap.J.*, in press.
[6] Falomo, R., et al., 1989, in *BL Lac objects*, eds. L. Maraschi, T. Maccacaro & M.H. Ulrich, (Springer–Verlag), p. 73.
[7] Punch, M. et al., 1992, *Nature*, **358**, 477.
[8] Ghisellini, G. & Maraschi, L., 1989, *Ap.J.*, **340**, 181.
[9] Makino et al., 1992, private communication.
[10] Thompson et al., 1992, submitted to *Ap.J.*
[11] Fink, H.H. et al., *A.A.*, **246**, L6.
[12] Hartman, R.C. et al., 1992, *Ap.J.*, **385**, L1.
[13] Lin, Y.C. et al., 1992, *Ap.J.*, in press.

Microquasars in the Galactic Centre Region

I.F. Mirabel *

Abstract

The two persistent soft gamma-ray sources in the galactic centre region are black hole candidates of stellar mass with comptonized accretion disks that radiate 10^{37-38} erg s^{-1}. They appear as microquasar stellar remnants from which emanate double-sided radio jets that extend over distances of a few parsecs.

1. The galactic centre in soft γ-rays

Contrary to the standard X-ray band (below 20 keV), where many sources have been detected at less than 5° from the galactic centre, at higher energies (35-500 keV) the field is dominated by only two persistent sources: 1E1740.7-2942 and GRS1758-258 ([1], [2]), located respectively at $\sim$ 50' and $\sim$ 5° from Sgr A (see Fig. 1). In the 30-500 keV band these sources radiate a few $\times 10^{37}$ erg s^{-1}, near the Eddington limit of collapsed objects of stellar mass. Although persistent, they are time variable. Since no γ-ray source has been detected from Sgr A, if there is a super-massive black hole at the dynamic centre of the Galaxy, at present it is in sepulchral silence.

The telescope SIGMA on board GRANAT detected a powerful burst around 420 keV from the strongest γ-ray source 1E1740.7-2942 which is interpreted ([3], [4]) as the redshifted (probably gravitationally) 511 keV line from the annihilation of e^+e^- pairs. Since this Einstein source can produce and annihilate 10 billion (10^{10}) tonnes of positrons in just one second, it is now known as the "Great Annihilator". No annihilation line has yet been detected from GRS1758-258, the second strongest soft γ-ray source in that region.

These two sources are likely to be black holes of stellar mass. In their standard state, the shapes of the X-ray spectra resemble that of the black hole candidate Cygnus X-1. The intrinsic X-ray luminosity of 1E1740.7-2942 is comparable to that of Cygnus X-1 at the distance of the galactic centre; GRS1758-258 is somewhat weaker.

2. Radio jets from the two soft γ-ray sources

To comprehend the nature of these sources we are carrying out ground-based multi-wavelength observations co-ordinated with observations by SIGMA from space. Due

---

*Service d'astrophysique. Centre d'etudes de Saclay. 91191 Gif sur Yvette. FRANCE.

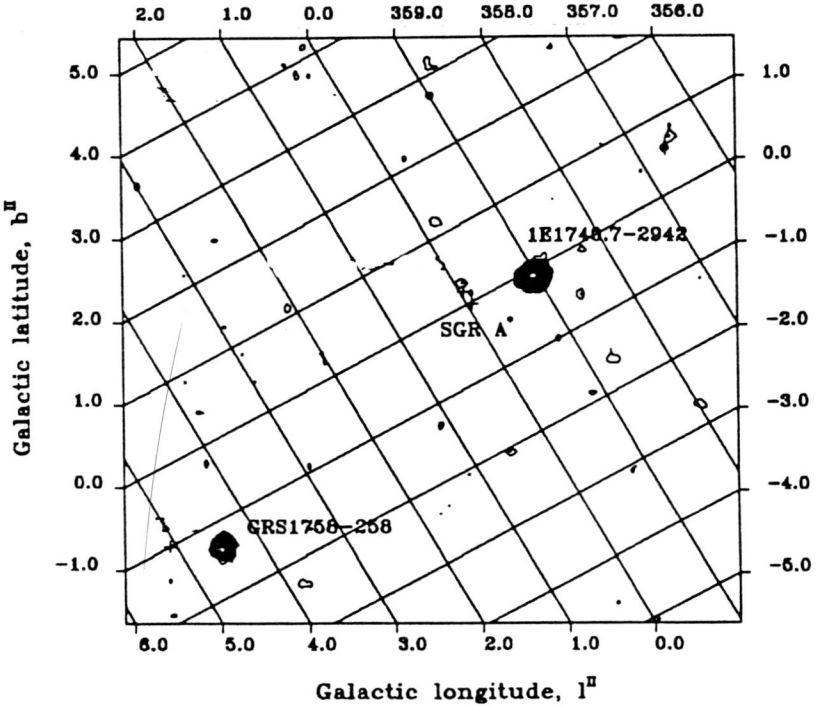

Figure 1. The hard (35–120 keV) X-ray image around the galactic centre obtained by the telescope SIGMA [2]. The two dominant persistent sources are 1E1740.7-2942 and GRS1758-258, respectively at $\sim 50'$ and $\sim 5°$ from Sgr A.

to the high interstellar absorption along the line of sight to the central region of the Galaxy, the optical identification of binary counterparts is very difficult. Therefore, we have undertaken observations at centimetre, millimetre, and infrared wavelengths.

The first result from co-ordinated VLA-SIGMA observations of these two sources was the identification of compact radio counterparts with position accuracy's $\leq 1''$, namely, 100 times better than the precision obtained with γ-ray telescopes. Inside the error circles of the X-ray and γ-ray telescopes we find compact radio sources, with time variations in the radio flux by factors greater than four, which are in the same sense as the soft γ-ray photon counts measured by SIGMA [5], [6].

The study of these radio counterparts took an unexpected turn. Radio jets whose centres coincide with the variable sources were soon discovered [5], [7] (see Fig. 2). Both sources are double-sided and the lobes appear to be aligned with the central, time variable source. The spectral index of the lobes in 1E1740.7-2942 is -0.8. These jets, a few parsecs long, are probably synchrotron emission from e^+e^- pairs streaming

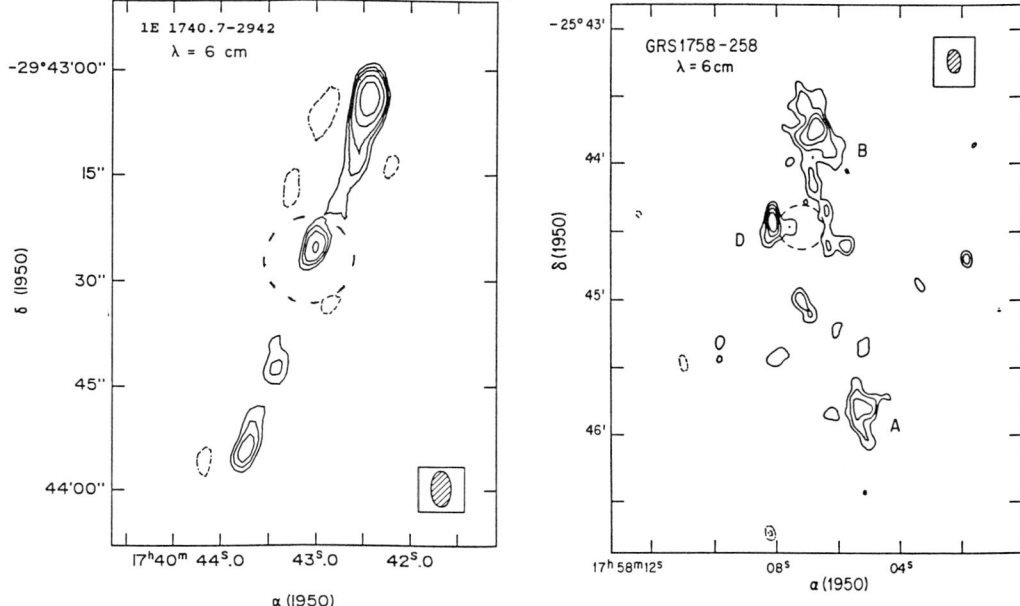

Figure 2. Radio jets associated to the soft γ-ray sources 1E1740.7-2942 and GRS1758-258 observed with the VLA at $\lambda 6$ cm [5], [7]. The small dashed circles indicate the position of these sources reported by the soft X-ray satellite ROSAT.

out at high velocities from the high energy sources. Although it is very unlikely that these are extragalactic radio sources in coincidental superposition [8], we have still to prove the contrary by the detection of time variability in the lobes.

No H, J, or K infrared band counterparts within 1" of the radio counterparts have been found down to a magnitude limit of K = 17. Assuming an optical absorption of 50 mag along the line of sight to the galactic centre, this infrared magnitude limit implies that these collapsed objects are not accompanied by a massive star, as is Cygnus X-1. It is estimated that no massive star with optical luminosity brighter than $M_V = -3$ mag is associated with these high energy sources.

3. The extragalactic analogy

There are indications that the analogy between these collapsed objects near the centre of the Galaxy and quasars is more than a morphological coincidence, and that perhaps the same basic physical mechanisms apply in both types of objects. The standard theories in extragalactic research propose that there are three basic ingredients in AGNs and QSOs: a black hole of 10^{6-8} $M_\odot$, an accretion disk, and radio jets extending to hundreds of kiloparsecs. The same ingredients, but scaled down, are present in 1E1740.7-2942 and GRS1758-258. In the context of the standard model [9] for hard X-ray sources, both sources are black holes of stellar mass with comptonized accretion disks having temperatures of a few tens of keV. Besides, we find that from both objects emit jets of high energy particles that reach distances of a few parsecs. Since these appear as quasars at stellar scales we call them "microquasars".

The two persistent soft γ-ray sources in the galactic centre region are somewhat different from the celebrated microquasar SS433. Although they are more luminous in the X-rays, their radio flux is $\sim 10^{-3}$ that of SS433. Furthermore, the jets from these sources extend up to a few parsec only, whereas the jets in SS433 appear to reach distances of 100 pc, where they interact with the supernova remnant W50.

Microquasars with the radio luminosity of SS433 must be rather uncommon since it is the only well collimated stellar radio jet with such a strong flux that is known within a distance of several kiloparsecs from the Sun [10]. However, weak radio jets on parsec scales may be commonly associated with hard X-ray sources when embedded in a high density interstellar environment such as that of the galactic centre region.

References

[1] Cordier, B. *Ph.D. thesis. Université de Paris VII (1992).*
[2] Sunyaev, R.A. et al. *A&A 247, L29 (1991).*
[3] Bouchet, L. et al. *ApJL 383, L45 (1991).*
[4] Sunyaev, R.A. et al. *ApJ 383, L49 (1991).*
[5] Mirabel, I.F., Rodríguez,L.F., Cordier,B., Paul,J. and Lebrun, F. *Nat. 358, 215 (1992).*
[6] Mirabel, I.F., Rodríguez,L.F., Cordier,B., Paul,J., Lebrun, F., Duc, P.A. *IAU Circ. 5655 (1992).*
[7] Rodríguez, L.F., Mirabel, I.F., Martí. J, 1992, *ApJ 401, L15 (1992).*
[8] Mirabel, I.F., Rodríguez,L.F., Cordier,B., Paul,J. and Lebrun, F. *in Sub-arsec radio astronomy in press (1993).*
[9] Sunyaev, R.A. and Titarchuk, L.G. *A&A 86, 121 (1980).*
[10] Margon, B. *Ann. Rev. A&A 22, 507 (1984).*

A Comparison of the Ultraviolet Continuum Variability Properties of Blazars and Seyfert 1s

R. Edelson, G. Pike, J. Saken, A. Kinney, M. Shull, J. Krolik

Abstract

Long time scale ultraviolet light curves of blazars and Seyfert 1s both show very strong continuum variations, but this similarity vanishes when short time scales, spectral variability and correlations between variability and luminosity are studied. For instance, blazars show much more rapid variations than Seyfert 1s. Also, the spectra of Seyfert 1s harden as the source brightens, while blazars show little spectral variability. Third, the most luminous blazars tend to be the most strongly variable, while for Seyfert 1s, the strongest variations are seen in the least luminous sources. These differences suggest that in spite of some overall similarities, the observed emission from blazars and Seyfert 1s have different physical origins. These results are consistent with models which hold that the ultraviolet emission from blazars is incoherent synchrotron emission from a jet, while that from Seyfert 1s is dominated by thermal emission from an accretion disk.

1. Background

In the 13 years since its launch, IUE has obtained over 5000 ultraviolet spectra of active galactic nuclei (AGN). In this paper, we use ∼2500 spectra of 16 objects to survey the ultraviolet variability properties of Seyfert 1s (defined to include quasars as well; [2]) and blazars (BL Lacs and OVV quasars; [6]).

2. Long and Short Time Scale Variability

Although it is a common prejudice that blazars are more strongly variable than Seyfert 1s, the long term variability properties of the two types of object are actually very difficult to distinguish. For example, Fig. 1 shows the long term ultraviolet light curves of the two best-observed AGN; NGC 4151 (a Seyfert 1; [2]) and PKS 2155-304 (a blazar; [4]). In fact, the Seyfert 1 shows the larger variation amplitude.

This ambiguity vanishes when shorter time scales are probed. Fig. 2 shows short term variability in the Seyfert 1 NGC 5548 ([1]; no similar data for NGC 4151 have been published) and PKS 2155-304 ([11]). Note the different time scales: The blazar data show many significant variations in a 4.3 day period, while the Seyfert 1 data, taken over 8 months, show no significant variations on time scales shorter than ∼20

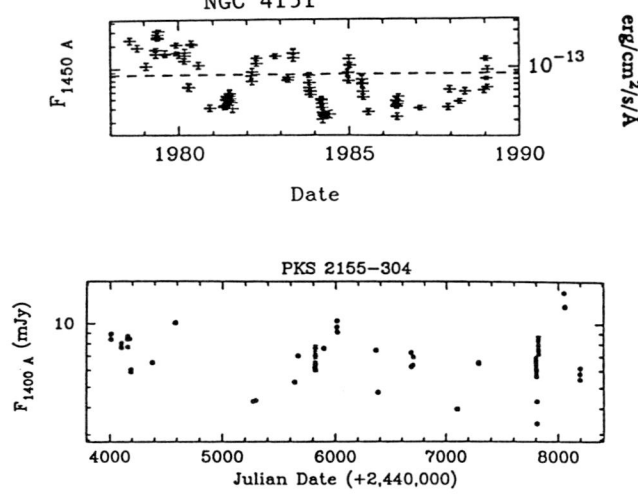

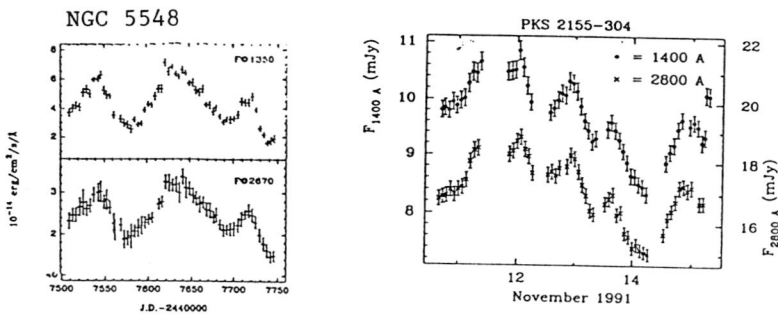

Figure 1. Long term variability in NGC 4151 and PKS 2155–304

Figure 2. Short term variability in NGC 5548 and PKS 2155–304

days. Although NGC 4151, the most rapidly variable of the well-studied Seyfert 1s, can vary on time scales as short of a few days, even it has never shown the rapid microvariability seen in blazars. Thus, although both types of objects show strong ultraviolet variability on long time scales, blazars appear to show more rapid variations than Seyfert 1s.

3. Spectral Variability

Fig. 3 compares the relationship between ultraviolet continuum spectral index (defined as $F_\nu \propto \nu^\alpha$, measured between 1400 and 2800 Å) and flux density for Seyfert 1s and blazars ([2]; [6]). For NGC 4151, there is a strong correlation in the sense that the spectrum hardens as the source brightens, while for PKS 2155–304, there is no evidence for a correlation between flux and spectral index. This distinction is typical of Seyfert 1s and blazars.

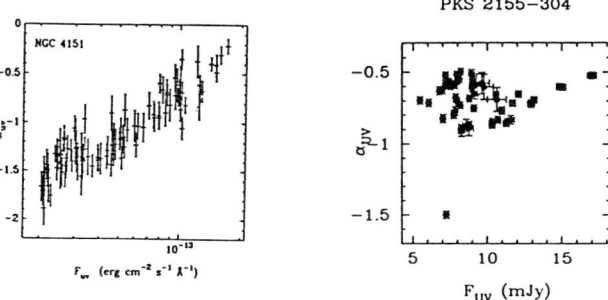

Figure 3. Correlation of flux with spectral index in NGC 4151 and PKS 2155-304

4. Variability and Luminosity

The previous plots referred to individual objects. Fig. 4 contains data from two large surveys of Seyfert 1s and blazars ([2]; [6]). For each object, the fractional rms variability and the apparent luminosity were computed and plotted. The circles refer to 1400 Å data and the triangles, to 2800 Å data. Only objects with more than 5 years of IUE observations were included. The blazar data (on the left) shows a positive correlation, with $r_S = 0.60$ (corresponding to $p = 0.04$ for $N = 10$ independent points), while the Seyfert 1s show an anticorrelation between degree of variability and luminosity, $r_S = -0.85$ ($p = 0.02$ for $N = 6$).

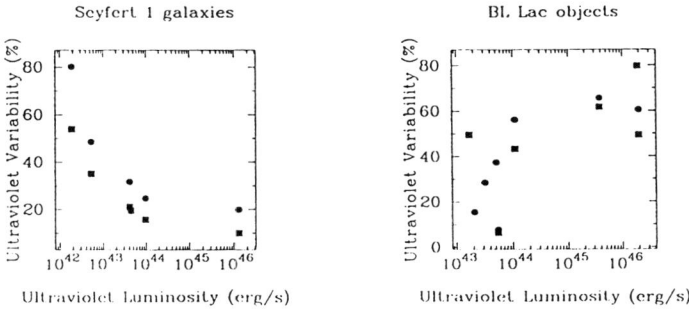

Figure 4. Correlation of degree of variability with Luminosity in Seyfert 1s and blazars

5. Theoretical Implications

These data show that although they have some superficial similarities, upon careful examination, blazars and Seyfert 1s show very different variability characteristics. This has important implications for the physics underlying the observed emission.

Accretion disk models have long been fitted to the ultraviolet spectra of Seyfert 1s (e.g., [9]). Variability data support this model. For instance, the anticorrelation between the degree of variability and the luminosity would be expected if the luminosity was proportional to the size, and therefore the light-crossing time, of isotropically-emitting sources. Also, the spectra of individual variable Seyfert 1s tend to harden as the flux increases, suggesting that the observed spectrum is a combination of a relatively hard, variable component (the putative accretion disk) and a softer, non-variable one. Finally, Seyfert 1s show relatively slow variations, consistent with the general picture, although perhaps not the details, of accretion disk models ([8]).

The ultraviolet variability behaviour of blazars provides the most direct evidence to date that the emission is beamed synchrotron emission from a relativistic jet. In this model, the most strongly beamed objects will tend to have both the highest apparent luminosities and the most compressed variability time scales. Furthermore, the most rapid variations seen in some blazars (3C 279 and PKS 2155–304) appear to violate the Eddington limit, directly indicating that the ultraviolet emission from these sources is not isotropically emitted. These results are consistent with the standard SSC model (e.g., [7]). On the other hand, the correlations and the rapid variability are incompatible with the accretion disk model proposed by [10].

Since the long time scale variability properties of Seyfert 1s and blazars are so similar, while the short term and spectral variability properties are very different, future observing campaigns should stress high density, evenly-sampled monitoring over a large bandwidth. In fact, both the intensive NGC 5548 and PKS 2155–304 campaigns involved other wavelength bands, leading to the setting of a lower limit of $\sim 0.1c$ on the speed of propagation in an accretion disk in NGC 5548 ([8]), and clear evidence that Compton scattering cannot dominate the soft-X-ray emission from PKS 2155–304 ([5]). Future approved or planned campaigns to monitor the blazars OJ 287, Mkn 421 and 3C 279, and the Seyfert 1 NGC 4151 have even more regular and intensive sampling and better multi-wavelength coverage than the previous ones, and will severely test both the accretion disk and synchrotron self-Compton model.

References

[1] Clavel, J. et al. 1991, ApJ, 366, 64
[2] Edelson, R., Krolik, J., and Pike, G. 1990, ApJ, 359, 86
[3] Edelson, R. et al. 1991, ApJL, 372, L9
[4] Edelson, R. et al. 1992, ApJ, in press
[5] Edelson, R. et al. 1993, in preparation
[6] Edelson, R. 1992, ApJ, in press
[7] Ghisellini, G. Maraschi, L., and Treves, A. 1985, AA, 146, 204
[8] Krolik, J. et al. 1991, ApJ, 371, 541
[9] Malkan, M. and Sargent, W. L. W. 1982, ApJ, 254, 22
[10] Wandel, A., and Urry, C. M. 1991, ApJ, 367, 7
[11] Urry, C. M. 1993, ApJ, in preparation

Simultaneous Optical and IR Monitoring of the Seyfert Nucleus NGC 7469

D. Dultzin-Hacyan * *A. Ruelas-Mayorga* * *R. Costero* *
M. Alvarez *

Abstract

We present results of four nights of optical monitoring of the Seyfert nucleus NGC 7469 in search for microvariability. At visible wavelengths we used the same instrumentation, data and error analysis as in a previous monitoring programme (Dultzin-Hacyan et al. 1992 – paper 1) but found (contrary to the result of paper 1) a negative result: no microvariability. During two nights, simultaneous monitoring was done at $1.65\mu m$ (H band) and variations of 10% and 40% were observed in $\Delta t \sim 2$ hours. Although this observational result should be taken with care, if confirmed it implies that near IR light is dominated by non-thermal emission. We stress the transient character of microvariability and briefly discuss models of local disk instabilities for Seyferts.

1. Introduction

The existence of optical microvariability (small amplitude variations in less than a day) is considered strong observational evidence in favour of a very compact object (black hole?) in the centre of active nuclei. The presence of microvariability has been quite firmly established for several blazars (e.g., Carini et al. 1991 and also Miller in this volume). There are also reports of observations at near IR and radio wavelengths (e.g., Kidger & De Diego, 1991; Wagner et al. this volume). In the case of Seyfert galaxies, there are, to our knowledge, 3 Seyfert 1 galaxies studied for microvariability: NGC 4151, for which Lawrence et al. (1981) found a negative result, while Lyutyi et al. (1989) found microvariability with $\Delta V \sim 0\overset{m}{.}05 - 0\overset{m}{.}10$ in $\Delta t \sim 15 - 30$ min during 5 nights out of 13; for NGC 4051 no optical nor IR variability was found while it varied strongly in X rays (Done et al. 1990); and finally, Aslanov et al. (1989) first reported optical microvariability for NGC 7469 and this result was recently confirmed by Dultzin-Hacyan et al. 1992 (paper 1). After performing a very thorough error analysis, the latter authors found variations with mean amplitudes of $\sim 0\overset{m}{.}40$ in mean timescales of $\Delta t \sim 13$ min during 4 nights out of 5; during the last night the flickering

---

*Instituto de Astronomía, UNAM, Mexico. D. D-H., received support from DGAPA, UNAM through grant IN 104591 and partial travel support from CONACYT.

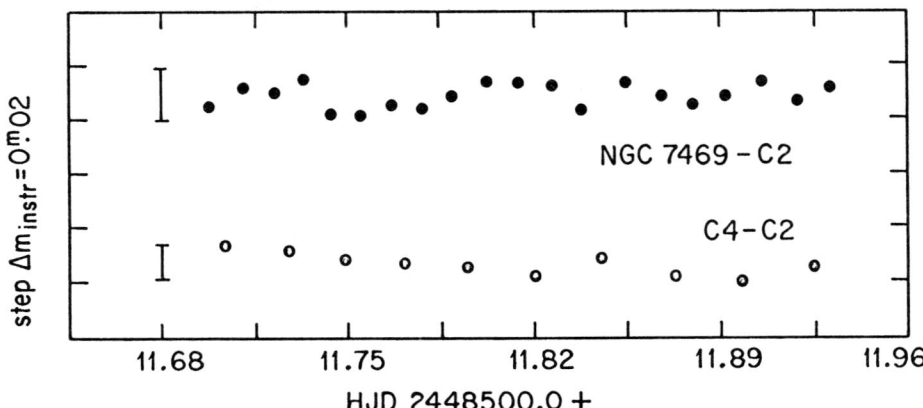

Figure 1. Differential light curves in instrumental magnitudes. The error bars include the mean photometric error, errors due to the contribution of underlying galaxy and microscale seeing and atmospheric transparency changes. C_2 and C_4 are constant brightness control stars.

was marginal within the error bars. The implied change in luminosity was found to be $\Delta L_{var}/\Delta t \sim 6 \times 10^{37}\,\mathrm{erg\,s^{-2}}$.

Here we report the results of a second monitoring programme on NGC 7469 carried out in 1991 at optical and IR (1.65μm) wavelengths.

2. Observations

The optical observations were made with the 1.5 m telescope at San Pedro Martir, Baja California, Mexico, using a Strömgren photometer (Schuster & Nissen 1988) and the IR observations with the 2.12 m telescope at the same site, coupled with an InSb IR photometer (Roth, *et al.* 1984).

The optical observations (11-14 Sept.) were made using exactly the same setting, acquisition and reduction techniques as well as the same error analysis as in the previous (1990) monitoring programme (paper 1). No variability was found during the entire $\sim$ 19 hrs of monitoring. In Fig. 1 we show an example of the differential photometry for one night (September 11).

During the nights 11 and 13 September (1991) we were also able to do IR (H band) monitoring. During both nights a strong (10 and 40%) decrease in brightness followed by an increase to the previous level was observed within about 2 hours (Fig. 2).

The time resolution of the IR observations is about 50 min, because in order to attain photometric errors of typically $\sim 0.^m03$, integrations of 20–30 min were needed.

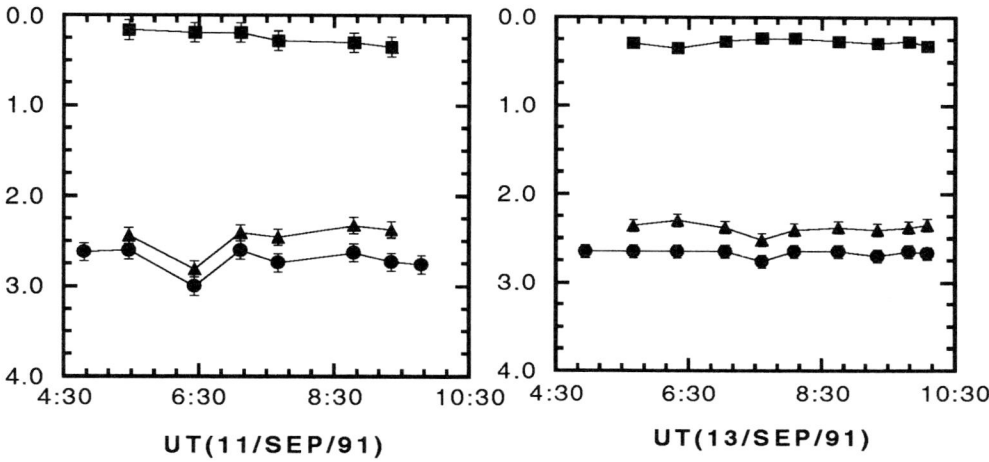

Figure 2. We plot 3 differential light curves for each night: Squares trace the difference between the control stars, triangles and circles trace the difference between the nucleus and each control star. The observations of the nucleus and control stars were not simultaneous (but close) and thus an interpolation was made to adjust the curves. The vertical axis is ΔH.

In order to avoid guiding errors grater than 4" (the diaphragm was 14"), which may cause variations in the contribution of the underlying galaxy (see paper 1, also Kidger & De Diego 1991), careful re-centring was done frequently. Also the frequent observation of two control stars enabled us to check the photometric stability of the night.

Variability results based on just a single point should, evidently, be taken with care (and even scepticism); perhaps only as an indication that further monitoring (with better time resolution) is needed. If confirmed, however, this result implies that changes in the near IR may be larger than in the visible!, and also that the near IR light is probably dominated by non-thermal emission in the nucleus of this galaxy.

3. Final Considerations

Microvariability is a transient phenomenon within timescales of one day. Models involving local disk instabilities which require that we "see" the innermost part of the disk (e.g., Lesch & Pohl, 1992) are valid only for blazars; in the model of Wiita et al. (1992) flares or hot spots occur in the outer part of the disk, but still, in the case of most Seyferts, we "see" through the whole or part of the galactic disk, perhaps for this reason microvariability is harder to observe in Seyferts.

References

Aslanov A.A., Kolosov D.E., Lipunova N.A. and Lyutyi V.M. 1989, SvAL, 15, 132.

Carini M.T., Miller H.R., Noble J.C. and Sadun A.C. 1991, AJ, 101, 1196.

Done C., Ward M.J., Fabian A.C., Kuneida H., Tsuruta S., Lawrence A., Smith M.G. and Wamsteker W. 1990, MNRAS, 243, 713.

Dultzin-Hacyan D., Schuster W.J., Parrao L., Peña J., Peniche R., Benitez E. and Costero R. 1992, AJ, 103, 1769.

Kidger M. and De Diego J. 1992, A&AS, 93, 1.

Lesch H. and Pohl M. 19192, A&A, 254, 29.

Lawrence A., Giles A.B., McHardy I.M. and Cooke B.A. 1981, MNRAS, 195, 149.

Lyutyi V.M., Aslanov A.A., Khruzina T.S., Kolosov D.E. and Volkov I.M. 1989, SvAL, 15, 247.

Roth M., Iriarte A., Tapia M. and Resendiz G. 1984, Rev Mex AA, 9, 25.

Schuster W.J. and Nissen P.E. 1988, A&AS, 73, 225.

Wiita P.J., Manglam A.V. and Chakrabati S.K. 1992, in *Testing the AGN Paradigm*, edited by S.S. Holt, S.G. Neff and C.M. Urry (AIP, New York), p. 251.

Broad-band Spectra and Polarization Properties of Variable Flat-Spectrum Radio Sources

Stefan J. Wagner *   *Arno Witzel* [†]

Abstract

New results from multi-frequency monitoring campaigns of variable flat-spectrum radio sources are reported. They strengthen the assumption that the intraday variability occurs in a correlated fashion throughout the radio, optical and X-ray wavebands. Various properties of the behaviour exclude propagation effects as the dominant cause of the variations. This implies excessive brightness temperatures. A large fraction of the primary synchrotron radiation may be upscattered into the Gamma ray regime. We also discuss first results from polarization studies in different frequency regimes. At least the BL Lac object S5 0716+714 exhibits variations of the polarized flux which are correlated with variations of the total flux. Neither simple two component models nor "christmas-tree" scenarios of a large number of individual emitters seem able to explain the polarization data.

1. Introduction

Compact, flat-spectrum radio sources are well known to be variable at optical and radio frequencies. It was generally assumed that the typical timescales increase with wavelength (from hours at x-ray energies to days/weeks at optical wavelength and months/years at radio frequencies). During the last few years we have performed several simultaneous multi-frequency campaigns with high temporal resolution to study the nature of the fastest variations which probe the smallest regions in these AGN. Variability on timescales of hours to a few days were found to be common at optical and radio frequencies in a sample of bright radio-sources selected to have flat spectra in the 3–5 GHz regime. The mere existence of rapid variations requires a very compact source structure either due to light travel time arguments if the variations are intrinsic or in a more indirect way if the flickering is introduced by propagation effects (microlensing or refractive interstellar scattering). It was found that the observed timescales are so short that microlensing cannot induce the variations (e.g. Wagner, 1992). During a simultaneous campaign of four weeks in February 1990, we found

---

*Landessternwarte Königstuhl, 6900 Heidelberg, Germany
[†]MPIfR, Auf dem Hügel 69, 5300 Bonn, Germany

a close correlation of the variations at optical and radio frequencies (Wagner 1991, Quirrenbach et al. 1991). Several statistical tests strengthen the assumption that the variations are similar and separated by a short time-lag of less then two days (Wagner & Witzel, 1992). This implies that the radio variability is not dominated by scintillation of the galactic interstellar medium and hence must be intrinsic to the source. This in turn, implies linear dimensions smaller than a light-day either for the entire source, or at least a region within which a large fraction of the radio luminosity is emitted. Assuming a distance of $z > 0.4$, this implies extremely high brightness temperatures, which are in excess of the Compton-limit for incoherent synchrotron emission by several orders of magnitude. This can only be resolved by relativistic beaming if the Doppler factors are of the order of 10^2, an order of magnitude larger than observed in the most extreme superluminal sources and in clear disagreement with independent indications for smaller Doppler factors for BL Lac sources (see e.g. Padovani, this volume). Further campaigns at radio and optical frequencies, as well as a few additional simultaneous campaigns have revealed intraday variability (IDV) to be common and suggest correlations in other sources as well (Quirrenbach et al., 1992, Heidt et al., 1991, Wagner et al., 1992).

It has been suggested that modifications to the standard synchrotron self Compton (SSC) model of incoherent synchrotron emission may modify the constraints given by the Compton catastrophe and may help to explain the rapid flares, observed almost simultaneously over 5 decades in frequency. One promising suggestion is maser-induced coherent radiation (e.g. Benford, 1992).

2. Multi-frequency observations

Most of the sources which we found to exhibit IDV at radio and optical frequencies have now been studied at many wavelengths. Fig. 1 shows the overall spectrum of one of the best-studied objects, the BL Lac candidate 0716+714. Here we give a brief description of the observations carried out in different wavebands.

a) Radio Measurements
The spectrum of 0716+714 is flat in the range of 21 cm to 7 mm. Several campaigns of intense monitoring have been carried out. Some of the results have been described in Quirrenbach et al., 1991, Wagner 1991 and Witzel and Wagner, 1992. In these papers we mainly discuss the correlation between optical and radio observations. Variations of the radio spectral index are described e.g. by Wegner et al., 1992. In particular, we would like to stress that we found suggestive evidence that individual outbursts seem to appear first at 6 cm and later at 3.6 cm. This is in contradiction to the prediction of classical synchrotron models which would imply the opposite lag due to optical depth effects.

b) Optical/UV measurements
Several simultaneous campaigns have shown that the variations at optical frequen-

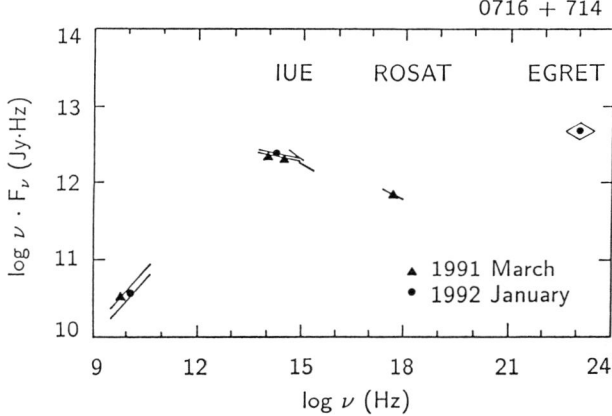

Figure 1. The overall spectrum of the permanently variable BL Lac object 0716+714 during two simultaneous campaigns in 1991 and 1992.

cies are closely correlated to the intraday variations observed at 6 cm. Within the optical regime no variations of the spectral index were observed. During February 1991 and January 1992 we observed 0716+714 simultaneously at optical and UV wavelengths. We found a smooth curvature from the optical to the UV spectral index ($\alpha_{550\,nm-200\,nm} = 1.4$). Again no temporal variations of the spectral index were observed, neither within the wavelength range covered by the IUE satellite nor of α_{O-UV}.

c) X-ray observations

In March 1991 0716+714 was observed for 20 000 s with ROSAT. The 1 keV flux varied by a factor of 2 within 24 hours. No significant variations were detected on short ($\leq 1000\,\text{sec}$) timescales. Again no significant variations of the spectral index within the range 0.3 keV–2 keV were detected. The x-ray spectral index, however, is remarkably steep ($\alpha_{1keV} = -1.7$).

As discussed in more detail by Witzel & Wagner (1992) we found evidence for a correlation between the X-ray variations and optical fluctuations. although poorly sampled, our observations are consistent with a small ($< 5\,\text{h}$) time-lag. This, as well as the smooth transition of the spectral index throughout the optical-UV-X-ray regime, indicates that the X-ray and optical photons emerge from the same physical process within the same volume. The data are consistent with a classical synchrotron-source whose high-energy cut-off is responsible for the steep X-ray spectral index. Unexpectedly, the Compton-scattered radio emission (from the luminous and compact radio source) would still be suppressed in the soft X-ray regime.

d) Gamma-ray observations

During the pointing in January 1992 GRO detected γ-ray emission from 0716+714

(Michelson et al., 1992). The EGRET flux is shown in Fig. 1. No simultaneous X-ray observations of the source were carried out during this period. We have, however, monitored the synchrotron flux during this period and found it to be identical within the usual 50 % fluctuations at optical frequencies to the flux level recorded during the simultaneous optical–X-ray monitoring in March 1991. If the soft X-ray flux is indeed dominated by synchrotron radiation the short-term and long-term stability of α_{ox} may allow an extrapolation of the X-ray flux density in January 1992 from the contemporary optical and UV measurements.

We want to stress that the extraordinary γ-ray luminosity is a common phenomenon of intraday variable sources. All of the EGRET sources discovered to date are rapidly variable at optical frequencies. In some cases the γ-ray detection may be directly related to an outburst at synchrotron frequencies (von Linde et al., 1992).

3. Polarization Properties

All of the intraday variable sources have a detectable amount of linearly polarized emission. We have monitored the polarization properties of some of these sources in several campaigns, a few of which have been carried out simultaneously at radio and optical frequencies. The results can be summarized as follows:

All sources with rapidly variable flux densities of the total flux also show changes of the flux density of polarized light. In all cases observed so far, the percentage polarization is variable also. The changes of the degree of linear polarization are accompanied by changes of the polarization angle. All of these statements are valid within the radio regime and the optical regime separately.

We have observed both direct and anti-correlations between the changes of the total and the polarized flux. Since the best observed sources generally show oscillatory variations with a duty cycle close to one, anti-correlations may be interpreted as phase-shifted correlations. We have not yet observed correlations with phase shifts clearly different from multiples of 0.5, but we found sources to change from a correlated to an anti-correlated behaviour within a single campaign. Again, all of these statements are true for optical and radio regimes separately.

Not all of the sources always show measurable amounts of intraday variability of the polarized light. Since typical variations of the polarized component rarely exceed about 5 to 8σ of individual measurements, this may also be due to modest changes of the amplitude of the variations.

The long term trends of polarization angle have preferred values, correlated with the position angle of the structural axes of the VLBI jets. While the fractional variations of the total and polarized flux densities are of similar amounts (20%) in the optical, $\Delta S/S \sim 0.1$ and $\Delta P/P \sim 0.8$ in the radio regime.

4. Emission Processes

The short time-scales of the variations and the fact that the lag of the radio-optical correlations is larger then the emission regions in either of these spectral domains (measured from the widths of the autocorrelation functions) both exclude microlensing as the dominant cause of the rapid variability. The close correlation of the variations at optical and radio frequencies likewise exclude RISS as a *dominant* mechanism in 0716+714 (Quirrenbach et al., 1991).

We conclude that IDV is intrinsic. The classical SSC models of radio sources are compatible with the observed spectra and implied linear dimensions only if the sources are strongly beamed. Doppler factors of ~ 100 are required. This would be possible in jet models if the line of sight is inclined by only $< 1°$ to the velocity vector of the emitting component. SSC models would also explain the correlation between the occurrence of IDV and the high luminosities in the γ-ray regime. This relation, which is now emerging from the growing number of Blazars discovered by EGRET indicates that second order Compton upscattering is an efficient process in the most compact sources (which show the shortest time-scales). While the multiplicity of the bursts could be due to frequent, rapid injections and accelerations of e $-$ within the jets, the quasi-periodic nature (Wagner & Witzel, 1992) suggests a geometric scenario, as predicted e.g. by Camenzind & Krockenberger, 1991.

The polarization behaviour cannot be explained by superposition of two components with variable amounts of total flux and percentage polarization but constant polarization angles. This rules out simple ("few candle") christmas-tree models and stresses the need for very efficient acceleration mechanisms. Lesch (1991) suggested magnetic reconnection driven by plasma instabilities as a possible accelerator.

The extreme Doppler factors may be avoided if the radiation process is not incoherent synchrotron radiation. Coherent radiation mechanisms have been suggested e.g. by Benford (1992) as a product of an electron maser mechanism (Weatherall & Benford, 1991). Certain amounts of coherent synchrotron radiation could be caused by anisotropies of the distribution functions of the primary particles. It is unclear, however, if such anisotropies can lead to sufficient amounts of coherent radiation with the observed spectral characteristics of the radiation emitted from the IDV component. Rees (1992) also pointed out that an additional constraint is imposed on scenarios involving coherent radiation mechanisms by the requirement that the Thomson depths has to be extremely low to avoid scattering.

Irrespective of the importance of coherent synchrotron mechanisms, we want to point out that homogeneous synchrotron sources cannot explain the spectral behaviour in the radio regime (Wagner & Witzel, 1992, Wegner et al., 1992) The rapid variability of the spectral index cannot be explained by synchrotron radiation of a homogeneous cloud and ordinary optical depth effects. This problem also arises if non-SSC models (such as synchrotron radiation by the decay products of high energy primary particles

(Mannheim & Biermann, 1989) are taken into account.

The long-term stability of the spectrum shown in Fig. 1 suggests that most of the radiation emerges from variable components with saturated variability (i.e. duty cycles close to 1) of similar spectral properties *or* the superposition of a stable source and variable components which have very similar spectral properties. In classical SSC models the particle distribution functions would be fairly well constrained by the steep cut-off of the synchrotron emission in the soft X-rays, which are apparently primarily synchrotron, and by the X-ray/γ-ray flux ratios.

The alternative suggestion by Mannheim & Biermann that the synchrotron branch is emitted by the decay products of high energy particles of γ-ray emitting primaries would likewise be well constrained by the high-energy cut-off in the soft X-ray regime.

Although both the microphysics of the radiation mechanisms as well as the overall geometry are still controversial, the mere existence of IDV is promising as a new tool of investigating AGN on scales as small as 10^{14} cm.

Acknowledgements We would like to thank all of our collaborators, especially Jochen Heidt, Dr. Thomas Krichbaum and Ralf Wegner for continuously successful collaboration, Dr. Max Camenzind for stimulating discussions, the staff of the Calar Alto observatory and the Effelsberg telescope as well the DFG (SFB 328) for financial support.

References

Benford, G.: 1992, in *Extragalactic Radio Sources - from Beams to Jets"*, J. Roland et al. (Eds.), Cambridge University Press, p. 85
Camenzind, M. and Krockenberger, M.: 1992, AA 255, 59
Heidt et al.: 1992, AA submitted
Lesch, H.: 1991,in *Variability of Active Galaxies*, W. Duschl et al. (Eds.), Springer, Lecture Notes in Physics, Vol. 377, p. 211
Mannheim, K. and Biermann, P.: 1989, AA 221, 211
Michelson et al.: 1992, IAUC 5470
Quirrenbach, A. et al.: 1991, ApJ 372, L71
Quirrenbach, A. et al.: 1992, AA 258, 279
Rees, M., 1992, in *X-ray Emission from Active Galactic Nuclei and the Cosmic X-ray Background*, MPE Report 235, W. Brinkmann and J. Trümper (Eds.), p. 255
von Linde, J. et al.: 1992, AA submitted
Wagner, S.J.: 1991, in *"Variability of Active Galaxies"*, W. Duschl et al. (Eds.), Springer, Lecture Notes in Physics, Vol. 377, p. 163
Wagner, S.J.: 1992, in *"Gravitational Lenses"*, R. Kayser et al. (Eds.), Springer Verlag
Wagner, S.J. and Witzel, A.: 1992, in *"Extragalactic Radio Sources - from Beams to Jets"*, J. Roland et al. (Eds.), Cambridge University Press, p. 59
Wagner, S.J. et al.: 1992, AA submitted
Weatherall, J. and Benford, G.: 1991, ApJ 378, 543

Wegner, R. et al.: 1992, in *Sub-Arcsecond Radio Astronomy*, R. Davis and R. Booth (Eds.), Cambridge University Press

Witzel, A. and Wagner, S.J.: 1992 in *Sub-Arcsecond Radio Astronomy*, R. Davis and R. Booth (Eds.), Cambridge University Press

The Radio to Optical Variability of the BL Lac Object ON 231

E. Massaro, R. Nesci and G.C. Perola [*]
D. Lorenzetti and L. Spinoglio [†]   M. Felli and F. Palagi [‡]

Abstract

We report new measurements of the BL Lac object ON 231 (W Com) from radio to optical wavelengths. This source was found to be at its highest brightness in the near IR and optical bands for many years.

1. Introduction

ON 231 is one of the targets in the observational programme in the near IR that we have been carrying out since 1986 at the 1.5 m Italian IR Telescope at Gornergrat (TIRGO, 3150 m a.s.l.). The results of the measurements performed up to the spring of 1988 are given in [3]. At the beginning of February 1992 we found the source in the highest state for about 20 years and therefore, in addition to the IR and optical measurements, we observed ON 231 at radio frequencies to obtain a more complete picture of its spectral distribution. Radio observations were carried out with the 32 m dish of Istituto di Radioastronomia (CNR - Bologna, Medicina) at a frequency of 22.2 GHz.

2. Results

The results of the new measurements are presented in Fig. 1 (open circles) together with those of Landau et al. [2] performed in the spring 1983 (crosses). Some remarkable facts are evident: i) the flux in the near IR and optical bands in 1992 is generally higher than in 1983; ii) at variance with the above, the flux at 22.2 GHz does not show a significant change with respect to the previous measurement; iii) the maximum power emitted by ON 231 in the high state lies between 250 and 500 THz, while in 1983 it was at a lower frequency, likely below 100 THz; iv) the spectral slopes in the optical and near IR bands are flatter when the luminosity is higher. We would like to point out the similarity of the behaviour of ON 231 with that of the

---

[*] Istituto Astronomico, Universita' "La Sapienza", Roma, Italy
[†] Istituto di Fisica dello Spazio Interplanetario, CNR, Frascati, Italy
[‡] Osservatorio Astrofisico di Arcetri, Firenze, Italy

OVV quasar 3C 279, recently detected at photon energies above 100 MeV by EGRET-CGRO [1] and having a luminosity more than one order of magnitude higher than that of ON 231. Both sources exhibited a large increase of the flux in the near IR-optical, while at the radio frequencies the flux remained unchanged. It will be important, therefore, to verify if the high frequency flux of ON 231 undergoes a significant increase. In this respect we stress that ON 231 must be considered an interesting target for X and γ-ray observatories in orbit and scheduled in the near future.

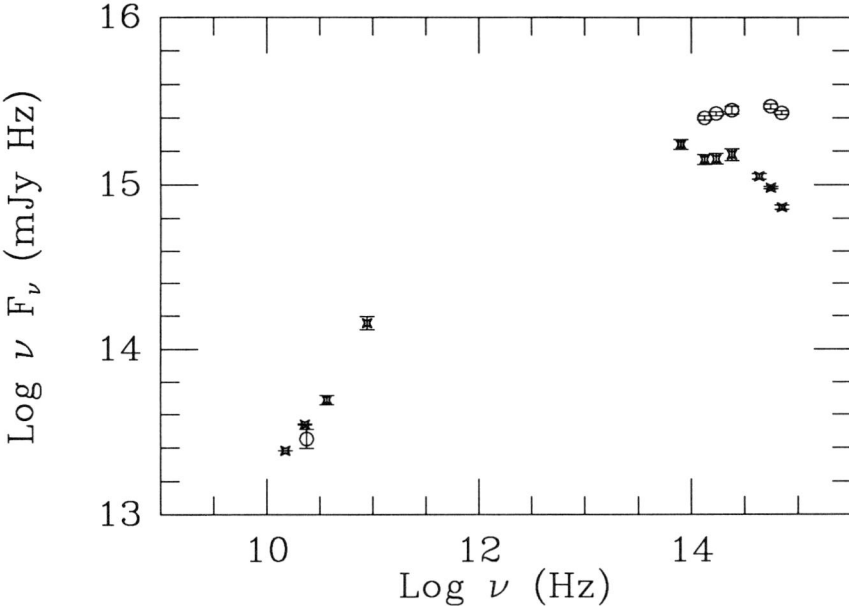

Figure 1. The radio to optical spectra of ON 231 in 1983 spring [2] (crosses) and 1992 winter (open circles).

References

[1] Hartman R.C. et al., *Astrophys. J. Letters* **385** (1992) L1.
[2] Landau R. et al., *Astrophys. J.* **308** (1986) 78.
[3] Lorenzetti D., Massaro E., Perola G.C., Spinoglio L., *Astron. Astrophys.* **235** (1990) 35.

January 1992 Microvariability Campaign on OJ 287

A. Sillanpää, L.O. Takalo and K. Nilsson [*]   S. Kikuchi [†]
Yu. S. Efimov and N.H. Shakhovskoy [‡]
D. Dultzin-Hacyan, R. Costero, and E. Benitez [§]
M.R. Kidger and J.A. de Diego [¶]

We present preliminary results from a multi-site microvariability campaign on the blazar OJ 287, conducted during January 10–15th, 1992. We observed OJ 287 using both CCD's (at Mexico and La Palma) and photopolarimeters (at Crimea, Japan and La Palma) in optical regions and also with an infrared photometer (Tenerife). No clear microvariability was detected during this campaign. On the other hand we detected small (0.5 mJy) amplitude variability in timescales of hours. The average V magnitude was 15.5.

In polarization we detected small night-to-night variations with the average polarization being 18%. The position angle was quite stable at 106°. During the first night of observations we detected a possible rotation of the position angle. The variations are shown in Fig. 1, where we have plotted the results from the most intense part of monitoring campaign.

---

[*]University of Turku, Tuorla Observatory, Tuorla, SF-21500 Piikkiö, Finland
[†]National Astronomical Observatory, Tokyo, Japan
[‡]Crimean Astrophysical Observatory, Nauchny, Russia
[§]Instituto de Astronomia, UNAM, Mexico
[¶]Instituto de Astrofisica de Canarias, 38200 La Laguna, Tenerife, Spain

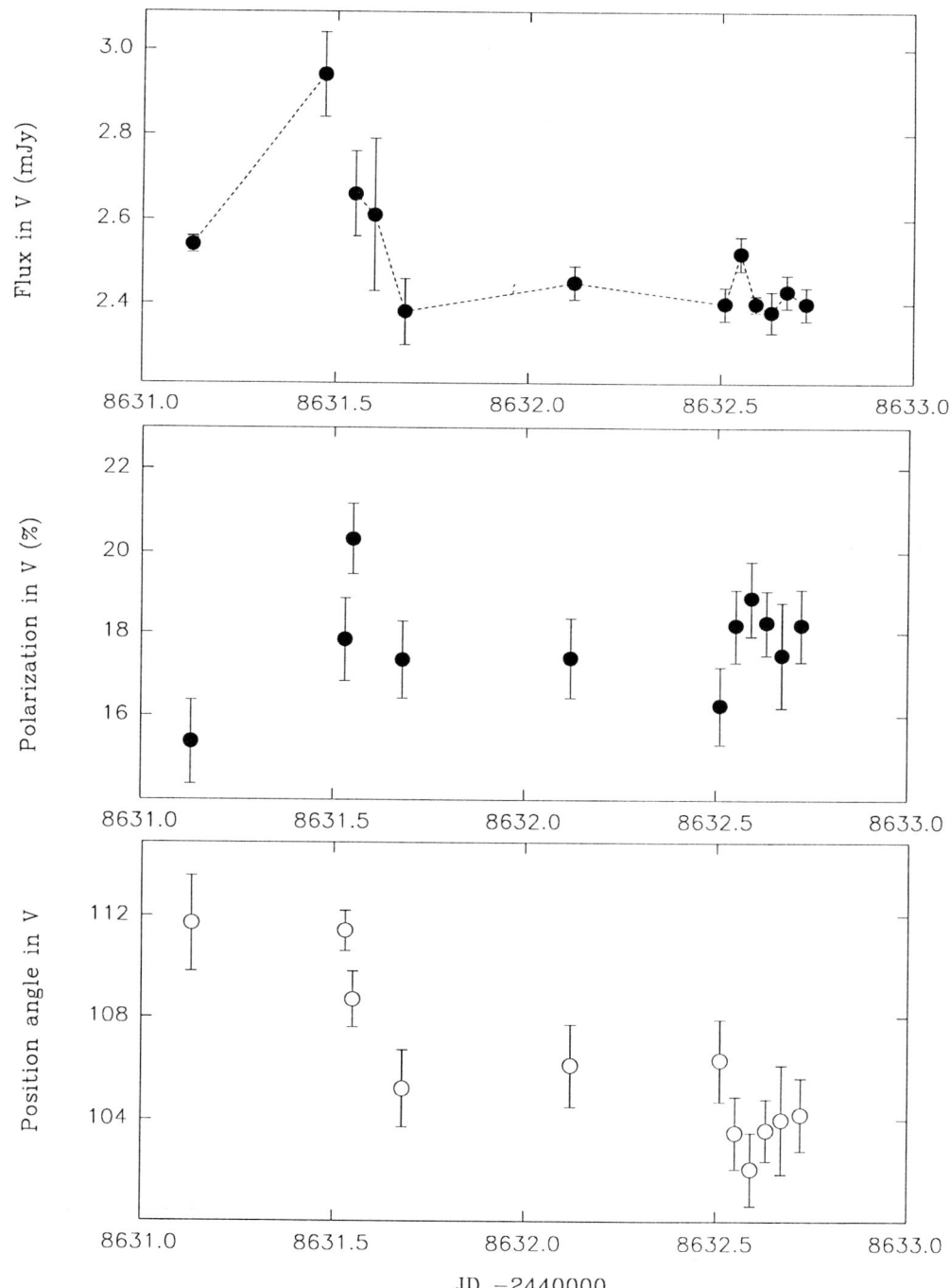

Figure 1. Photometric and polarimetric behaviour observed in OJ 287 during the monitoring campaign.

Blazar Microvariability: A Case Study of AO 0235+164

John C. Noble and H. Richard Miller [*]

1. Introduction

The existence of microvariabilty at optical wavelengths has been clearly demonstrated for BL Lacertae objects by a number of groups during the past several years (Miller & Carini 1991, Carini & Miller 1992, Wagner *et al.* 1991). However, in no instance has the nature of the microvariability been investigated when a blazar was near a minimum in brightness in its long-term variability. Thus, the blazar AO 0235+164 was selected to be monitored with the goal of determining whether or not rapid variations are present when the object is near its minimum brightness level, based on its known historical variability.

2. Observations

The observations of AO 0235+164 reported here were obtained with the 42-inch telescope at Lowell Observatory equipped with a direct CCD camera. The observations were made through an R filter with an RCA CCD. Repeated exposures of typically 300 seconds were obtained for the star field containing AO 0235+164 and several standard stars. These standard stars, located on the same CCD frame as AO 0235+164, provided comparisons for use in the data reduction process. The observations were reduced using the method of Howell & Jacoby (1986). Each exposure is processed through an aperture photometry routine which reduces the data as if it were produced by a multi-star photometer. Differential magnitudes can then be computed for any pair of stars on the frame. Thus, simultaneous observations of AO 0235+164, several comparison stars and the sky background will allow one to remove variations which may be due to fluctuations in either atmospheric transparency or extinction.

3. Discussion

The observations of AO 0235+164 obtained during 1991 November 5-8 indicated the presence of unusually rapid variations on all four of those nights. Well-defined events in which the object fades and recovers or undergoes an outburst and returns to its original state were detected on the nights of November 6 and November 7. However,

---

[*] Georgia State University

very rapid variations were detected on the nights of November 5 and November 8, with the most rapid rate of change of 0.05 mag/hour detected on the latter night. The variations observed for this object are among the most violent ever detected on these timescales for any blazar. This suggests that the radiation is generated in a small spatial volume. However, since the microvariations are of relatively small amplitude (typically on the order of 10%), One should be cautious not to necessarily interpret these changes as the result of global variations of the source.

One model (Wiita et al. 1991) would explain the observed microvariability as the result of excess emission produced by *flares* or *hot spots* randomly appearing and disappearing on the accretion disks around supermassive black holes. A similar model, stressing the X-ray fluctuations in AGNs, has also been proposed by Abramowicz et al. (1991). The lack of observed periodicities is in accord with this class of models, which assumes contributions from, typically, 10–100 flaring regions at any given time. Preliminary results reported by Wiita et al. (1991) yield good matches to the power density spectra (PDS) observed for BL Lacertae objects where the data has been taken with high sampling frequencies over several consecutive nights. It is important for future studies to obtain sufficiently well-sampled light curves for a number of blazars so that a similar analysis of the power density spectra for these variations can be compared with models such as those described above.

References

Abramowicz, M.A., Bao, G., Lanza, A., and Zhang, X.-H. 1991, *Astr. Ap.*, **245**, 454
Howell, S.B. and Jacoby, G.J. 1986, *P.A.S.P.*, **98** 802.
Wagner, S. J., Sanchez–Pons, F., Anton, K., Quirrenbach, A., and Witzel, A. 1991, Krichbaum, T., Hummel, C.A., Alberdi, A., *Variability of Active Galactic Nuclei*, ed. H.R. Miller and P.J. Wiita (Cambridge Univ. Press), p. 123.
Wiita, P.J., Miller, H.R., Gupta, N., and Chakrabarti, S.K. 1992, in *Variability in Blazars*, ed. E. Valtaoja and M. Valtonen (Cambridge: Cambridge U. Press) p. 311

The Timescales of the Optical Variability of the BL Lacertae Galaxy PKS 2201+044

J.W. Wilson [*]   H.R. Miller [*]   J.C. Noble [*]   M.T. Carini [†]

1. Introduction

BL Lacertae objects have been characterized by rapid and large amplitude optical variability, by a highly variable and polarized optical continuum which is featureless, or one in which any discrete features are found only in low contrast to the continuum. In the present investigation, the BL Lacertae object with the most significant galaxy component detected to date, PKS 2201+044, has been studied. The purpose of the investigation is to present the results of eighteen years of photometric monitoring of this BL Lacertae galaxy.

2. Observations

The observations of PKS 2201+044 were obtained with the 0.9 m and 1.3 m telescopes at KPNO and the 42–in. telescope at Lowell Observatory Observatory, all of which were equipped with a direct CCD camera. The details of the observations, data reduction, and analysis are the same as those described by Noble and Miller (this volume).

3. Discussion

Multiple aperture photoelectric and CCD observations were obtained on several nights. These were used to derive an aperture correction using the method outlined by Sandage (1973), which was applied to all the observations to derive V^*, the V-magnitude in a standard aperture of 15.42 arcseconds. Over the eighteen-year period of the observations, we see a general increase in brightness reaching a maximum in 1987, followed by a decline until the fall, 1991. A major exception to this is the observation of 1981 June 7 when the object was observed at $V = 16.31$. PKS 2201+044 was also observed at a similar brightness of $V = 16.43$ on 1987 November 9. The magnitude measured in these two instances is very close to the expected magnitude of the underlying galaxy. This indicates that the nonthermal component has essentially vanished at these times. If an accretion process is responsible for the nonthermal

---

[*]Georgia State University
[†]Science Programs, Computer Sciences Corporation & IUE Observatory

component, then these observed minima suggest either that the accretion has ceased or that the nonthermal central source has been obscured. This is similar to recent behaviour observed in BL Lac itself (Miller, 1988). Although this object exhibits a long-term brightening trend, on at least two occasions moderate but significant fluctuations in brightness were observed to take place. The first of these was noted on 1978 September 24 and 26 when the object was observed to have $V = 15.46$ and $V = 15.43$ respectively, compared to $V = 15.22$ on 1978 August 6. The second of these was observed on 1984 May 29. PKS 2201+044 is 0.41 magnitudes fainter than the observations obtained on 1983 June 6 and is 0.60 magnitudes fainter than the observations obtained 1985 June 21. In addition, evidence for microvariability was seen on 1991 September 21 when a variation of more than 0.1 magnitudes was observed during one night.

The observations presented in this paper, when compared with the historical light curve for this object, indicate a total range of 2.3 magnitudes, which is somewhat greater than that reported previously (Miller, 1977). However, the central source displays a much larger range, with an overall change observed in excess of 4 magnitudes. This is comparable to the historical variations observed for a number of BL Lacertae objects. The most rapid variation which was observed for the central source was the 0.4 magnitude change which was detected between 1979 November 13 and 1979 November 14.

References

Miller, H.R. 1988, in *Active Galactic Nuclei*, ed. H.R. Miller and P.J. Wiita (Berlin: Springer Verlag), p. 146.

Dynamics of Quasar Variability

S. Cristiani and R. Vio [*]

1. Introduction

We have heard many times during this conference that variability is a powerful tool to investigate the nature of AGN. The goal of this talk is to demonstrate that the clues derived from variability studies may be partial, disappointing and even misleading, if some characteristics of variability are overlooked and the proper statistical tools are not adopted. To convince you that such a caveat, although obvious, is important and productive, we should like to focus on a few issues, particularly important from our point of view.

2. The stochastic nature of the light curves

First of all, let's draw your attention to the fact that light curves of quasars have generally random behaviour, i.e., that the knowledge of the value of a time series at a given instant does not allow (in a broad sense) one to forecast the future evolution of the light curve. What does it mean from a physical point of view? It means that *quasars are dynamical systems of high dimensionality*, systems whose temporal evolution is described by an extremely large set of differential equations (or by partial differential equations). In other words, quasars are dynamical systems whose evolution is determined by a large number of external factors. To understand this point let us consider an example close to common sense: a pendulum. As is well known, the temporal evolution of such a system is described by a second order ordinary differential equation. However, if the pendulum is in open air, exposed to the action of a wind, its evolution depends also on a large set of differential equations describing the meteorological conditions around the system. It can be demonstrated that the superposition of the actions of a very large number of independent factors has a stochastic resultant, implying that the temporal behaviour of the pendulum will be random (unless somebody is able to forecast the intensity and the direction of the wind at any given instant).

3. The non-linear nature of the light curves

Failures of variability studies are partly due to the fact that light curves have been usually studied using "classical" statistical techniques, such as power spectrum analysis, the structure function and so on. These are all methods suited for analysing

---

[*]Department of Astronomy, University of Padova, Vicolo dell'Osservatorio 5, I-35122 Padova, Italy.

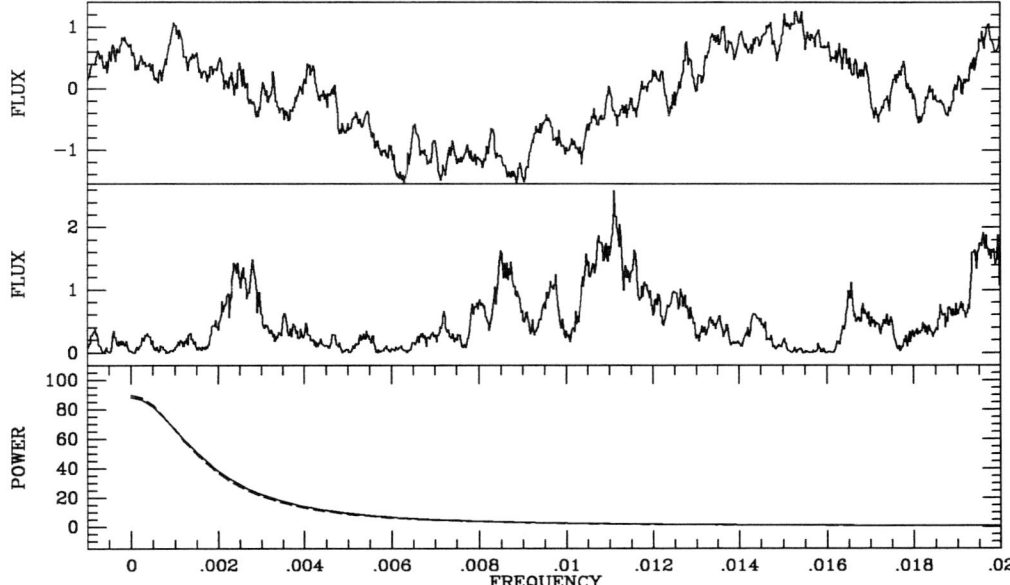

Figure 1. a) linear time series, b) non-linear time series, c) corresponding power spectra (process a = continuous line).

linear time series, e.g. signals characteristic of systems described by linear differential equations. In the case of quasars there is no *a priori* indication that things go in this way, the scenario involves a mixture of highly non-linear phenomena and the time series are expected to be non-linear. And indeed in a number of cases for which we have sufficient data our analysis shows that we are dealing with non-linear time series [2], [3]. Therefore other statistical techniques must be used.

To show you how unsatisfactory it can be to apply a linear technique to non-linear data let us show two simulated signals from two very different processes (Fig. 1), one linear and one non-linear, and look at their power spectra: they are indistinguishable. Which means that there is still a lot of information we are not using. And it would be easy to show you similar results for other linear indicators like the structure function.

To solve this problem it is necessary to adopt new methods, which are able to recognize the non-linearity of the signal under study, techniques like the bispectrum, state dependent models (SDM) and multifractal analysis. A bispectrum is able to distinguish between the two signals shown in Fig. 1 [3]. The bispectrum of a light curve from quasars like 3C 345 also shows the non-linear signature and this, together with other techniques, prompted us to a further characterization. The SDM led us to model the variability of 3C 345 with the simplest form of non-linear time series, a bilinear model which reproduces satisfactorily most of the characteristics of the observed light curve. This exercise showed us that the time series of 3C 345 is close to instability, non-stationary, non-invertible, which in turn explained a number of things we could not understand at the beginning, for example the issue of the "transient periodicities", which come out naturally, and have no special physical implication, for a system like this [2].

4. The stochastic approach

Is the problem solved by adopting more refined statistical techniques? Not at all. Although a correct statistical analysis is an essential step towards understanding, it has to be considered only a preliminary step. The main problem is that quasars are continuous stochastic systems, while most of the statistical techniques implicitly assume discrete evolution, so, for example, the results depend critically on the sampling frequency of the signal. The outcome is that in practice it is impossible to reconstruct the dynamical equations of the system on the basis of the observed signal only. *Statistics cannot substitute Physics*. How to overcome this drawback? With a tighter connection between statistical analysis of the data and theory:

1) From theory, and from information independently available, a physical scenario is derived. There will be a set of characteristic equations, a large set due to the complexity, the high dimensionality mentioned in Sec. 1. Think of our pendulum: there will be gravity, air resistance, the local atmosphere, the global atmosphere ... Of course, it is probably too much to hope to solve or even write down all the equations, we have to limit ourselves to a small set and put our ignorance somewhere, not simply ignore it.

2) The restricted set of equations has to be "stochasticized", i.e. some parameters have to become random variables or some random variables have to be added to account for our ignorance (remember that the superposition of a large number of independent processes gives a stochastic result). In this way complex systems are treated with a limited set of equations. Synthetic time series are produced by means of numerical integration.

3) The synthetic time series are compared with the observed ones on the basis of their statistical properties.

5. An example: NGC 6814

To illustrate how this simple but powerful approach works, let us show an exercise based on NGC 6814, the famous Seyfert galaxy mentioned several times during this conference. One of the models for its recurrent activity in the X-rays involves a pair production scenario [1]. The basic equations describe pair production and annihilation, photon production and escape, electron and positron heating and cooling, the proton density changes. Adopting a deterministic scenario, while ignoring the real complexity of the system, an interesting result is obtained, but clearly different from

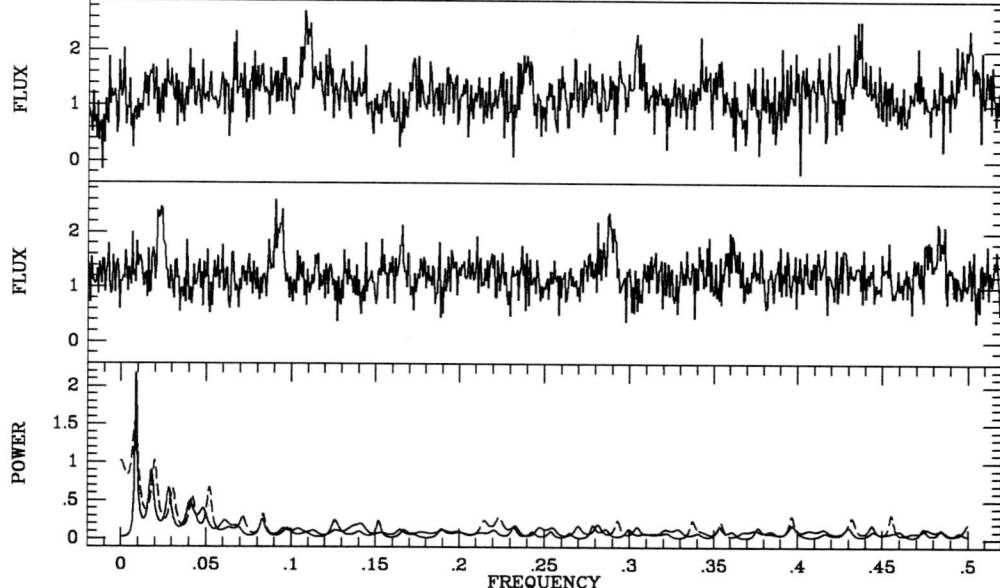

Figure 2. a) The observed X-ray EXOSAT light curve of NGC6814, b) simulation of the stochastic version of the pair production model, c) corresponding power spectra (process b = continuous line).

what is observed. Where should we put our ignorance? We have chosen the accretion rate (the only "external" parameter) as the variable to be stochasticized [4]. The result is shown in Fig. 2. The observed quasi-periodic behaviour, with large variations of the burst luminosity is naturally reproduced. Not too bad considering the extreme simplicity of the model!

The next step will be to apply this procedure to more complicated cases like perturbed accretion disks, for which partial differential equations have to be treated and life becomes much harder ...

References

[1] P. Moskalik, M. Sikora, 1986, *Nature* **319**, 649
[2] R. Vio, S. Cristiani, O. Lessi, L. Salvadori, 1991, *Astrophys. J.* **380**, 351
[3] R. Vio, S. Cristiani, O. Lessi, A. Provenzale, 1992, *Astrophys. J.* **391**, 518
[4] R. Vio, R. Turolla, S. Cristiani, C. Barbieri 1993, *Astrophys. J.* in press

The Variability of a Large Sample of Quasars

I. M. Hook [*] *R. G. McMahon* [*] *B. J. Boyle* [*] *M. J. Irwin* [†]

Abstract

The variability properties of a sample of over 300 optically–selected quasars near the South Galactic Pole (SGP) have been studied using a series of eleven UKST B_J plates at seven epochs, spanning 16 years. Quasars of high luminosity show significantly less variation than those with low luminosity. A similar, though much weaker, trend with redshift was found; lower redshift quasars varying proportionally more than high redshift quasars. The observed trends are a consequence of an intrinsic dependence of quasar variability on luminosity combined with the effects of time–dilation and have strong implications for quasar samples selected solely on variability.

1. Introduction

Variability provides a simple yet powerful means for investigating the physical processes at work in the inner regions of AGN. The primary diagnostics for optical variability are: the dependence on absolute magnitude and redshift, the timescale of variations in the quasar rest frame and the degree of coherence of individual quasar light curves — in our case taken as an ensemble. In addition to providing insight into quasar models an important feature of such a study is the ability to predict selection effects for quasar samples chosen purely on the basis of variability (*e.g.*, Hawkins 1986). In this paper we summarise our method and results: a more detailed account is given in Hook *et al.* (1991) and Hook *et al.* (1992).

2. Data

The sample of quasars was taken principally from the catalogue of Hewitt & Burbidge (1989) with additional objects from other surveys. All the quasars lie in one UK Schmidt field centred at the South Galactic Pole (SGP, $00^h53^m - 28°03'$) and were originally selected by UVX, objective prism and multicolour techniques — but not by their variability. The final working sample consists of 316 quasars in the range $0.3 < z < 4.1$, $m_B < 21.0$, $-29 < M_B < -23$. We have used $H_0 = 50\,kms^{-1}Mpc^{-1}$, $q_0 = 0.5$ and a $k-$correction with spectral index $\alpha = 0.5$ throughout.

---

[*] Institute of Astronomy, Madingley Road, Cambridge, CB3 OHA
[†] Royal Greenwich Observatory, Madingley Road, Cambridge, CB3 0EZ

The plate material consists of eleven plates at seven epochs, spanning a range of 16 years and having a minimum spacing between epochs of 1 month. The APM facility at Cambridge was used to measure the plates and to produce magnitudes for all the objects on the plates. The measuring errors as a function of apparent magnitude were then estimated. Over most of the range of interest ($17.5 < m_B < 20.5$) the rms measuring errors were approximately constant at $\sim 0.07 mag$.

3. Analysis and Results

The variability index σ_v for any particular quasar was defined by:

$$\sigma_v = \frac{1}{N-1} \sum_{i=1}^{N} | m_{Bi} - \overline{m}_B | \sqrt{\frac{\pi}{2}}$$

where N is the number of epochs at which the quasar was visible on the plate and m_{Bi} is the apparent magnitude measured from plate i. The factor $\sqrt{\frac{\pi}{2}}$ scales σ_v to be equal to the rms for a Gaussian distribution.

There is a highly significant correlation of σ_v with absolute magnitude (Fig. 1a) in the sense that more luminous quasars tend to have lower variability (rank correlation coefficient $R = 0.3$, significant at $> 5\sigma$ level). A weaker correlation can be seen with redshift ($R = -0.19$, at $\sim 3.3\sigma$ level, Fig. 1b). No correlation with apparent magnitude was found at the 3σ level.

The observed correlations found between variability index, absolute magnitude and redshift (Fig. 1) imply that surveys using a selection criterion of variability above some set threshold will be relatively incomplete for high luminosity quasars and for quasars at high redshift.

We have investigated the time dependence of the variations by computing the magnitude differences $| \Delta m |$ and corresponding time differences Δt_{obs} for each pair of plates for each quasar. The median values in bins of time-interval define the structure function for the ensemble. The structure function shows a steady rise in variation with time interval, indicating that quasars vary on all time–scales up to $\sim$ ten years (see Hook et al. 1991), and the dependence of variability with absolute magnitude (or redshift) is again apparent. To account for the contribution to this dependence from the effects of time-dilation $\Delta t_{rest} = \Delta t_{obs}/(1+z)$ was substituted for Δt above. Fig. 2 shows that, even with the effects of time–dilation removed, there is a clear trend for more luminous quasars to vary less. No significant trend with redshift was found.

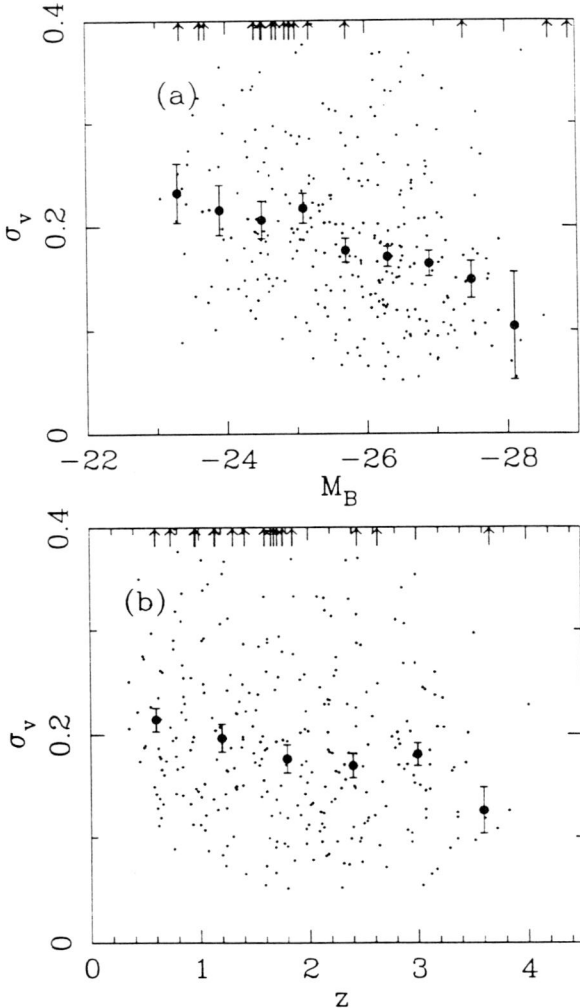

Figure 1. Variability as a function of (a) absolute magnitude and (b) redshift, with measuring errors included. Dots represent the data, circles mark the median values in each bin. The errors bars are 1σ errors from the median values.

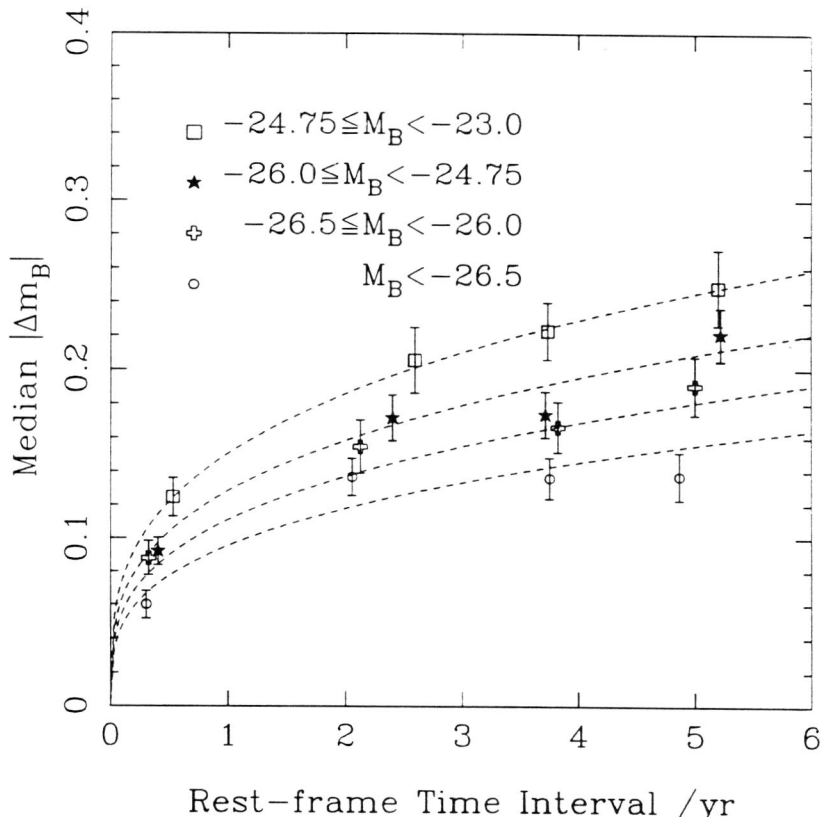

Figure 2. Ensemble changes in magnitude as a function of rest-frame time interval. The contribution to $|\Delta m|$ due to measuring errors (dependent on apparent magnitude) has been subtracted out. The curves represent the best fit to the data (using χ^2 minimisation) of the functional form $|\Delta m| = (A + BM_B)\Delta t_{rest}^p$, i.e., with $p = 0.30$, $A = 0.64$ and $B = 0.02$.

References

Hawkins, M. R. S. 1986, Mon. Not. R. astr. Soc. **219**, 417.
Hewitt, A. and Burbidge, G. 1987, Astrophys. J. Supp. **63**, 1.
Hook, I. M., McMahon, R. G., Irwin, M. J., Boyle, B. J. 1991, Publs. astr. Soc. Pacif. conference series vol. 21, *The Space Density of Quasars*, ed. David Crampton.
Hook, I. M., McMahon, R. G., Irwin, M. J., Boyle, B. J. 1992, Mon. Not. R. astr. Soc. *in preparation*

The Fate of Central Black Holes in Merging Galaxies

F. Governato, M. Colpi and L. Maraschi [*]

Abstract

We present results of a first series of numerical N-body experiments that describe the merging of galaxy pairs containing a massive black hole in their core.

If the nuclei of most galaxies harbour a massive black hole (BH) in their core, and most galaxies undergo a merging process, massive black hole binaries may form [1], [2], [3]. In this process, do the BHs always approach so closely as to become a bound pair? We report here on a first series of numerical examples in which we find that a key parameter controlling the process of BH binary formation is the initial density contrast of the two merging galaxies.

We describe the galaxies as isotropic King spheres with equal masses and dimensionless value of the central potential $\Psi(0)/\sigma^2$ equal to 7 and 1 (K7 and K1, hereafter). Model K7 is more centrally condensed than K1, the ratio of the two mean densities $\bar{\rho}_{K7}/\bar{\rho}_{K1}$ being ~ 13. To each galaxy model a BH is added with a mass equal to 1% of the total mass. We then simulated the circular encounter between two King spheres using the TREECODE with variable time step kindly provided us by Hernquist.

The present analysis shows that in encounters between identical galaxies dynamical friction is always effective in driving each black hole to the centre of the remnant, eventually forming a bound binary (Fig. 1). Conversely, if the encounter is between a loose (K1) and a centrally condensed (K7) galaxy, the merging process does not lead to a black hole binary (Fig. 2).

In the centrally condensed galaxy, the core survives the initial stage of merging and rapidly settles toward the centre of the system together with its black hole. In the loose low density galaxy, the central core is rapidly disrupted by the tidal field, and the naked black hole remains in the external region of the remnant where dynamical friction is weak. This behaviour is likely to be caused by the transfer of orbital energy from the surviving core of the condensed galaxy to the rest of the system.

We are now performing simulations including also disk galaxies, with the aim of selecting the conditions for which galaxy mergers do lead to bound BH pairs.

---

[*] Department of Physics, Universitá degli Studi di Milano, Via Celoria 16, 20133 Milano, Italy.

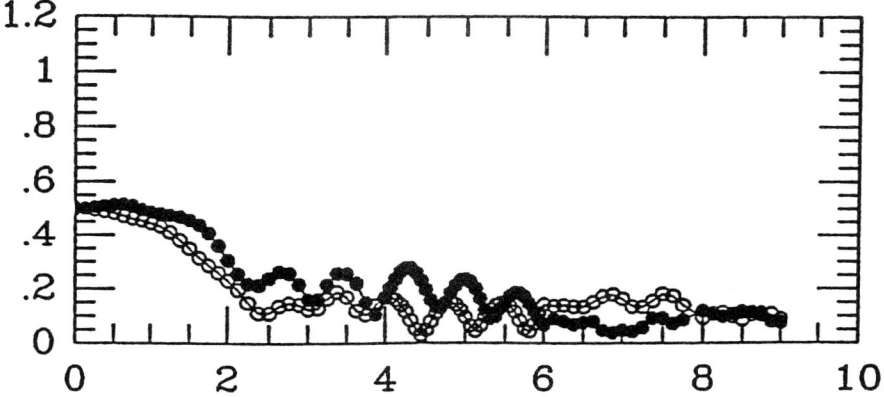

Figure 1. Circular encounter between two identical King spheres: merger between model K1 and K1; distance of the two BHs from the centre of mass of the system, as a function of time (in units of the crossing time).

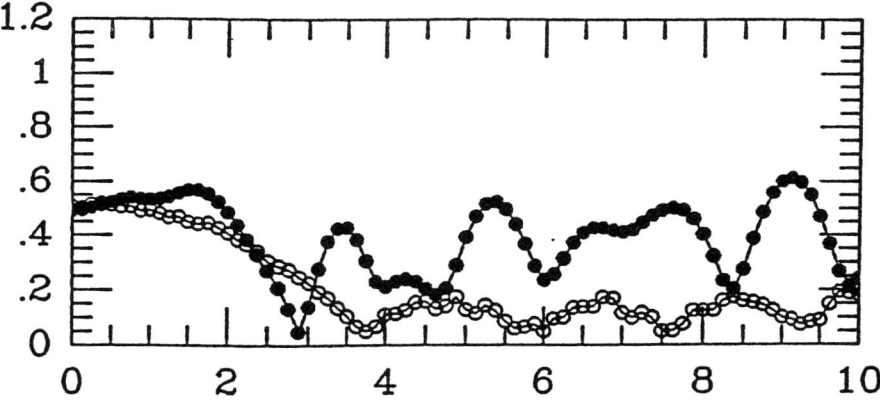

Figure 2. Circular encounter between two King spheres: merger between model K1 and K7; distance of K1's BH (*filled circles*) and K7's BH (*open circles*) from the centre of mass of the system, as a function of time (in units of the crossing time).

References

[1] Begelman, M.C., Blandford, R.D., & Rees, M.J. 1980, Nature 287, 307
[2] Rees, M.J. 1990, Science 247, 817.
[3] Gaskell, C.M. 1985, Nature 315, 386.

Polarimetric Searching for Goldstone Bosons from AGNS

Yu. N. Gnedin [*,†]   *S. V. Krasnikov* [†]

Abstract

The oscillations between photon and arion in intercluster galaxies magnetic field should produce a noticeable amount of polarization of AGN radiation. The radiation flux and degree of polarization can oscillate with respect to cosmological red-shift. This process is dependent on the values of the intercluster magnetic field and of the arion-photon coupling constant.

Astrophysical and cosmological considerations have led to numerous constraints on new particle physics phenomena. Observational and experimental searches for axions (arions or omions which are massless bosons) depend crucially on the symmetry-breaking scale F and couplings of axions with matter g.

The detection of axions is possible through their coupling to photons. The direct annihilation of an axion into two photons (Fig. 1a) has a finite although extremely low cross-section. However, an interaction with the magnetic field via the Primakoff process can lead to the production of photons whose energy is comparable to the axion (arion) total energy (Fig. 1b). The reverse process: photon ↔ arion (axion) transition in a magnetic field is also very important.

The probability of the transformation commonly depends on the refractive index of the medium n, the angle between the directions of the magnetic field B and the photon momentum and on the constant of arion-photon coupling g. If radiation with initial intensity I_0 and frequency ω travels a distance b in a magnetic field B, it's final intensity will be equal to:

$$I_* = I_0 \left[1 - \frac{1}{1+x^2} \sin^2 \left(\tfrac{1}{2}\sqrt{x^2+1}\, Bgb \right) \right] \tag{1}$$

where: $\hbar = c = 1$, $x = \ell_p/\ell_a = |1 - n|\omega/Bg$ (for full details see Raffelt and Stodolsky, 1988; Anselm, 1988 and Gnedin and Krasnikov, 1992).

---

[*] The Central Astronomical Observatory at Pulkovo, St - Petersburg, Russia
[†] Visiting Astronomer of the Royal Greenwich Observatory

For simplicity we restrict ourselves to the case when $B_\perp = B$. If photons propagate through the plasma with electron number density N_e, then $|x| \leq 1$ for

$$\omega \in \left(10^{-7} N_e / Bg, \; 4 \times 10^{26} g / B\right) \tag{2}$$

Since photons polarized perpendicularly to the magnetic field are not affected by arions the process of photon-arion conversion acts as an additional absorption process strongly depending on polarization state. Therefore it can lead to noticeable photometric and polarimetric effects for AGN radiation which propagates across magnetic field lines as in the vicinity of AGN so and in the intergalactic medium (IGM). It is possible to use AGN as probes of the IGM.

The oscillation lengths in the IGM can be estimated as (Gnedin, 1992):

$$\begin{aligned} \ell_a &= 60 \left(10^{-11} GeV^{-1}/g\right)\left(10^{-9} b/B\right) Mpc \\ \ell_p &= 8 (\omega/3 eV)(0.005/\Omega h_{50}) \; Mpc \end{aligned} \tag{3}$$

Equations 3 are valid for the medium between galaxy clusters. We use the estimate of the IGM magnetic field from Fujimato (1990). The expected level of polarization is

$$p_e \approx \left(\frac{\omega}{3eV}\right)^2 \left(\frac{0.005}{\Omega h_{50}}\right)^2 \left(\frac{g}{10^{-11} GeV^{-1}}\right)^2 \left(\frac{B}{10^{-9}}\right)^2 \% \tag{4}$$

The expected level of polarization is $p_e \sim 0.5\%$, and it should increase towards higher frequencies according to $p_e \sim \omega^2$. The polarization will oscillate with respect to cosmological redshift, with a maximum at

$$z \approx \frac{50(\pi h_{50})}{(gcB_{IGM})} \tag{5}$$

This effect can be tested by polarimetry of a sample of AGN.

References

Anselm A.A. 1988. *Phys. Rev.*, **D37**, 2001.
Fujimato, M. 1990. *Publ. Astron. Soc. Japan*, **43**, L39.
Gnedin, Yu. N. 1992, in *Frontiers of X-ray Astronomy*, Proc. of the 28th Yamada Conference, ed. Y. Taranaka and K. Koyama, p715.
Gnedin, Yu. N. & Krasnikov, S.V. 1992. JETP, in press.
Raffelt, G. & Stodolsky, L. 1988. *Phys. Rev.*, **D37**, 1237.

Concluding Talk

Unification of AGNs, and the Starburst Hypothesis

Alexei V. Filippenko [*]

1. Unification of AGNs

There is now ample evidence for partial unification of (a) Seyfert 1 and Seyfert 2 galaxies, in terms of obscuring tori and orientation effects; (b) steep-spectrum (lobe dominated) radio quasars and flat-spectrum (core dominated) radio quasars, in terms of relativistic beaming and orientation effects; and (c) blazars, radio-loud quasars and broad-line radio galaxies, and narrow-line radio galaxies, in terms of obscuring tori, relativistic beaming, and orientation effects. Variants or subsets of these basic sets have also been proposed — e.g., BL Lac objects and Fanaroff-Riley (1974) Class I galaxies. Much of the observational evidence was discussed at this conference, primarily by Bob Fosbury, Clive Tadhunter, Neal Jackson, Thaisa Storchi-Bergmann, Paul Alexander, Paolo Padovani, Dave Axon, and Alec Boksenberg, as well as in several posters; it involves spectropolarimetry, superluminal motion, projected sizes, relative numbers of different objects, lobe depolarization asymmetry, the inverse-Compton limit, ionization cones, variability, and other phenomena. A reasonable conclusion is that unification schemes *must* be correct at least to *some* degree.

It is not so clear, however, that the *strong forms* of the unification schemes are valid. Specifically, let me focus on the unification of Seyfert 1 and Seyfert 2 galaxies (Antonucci & Miller 1985). The strong form states that these are *exactly* the same objects, simply viewed from different directions; *no* other factors (such as differences in the thickness or opening angle of the obscuring torus) are involved.

To keep this idea firmly in mind, I present the following analogy. One notices that, observationally, there are two main types of pretzels (Fig. 1): long and skinny, which I call Type 2, and the more complicated "loopy" kind, which I call Type 1. [The Type 1s, at least, come in two sizes — small ("dwarf") and large — but this is not central to my argument.] Remarkably, these particular pretzels are made by Seyfert Foods-Y (part of Borden, Inc.); they are, in fact, "Seyfert's[R] Butter Pretzels." (Incidentally, I think they taste quite good.) Now, if we were observing these pretzels from a great distance, a natural question would arise: are the Type 2s really just Type 1s seen edge-on? In this case it is obvious that a slight deviation from the edge-on orientation would immediately reveal that the object is actually a Type 1. Thus, we would expect to see very few Type 2s if *all* the pretzels are intrinsically Type 1 and

---

[*]Department of Astronomy, University of California, Berkeley, CA, USA.

have random orientations with respect to our line of sight. A large *observed* fraction of Type 2s relative to Type 1s in a volume-limited sample would, consequently, imply that most (but not all) Type 2s differ intrinsically from Type 1s.

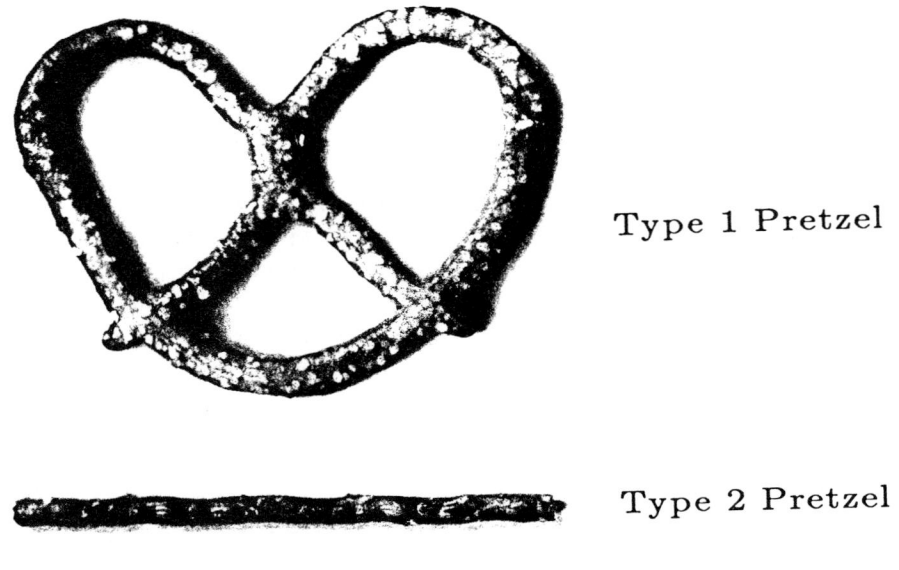

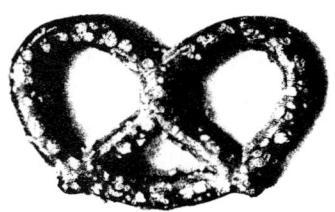

Figure 1. Pretzel classification. Partial unification of Types 1 and 2 is achieved by viewing Type 1 from the side. Made with "Seyfert's^R Butter Pretzels."

This kind of test can, in principle, be applied to Seyfert nuclei. Assuming *all* Seyfert 2 galaxies do indeed harbour Seyfert 1 nuclei, we can use the relative number densities of Type 1 and Type 2 Seyferts (e.g., Rush & Malkan 1992) to conclude that the obscuring tori have a sizeable physical thickness relative to the broad-line region (BLR) and the source of the continuum emission. A considerable deviation from an edge-on orientation is necessary for one to see the Seyfert 1 nucleus directly. If all obscuring tori were the same size, we could determine the opening angle of the torus (assuming a simple geometry) from the observed number densities.

There is, however, some intriguing evidence that the tori are *not* all identical in different Seyfert galaxies; this argues against the *strong* form of the Seyfert unification hypothesis. For example, Seyfert 2 galaxies seem to be more luminous than Seyfert 1s in CO emission and at infrared wavelengths (Heckman et al. 1989), suggesting that their molecular tori are larger on average. If Seyfert galaxies have tori with a range of sizes, those with large, thick tori are more likely to appear observationally as Seyfert 2s than as Seyfert 1s.

It is also important to note the existence of Seyfert 2 nuclei with prominent blue continua (in some cases polarized, in others unpolarized), yet no sign of broad Balmer lines in either their total flux or polarized flux spectra (Miller & Goodrich 1990). These objects might be "true blue Seyfert 2s" with no BLRs. Cid Fernandes & Terlevich (1992) suggest that a young stellar population is responsible for the blueness of the observed continua in Seyfert 2s showing low polarization. This would, in addition, explain why no scattered broad emission lines are seen. Direct tests of the starburst hypothesis for Seyfert 2s can be made with the Hubble Space Telescope, but in any case it seems likely that at least *some* Seyfert 2s do not harbour Seyfert 1 nuclei with BLRs. If so, Seyfert unification only works for a subset of objects.

Preliminary evidence for another important difference between Seyfert 1s and 2s was presented at this conference by Ed Moran and Jules Halpern (see also Moran et al. 1992). When compared with a plot of the $\lambda = 20$ cm luminosity function of Seyfert galaxies (Types 1 and 2 together, from the complete distance-limited sample of Ulvestad & Wilson 1989), the ten known Seyfert 2s with published detections of broad Balmer emission lines in their polarized-flux spectra ("hidden Seyfert 1s") fall in the top 15%, and eight of the ten fall in the top 4%. Thus, hidden Seyfert 1 nuclei are more radio loud than typical Seyfert 1s or 2s! If the radio luminosity in Seyferts is an isotropic property, then high radio luminosity is an intrinsic property of hidden Seyfert 1s. The same conclusion holds even when the radio radiation is significantly beamed; if anything, the hidden Seyfert 1 nuclei should appear *fainter* at radio wavelengths than normal Seyfert 1s, since the former are beamed away from us. Taken at face value, the discovery of Moran and Halpern therefore eliminates the strong form of Seyfert unification. The most radio-loud Seyfert 1 nuclei seem to "clear out" potentially obscuring material in the direction of their radio jets, thereby establishing conditions that are particularly favourable for detection of *scattered* radiation from the nucleus and the BLR. This is also consistent with Pogge's (1989) conclusion that there are significant differences in the circumnuclear regions of many Type 1 and Type 2 Seyfert galaxies (as determined from narrow-band optical images) which are not easily understood in the context of simple unification schemes.

We must, of course, be wary of selection effects. One disturbing problem is that only about half of the known hidden Seyfert 1s were taken from the Ulvestad & Wilson (1989) distance-limited sample; the others are more distant, and hence are probably more luminous at all wavelengths including radio. Moreover, the selection of objects has not been random even within the distance-limited sample; those chosen for spectropolarimetric observations have generally been the most luminous at UV and other wavelengths. This limitation could be overcome with an extensive, high-quality spectropolarimetric survey of *all* objects in the Ulvestad & Wilson (1989) list, preferably at Hα. [In her thesis, Kay (1990) studied a large sample of Seyfert 2s at Hβ, which is considerably more difficult to detect than Hα.] Finally, not all hidden Seyfert 1s may reveal themselves through polarized Balmer emission lines, due to unsuitable physical conditions (optical depth to electron scattering, etc.); other characteristics for which to search include broad unpolarized Paschen lines and ionization cones, although in my opinion the latter are not *necessarily* indicative of hidden Seyfert 1 nuclei.

Despite the above caveats, I am intrigued by the observed difference in radio luminosities between hidden Seyfert 1s and normal Seyferts. My gut feeling is that at least part of the effect is due to intrinsic differences, rather than to selection biases.

One definitive way to test the unification scheme for Type 1 and Type 2 Seyferts is to measure their hard X-ray to gamma-ray continua (100 keV to 10 MeV). The obscuring torus should be transparent to MeV photons (the Klein-Nishina cross section is low), so one would expect to see no differences in the continuum shape if Seyfert 2s are actually Seyfert 1s seen from the side. Although few concrete results are available to the public at this time, there are rumours that the OSSE instrument aboard the Gamma Ray Observatory detected some Seyfert 1s but not Seyfert 2s — suggesting that the two classes are not identical.

This is clearly an area ripe for further exploration, and I was happy to see that many of the participants at this conference are involved in it. There are numerous potentially interesting questions to answer. For example, do *all* Seyfert 2s with ionization cones show a polarized blue continuum? What fraction of these exhibit broad Hα emission? How do narrow-line radio galaxies (some of which, like Cyg A, have extended blue light) fit in? What can account for the highly polarized blue continuum in Seyfert 2s whose polarized-flux spectra are devoid of broad Balmer lines? Is there any evidence for absorption lines from hot stars in the UV continua of Seyferts? Are there compact, variable, radio or X-ray cores in "pure" Seyfert 2s? How frequently can one detect broad Paschen lines in high-quality infrared spectra of Seyfert 2s? We have a lot to learn — not just about unification schemes, but also about the fundamental nature of Seyfert nuclei.

2. The Starburst Hypothesis for AGNs

This leads me to the second main topic that I want to cover: the starburst hypothesis of AGNs, especially that of Roberto Terlevich and his collaborators (Terlevich 1992; Terlevich *et al.* 1992, and references therein). Interesting presentations were made at this conference by Roberto Terlevich and Guillermo Tenorio-Tagle, and several posters further explored the idea. I feel that great strides have been made by Terlevich's team, and the idea merits serious consideration. At least partial success has been achieved on a number of topics in very few years of work: the starburst model has addressed issues such as energetics, optical light curves (flares, undulations), optical and UV emission-line intensity ratios, continuum/emission-line lags, QSO formation in terms of the young cores of elliptical galaxies, sizes and velocity dispersions of AGNs, and "Warmer" models of Seyfert 2 galaxies and low-ionization nuclear emission-line regions (LINERs; Heckman 1980). At this conference, I was particularly impressed with the possibility that Brian Boyle's QSO luminosity function might be explained as a natural consequence of core formation in elliptical galaxies, and with the discovery by Gary Ferland and Fred Hamann of a very high nitrogen abundance in high-redshift QSOs (suggestive of early starbursts). It should also be noted that there is direct evidence of at least *some* Type II (hydrogen-rich) supernovae in which interaction of the ejecta with a dense circumstellar wind ($n \approx 10^7$ cm^{-3}) leads to spectroscopic properties resembling those of Type 1 Seyferts and QSOs (Filippenko 1989, 1992).

However, as discussed by Heckman (1991), Filippenko (1992), and in the written version (Filippenko *et al.* 1993) of the panel discussion held at the May 1992 Madrid meeting on *The Nearest Active Galaxies*, the starburst hypothesis still faces many major challenges. I believe the most serious problem is to explain the high-energy radiation from AGNs. This includes the gamma rays seen in at least a few luminous quasars, as well as the rapid X-ray variability observed in many objects. Although the details of the production of high-energy radiation and rapid variability have not yet been determined in the context of the standard black hole model, I don't consider this to be a fundamental difficulty; black holes have deep potential wells that should, in principle, give rise to the observed phenomena. Indeed, it is worth noting the class of Galactic X-ray binaries whose properties appear to require the presence of a black hole. By contrast, it is difficult to see how supernovae in dense environments could produce rapid X-ray variability and significant radiation above 100 keV.

I am also quite concerned by the fact that the starburst scenario does not include radio-loud AGNs, which in most other respects are nearly identical to radio-quiet AGNs. It seems that the existence of superluminal motion, narrow jets extending tens of Mpc in a constant direction, and extremely compact cores cannot be explained by a collection of supernovae. These properties constitute perhaps the best indirect evidence for the existence of a spinning, supermassive black hole. BL Lac objects and optically violent variables, which are consistent with unified schemes of AGNs, also find no natural explanation in the starburst picture. I think it is likely that radio-loud and radio-quiet AGNs constitute fundamentally the *same* phenomenon, with somewhat different manifestations due to differences in certain physical parameters (e.g., the mass, spin, and accretion rate of the black hole; the density of the interstellar medium; see Blandford 1990).

Another important problem is the fact that we have not yet identified the "progenitor starbursts" of QSOs. The starburst preceding the QSO phase should be 10 to 100 times more luminous than the QSO itself. Where are these objects? They might be ultraluminous IRAS galaxies, but we are not yet sure, and in any case difficulties still remain. Related to this is the deduction that the central masses of QSOs should be enormous if QSOs are fundamentally a starburst phenomenon; the conversion efficiency of mass into energy is very much lower for stars than for accretion into a black hole, meaning that a large number of stars is required. The resulting central velocity dispersion is greater than measured in any nearby galaxies *unless* the starburst extended over a reasonably large radius (Terlevich 1992). If so, however, I would think that at least some QSOs should have been spatially resolved by the Hubble Space Telescope at the 0.1" scale, yet none has been to my knowledge.

It is also not clear that the starburst scenario can reproduce the overall spectra of AGNs together with their variability. Fairall 9, for example, has shown UV variability of a factor of 33 in 6 years (Clavel, Wamsteker, & Glass 1989) — yet a substantial fraction of the UV light in the starburst model is supposed to come from hot stars, and variability of such high amplitude should therefore not be present.

In Terlevich's model, supernovae should produce the broad Hα seen in Seyfert 1 nuclei and some LINERs; an AGN with $M_B \approx -19.5$ mag is powered by about one supernova per year. Thus, NGC 4051 ($M_B \approx -16$ mag) and M81 ($M_B \approx -13$ mag) require a supernova every 25 and 400 years, respectively. One therefore predicts the absence of variations in the broad Hα flux on short time scales. Filippenko & Sargent (1988) did not detect any Hα variability in several spectra of M81 spanning a baseline of a few years, but broad-line variability *has* been detected in NGC 4051 over months or years (e.g., Peterson, Crenshaw, & Meyers 1985). Moreover, quite a few low-luminosity Seyfert 1 nuclei (including NGC 4051 and M81) are known to have exhibited very rapid, high-amplitude X-ray variability (e.g., Barr & Mushotzky 1986). I consider this to be a significant problem for the starburst hypothesis.

There are, in addition, numerous details that remain to be explained by the starburst hypothesis. For example, it appears that the characteristic size of the broad-line region in AGNs scales roughly as the square root of the continuum luminosity. Why should this be the case with supernovae? Profile variability is also a difficult question: why are the *wings* of line profiles more stable than the cores in at least some well-observed AGNs (Mike Crenshaw, this conference; see also Ferland, Korista, & Peterson 1990), when the simple supernova hypothesis would predict the opposite trend? Why are the profile shapes in a given AGN approximately stable over long time intervals, despite the presence of substantial short-term variability? Moreover, why do certain spectral details (such as the profiles of the broad emission lines) of known "Seyfert 1 supernovae" (Filippenko 1989) differ from those of Type 1 Seyferts? The amplitude of QSO variability as a function of luminosity also seems to be a problem for the starburst scenario, as emphasized by Isabel Hook at this conference. Another potential thorn may be the upper limit to the size of one QSO determined through studies of microlensing (Rauch & Blandford 1991). Of course, at present not all of these problems have unambiguous solutions in the context of the black hole model, so they shouldn't be used as strong evidence *against* the starburst hypothesis. However, starburst proponents would do well to keep them in mind, because eventually they will have to be explained.

Despite being somewhat sceptical about the general applicability of the starburst hypothesis to AGNs, especially very luminous ones, I think it may indeed explain certain types of low-luminosity AGNs. It is possible, for example, that some fraction of Seyfert 2 galaxies have an essentially stellar origin for their apparent "activity" (Cid Fernandes & Terlevich 1992). Furthermore, a majority of LINERs having spatially extended emission ($r > 200$ pc) are probably powered by supernova-driven winds, galaxy interactions, or mergers, rather than by a nonstellar ionizing continuum emerging from the nucleus. Turning now to compact LINERs, Filippenko & Terlevich (1992) demonstrate that those with weak [O I] $\lambda 6300$ emission (relative to Hα) could harbour nothing more than a collection of hot O-type stars in their nuclei, rather than massive black holes. Only about 30 O-type stars are needed to explain an Hβ luminosity of 10^{38} erg s^{-1}. Extending this argument, Shields (1992) suggests that many LINERs with considerably higher [O I]/Hα intensity ratios may be powered by starbursts; gas spanning a wide range of densities is necessary in the galactic nuclei, but there is ample observational evidence for this (Filippenko & Sargent 1988, and references therein).

3. Questions

Let me end with a list of questions I prepared for the discussion following the three summary presentations. Some of these were directed at specific participants, while others were open to everyone. There was insufficient time to cover all topics, but perhaps the reader will now be motivated to think about them. I do not repeat some of the questions listed in the previous two sections of this summary.

(1) In their starburst hypothesis, Terlevich and collaborators have eliminated (at least temporarily) AGNs with evidence for relativistic jets. Should they also ignore objects with high gamma-ray and X-ray luminosities, and AGNs with rapid X-ray variability? Is there any realistic hope that such objects will be explained with starbursts?

(2) Do the results of reverberation mapping of Type 1 Seyferts (as discussed at this conference by Brad Peterson, Ernst van Groningen, Keith Horne, and others) prove the existence of a *single* compact energy source, or is it still possible that the source position could vary for different outbursts? Also, could the evidence for anisotropic emission be best interpreted in terms of a stellar origin for the broad emission lines?

(3) Is there any evidence for mass redistribution in Seyfert 1 nuclei, or does the trans-

fer function change for other reasons (e.g., changes in the gas density or ionization parameter)?

(4) Is there strong evidence that *all* Type 2 Seyfert nuclei are Seyfert 1s viewed from the side? Are there any "pure Seyfert 2s"? What factors other than orientation are important? Are all Seyfert 2s with ionization cones actually hidden Seyfert 1s?

(5) Are the proposed explanations for the long-term periodic variability of X-rays in NGC 6814 realistic? Do they strongly support the black hole model, and can they be used to definitively argue against the starburst hypothesis?

(6) Why aren't the heavy-element abundances (specifically nitrogen) of low-redshift QSOs clearly *higher* than those of high-redshift QSOs? How do the derived abundances scale with luminosity for QSOs of a given redshift, and how do they scale with redshift for QSOs of a given luminosity? What sorts of starbursts were involved?

(7) In the context of the starburst model, where does the highly variable UV continuum come from in Fairall 9? Can the energy budget of the largest observed outbursts ($\sim 10^{53}$ erg over a thousand days) be explained with supernovae?

(8) What is the direct observational evidence for accretion disks in AGNs? Do present data (e.g., absence of Lyman edges and polarization) greatly threaten a thermal interpretation of the big blue bump? Could hot stars be responsible for the bump?

(9) How are broad absorption line (BAL) QSOs produced? Why are BAL QSOs rarely observed as radio sources, and what is their relationship to other AGNs? Can BAL QSOs be explained with starburst models?

(10) How well do the various unification schemes explain the relative number densities of different types of AGNs? Is there strong evidence for large differences in the Lorentz factor (γ) for radiation at different wavelengths?

Acknowledgements I thank the organizers for a stimulating conference and financial assistance, and Brian Boyle for setting up a computer account to be used by participants. Additional travel support was provided by the Committee on Research at U. C. Berkeley. My work on AGNs is funded by NSF Grant AST–8957063. I am grateful to Ed Moran, Jules Halpern, Roberto Terlevich, and Guillermo Tenorio-Tagle for particularly informative discussions, and to Joe Shields for my first package of Seyfert's^R Butter Pretzels.

References

Antonucci, R.R.J., & Miller, J S. 1985, *ApJ*, **297**, 621.
Barr, P., & Mushotzky, R.F. 1986, *Nature*, **320**, 421.
Blandford, R.D. 1990, in *Active Galactic Nuclei*, by R.D. Blandford, H. Netzer, & L. Woltjer (Berlin: Springer-Verlag), p. 161.

Cid Fernandes, R., Jr., & Terlevich, R. 1992, in *Relationships Between Active Galactic Nuclei and Starburst Galaxies*, ed. A. V. Filippenko (San Francisco: Astron. Soc. Pacific), p. 241.
Clavel, J., Wamsteker, W., & Glass, I.S. 1989, *ApJ*, **337**, 236.
Fanaroff, B.L., & Riley, J.M. 1974, *MNRAS*, **167**, 31P.
Ferland, G.J., Korista, K.T., & Peterson, B.M. 1990, *ApJL*, **363**, L21.
Filippenko, A.V. 1989, *AJ*, **97**, 726.
Filippenko, A.V. 1992, in *Physics of Active Galactic Nuclei*, ed. W.J. Duschl & S.J. Wagner (Berlin: Springer-Verlag), p. 345.
Filippenko, A.V., Conti, P.S., Genzel, R., Heckman, T.M., Mushotzky, R.F., & Terlevich, R.J. 1993, in *The Nearest Active Galaxies*, ed. J. Beckman (Madrid: CSIC), in press.
Filippenko, A.V., & Sargent, W.L.W. 1988, *ApJ*, **324**, 134.
Filippenko, A.V., & Terlevich, R. 1992, *ApJL*, **397**, L79.
Heckman, T. . 1980, *A&A*, **87**, 152.
Heckman, T.M. 1991, in *Massive Stars in Starburst Galaxies*, ed. C. Leitherer, *et al.* (Cambridge: Cambridge Univ. Press), p. 289.
Heckman, T.M., Blitz, L., Wilson, A.S., Armus, L., & Miley, G.K. 1989, *ApJ*, **342**, 735.
Kay, L.E. 1990, *Ph.D. Thesis*, University of California, Santa Cruz.
Miller, J.S., & Goodrich, R.W. 1990, *ApJ*, **355**, 456.
Moran, E.C., Halpern, J.P., Bothun, G., & Becker, R. 1992, *AJ*, in press.
Peterson, B.M., Crenshaw, D.M., & Meyers, K.A. 1985, *ApJ*, **298**, 283.
Pogge, R. 1989, *ApJ*, **345**, 730.
Rauch, K.P., & Blandford, R.D. 1991, *ApJL*, **381**, L39.
Rush, B., & Malkan, M.A. 1992, in *Relationships Between Active Galactic Nuclei and Starburst Galaxies*, ed. A.V. Filippenko (San Francisco: Astron. Soc. Pacific), p. 21.
Shields, J.C. 1992, *ApJL*, **399**, L27.
Terlevich, R. 1992, in *Relationships Between Active Galactic Nuclei and Starburst Galaxies*, ed. A.V. Filippenko (San Francisco: Astron. Soc. Pacific), p. 133.
Terlevich, R., Tenorio-Tagle, G., Franco, J., & Melnick, J. 1992, *MNRAS*, **255**, 713.
Ulvestad, J.S., & Wilson, A.S. 1989, *ApJ*, **343**, 659.